Origins and Evolution of the Antarctic Biota

Geological Society Special Publications
Series Editor K. COE

GEOLOGICAL SOCIETY SPECIAL PUBLICATION NO 47

ORIGINS AND EVOLUTION OF THE ANTARCTIC BIOTA

EDITED BY
J. A. CRAME
British Antartic Survey
Natural Environment Research Council
Cambridge

1989
Published by
The Geological Society
LONDON

THE GEOLOGICAL SOCIETY

The Geological Society of London was founded in 1807 for the purposes of 'investigating the mineral structures of the earth'. It received its Royal Charter in 1825. The Society promotes all aspects of geological science by means of meetings, special lectures and courses, discussions, specialist groups, publications and library services.

It is expected that candidates for Fellowship will be graduates in geology or another earth science, or have equivalent qualifications or experience. All Fellows are entitled to receive for their subscription one of the Society's three journals: *The Quarterly Journal of Engineering Geology*, the *Journal of the Geological Society* or *Marine and Petroleum Geology*. On payment of an additional sum on the annual subscription, members may obtain copies of another journal.

Membership of the specialist groups is open to all Fellows without additional charge. Enquiries concerning Fellowship of the Society and membership of the specialist groups should be directed to the Executive Secretary, The Geological Society, Burlington House, Piccadilly, London W1V 0JU.

Published by the Geological Society from:
The Geological Society Publishing House
Unit 7
Brassmill Enterprise Centre
Brassmill Lane
Bath
Avon BA1 3JN
UK
(*Orders*: Tel. 0225 445046)

First published 1989

British Library Cataloguing in Publication Data
Origins and the evolution of the Antarctic
   1. Antarctic. Fossils
   I. Crame, J. A. (James Alistair) *1949−*
   560.998'9

   ISBN 0−903317−44−3

Printed in Great Britain at the Alden Press, Oxford

Contents

Origins and evolution of the Antarctic biota: an introduction

J. A. CRAME

British Antarctic Survey, Natural Environment Research Council, High Cross, Madingley Road, Cambridge CB3 OET, UK

Within the last 25 years there has been a dramatic increase in our knowledge of the fossil record of Antarctica. Improved access to the remotest parts of the continent, the advent of offshore drilling and intensive study of early expedition collections have all led to the accumulation of a vast amount of data that stretches back nearly 600 Ma to the beginning of the Cambrian period. No longer can Antarctica be dismissed from our view of the history of life on earth simply because so little is known about it; it is fast becoming another crucial reference point for global palaeontological syntheses.

If, today we have an image of Antarctica as a remote, inhospitable continent that supports little life, we now know that such a view cannot be projected back indefinitely through time. Abundant plant and animal fossils from a variety of periods point to much more benign climates and immediately raise a series of interconnected questions: where did such organisms come from, how long did they persist, and precisely when (and how) did they become extinct? Can our most southerly continent throw further light on the long-term role of climate in driving evolutionary trends (e.g. Valentine 1967; Vrba 1985)?

It was with points such as these in mind that a mixed group of palaeontologists, biologists, geologists and geophysicists gathered together for an international discussion meeting on the 'Origins and evolution of the Antarctic biota' at the Geological Society, London on 24 and 25 May 1988. A further workshop day (26 May 1988) was held at the British Antarctic Survey, Cambridge and altogether over 100 scientists from some 15 different countries took part in the discussions.

Obviously, in such a short space of time, it was not possible to cover all aspects of Antarctic palaeontology, and some subjects were necessarily reviewed in greater detail than others. As Palaeozoic biotas and biogeography were to be the subject of a complementary symposium ('Palaeozoic biogeography and palaeogeography'; Oxford, 14−19 August 1988), they were not examined exhaustively here. Nevertheless it was felt to be important to spend some time considering Antarctica's first macrofossils, and

in this volume **L. R. M. Cocks** provides a brief but succinct review of Antarctica's place within Cambrian to Devonian Gondwana. It is perhaps not generally appreciated that the greater part of Antarctica's extensive Early and Middle Cambrian sedimentary record was laid down under tropical conditions. The extensive Shackleton Limestone of the Transantarctic Mountains, and its correlatives, contains abundant shelly faunas with Australian and Chinese affinities, and part of the margin of what we now know as the Gondwana supercontinent formed a reef belt that protruded $10°−20°$ into the Northern Hemisphere. The most abundant and characteristic early Palaeozoic fossils of Antarctica are in fact archaeocyaths and **F. Debrenne** & **P. D. Kruse** show how, in conjunction with certain skeletal cyanobacteria and algae, these primitive coral-like organisms were the primary framework-builders of the reefs. Taxonomic revisions presented by these authors confirm the particularly strong faunal links between Antarctica and South Australia in the Early Cambrian.

Amongst the inarticulate brachiopods, trilobites, primitive molluscs and other marine invertebrates that inhabited the Early Cambrian reefs and carbonate platforms, a particularly important series of monoplacophorans and gastropods has been identified by **G. F. Webers** & **E. Yochelson**. These occur in the Late Cambrian (Dresbachian) Minaret Formation of the Ellsworth Mountains and comprise some 20 taxa. The most important of these is the monoplacophoran, *Knightoconus antarcticus*, which could well have been a direct ancestor of the cephalopods. Its 7 cm long, curving, cap-like shell bears multiple internal septa along at least one-third of its length and these are perforated by a rudimentary connecting tube (siphuncle). The stratigraphic position of this species can be demonstrated to be slightly beneath that of *Plectronoceras*, the first true cephalopod (from China).

In his review, **Cocks** points out that there is a very limited Ordovician and Silurian sedimentary and palaeontological record in Antarctica; it is likely that this was a time of extensive uplift and erosion. However, it is apparent that,

From Crame, J. A. (ed.), 1989, *Origins and Evolution of the Antarctic Biota*, Geological Society Special Publication No. 47, pp. 1−8.

1

during the later part of early Palaeozoic time, profound changes were occurring in the distribution of supercontinents and the large Gondwana plate slid progressively over the contemporary south pole. By the early Middle Devonian the pole lay under either S. Africa or S. America, and Antarctica came to occupy a mid- to high-latitude position (35°–75°S depending on the reconstruction used). It has essentially maintained this ever since.

Although palynomorphs of probable Early Devonian age are known from the Beacon Supergroup of southern Victoria Land (Kyle 1977), the most characteristic fossils of this age comprise a series of marine invertebrates from the Horlick Mountains, Ohio Range, Transantarctic Mountains. These clearly have Malvinokaffric Province (i.e. cool-water) affinities (Doumani *et al.* 1965) and testify to the comparatively high-latitude position of Antarctica in the Early Devonian. Some Lower Devonian fish fossils have been recorded from both the Horlick and Ellsworth mountain ranges and Middle–Late Devonian plant remains are known from the Ruppert Coast of Marie Byrd Land (Grindley *et al.* 1980); however, these are overshadowed in both stratigraphical and biogeographical importance by the Givetian (late Middle Devonian) Aztec Siltstone fish fauna from southern Victoria Land.

In a comprehensive review, **G. C. Young** shows that the Aztec fauna comprises not only the oldest but also one of the most diverse fossil vertebrate assemblages yet found in Antarctica. Over 30 taxa have so far been identified and these belong to four major groups of jawed fishes — the placoderms (armoured fishes), acanthodians (spiney-finned fishes), chondrichthyans (sharks and their relatives) and osteichthyans (boney fish) — and one group of agnathans (primitive jawless vertebrates; the thelodontids). The fauna has its strongest biogeographic links with eastern Australia and indicates that a distinctive East Gondwana vertebrate faunal province persisted until at least the end of the Middle Devonian (Young 1981). By the latest Devonian the vertebrate faunas of Euramerica and East Gondwana are much more similar, particularly in the composition of their respective freshwater assemblages. **Young** links this change to major palaeogeographic reconstructions and suggests that this may have been the time when the ocean separating these two supercontinents disappeared. The presence of a significant number of primitive taxa (or paraphyletic stem groups) in Antarctica and Australia suggests that this extensive continental region was the site of origin of major groups of

phyllolepid placoderms, xenacanth sharks and bothriolepid antiarchs.

Collision of the Gondwana and Baltica/Laurentia blocks in the latest Devonian to form the single supercontinent of Pangea radically alters our view of the subsequent evolution of life in the southern high latitudes. It has traditionally been felt that the classic 'Gondwana' sequences in these regions, comprising Permo-Carboniferous non-marine glacial facies overlain by coal measure facies containing a *Glossopteris* flora and Triassic non-marine facies with a *Lystrosauras* fauna and *Dicroidium* flora, were the product of supercontinent isolation. However, viewed in terms of the single supercontinent model, it would now appear that these classic sedimentary facies and biotas can be re-interpreted as austral components of a Pangean super-sequence (Veevers 1988). The basal glacial successions reflect the mid- to high southern latitudes and succeeding beds contain a provincial biota separated from boreal Pangea on the east by the Tethyan gulf and on the west by a tropical land zone (Veevers 1988, fig. 2c). Although Veevers dates the coalescence of Gondwana and Euramerica slightly later (mid-Carboniferous, *c.* 320 Ma), it is clear from his reconstruction that the classic *Glossopteris* flora now has a distinctly austral aspect.

Dicynodonts, a widespread, morphologically diverse group of mammal-like reptiles prevalent during the Permian and Triassic, are the subject of a comprehensive review by **S. L. De Fauw**. Essentially composed of herbivores, this group appears to have undergone two significant adaptive radiations: one in the Late Permian (Tatarian) and the other in the early Middle Triassic (Anisian). Both radiations were undoubtedly aided by the development of homogeneous Gondwana floras (the *Glossopteris* and *Dicroidium* floras, respectively), but it is important to emphasize that many genera achieved very widespread distributions in Pangea. The equatorial barrier does not appear to have been impenetrable to terrestrial vertebrates (see also Chatterjee 1987), and of the 35 known dicynodont genera approximately 11 are known from two or more cratonic regions. *Lystrosaurus*, the famous dicynodont from the Fremouw Formation of Antarctica, is now known to have been particularly widely dispersed. Semi-aquatic and probably able to forage for a variety of food types, it may have been the most successful mammal-like reptile of all time.

One of the most significant palaeontological discoveries yet made in Antarctica is that of a series of fossil forests. Characterized by both petrified logs and in situ stumps bearing clearly-defined annual growth rings, they are now known from the Permian and Lower Triassic of the Transantarctic Mountains, the Lower Cretaceous of Alexander Island and the Upper Cretaceous–lower Tertiary of the nothern Antarctic Peninsula region (e.g. Creber & Chaloner 1984). So striking is their abundance and diversity, and so mild the climatic deductions made from them, that it was thought at one time that they may indicate periods of fundamental change in the growing conditions in polar regions (perhaps related to shifts in the tilt of the earth's axis; e.g. Jefferson 1983). However, **W. G. Chaloner & G. T. Creber** have been interested in this problem for some time, and in their review of the phenomenon of forest growth in Antarctica they suggest that the only factor which inhibits tree growth in the highest latitudes at the present day is temperature. It is now apparent that the annual input of solar energy to the polar regions is sufficient to produce large annual growth rings and it has been demonstrated that many evergreen trees can tolerate long, dark winters. Studies on the composition of both living and fossil forests have shown that the low angle of sunlight does not cause excessive mutual shading of trees.

It has long been suspected that many of the key elements of today's Southern Hemisphere humid and perhumid forests could be traced well back into the Mesozoic period. This is particularly so for certain lycopods, ferns, araucarians and podocarps which may belong to lineages originating as early as the early Jurassic, or even the late Triassic (Fleming 1963, 1975). Certainly, elements of the extensive Late Jurassic evergreen coniferous rainforests that covered much of the southern Gondwana margins have persisted to the present day (Jefferson 1983; Dettman 1986).

If much of the early evidence for the antiquity of austral forest floras came from macropalaeobotany, much of the latest information stems from a proliferation of palynological investigations. Utilizing both on- and offshore successions, and a wide variety of palynoflora taxa, it is now possible to establish at least the Cretaceous ancestry of a number of living groups. **M. E. Dettmann**, for example, shows how the fern *Lophosoria*, which is restricted to South and central America, had a much wider distribution in the Cretaceous and Tertiary. Traced by its unique spore, *Cyatheacidites annulatus* (Dettman 1986), this genus can be shown to have its earliest stratigraphic occurrence (in basal Cretaceous strata) in the Antarctic Peninsula–South America region; it then migrated eastwards to Australia (where it

occurs in later Cretaceous beds) and northwards through South America. Similarly, the gymnosperm *Dacrydium balansae/D. bidwillii* alliance is now distributed from South America, through New Zealand, to certain Pacific islands and Malaysia. Its history can be traced by the pollen genus *Lygistepollenites* and the oldest occurrences (Coniacian–Santonian) shown to be on the Antarctic Peninsula and in southeast Australia.

Particularly important in confirming the high-latitude origins of a number of angiosperm lineages has been the elucidation of the Cretaceous–Tertiary boundary section on Seymour Island by **R. A. Askin**. Data from here are important in confirming that species from a range of families, which includes the Fagaceae, Myrtaceae, Proteaceae, Winteraceae, Casuarinaceae, Gunneraceae, Bombacaceae and Loranthaceae, have late Cretaceous–early Tertiary first appearances somewhere along the southern Gondwana margins. It should be emphasized, however, that not all these families necessarily originated in this region. Myrtaceous pollen, for example, is known from early Campanian and Maastrichtian sediments on the Antarctic Peninsula; this occurrence clearly predates Paleocene records in Australia and New Zealand, but postdates those from undifferentiated 'lower Senonian' in Borneo and Santonian in Gabon. In her paper, **Dettmann** speculates that a number of austral angiosperm taxa may actually have had northern Gondwanan or even Laurasian origins and then migrated via a South America–Antarctica–Australia route.

It seems to be possible to conclude that Antarctica was both an important source area and dispersal corridor (during the early Cretaceous–early Tertiary period) for austral plants now living at mid- to low latitudes. Quite why a number of major groups should either originate in, or disperse through, the highest southern latitudes is not readily apparent, but, as **Askin** points out, the climate during this period was for the most part mild and equable. Indeed, it could well be that an unusual combination of climatic conditions (comprising a polar winter/summer light cycle, low light angles and relatively mild and wet weather at the poles) provided an unique stress impetus for diversification of a wide variety of taxa. We will return to polar origination events and the time-discrepancy of high- and lower-latitude occurrences of taxa (often referred to as heterochroneity) below.

We know that for much of the Cretaceous Antarctica was covered by thick forests, but which animals may have inhabited them? Although the Cretaceous terrestrial and freshwater faunal record from the continent is still poor, **T. H. Rich, P. V. Rich, B. Wagstaff, J. McEwen-Mason, C. B. Douthitt & R. T. Gregory** maintain that it is possible to speculate on the composition of at least the Early Cretaceous biotas by close comparison with southeastern Australia. At this time the two continents were of course still joined and it is estimated that the terrestrial vertebrates preserved within the Otway and Gippsland basins could have lived at as high a palaeolatitude as 85°S! Hypsilophodont dinosaurs are particularly common in these Victoria coast assemblages, and there is at least one theropod, *Allosaurus* sp.. Turtles are common too, although they seem to belong to primitive types, and there is a variety of fish (including ceratodont lungfish and the unique Australian koonwarrids), lepidosaurs, pterosaurs, plesiosaurs (presumably freshwater) and birds, together with a possible labyrinthodont amphibian. That dinosaurs could have lived in southeastern Australia (and thus Antarctica too) is not altogether surprising for they have been known for some time from Late Cretaceous deposits on the North Slope of Alaska. **Rich et al.** believe it unlikely that the herbivorous hypsilophodontids would have migrated vast distances to areas of winter daylight and suggest instead that a large brain and eyes may have pre-adapted them to the low light conditions of polar habitats.

By comparing assemblages from both Australia and southern South America, **R. E. Molnar** has suggested that the one dinosaur known from the Late Cretaceous of Antarctica (an ankylosaur from James Ross Island; Olivero et al. in press) was part of a terrestrial fauna that included hypsilophodontids, pterosaurs, ratites, sphenodontians, leiopelmatid frogs and ceratodontid lungfish. In reviewing austral Mesozoic terrestrial vertebrate faunas, he was struck by the fact that three of the fifteen known genera from the Jurassic–Cretaceous of Australia (*Allosaurus*, *Austrosaurus* and *Siderops*) seem to have been relicts. Coupled with the occurrence of the apparent labyrinthodontid amphibian, this could be taken to indicate one of two things: either there was some form of geographical barrier protecting the Australian region (and Antarctica too?), or certain taxa may have preferentially survived in polar regions (cf. Vermeij 1987).

Returning to the marine realm, **G. R. Stevens** indicates that a distinctive cool-temperate Maorian province can be detected within the Triassic invertebrate assemblages known from New Caledonia, Papua New Guinea and South America. However in the Jurassic, eastern Gondwana rotated in a direction away from the South Pole and this led directly to a substantial improvement in global climates. Simultaneously, new shallow-water migration routes were established around the southern Gondwana margins and by the Middle Jurassic warm-temperate Tethyan faunas were able to spread to Australia, New Caledonia, New Zealand and West Antarctica. Such a scenario is certainly borne out by the belemnite assemblages described by **P. Doyle & P. J. Howlett**, for in the Middle–Late Jurassic a distinctive *Belemnopsis–Hibolithes–Duvalia* fauna can be traced southwards from European Tethys to the Gondwana margins. Local endemic centres can be identified at this time, such as in the embryonic trans-Gondwana seaway, but they are only discernible at the species level.

By the Early Cretaceous, extensive rifting had occurred across Gondwana and the proto-Indian and South Atlantic oceans were beginning to form. This in turn led to a reversal of the rotation of the eastern Gondwana block and the re-introduction of its southern margins into the highest latitudes. Here, a cool-temperate molluscan fauna developed, with one of its most important components being the endemic belemnite family, the Dimitobelidae. Further opening of the Atlantic and Indian oceans occurred throughout the Late Cretaceous and it was at this time that New Zealand's land links with the rest of Gondwana were severed by the formation of the Tasman Sea. However, as **Stevens** points out, there were still good Late Cretaceous shallow-marine connections between New Zealand, West Antarctica and South America. A cool-temperate Weddellian Province can be established between these localities until well into the Paleogene (Zinsmeister 1982).

The abundant early Campanian–late Eocene marine invertebrate faunas of the James Ross Island region have been investigated by **R. M. Feldmann & D. M. Tschudy**. They are particularly interested in an unusually rich decapod crustacean fauna which occurs in association with ammonites, bivalves, gastropods, echinoderms and brachiopods. Concentrating on the macrurous decapods (lobsters), they have been able to identify four species in the shallow-water Campanian–Paleocene Lopez de Bertodano Formation. *Hoploparia stokesi* is extremely abundant, but shows no clear-cut evolutionary trends in shape over a 20 Ma period. *Metanephrops jenkinsi*, a probable derivative of *H.*

stokesi, is particularly noteworthy as it extends the range of the genus back from the Pliocene to the Campanian. *Metanephrops* today is in fact only known from outer shelf/slope habitats at lower latitudes, and **Feldmann** & **Tschudy** believe that this may be another example of a marine invertebrate taxon that originated in the Late Cretaceous–early Tertiary of the James Ross Island region and then dispersed slowly to lower latitudes through the Cenozoic. So far, at least five bivalve, one gastropod, three echinoderm, seven decapod and two brachiopod genera known previously from either the late Cenozoic or Recent of mid- to low-latitude regions can be shown to have their earliest stratigraphic occurrences at this Antarctic locality (Zinsmeister & Feldmann 1984; Crame 1986; Wiedman *et al.* 1988). This is another striking example of high-latitude heterochroneity.

Marine vertebrates from the Lopez de Bertodano Formation include teleost fish, sharks, mosasaurs and plesiosaurs. The last two groups have been studied by **S. Chatterjee** & **B. J. Small** as they almost certainly occupied the top-predator niches filled at the present day by seals and whales. Material belonging to two separate plesiosaur families, the Plesiosauridae and Cryptoclididae, has now been identified, and a new species within the latter taxon is described (*Turneria seymourensis* sp. nov.). Interestingly enough, it would appear that cryptoclidids fed by sieving food particles through a mesh formed by their slim, delicate, interlocking teeth. **Chatterjee** & **Small** speculate that they may even have fed on *Hoploparia*, in a manner analogous to that of crabeater seals feeding on krill.

In a stimulating new hypothesis to account for the origin of the Australian marsupials, **J. A. Case** suggests that Zinsmeister's (1982) Weddellian Province can also be identified in the terrestrial realm. Here it is characterised, in the Paleocene and Eocene of South America, south-east Australia and Antarctica (Seymour Island), by two key elements: *Nothofagus* and marsupials. **Case** points out that, just as the concept of high-latitude heterochroneity can be used to account for the origin of certain plant and marine invertebrate taxa within the province, so it may also be applicable to terrestrial vertebrates too. In its original formulation (Zinsmeister & Feldmann 1984), the concept was split into two distinct components: first, the high latitudes may serve as centres of origin for taxa which can 'escape' under existing climatic conditions, and second, they may serve as 'holding tanks' for new taxa which remain isolated

until suitable conditions develop for their dispersal.

It is this second aspect of heterochroneity which may have been crucial here, with Australian marsupial radiation being closely linked to habitat diversification. New stem taxa are held to have evolved in the latest Cretaceous–early Tertiary Weddellian Province but these could not proliferate in the uniform cool-temperate closed rainforests. However, by the mid- to late Eocene, Australia had separated sufficiently from Antarctica to generate much more diverse open-forest habitats and this provided the trigger for extensive family-level cladogenesis to occur.

Although it is generally assumed that the best record of the deterioration of Cenozoic climates and onset of glaciation in Antarctica is contained in marine cores (e.g. Leg 113 scientific party 1987; Leg 119 scientific party 1988), there are also some important terrestrial sequences to be considered. **K. Birkenmajer** & **E. Zastawniak** indicate that a long sequence of late Mesozoic–Tertiary floras is contained within the volcaniclastic sediments interbedded with a thick volcanic pile on King George Island, South Shetland Islands. Radiometric dating of the volcanics has enabled the plants to be grouped into a series of discrete assemblages, which can then be analysed palaeoecologically. The results of this study, in conjunction with sedimentological investigations of intercalated glacial deposits, have been used to construct a 'climatostratigraphy' for the South Shetland Islands. Warm phases are identified in the Late Cretaceous–Paleocene, middle Eocene–early Oligocene, late Oligocene (in part) and at the Oligocene–Miocene boundary; intervening cold ones are identified in the early Eocene (the Krakow Glaciation) and late Oligocene (Polonez and Legru glaciations, separated by the Wesele Interglacial).

In a review of the evolution of the Antarctic fish fauna, **J. T. Eastman** & **L. Grande** demonstrate that the dominant living group, the notothenioids, have no fossil record. Their rise to prominence seems to have been a consequence of the thermal isolation of Antarctica, which was partially achieved after the final (deep water) separation of Australia in the late Eocene–early Oligocene (38 Ma) and completed on the formation of the Drake Passage at the Oligocene–Miocene boundary (23 Ma) (e.g. Kennett 1977). With the formation of the Antarctic Convergence, it is likely that the southward migration/colonization by most pelagic fishes would have been prohibited; the ancestral notothenioid stocks could then develop unchallenged in the vast Southern Ocean.

Eastman & Grande point out that the most logical explanation for the replacement of the Seymour Island Tertiary fish fauna by a relatively depauperate Recent one is the 15°C decline in temperature over a 50 Ma period. Nevertheless this should not have been an insurmountable evolutionary problem, for this averages out to a decrease of only 0.03°C per 100 000 years. Adaptation to low temperature as such presents few major biological obstacles and literally thousands of species have accomplished it (Dunbar 1968; Clarke 1988). Other ecological restraints, such as limited habitat space and trophic resources, are far more likely to have inhibited the development of the Antarctic fish fauna.

A more general survey of the Southern Ocean marine fauna by **A. Clarke & J. A. Crame** reveals that a substantial part of it may have evolved in situ over a long period of time. A number of living invertebrate groups, such as the pycnogonids, certain gastropods (trochids, littorinids, trichotropids and buccinaceans), echinoderms (ophiuroids, ctenocidarid cidaroids and schizasterid spatangoids) and ascidians, appear to be the products of adaptive radiations that began in the Cenozoic, or even, in some instances, the late Mesozoic. Far greater areas of shallow-water habitats were available for colonisation in the geological past; even after the onset of the main ice caps, substantial de-glaciations occurred in interglacial periods.

Turning to more physiological matters, **Clarke & Crame** consider the slow growth rates, extended development times and low metabolic rates of cold-water ectotherms (cold-blooded organisms). Low temperature is of course the traditional explanation for all these phenomena, but it seems somewhat paradoxical to suggest that it should hinder some processes such as growth, but not others such as locomotion. There is in fact a great energetic benefit to be derived from living in cold water, because basic maintenance costs are so low (Clarke 1988). The real problem for organisms in polar waters seems to be adapting to a severely pulsed food supply.

Using evidence gleaned from both the austral fossil record and the adaptations and distribution of living organisms, **R. E. Fordyce** suggests that the formation of the Southern Ocean may have had a seminal influence on the development of certain groups of marine mammals. The formation of the circum-Antarctic current, psychrosphere and Atlantic Convergence almost certainly provided the stimulus to develop new feeding strategies.

Fossil evidence from both Antarctica (Seymour Island) and New Zealand supports the conclusion that, within the Cetacea (whales and dolphins), the filter-feeding Mysticeti and echolocating Odontoceti arose from the primitive Archeoceti in the latest Eocene–earliest Oligocene interval. Both groups certainly appear to have then proliferated in the mid- to high southern latitudes through the Oligocene. Lobodontine phocids (true seals) radiated in the Southern Ocean region in approximately the late Miocene, and perhaps excluded certain otariids (fur seals and sea lions), such as *Arctocephalus*, from high-latitude pagophilic lifestyles.

E. Thomas considers the fact that polar cooling had major effects on the formation of bottom water, and this in turn is reflected in the composition of benthic foraminiferal assemblages. She describes a new sequence of Late Cretaceous (Maastrichtian) to late Miocene taxa that were obtained during ODP Leg 113 drilling on Maud Rise. They are all lower bathyal to upper abyssal types and analysis of them reveals distinct compositional shifts at seven levels (early/late Paleocene boundary, latest Paleocene, early Eocene, early middle Eocene, middle middle Eocene, earliest Oligocene, middle Miocene). At two of these, the early Eocene and early middle Eocene, the changes are particularly striking and suggest that the formation of bottom water may have been very different from preceding and succeeding periods. The marked lack of spiral forms and dominance of buliminids can be linked to lower oxygen levels and higher nutrient supply. These may have been times when deep waters were formed by evaporation at low latitudes (to form warm saline bottom water) rather than in polar regions.

Finally, **L. E. Watling & M. H. Thurston** consider a method of investigating the origins and evolution of an Antarctic amphipod family, the Iphimediidae, without recourse to the fossil record. Using a combination of phylogenetic (cladistic) and biogeographic data, they show that the most primitive members of the family are distributed primarily outside Antarctica; these are inferred to be the relicts of a former global distribution. The evolution of the family is then marked by a basic reorganization of the mouth field appendages which led directly to a major radiation of taxa. It is suggested that this occurred in the Antarctic region at approximately the Eocene–Oligocene boundary, with the cool waters of the embryonic Southern Ocean acting as an incubator for the evolutionary advance. With the possession of a new scissors-like mandible, the family spread swiftly outwards from the Antarctic (principally through the genus *Iphimedia*; approximately 35 species worldwide) and successfully colonised the thermally changing global ocean.

Summary

The new, but still fragmentary, fossil record that we have from Antarctica indicates that a surprisingly wide variety of organisms has lived on or around our southernmost continent. Since the early Middle Devonian (385 Ma ago), when Antarctica (within Gondwana) reached approximately its present position, it has been colonized by a succession of plants and animals that shows signs of being as complex as those on any other continent.

Vast habitat areas available for occupation by both terrestrial and marine organisms, coupled with equable climates for very long periods of time, certainly contributed towards this proliferation of life. The southern margins of Antarctica (and Gondwana) seem to have been particularly important dispersal routes and undoubtedly served as a major corridor for floral and faunal interchange between the high and low southern latitudes.

What is perhaps more surprising is to find that a number of plant and animal groups seem to have both originated in, and then radiated from, the high southern latitudes. Even after the marked deterioration of climates and loss of habitats through the Cenozoic, there still appear to have been adaptive radiations (especially in the marine realm).

Could it be that temperate, cool-temperate and even cold-temperate regions of the world have been more effective in the process of species diversification than hitherto recognized? The improved fossil record from Antarctica may be crucial in determining the contribution of high-latitude regions to the global species pool.

References

CHATTERJEE, S. 1987. A new theropod dinosaur from India with remarks on the Gondwana — Laurasia connection in the Late Triassic. *In*: McKENZIE, G. D. (ed.) *Gondwana six : stratigraphy, sedimentology and paleontology.* American Geophysical Union Monograph, **41**, 183−189.

CLARKE, A. 1988. Seasonality in the Antarctic marine environment. *Comparative Biochemistry & Physiology*, **90B**, 461−473.

CRAME, J. A. 1986. Polar origins of marine invertebrate faunas. *Palaios*, **1**, 616−617.

CREBER, G. T. & CHALONER, W. G. 1984. Climatic indications from growth in fossil woods. *In*: BRENCHLEY, P. J. (ed.) *Fossils and climate.* Wiley, Chichester, 49−74.

DETTMANN, M. E. 1986. Significance of the Cretaceous — Tertiary spore genus *Cyatheacidites* in tracing the origin and migration of *Lophososria* (Filicopsida). *Special Papers in Palaeontology*, **35**, 63−94.

DOUMANI, C. A., BOARDMAN, R. S., ROWELL, A. J., BOUCOT, A. J., JOHNSON, J. G., McALESTER, A. L., SAUL, J., FISHER, D. W. & MILES, R. S. 1965. Lower Devonian fauna of the Horlick Formation, Ohio Range, Antarctica. *Antarctic Research Series*, **6**, 241−281.

DUNBAR, M. J. 1968. *Ecological development in polar regions. A study in evolution.* Prentice-Hall, Englewood Cliffs, N. J.

FLEMING, C. A. 1963. Paleontology and southern biogeography. *In*: GRESSITT, J. L. (ed.) *Pacific basin biogeography.* Bishop Museum Press, Honolulu, 369−385.

—— 1975. The geological history of New Zealand and its biota. *In*: KUSCHEL, G. (ed.) *Biogeography and ecology in New Zealand.* Monographiae Biologicae, **27**, W. Junk, The Hague, 1−86.

GRINDLEY, G. W., MILDENHALL, D. C. & SCHOPF, J. M. 1980. A mid-late Devonian flora from the Ruppert Coast, Marie Byrd Land, West Antarctica. *Royal Society of New Zealand Journal*, **10**, 271−285.

JEFFERSON, T. J. 1983. Palaeoclimatic significance of some Mesozoic Antarctic fossil floras. *In*: OLIVER, R. L., JAMES, P. R. & JAGO, J. B. (eds) *Antarctic earth science.* Australian Academy of Science, Canberra, 593−598.

KENNETT, J. P. 1977. Cenozoic evolution of Antarctic glaciation, the circum-Antarctic Ocean, and their impact on global paleooceanography. *Journal of Geophysical Research*, **82**, 3843−3860.

KYLE, R. A. 1977. Devonian palynomorphs from the basal Beacon Supergroup of south Victoria Land, Antarctica. *New Zealand Journal of Geology and Geophysics*, **20**, 1147−1150.

LEG, 113 SHIPBOARD SCIENTIFIC PARTY 1987. Glacial history of Antarctica. *Nature*, **328**, 115−116.

LEG, 119 SHIPBOARD SCIENTIFIC PARTY 1988. Early glaciation of Antarctica. *Nature*, **333**, 303−304.

OLIVERO, E. B., GASPARINI, Z., RINALDI, C. A. & SCASSO, R. In press. First record of dinosaurs in Antarctica (Upper Cretaceous, James Ross Island) : palaeogeographical implications. *In*: THOMSON, M. R. A., CRAME, J. A. & THOMSON, J. W. (eds). *Geological evolution of Antarctica.* Cambridge University Press, Cambridge.

VALENTIVE, J. W. 1967. The influence of climatic fluctuations on species diversity within the Tethyan provincial system. *In*: ADAMS, C. G. & AGER, D. V. (eds) *Aspects of Tethyan biogeography.* Systematics Association Publication, **7**, 153−166.

VEEVERS, J. J. 1988. Gondwana facies started when Gondwanaland merged in Pangea. *Geology*, **16**, 732–734.

VERMEIJ, G. J. 1987. *Evolution and escalation*. Princeton University Press, Princeton.

VRBA, E. S. 1985. Environment and evolution : alternative causes of the temporal distribution of evolutionary events. *South African Journal of Science*, **81**, 229–236.

WIEDMAN, L. A., FELDMANN, R. M., LEE, D. E. & ZINSMEISTER, W. J. 1988. Brachiopoda from the La Meseta Formation (Eocene), Seymour Island, Antarctica. *In*: WOODBURNE, M. O. & FELDMANN, R. M. (eds). *Geology and paleontology of Seymour Island*. Geological Society of America Memoir, **169**, 449–457.

YOUNG, G. C. 1981. Biogeography of Devonian vertebrates. *Alcheringa*, **5**, 225–243.

ZINSMEISTER, W. J. 1982. Late Cretaceous — early Tertiary molluscan biogeography of the southern circum-Pacific. *Journal of Paleontology*, **56**, 84–102.

— — & FELDMANN, R. M. 1984. Cenozoic high latitude heterochroneity of southern hemisphere marine faunas. *Science*, **224**, 281–283.

Antarctica's place within Cambrian to Devonian Gondwana

L. R. M. COCKS

Department of Palaeontology, British Museum (Natural History)
Cromwell Road, London SW7 5BD, UK

Abstract: Analysis of various fossil groups, in particular brachiopods, graptolites and trilobites, from Gondwana as a whole, which included much of what is now southern Europe, southern Asia, Africa, South America, Australasia and Antarctica, has determined not only the margins of the Gondwanan palaeocontinent in the early and middle Palaeozoic, but also the relative palaeotemperatures and the probable palaeo-latitudes. The patchy faunal record from Antarctica is reviewed from Cambrian to Devonian times, and this consists of well-dated Early, Middle and Late Cambrian shelly faunas with eastern Gondwanan affinities and deposited under tropical conditions. By contrast, there are no proven Ordovician fossils known and only one disputed Silurian record of two coral specimens from the Beardmore Glacier moraine. In early Devonian times, well-dated Emsian shelly faunas of Malvinokaffric affinity are known from the Horlick Range and there is a diverse Middle Devonian fish fauna from Victoria Land. From these records it can be deduced that, as the large Gondwana plate slid over the contemporary South Pole, the palaeo-latitude of Antarctica slowly increased from tropical to high temperate from Cambrian to Devonian times.

Excellent summaries of Antarctic biostratigraphy in Cambrian to Devonian times have been published (e.g. Laird 1981), but, whilst recognizing that Antarctica formed part of the large Gondwana palaeocontinent, these have often taken a relatively narrow view, which is not surprising in view of the difficult task of winning information and fossils from the inhospitable terrain of Antarctica today. This paper views Antarctica in a wider context and follows other reviews of Lower Palaeozoic Gondwana as a whole (e.g. Cocks & Fortey 1988). The chief faunal records and correlation are reviewed first (Fig. 1) and then the evidence for the changing palaeolatitudes of Gondwana is adduced in the closing section of the paper.

Cambrian

Proven Cambrian rocks are confined to the Transantarctic Mountains and the Ellsworth Mountains, and come from a large variety of formations chiefly summarized by Cooper *in* Shergold *et al.* (1985). There is no unbroken sequence through the entire Cambrian but both in the Bowers Mountains of northern Victoria Land and in the Ellsworth Mountains there are extensive deposits of parts of the system. In the Bowers Mountains, the Middle Cambrian tholeiitic island arc rocks of the Glasgow Formation (Weaver *et al.* 1984) confirm Antarctica's place near the edge of the palaeocontinent and suggest the possible accretion of one or more suspect terranes at that time. The richly fossiliferous Spurs Formation of the succeeding Mariner Group has yielded several late Middle to early Late Cambrian faunas (Shergold *et al.* 1976) from shallow-water limestones, which can be identified with trilobite genera and even species from South China such as *Prochuangia* aff. *granulosa* Lu. Above this unit in the Mariner Group is the regressive Eureka Formation. Although the upper part of this formation contains possible mud-cracks and was presumably deposited under relatively shallow water, the middle and lower parts have yielded a middle late Cambrian fauna including trilobites such as *Olentella* and *Apheloides*, which are characteristic of deeper-water marginal facies comparable to contempory faunas from Siberia. This latter occurrence demonstrates how essential it is to compare like with like; if that fauna had been thought of as cratonic (like the Spurs Formation fauna) then mistaken conclusions might have been drawn on Siberia's relationship with Antarctica.

Lower Cambrian (Atdabanian) rocks are known from two broad regions in Antarctica; firstly in situ from the Shackleton Limestone of the central Transantarctic Mountains, in particular at the head of the Beardmore Glacier and in the Churchill Mountains, where a shelly fauna of archaeocyathans and 17 other species of brachiopods, trilobites, monoplacophorans, hyolithids and problematica has been recovered (Rees *et al.* 1985; Rowell *et al.* 1988 *a b*). Most

From Crame, J. A. (ed.), 1989, *Origins and Evolution of the Antarctic Biota*,
Geological Society Special Publication No. 47, pp. 9–14.

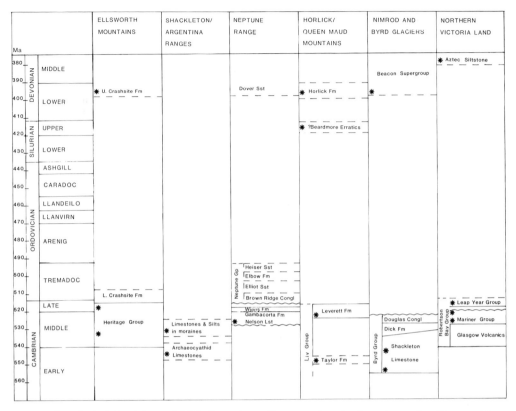

Fig. 1. The principal Cambrian to Middle Devonian sedimentary deposits from Antarctica. The asterisks denote fossil records which have proved useful for correlation; other fossil records are omitted. The time-scale on the left follows McKerrow *et al.* (1985).

of the fauna is either endemic or of Australian affinity, but Chinese and more cosmopolitan species are also present. The second fauna is from limestone erratic blocks in the Miocene Cape Melville Formation of King George Island off the Antarctic Peninsula and consists of small problematica such as *Camenella*, *Lapworthella*, *Chancelloria*, *Halkieria* and *Mongolitubus*, indicating that these early Cambrian limestones were or are much more widespread than shown by the outcrop as known at present (Gazdzicki & Wrona 1986). In addition fossiliferous early and middle Cambrian rocks, many including abundant archaeocyathids, are known from a variety of areas over many parts of the continent.

Latest Cambrian rocks are present at two localities, one at Handler Ridge, where the Robertson Bay Group has yielded seven different conodonts, including *Proconodontus posterocostatus* Miller and *Iapetognathus* sp. (Burrett & Findlay 1984). A second fauna from Handler Ridge (Wright *et al.* 1984)

yielded another conodont fauna, including *Hirsutodontus hirsutus* Miller, and four trilobite genera, *Harpides?*, *Pseudokainella*, *Tsinania* and a saukiid. The second locality is at Reilly Ridge (Schmidt-Tomé & Wolfart 1984), about 250 km from Handler Ridge, from where shallow-water trilobites and brachiopods are recorded. All these faunas indicate an age extremely close to, but probably just below, the Cambro-Ordovician systematic boundary, and the trilobites reinforce the Chinese provincial affinities of that fauna. However, recent recollecting at Reilly Ridge by R. A. Cooper (pers. comm.) has yielded new fossils, including *Brassicicephalus*, *Catillicephalus* and *Distazeris*, which indicate a Mindyallan age (early Upper Cambrian) and North American affinities.

Ordovician

No fossils of unequivocally Ordovician age have yet been identified and published from the Antarctic Continent. However, there are rocks which overlie latest Cambrian sequences with definite fossils, for example the lower and middle members of the Crashsite Quartzite Formation in the Ellsworth Mountains, which contain unidentified and fragmentary inarticulate brachiopods and a variety of trace fossils which seem most likely to be of early Ordovician age. Clarkson *et al.* (1981) reviewed the most probable Ordovician localities which, in addition to the Crashsite Quartzite, are: (a) the Camp Ridge Quartzite of northern Victoria Land, which overlies well-dated late Cambrian beds unconformably, but which appears to have been affected by a late Silurian metamorphic event; (b) the Cocks Formation and the underlying Skelton Group of southern Victoria Land which lies unconformably on middle Cambrian beds and is intruded by middle Ordovician plutonic rocks, although these metamorphosed rocks might be older; (c) the Brown Ridge Conglomerate and other lower rocks of the Neptune Group in the Pensacola Mountains and Neptune Range which rest unconformably on late Cambrian rocks and underlie beds of probable middle Devonian age. The same authors also review several other less reliable reports from various parts of the continent.

There are also a large number of largely Ordovician (but ranging up to Devonian) radiometric age dates reported from Antarctica (e.g. Clarkson *et al.* 1981, fig. 3), but these represent tectonic events outside the scope of this paper.

Silurian

The only reported Silurian faunas from the entire Antarctic continent are in four microscope slides containing the upper Silurian corals *Acanthohalysites* sp. and *Syringopora* sp. identified by Jell (1981), from a block reputedly collected from the Cloudmaker moraine of the Beardmore Glacier by the Shackleton expedition of 1908 and now in the Sedgwick Museum, Cambridge. Despite attempts to recollect, no comparable material has been found in the Beardmore Glacier or elsewhere in Antarctica, which led Rowell *et al.* (1987) to write a note suggesting that the original specimens may not have come from Antarctica at all, particularly since there was an interval between their collection in 1908 and their museum registration in 1930. However, since there are also undoubtedly other specimens from the Cloudmaker moraine in the Sedgwick Museum, I think that the stated provenance may be correct. There seems no good reason why marine rocks of late Silurian age should not have been present in the Trans-Antarctic

Mountains–Ross Sea area. On the other hand, the analysis by Rowell *et al.* (1988*b*) of the origin and depositional setting of the Douglas Conglomerate concluded that those parts of the Transantarctic Mountains may have been subjected to tectonism and erosion during much of the Ordovician and Silurian, but this would still have left time for a late Silurian marine interval.

Devonian

Faunas of Devonian age may be divided into two groups, firstly the abundant and well-preserved fossil fish material, which has been described by Young (1982) from the late Givetian to early Frasnian Aztec Siltstone of southern Victoria Land and which are reviewed by Young elsewhere in this volume, and secondly the more varied fossil material from the Ellsworth Mountains and Horlick Mountains. This has been best documented from the Horlick Mountains (Doumani *et al.* 1965), where the Horlick Formation has yielded dominant brachiopods (*Pleurothyrella*, *Australospirifer* and others) as well as many other marine groups, including fish (Young 1986). Boucot *et al.* (1963) determined these fossils as being of Emsian age and of distinct Malvinokaffric faunal affinities, and Bradshaw & McCartan (1983) established their shallow-water marine environment. Craddock (1969) also reported brachiopods of 'probable Devonian' age from the upper Dark Member of what he termed the Crashsite Formation in the Sentinel Range of the Ellsworth Mountains, but it seems most probable that there is an unrecognized unconformity below the Dark Member and above the main part of the underlying two other members of the Crashsite Quartzite which appear to be of earliest Ordovician age. Grindley *et al.* (1980) have also recorded some middle to late Devonian plants from the Ruppert Coast of Marie Byrd Land, and Kyle (1977) has recorded palynomorphs of probable early Devonian age from the Beacon Supergroup at Table Mountain, southern Victoria Land.

Regional setting within Gondwana

The general place of Antarctica within the Gondwana continent is now well established, lying as it did with Greater (East) Antarctica against southeastern Africa, the Indian subcontinent and Australia and Lesser (West) Antarctica partly against the New Zealand block and partly on the edge of the palaeocontinent (Fig. 2). However, what has become clearly

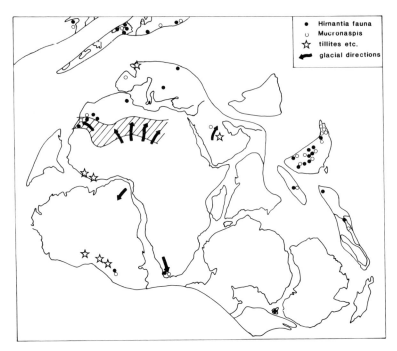

Fig. 2. Palaeogeographic reconstruction of Gondwana in latest Ordovician times, showing the late Ashgill glacial evidence and the distributions of the distinctive *Hirnantia* Fauna and the trilobite *Mucronaspis* (updated from Cocks & Fortey 1988). The contemporary South Pole was in central Africa and the equator in Papua New Guinea. This demonstrates that, although there is virtually no evidence from Antarctica itself for this geological period, the continental distributions and palaeolatitudes are known fairly accurately by the evidence from the adjacent landmasses.

focussed only in recent years (e.g. Cocks & Fortey 1988) is the way in which the whole Gondwanan continent moved during early and middle Palaeozoic times. In the Cambrian and early Ordovician the South Pole lay under what is now North Africa, and the Australian edge on the Equator, and during the succeeding Silurian and Devonian, Gondwana drifted across the pole, so that by the early middle Devonian the pole lay under southern Africa or perhaps southern America.

These movements have been deduced and checked using two completely independent methods, firstly by palaeomagmatism and secondly by an analysis of the faunas to be found in all of the various Gondwanan components. The faunas have produced a very coherent picture. Once initial ecological analysis had been made to determine the general type of fauna (mistakes have been made in the past for example by trying to compare shallow and deep-water faunas with each other to establish 'provincial' differences), then the shallower-water faunas on the cratons were considered and compared with each other. In, for example, the Lower

Ordovician, two clear ends of a continuous spectrum became apparent. In 'West' Gondwana, chiefly in North Africa and western Europe, the cratonic faunas were dominated by a coldwater fauna of large inarticulate brachiopods and the dalmanitid–calymenacean trilobite fauna, and there is a geographical cline from there to the tropical faunas of 'East' Gondwana, known best today from South China and Australia. A mixture between the two extremes can be found in the contemporary faunas of Turkey, the Middle East, and western South America (Cocks & Fortey 1988, Fig. 5).

Thus Antarctica can be inferred to have been oriented so that northern Victoria Land was the closest to the then equator (the 'north' in the following sentences) and Queen Maud Land and the Weddell Sea area closest to the South Pole (the 'south'). In the early Cambrian the north was tropical and the south at a latitude of about 35°. During late Cambrian, Ordovician, Silurian and early Devonian times, as the whole Gondwanan plate slid progressively across the South Pole, the palaeolatitude of Gondwana steadily increased, until by the middle Devonian the northern Victoria Land fish beds would have been deposited at about 40°S and Queen Maud Land would have been well within the Antarctic circle. Therefore the faunas from Antarctica fit very well into a wider pattern, with the expected links between the subjacent Chinese and Australian subcontinents, and the rich tropical early Cambrian faunas of the Shackleton Limestone and elsewhere slowly giving way to the poorly diverse cold water faunas of the Horlick Mountains which form a natural part of the temperate to subpolar Malvinokaffric faunas so well known from elsewhere in the Devonian.

During this 200 million year period, Antarctica was by no means passive. Active accretion of island arcs to the northeast in the early to middle Cambrian to form what is now

northern Victoria Land might have been paralleled under what is now Marie Byrd Land and the southern part of the Antarctic Peninsula which formed the edge of the Gondwanan palaeocontinents (Fig. 2). Tectonic activity at intervals during the whole Lower Palaeozoic is reflected in the unconformities, the relative lack of Ordovician and Silurian rocks, and metamorphic events and the radiometric dating from plutons and other igneous rocks. Further work to establish more data and focus this long period more sharply will be an intriguing challenge for future years.

I am most grateful to R. A. Cooper and A. J. Rowell for discussion and to R. A. Fortey for commenting on a draft of the manuscript.

References

BOUCOT, A. J., CASTER, K. E., IVES, D. & TALENT, J. A. 1963. Relationships of a new Lower Devonian terebratuloid (Brachiopoda) from Antarctica. *Bulletins of American Paleontology*, 46, 81–151, pls. 16–41.

BRADSHAW, M. A. & McCARTAN, L. 1983. The Depositional environment of the Lower Devonian Horlick Formation, Ohio Range. *In*: OLIVER, R. L., JAMES, P. R. & JAGO, J. B. (eds) *Antarctic Earth Science.* Cambridge University Press, Cambridge. 102–106.

BURRETT, C. F. & FINDLAY, R. H. 1984. Cambrian and Ordovician conodonts from the Robertson Bay Group, Antarctica and their tectonic significance. *Nature*, 307, 723–5.

CLARKSON, P. D., COOPER, R. A., HUGHES, C. P., THOMSON, M. R. A. & WEBERS, G. F. 1981. Ordovician of Antarctica — A review. *In*: WEBBY, B. D. (ed). The Ordovician system in Australia, New Zealand and Antarctica. *International Union of Geological Sciences Publication*, 6, 46–50.

COCKS, L. R. M. & FORTEY, R. A. 1988. Lower Palaeozoic facies and faunas around Gondwana. *In*: AUDLEY-CHARLES, M. G. & HALLAM, A. (eds) *Gondwana and Tethys* Geological Society, London, Special Publication, 37, 183–200.

CRADDOCK, C. 1969. *Sheet 4* — Geology of the Ellsworth Mountains *In*: BUSHNELL V. C. (ed). *Geologic Map of Antarctica 1:1 000 000, Antarctic Map Folio Series Folio 12, Plater IV*. American Geographical Society, New York.

DOUMANI, G. A., BOARDMAN, R. S., ROWELL, A. J., BOUCOT, A. J., JOHNSON, J. G., McALESTER, A. L., SAUL, J., FISHER, D. W. & MILES, R. S. 1965. Lower Devonian fauna of the Horlick Formation Ohio Range, Antarctia. *Antarctic Research Series*, 6, 241–81, pls. 1–18.

GAZDZICKI, A. & WRONA, R. 1986. Polskie Badania paleontologiczne w Antarktyce zachodniej (1986). *Przeglad Geologiczny*, 11, 609–17.

GRINDLEY, G. W., MILDENHALL, D. C. & SCHOPF, J. M. 1980. A mid–late Devonian flora from the Ruppert Coast, Marie Byrd Land, West Antarctica. *Royal Society of New Zealand Journal*, 10, 271–85.

JELL, J. S. 1981. Silurian tabulate corals from the Cloudmaker Moraine, Beardmore Glacier, Antarctica. *Alcheringa*, 5, 311–6.

KYLE, R. A. 1977. Devonian palynomorphs from the basal Beacon supergroup of south Victoria Land, Antarctica. *New Zealand Journal of Geology and Geophysics*, 20, 1147–50.

LAIRD, M. G. 1981. Lower Palaeozoic rocks of Antarctica. *In*: HOLLAND, C. H. (ed.) *Lower Palaeozoic Rocks of the World*, 3. John Wiley & Sons, Chichester, 257–314.

McKERROW, W. S., LAMBERT, R. St. J. & COCKS, L. R. M. 1985. The Ordovician, Silurian and Devonian periods. *In*: Snelling, N. J. (ed.) *The Chronology of the Geological Record*. Geological Society, London, Memoir, 10, 73–80.

REES, M. N., ROWELL, A. J. & PRATT, B. R. 1985. The Byrd Group of the Holyoake Range, central Transantarctic Mountains. *Antarctic Journal of the United States*, 20(5), 3–5.

ROWELL, A. J., EVANS, K. R. & REES, M. N. 1988a. Diversity and significance of the Early Cambrian Shackleton Limestone fauna, central Transantarctic Mountains. *Origins and evolution of the Antarctic Biota Abstracts*, 37.

— —, REES, M. N. & BRADDOCK, P. 1987. Silurian marine fauna not confirmed from Antarctica. *Alcheringa*, 11, 137.

— —, — — & — — 1988b. Pre-Devonian Paleozoic rocks of the central Transantarctic Mountains. *Antarctic Journal of the United States*, **21**, 48−50.

SCHMIDT TOMÉ, M. & WOLFART, R. 1984. Tremadocian faunas (trilobites, brachiopods) from Reilly Ridge, North Victoria Land, Antarctica. *Newsletters on Stratigraphy*, **13**, 88−93, pl. 1.

SHERGOLD, J. H., COOPER, R. A., MACKINNON, D. I. & YOCHELSON, E. L. 1976. Late Cambrian Brachiopoda, Mollusca and Trilobita from Northern Victoria Land, Antarctica. *Palaeontology*, **19**, 247−291, pls. 38−42.

−−, JAGO, J. B., COOPER, R. A. & LAURIE, J. 1985. The Cambrian System in Australia, Antarctica and New Zealand. *International Union of Geological Sciences Publication*, **19**, 1−85, charts 1−8.

WEAVER, S. D., BRADSHAW, J. D. & LAIRD, M. G. 1984. Geochemistry of Cambrian volcanics of the Bowers Supergroup and implications for the Early Paleozoic tectonic evolution of northern Victoria Land. *Earth and Planetary Science Letters*, **68**, 128−14.

WRIGHT, T. O., ROSS, R. J. & REPETSKI, J. E. 1984. Newly discovered youngest Cambrian or oldest Ordovician fossils from the Robertson Bay terrane (formerly Precambrian), northern Victoria Land, Antarctica. *Geology*, **12**, 301−5.

YOUNG, G. C. 1982. Devonian sharks from south-eastern Australia and antarctica. *Palaeontology*, **25**, 817−43, pls. 87−89.

YOUNG, V. T. 1986. Early Devonian fish material from the Horlick Formation, Ohio Range, Antarctica. *Alcheringa*, **10**, 35−44.

Cambrian Antarctic archaeocyaths

FRANÇOISE DEBRENNE[1] & PETER D. KRUSE[2]

[1] CNRS UA, 12, Institut de Paléontologie, 8 rue de Buffon, 75005 Paris, France
[2] Northern Territory Geological Survey, PO Box 2901, Darwin, NT, 5794 Australia

Abstract: Most Antarctic archaeocyaths have been collected as allochthonous blocks, but in situ collections from the Shackleton Limestone in the vicinity of Nimrod and Byrd Glaciers (Transantarctic Mountains) provide stratigraphic control. Collections from King George Island, the Weddell Sea, Whichaway Nunataks, Argentina Range, Beardmore Glacier, Nimrod Glacier and Byrd Glacier (Cracking Cwm) are all of Botomian age according to the Siberian stage scale for the Early Cambrian. One collection from the Byrd Glacier (Mt Egerton) may be of slightly younger, Toyonian age. A unique species from the Minaret Formation in the Heritage Range has a Late Cambrian (Idamean) age, and is the youngest archaeocyath known. All existing collections are taxonomically revised here. Strong faunal affinities are evident at the species level with in situ Early Cambrian faunas in South Australia and allochthonous faunas in the Permo-Carboniferous Dwyka Tillite of South Africa. This allows the recognition of an Early Cambrian Gondwana province and confirms the existence of a Gondwana supercontinent throughout the Palaeozoic. Archaeocyatha are the most abundant and characteristic early Palaeozoic macrofossils of the Antarctic continent (Fig. 1). Most Antarctic collections have been recovered as allochthonous blocks in moraine or in dredgings at sea, but some in situ faunas are now known. To date, the only in situ collections studied in detail are from the Shackleton Limestone in the environs of the Nimrod and Byrd Glaciers, Transantarctic Mountains (Debrenne & Kruse 1986). On the basis of this latter work, we here make critical comparisons (and synonymies where necessary) of the faunal lists of previous authors to tentatively establish the composition of the Antarctic archaeocyathan fauna (Figs 2–4).

Discoveries of Antarctic archaeocyaths

The Scottish National Antarctic Expedition (1902–1904) recovered the first Archaeocyatha known from the Antarctic continent. Additional discoveries were made by later expeditions, either in erratics or more rarely in situ. Discoveries are listed below in chronological order (Figs 1–4).

(1) Weddell Sea: allochthonous, provenance unknown. Collected by Bruce (Scottish National Antarctic Expedition) in 1902–1904; fauna described by Gordon (1920).

(2) Beardmore Glacier: allochthonous, derived from Shackleton Limestone. Discovered by Wild and Shackleton of the British Antarctic Expedition (1907–1909). Fragments of cups were identified by Taylor & Goddard (1914), Regulares as well as Irregulares (the latter probably *Protopharetra*) being present. Skeats (1916) also indicated their presence and Scott (1911–1912) collected further allochthonous blocks from within which Archaeocyatha were recognized by Debenham (1921). From the head of the same glacier, R. L. Oliver in the 1960s collected some loose material at Plunket Point. Hill (1964a) described six forms of Archaeocyatha from the latter.

(3) Whichaway Nunataks: allochthonous,
provenance unknown, as no archaeocyath-bearing limestone outcrops were found in the area. Numerous boulders were collected by Fuchs, Pratt and Stephenson of the Transantarctic Expedition (1955–1958). The fauna was described in a monograph by Hill (1965).

(4) Heritage Range: allochthonous, White-out Conglomerate. An erratic from a Permo-Carboniferous tillite (Craddock & Webers 1964). The description of this small fauna is in press (Debrenne in Webers *et al.*, in press) Fig. 4A).

(5) Nimrod Glacier: in situ, Shackleton limestone (Botomian). The Nimrod Glacier region was investigated by M. G. Laird of the New Zealand Geological and Survey Antarctic Expedition in 1960–1961. The presence of Archaeocyatha was reported by Laird & Waterhouse (1962) and Hill (1964b) described the fauna, the first found in situ in Antarctica. Further collections by M. G. Laird and G. D. Mansergh (1964–1965), reported in Laird *et al.* (1971), were described by Debrenne & Kruse (1986).

(6) Heritage Range: in situ, Minaret Formation (Late Cambrian–Idamean). Discovered by Webers (1962–1963) and reported by Craddock & Webers (1964). These are the only Late Cambrian archaeocyaths known.

From Crame, J. A. (ed.), 1989, *Origins and Evolution of the Antarctic Biota*,
Geological Society Special Publication No. 47, pp. 15–28.

15

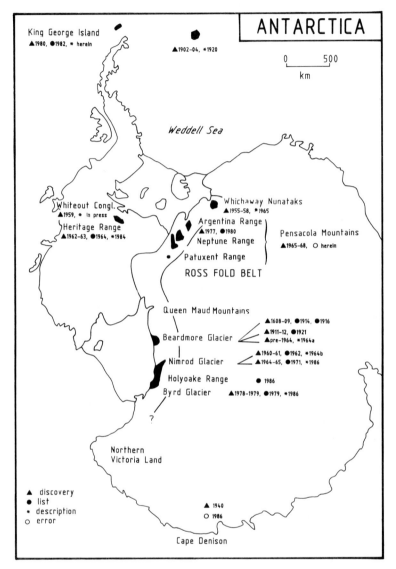

Fig. 1. Antarctic archaeocyathan discoveries.

Their attribution to the group was confirmed by Debrenne *et al.* (1984), who described *Antarcticocyathus webersi*, the sole archaeocyathan taxon present in this diverse fauna (Fig. 2B).

(7) Argentina Range: in situ, unnamed Early Cambrian unit. Collected by Shulyatin of the twenty second Soviet Antarctic Expedition (1977). A succinct faunal list was given by Konyushkov & Shulyatin (1980).

(8) Byrd Glacier: in situ, Shackleton limestone (Botomian and possible early Toyonian). Collected by Burgess (1978–1979) and briefly listed by Burgess & Lammerink (1979). The archaeocyathan fauna was described by Debrenne & Kruse (1986).

(9) King George Island: allochthonous, Polonez Cove Formation (late Oligocene). Discovered by the IV Polish Antarctic Expedition in 1980 and listed by Morycowa *et al.* (1982).

(10) Holyoake Range, Transantarctic Mountains: in situ, Shackleton Limestone. First Antarctic report of bioherms with an algal–archaeocyathan framework (Rees *et al.* 1986). The archaeocyaths have not yet been studied (Fig. 3A).

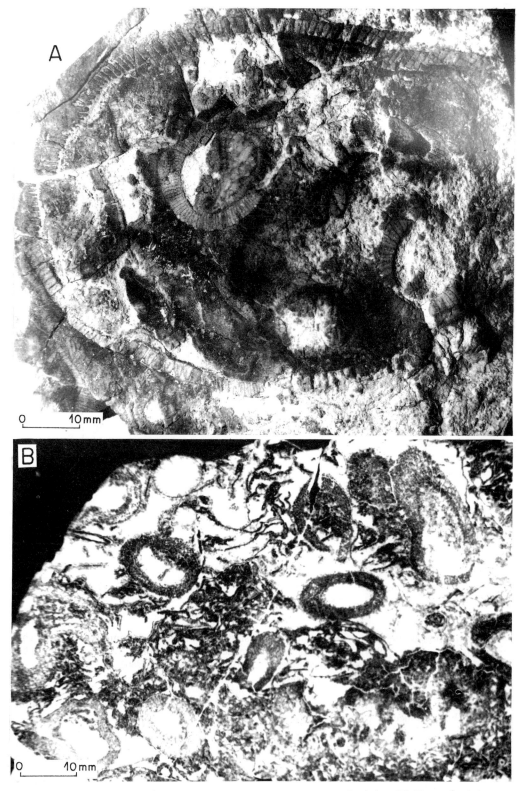

Fig. 2. (A) *Archaeocyathus* bearing rock, Shackleton limestone, Lower Cambrian. **(B)** Upper Cambrian limestone from Heritage Range with cups of *Antarcticocyathus webersi*.

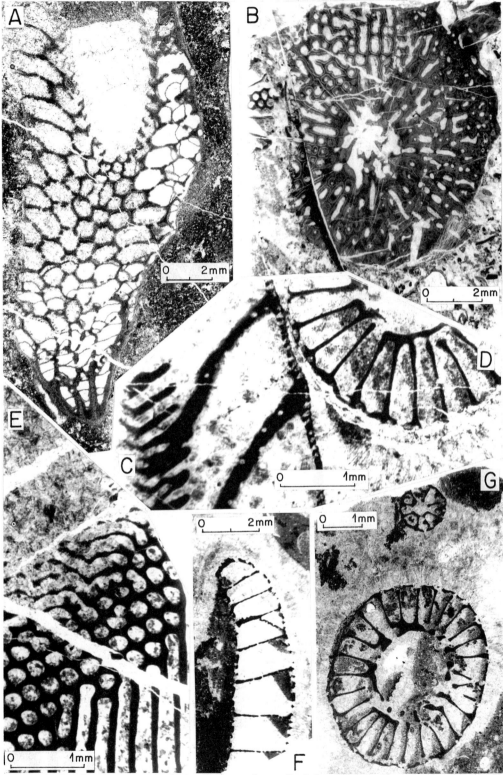

Fig. 3. (A) *Pseudosyringocnema* sp., Holyoake Range, Lower Cambrian. **(B)** *Pycnoidocyathus* sp., Beardmore Glacier, Lower Cambrian. **(C, D)** *Sanarkophyllum antarcticum* and *Flexanulus* cf. *oosthuizeni*, Beardmore Glacier, Lower Cambrian. **(E)** Detail of the inner wall with tubes and annuli of *Sanarkophyllum*, Beardmore Glacier, Lower Cambrian. **(F)** *Erugatocyathus* sp., King George Island, Lower Cambrian. **(G)** *Cyathocricus tracheodentatus*, King George Island, Lower Cambrian.

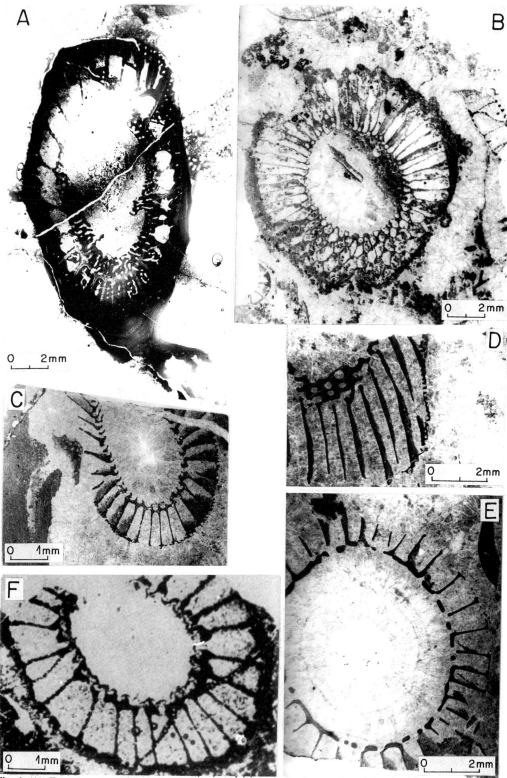

Fig. 4. (A) *Erismacoscinus endutus*, Whiteout conglomerate, Lower Cambrian. **(B)** *Syringocnema favus*,
Nimrod Glacier, Lower Cambrian. **(C)** *Thalamocyathus trachealis*, King George Island, Lower Cambrian.
(D) *Stapicyathus incisus*, tangential section of inner wall, King George Island, Lower Cambrian.
(E) *Stapicyathus incisus*, transverse section, King George Island, Lower Cambrian. **(F)** *Erugatocyathus
scutatus*, Whichaways Nunataks, Lower Cambrian.

Table 1. *Geographical distribution of archaeocyathan genera and species recorded from Antarctica. Comparisons with South Africa and Australia are given in the right-hand columns.*

	Antarctica									South Africa	South Australia
				Heritage Range							
	King George Island allochthonous	Weddell Sea allochthonous	Whichaway Nunataks allochthonous	Argentina Range in situ	Whiteout Conglomerate allochthonous	Minaret Formation in situ	Beardmore Glacier allochthonous	Nimrod Glacier in situ	Byrd Glacier in situ	allochthonous	in situ
EARLY CAMBRIAN											
Archaeolythus contractus (Hill)			+								□
Archaeolynthus sp.											
Dokidocyathus simplicissimus Taylor				?				+			■ □
Dokidocyathus sp.											□
Kymbecyathus avius Debrenne & Kruse									+		
Ajacicyathus ajax (Taylor)	+	+	+	+				cf.			■
Ajacicyathus cf. *araius* Debrenne			+	+				+			■ □
Ajacicyathus sp.	+	+	+	+	+						□
Nochoroicyathus sp.							+		+		
Stapicyathus incisus (Hill)	+	+	+					+		■	■
?Cadniacyathus curvatus (Hill)			+								
Gordonicyathus sp.				+	+						
Thalamocyathus trachealis (Taylor)	+	+	+	cf.				+	+	■	□ □
Thalamocyathus tectus Debrenne		+	+	?				+		■	■
Ehmocyathus lineatus Bedford & Bedford								+			■
Cyathocricus tracheodentatus (Bedford & Bedford)	+							+			■

Kiwicyathus nix Debrenne & Kruse

Kiwicyathus sp.

Sajanocyathus sp.

Ladaecyathus jagoi Debrenne & Kruse

Somphocyathus sp.

Flexanulus cf. *oosthuizeni* Debrenne

Aporosocyathus sp.

Tegerocyathella burgessi Debrenne & Kruse

Sanarkophyllum antarcticum (Hill)

Tegerocyathus biserialis (Hill)

Clathrithalamus mawsoni Debrenne & Kruse

?Didymocyathus ?hillae Debrenne & Rozanov

Erismacoscinus endutus (Gordon)

Erismacoscinus stephensoni (Hill)

Erismacoscinus sp.

?Veronicacyathus gravestocki Debrenne & Kruse

Veronicacyathus sp.

Anaptyctocyathus manserghi Debrenne & Kruse

Anaptyctocyathus sp.

Table 1. *cont.*

Regional groupings: Columns 1–9 = **Antarctica** (columns 4–6 belong to the **Heritage Range**); column 10 = **South Africa**; column 11 = **South Australia**.

Taxon	King George Island (allochthonous)	Weddell Sea (allochthonous)	Whichaway Nunataks (allochthonous)	Argentina Range (in situ)	Whiteout Conglomerate (allochthonous)	Minaret Formation (in situ)	Beardmore Glacier (allochthonous)	Nimrod Glacier (in situ)	Byrd Glacier (in situ)	South Africa (allochthonous)	South Australia (in situ)
Erugatocyathus scutatus (Hill)	+	+	+	+						□	□
Erugatocyathus sp.			+					+			□
Coscinoptycta sp.								?			
?*Porocoscinus fuchsi* (Hill)			+				+				
?*Sigmocoscinus* cf. *annulatus* (Bedford & Bedford)							+				■
Signocoscinus sp.								+			□
Mawsonicoscinus sigmoides Debrenne & Kruse								+			
Putapacyathus excavatus Hill			+					+			□
Protopharetra pauciseptata (Gordon)		+								■	
Protopharetra cf. *dubiosa* Taylor		+		+							■
Protopharetra sp.											□
Archaeofungia sp.				+			+	+		□	□
Archaeocyathus cf. *kuzmini* (Vologdin)		+		+							□
Archaeocyathus sp.					+						
Paranacyathus cf. *parvus* (Bedford & Bedford)				+	+						■
Paranacyathus sp.		?			+						□
Graphoscyphia sp.			?							□?	□
Pycnoidocyathus sp.			+				+				□

Pycnoidocyathus latiloculatus (Hill)

Pycnoidocyathus contractus (Hill)

?Pycnoidocyathus ecdemus Debrenne & Kruse

Metaldetes pratti (Hill)

?Metaldetes lairdi (Hill)

Metaldetes plicatus Gordon

?Metaldetes fortiseptatus (Hill)

?Palmericyathellus ?tabularis (Bedford & Bedford)

Palmericyathellus sp.

Pseudosyringocnema uniserialis (Hill)

Pseudosyringocnema gracilis (Hill)

Syringocnema favus Taylor

Syringocnema sp.

LATE CAMBRIAN

Antarcticocyathus webersi Debrenne & Rozanov

South Africa, South Australia
species present ■
genus present □

Archaeocyaths were also reported by Mawson (1940) from Cape Denison. Jago & Oliver (1985) have recently demonstrated that the specimen in question does not contain remnants of Archaeocyatha.

During the mapping programme of the twenty fifth Soviet Antarctic Expedition (1981–1982) in the Neptune Range, Pensacola Mountains, W. Weber collected samples from the Nelson Limestone, some containing a Cambrian fauna in which 'undefined' Archaeocyatha (fragments) were mentioned (Tröger & Weber 1985). Re-examination of the material by F. Debrenne invalidates this assertion. No archaeocyaths have been recognized within the Nelson Limestone.

Nomenclatural approach

Recent investigations of archaeocyathan systematics have led to some changes in generic taxonomy; the results of these studies (Debrenne et al., in press) are applied here to the taxa determined in the various Antarctic localities.

The major nomenclatural changes concern the systematics of tabulate Regulares. Ontogenetic studies have demonstrated that the tabulae appear either after the septa (Erismacoscinus) or before the septa (Coscinocyathus) in the course of ontogeny. In the former, tabulae are independent from the wall; in the latter they participate in the construction of one or both walls. The significance of pectinate tabulae has been reevaluated and they are now considered to be sporadic features with no systematic value. Consequently Thalamopectinus arterialis is a junior synonym of Thalamocyathus trachealis, and Ethmopectinus walteri of Ethmocyathus lineatus. At the genus level, among other synonymies, Aldanocyathus is a junior synonym of Nochoroicyathus, Cricopectinus of Cyathocricus and Glaessnericyathus of Stillicidocyathus.

In addition to these basic nomenclatural changes, F. Debrenne has recently had the opportunity to re-examine most of the Antarctic collections. These revised collections comprise Gordon's (1920) material from a dredging in the Weddell Sea (helded in King's College, London), Hill's (1965) specimens the Whichaway Nunataks (in the Bristih Museum (Natural History)), Hill's (1964a) Plunket Point collection (University of Adelaide), and the King George Island specimen of Morycowa et al. (1982) (on loan from the University of Jagellon, Krakow, Poland).

The Shackleton Limestone material from the Nimrod Glacier studied by Hill (1964b) is un-available, but this shortcoming is rectified by later collections studied by Debrenne & Kruse (1986). Unfortunately we could not gain access to the collections from the Argentina Range brought back to Leningrad by geologists of the twenty second Soviet Antarctic Expedition; we only have at our disposal a list of genera and species with no figures (Konyushkov & Shulyatin 1980). It is only by comparison with the known Antarctic species that we propose a revised list. The systematic composition of Antarctic fauna is given in the Appendix.

Intercontinental distribution of Antarctic archaeocyaths

Table 1 shows the geographical distribution of genera and species recorded from Antarctica; comparison with South Africa and Australia is given in the two right hand columns. The South African fauna (Debrenne 1975) has also been revised taking into account the recent developments in systematics. The Protopharetra species of South Africa, for instance, have been re-determined after examination of new Sardinian collections: Protopharetra densa Bornemann is now considered as Archaeocyathus sp., Protopharetra grandicaveata Vologdin and polymorpha Bornemann as Protopharetra pauciseptata (Gordon). Andalusicyathus cooperi is transferred to ?Graphoscyphia.

Furthermore, examination of Plunket Point material has disclosed the presence of Flexanulus which was to date known only in South Africa. Of the South African genera, only Statanulocyathus has not yet been found beyond South Africa: all the other genera are also known from Antarctica.

The South African archaeocyaths have not been found in situ, but within boulders in the Permo-Carboniferous Dwyka Tillite (in an analogous situation to the Anctic Whiteout Conglomerate). In the Permian and Carboniferous these continents were part of a larger Gondwana supercontinent, and the blocks now found in the South African tillites were probably derived from Cambrian archaeocyath-bearing rocks in Antarctica (Cooper & Oosthuizen 1974).

Similarly, the strong Antarctic faunal affinity with Australia favours the juxtaposition of Antarctica and Australia during the Early Cambrian. In South Australia, carbonates of the Wilkawillina and Ajax Limestones yield most of the species in common. In addition, a forthcoming article on the Sellick Hill archaeocyathan fauna (Debrenne & Gravestock in,

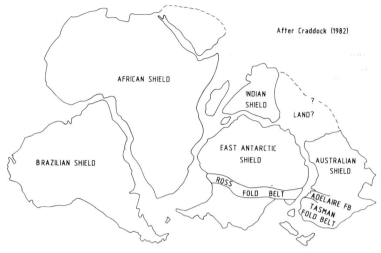

Fig. 5. Reconstruction of Gondwana in Cambrian time.

press) records the genus *Kymbecyathus*, to date known only in Antarctica, and South Australia.

Forty-three Early Cambrian forms have been determined at the species level in Antarctica. Fifteen of these are present also in South Australia, and nine in South Africa. None of the approximately 270 known Gondwana species is present on other continents, hence they may all be considered as endemic to an Australo-Antarctic (or Gondwana) province. At the generic level, however, 32 of the 43 Gondwana genera are found commonly in Siberia, the Far East, Altay Sayan, Baykal, Mongolia and Tuva, and more rarely, in North Africa and North America. Eleven genera are endemic.

Correlations based on these genera support a Botomian age for the Antarctic fauna, with a possible Toyonian-age fauna at Mt Egerton (Byrd Glacier) (Debrenne & Kruse 1986). The Late Cambrian age of the Heritage Range archaeocyath is attested to by the accompanying fauna (Debrenne *et al.* 1984).

Sedimentological approach

A detailed sedimentological analysis of the Antarctic Cambrian limestones has recently been carried out by Rees *et al.* (1986), who were the first to describe frameworks composed of an archaeocyath-cyanobacteria association. To be complete, such an analysis must be performed at localities where the limestones are in situ. Konyushkov & Shulyatin (1980) reported some algae (*Renalcis* and *Epiphyton*) and

stromatolite associations that support the assumption of a biohermal construction. An analogous interpretation may be given for some boulders containing *Epiphyton* (Gordon 1920).

In the Shackleton Limestone no archaeocyathan bioherms have yet been recognized. In the material studied by Debrenne & Kruse (1986), archaeocyaths, trilobites, molluscs and echinoderms flourished in shallow shelf waters, behind shelf margin ooid shoals; in addition, algal−archaeocyath bioconstructions are mentioned by Rees *et al.* (1986). Nevertheless it is difficult at the present stage of investigations, which are scarce and scattered, to reconstruct precise conditions of deposition within Antarctica, and the relative position of the build-ups and shallow shelves with respect to the Cambrian coast.

Conclusions

Most of the Antarctic collections come from allochthonous blocks and are based on few specimens. Good in situ collections have only been made within the Shackleton Limestone: these comprise Nimrod and Byrd Glacier faunas and recently-collected material from the Holyoake Range which remains to be studied. The in situ collection from the Argentina Range has also yet to be investigated in detail. Nevertheless a good overview of the Antarctic archaeocyathan fauna is available and the close relationships with South Australia and South Africa clearly established, confirming the reconstruction of the Gondwana continents at both the commencement and conclusion of Palaeozoic time (Fig. 5). Antarctica is the only place in the world where archaeocyaths survived until the Late Cambrian.

P. D. K. publishes with the approval of the Secretary, Northern Territory Department of Mines and Energy.

Appendix: *Systematic composition of Antarctic fauna*

Original name	Revised name
(A) Gordon (1920), Weddell Sea; allochthonous	
Archaeocyathus pauciseptatus Gordon	*Protopharetra pauciseptata* (Gordon)
Thalamocyathus flexuosus Gordon	*Ajacicyathus ajax* (Taylor)
Thalamocyathus ichnusae (Meneghini)	*Ajacicyathus ajax* (Taylor)
Thalamocyathus infundibulum (Bornemann)	*Ajacicyathus* sp.
Thalamocyathus trachealis (Taylor)	pars *Thalamocyathus trachealis* (Taylor)
	pars *Thalamocyathus tectus* Debrenne
Thalamocyathus tubavallum (Taylor)	*Stapicyathus incisus* (Hill)
Spirocyathus atlanticus (Billings)	*Archaeocyathus* cf. *kuzmini* (Vologdin)
Syringocnema gracilis Gordon	*Pseudosyringocnema gracilis* (Gordon)
Coscinocyathus endutus Gordon	pars *Erismacoscinus endutus* (Gordon)
	pars *Erugatocyathus scutatus* (Hill)
Coscinocyathus fultus Gordon	pars *Erismacoscinus endutus* (Gordon)
	pars? *Erugatocyathus scutatus* (Hill)
Protopharetra polymorpha Bornemann	*Protopharetra* sp.
Protopharetra radiata Bornemann	*Protopharetra pauciseptata* (Gordon)
Metaldetes plicatus Gordon	*Metaldetes plicatus* Gordon
Dictyocyathus sp.	*?Graphoscyphia* sp.
(B) Hill (1964a), Plunket Point (Beardmore Glacier); allochthonous, Shackleton Limestone	
Ajacicyathus sp.	*Ajacicyathus* sp.
Coscinocyathus sp.	*Erugatocyathus* sp.
Coscinoptycta bilateralis (Taylor)	*Coscinoptycta* sp.
Thalamocyathus trachealis (Taylor)	*Flexanulus* cf. *oosthuizeni* Debrenne (Fig. 30)
Formosocyathus antarcticus Hill	*Sanarkophyllum antarcticum* (Hill)
	(Fig. 3C—E)
Flindersicyathus sp.	*Pycnoidocyathus* sp. (Fig. 3B)
(C) Hill (1964b), Nimrod Glacier; in situ, Shackleton Limestone	
Porocyathus nimrodi Hill	indeterminate − no outer wall
Thalamocyathus trachealis (Taylor)	*Thalamocyathus tectus* Debrenne
Coscinocyathus sp.	*?Erugatocyathus* sp.
Flindersicyathus latiloculatus Hill	*Pycnoidocyathus latiloculatus* (Hill)
Protopharetra sp.	*Protopharetra* sp.
Bedfordcyathus lairdi Hill	*Metaldetes lairdi* (Hill)
	Syringoncnema favus (Taylor) (Fig. 4B)
(D) Hill (1965), Whichaway Nunataks; allochthonous	
Monocyathus contractus Hill	*Archaeolynthus contractus* (Hill)
Ajacicyathus sp.	*Ajacicyathus* sp.
Robustocyathus incisus Hill	*Stapicyathus incisus* (Hill)
Ethmophyllum biseriale Hill	*Tegerocyathus biserialis* (Hill)
Tumulocyathus curvatus Hill	*?Cadniacyathus curvatus* (Hill)
Ladaecyathus pratti Hill	*Metaldetes pratti* (Hill)
Ladaecyathus fortiseptatus Hill	*Metaldetes fortiseptatus* (Hill)
?Syringocyathus sp.	*?Graphoscyphia* sp.
Thalamocyathus trachealis (Taylor)	pars *Aporosocyathus* sp.
	pars *Thalamocyathus trachealis* (Taylor)
	pars *Thalamocyathus tectus* Debrenne

Appendix: *(cont.)*

Original name	Revised name
Coscinocyathus endutus Gordon	*Erismacoscinus endutus* (Gordon)
Coscinocyathus stephensoni Hill	*Erismacoscinus stephensoni* (Hill)
Coscinocyathus sp.	*Erugatocyathus* sp.
Torgaschinocyathus scutatus Hill	*Erugatocyathus scutatus* (Hill) (Fig. 4F)
Coscinoptycta fuchsi Hill	*?Porocoscinus fuchsi* (Hill)
Putapacyathus excavatus Hill	*Putapacyathus excavatus* Hill
Flindersicyathus latiloculatus Hill	*Pycnoidocyathus latiloculatus* (Hill)
Flindersicyathus uniserialis Hill	*Pseudosyringocnema uniserialis* (Hill)
Flindersicyathus contractus Hill	*Pycnoidocyathus contractus* (Hill)
Flindersicyathus sp.	*Pycnoidocyathus* sp.
Claruscyathus sp.	*Palmericyathellus* sp.
Syringocnema gracilis Gordon	*Pseudosyringocnema gracilis* (Gordon)

(E) Konyushkov & Shulyatin (1980), Argentina Range; in situ, Schneider Hills group

Dokidocyathus(?) sp.	*?Dokidocyathus* sp.
Aldanocyathus ex gr. *arteintervallum* (Vologdin)	*Ajacicyathus ajax* (Taylor)
Ajacicyathus sp.	*Ajacicyathus* sp.
Archaeofungia sp.	*Archaeofungia* sp.
Loculicyathus(?) sp.	*Paranacyathus* sp.
Ethmophyllum(?) sp.	*?Kiwicyathus* sp.
Erbocyathus sp.	*?Somphocyathus* sp.
Gordonicyathus sp.	*Gordonicyathus* sp.
Thalamocyathus(?) cf. *trachealis* (Taylor)	*Thalamocyathus* cf. *trachealis* (Taylor)
Coscinocyathus sp.	*?Erismacoscinus* sp.
Torgaschinocyathus(?) cf. *scutatus* Hill	*Erugatocyathus scutatus* (Hill)
Archaeocyathus cf. *atlanticus* Billings	*Archaeocyathus* cf. *kuzmini* (Vologdin)
Flindersicyathus cf. *latiloculatus* Hill	*Pycnoidocyathus* cf. *latiloculatus* (Hill)
Protopharetra cf. *dubiosa* Taylor	*Protopharetra* cf. *dubiosa* Taylor
Claruscyathus(?) sp.	*?Palmericyathellus* sp.
Syringocnema gracilis Gordon	*Pseudosyringocnema gracilis* (Gordon)
Syringocnema sp.	*Syringocnema* sp.

(F) Morycowa *et al.* (1982), King George Island; allochthonous, Polonez Cove Formation

Ajacicyathus sp.	*Ajacicyathus* sp.
Thalamocyathus trachealis (Taylor)	*Thalamocyathus trachealis* (Taylor)
new determinations:	*Stapicyathus incisus* (Hill) (Fig. 4D, E)
	Ajacicyathus ajax (Taylor)
	Cyathocricus tracheodentatus (Bedford & Bedford) (Fig. 3G)
	?Kiwicyathus sp.
	Erugatocyathus scutatus (Hill) (Fig. 3F)
	?Veronicacyathus sp.

(G) Debrenne & Kruse (1986); in situ, Shackleton Limestone. Only one change: *Thalamopectinus arterialis* Debrenne is a junior synonym of *Thalamocyathus trachealis* (Taylor). Examination of the type specimen of *Sanarkophyllum antarcticum* (Hill) shows that the inner wall pore tubes are of composite type: from each intersept arises a downwardly directed aporose tube bearing a long upwardly directed bushy louvre. Neighbouring louvres may coalesce to form a continuous annular structure, which was originally mistaken by Hill to be a porous membrane (Fig. 3E).

References

BURGESS, C. J. & LAMMERINK, W. 1979. Geology of the Shackleton Limestone (Cambrian) in the Byrd Glacier area. *New Zealand Antarctic Record*, **2**, 12–16.

COOPER, M. R. & OOSTHUIZEN, R. 1974. Archaeocyathid-bearing erratics from Dwyka Subgroup (Permo-Carboniferous) of South Africa, and their importance to continental drift. *Nature*, **247**, 396–398.

CRADDOCK, C. 1982. Antarctica and Gondwanaland. Review paper. *In*: CRADDOCK, C. (ed.) *Antarctic Geoscience*, International Union of Geological Sciences Series B, No. 4, University of Wisconsin Press, Madison, 3–14.

— — & WEBERS G. F. 1964. Fossils from the Ellsworth Mountains, Antarctica. *Nature*, **201**, 174–175.

DEBENHAM, F. 1921. The sandstone, etc., of the McMurdo Sound Terra Nova Bay, and Beardmore Glacier regions. *British Antarctic (Terra Nova) Expedition 1910, Natural History Report, Geology*, **1(4)**, 103–119.

DEBRENNE, F. 1975. Archaeocyatha provenant de blocs erratiques des tillites de Dwyka (Afrique du Sud). *Annals of the South African Museum*, **67**, 331–361.

— — & GRAVESTOCK, D. I., in press. Archaeocyatha from the Sellick Hill Formation and Fork Tree Limestone on Fleurieu Peninsula, South Australia. *Special Publication of the Geological Society of Australia*

— — & KRUSE, P. D. 1986. Shackleton Limestone archaeocyaths. *Alcheringa*, **10**, 235–278.

— —, ROZANOV, A. Yu. & WEBERS, G. F. 1984. Upper Cambrian Archaeocyatha from Antarctica. *Geological Magazine*, **121**, 291–299.

— —, — — & ZHURAVLEV, A. Yu., in press. *Regular Archaeocyatha*. Nauka, Moscow. (In Russian).

GORDON, W. T. 1920. Scottish National Antarctic Expedition 1902–04: Cambrian organic remains from a dredging in the Weddell Sea. *Transactions of the Royal Society of Edinburgh*, **52**, 681–714.

HILL, D. 1964a. Archaeocyatha from loose material at Plunket Point at the head of Beardmore Glacier. *In*: ADIE R. J. (ed.), *Antarctic Geology*. North Holland, Amsterdam, 609–622.

— — 1964b. Archaeocyatha from the Shackleton Limestone of the Ross System, Nimrod Glacier area, Antarctica. *Transactions of the Royal Society of New Zealand, Geology*, **2(9)**, 137–146.

— — 1965. Archaeocyatha from Antarctica and a review of the phylum. *Scientific Reports, Transantarctic Expedition 10, Geology 3*, 150 pp. 12 pls.

JAGO, J. B. & OLIVER, R. L. 1985. An alleged archaeocyath from Cape Denison, Antarctica. *Transactions of the Royal Society of South Australia*, **109**, 183–185.

KONYUSHKOV, K. N. & SHULYATIN, O. G. 1980. Ob arkheotsiatakh Antarktidy i ikh sopostavlenii s arkheotsiatami Sibiri. (On the archaeocyaths of Antarctica and their comparison with the archaeocyaths of Siberia). *In*: ZHURAVLEVA I. T. (ed.), *Kembriy Altae-Sayanskoy skladchatoy oblasty (Cambrian of the Altay Sayan Fold Belt)*. Nauka, Moscow, 143–150.

LAIRD, M. G. & WATERHOUSE, J. G. 1962. Archaeocyathine limestones of Antarctica. *Nature*, **194**, 861.

— —, MANSERGH, G. D. & CHAPPELL, J. M. A. 1971. Geology of the central Nimrod Glacier area, Antarctica. *New Zealand Journal of Geology and Geophysics*, **14**, 427–468.

MAWSON, D. 1940. Sedimentary rocks. *Australasian Antarctic Expedition, 1911–14, Scientific Reports, series A.*, **4(II)**, 347–367.

MORYCOWA, E., RUBINOWSKI, Z. & TOKARSKI, A. K. 1982. Archaeocyathids from a moraine at Three Sisters Point, King George Island (South Shetland Islands, Antarctica). *Studia Geologica Polonica*, **74**, 73–80.

REES, M. N., ROWELL, A. J. & PRATT, B. R. 1986. Lower Cambrian reefs and associated deposits, central Transantarctic Mountains. *Abstracts, 12th International Sedimentological Congress, Canberra, Australia* 255–256.

SKEATS, E. W. 1916. Report on the petrology of some limestones from the Antarctic. *Reports of the British Antarctic Expedition 1907, Geology* (**12**), 189–200, 2 pls.

TAYLOR, T. G. & GODDARD, E. J. 1914. Short notes on Palaeontology. *In*: DAVID T. W. E. & PRIESTLEY, R. E. (eds). *Glaciology, physiography, stratigraphy, and tectonic geology of South Victoria Land*. Reports on Scientific Investigations, British Antarctic Expedition 1907, Geology **1**, 319 pp., 95 pls.

TRÖGER, K. A. & WEBER, W. 1985. Description of a Cambrian fauna from the Neptune Range, Pensacola Mountains, Antarctica. *Zeitschrift Geologischer Wissenschaft, Berlin*, **13**, 359–367.

WEBERS, G. F., CRADDOCK, C. & SPLETTSTOESSER, J. F. in press. Geology and palaeontology of the Ellsworth Mountains, west Antarctica. *Geological Society of America Memoir*.

Late Cambrian molluscan faunas and the origin of the Cephalopoda

GERALD F. WEBERS[1] & ELLIS L. YOCHELSON[2]

[1] Macalester College, 1600 Grand Avenue, Saint Paul, Minnesota 55105, USA
[2] Department of Paleobiology, National Museum of Natural History, Washington, DC 20560, USA

Abstract: Late Cambrian gastropod and monoplacophoran faunas from western Antarctica, eastern and mid-western North America, and northern China, are diverse and provide some insight into palaeogeography and faunal distribution. The oldest of these, a well-preserved trilobite-mollusc fauna from the Ellsworth Mountains, Antarctica is dated as latest Dresbachian (Idamean); the Antarctic fauna is not unlike a slightly younger one (Franconian) from east-central Minnesota, USA. Slightly younger Franconian-age rocks from north China contain the first authentic cephalopods. A Trempealeauan fauna from eastern New York, USA, is closely comparable to another Trempealeauan fauna from north China.

Relatively high taxonomic diversity suggests tropical to sub-tropical marine environments for all these faunas. Plate tectonic position as interpreted from palaeomagnetism supports this interpretation. The occurrence of these molluscs on separate continents, and the presumed widely separated position of these continents in the Late Cambrian, makes it likely that the distribution mechanism for the molluscs was long-lived larval forms.

A key member of the Antarctic faunule is the monoplacophoran *Knightoconus* which has been considered representative of a group directly ancestral to the oldest cephalopod *Plectronoceras*. This view is reconsidered in the light of criticisms and alternatives proposed during the last 15 years; these occupy the bulk of the paper. Reaffirmation of evolution from the Antarctic *Knightoconus* is supported both by refutation of the criticisms and by ontogenetic studies of new topotypical material of the genus.

Although gastropods are seldom used for precise correlation within the Palaeozoic, they do have considerable biostratigraphic potential. It has been demonstrated in the Late Mississippian of Utah that well-preserved gastropods provide a more precise local zonation than associated cephalopods (Gordon & Yochelson 1987). Even though early Palaeozoic gastropod and monoplacophorans are not well studied, relative to trilobites and brachiopods of that time interval, it should not be assumed that overall similarities among molluscan assemblages are simply the result of these fossils having long stratigraphic ranges, relative to the fossil forms more commonly used for correlation.

Insofar as the Late Cambrian is concerned, it seems possible to differentiate, at the species level, Trempealeauan-age gastropods and monoplacophorans from those in older Late Cambrian strata. Too little is yet known of Dresbachian mollusc assemblages to be certain they can be distinguished from those in the Franconian. The utility of a fossil group is related directly to the amount of study it has received and the Late Cambrian molluscs have been neglected. Berkey (1898) was both the first and the last to describe the Franconian molluscs of Minnesota. The Trempealeauan

molluscs of New York were last described by Walcott (1912). With more investigation, these molluscs will become increasingly useful for biostratigraphic correlation.

Dresbachian and Franconian molluscs

The mollusc fauna of the Ellsworth Mountains, West Antarctica (Fig. 1, locality A), is assigned a latest Dresbachian (Idamean) age with considerable confidence. Associated trilobites are common and include species of *Homagnostus*, *Pseudagnostus* and *Onchopeltis* along with a number of other genera (Shergold & Webers, in press). The fossils are well-preserved, suggesting little transport, but are crowded into a coquina. The rock is considered to have been deposited in very shallow water (probably less than 10 m in depth) and probably nearshore. This inference is reinforced by the presence of oolites in the metamorphosed limestones of the Minaret Formation which contain the fauna; it is further supported by wave ripples and mudcracks in the overlying Howard Nunataks Formation.

The Ellsworth Mountains molluscs are diverse; 19 genera and 20 species (six gastropod, seven monoplacophoran, three rostroconch,

From Crame, J. A. (ed.), 1989, *Origins and Evolution of the Antarctic Biota,*
Geological Society Special Publication No. 47, pp. 29–42.

29

and four hyolithid) are represented (Webers *et al.* in press). Late Cambrian continental reconstructions (Fig. 1) place the fauna in an equatorial belt, a biogeographical position fully in accord with the high diversity. Outside of the Ellsworth Mountains, early Late Cambrian molluscan faunas are very poorly known.

The molluscan fauna that is most similar in both generic composition and relative abundance of high-coned monoplacophorans to that of western Antarctica is from the Taylors Falls area of east-central Minnesota. In terms of present-day geography, these two localities are separated by approximately one-third of the circumference of the earth. The Minnesota locality (Fig. 1, locality B) is at the present-day southern margin of the Canadian Shield. Perhaps coincidentally, this locality was also nearshore or even at the shoreline during the Franconian. That interpretation is based on an exposure of the 'Millstreet Conglomerate' containing boulders as large as 1.5 m in diameter. The conglomerate is in unconformable contact with basalt bedrock which formed a seacliff at the time of the conglomerate deposition.

Trilobites and molluscs occur in the matrix between the basalt boulders. Presumably it is in situ, for otherwise shells would have been broken to fragments in this high energy environment. A second exposure of the conglomerate containing the fauna was about 50 m from the shoreline. The Millstreet Conglomerate, forming a transgressing veneer over the unconformity surface, ranges in age from Dresbachian to Franconian (Cavaleri *et al.* 1987). The beds of the Millstreet Conglomerate which yield the

molluscan fauna are equivalent to the lowest part of the Franconia Formation, as exposed at Franconia, Minnesota. As noted, this fauna has not been critically reevaluated since Berkey (1898).

The monoplacophorans and gastropods of the Late Cambrian are judged to have been epifaunal herbivores. This view is based on morphological considerations of the forms, and on data from the life habits of most living polyplacophorans and gastropods. Presumably, in both areas they lived by rasping algal films or algal mats.

There are some differences at the species level, and absence of a few genera between these two areas, but the overall similarity is high. It is so high as to leave little doubt as to their near contemporaneity, a point confirmed by study of the associated trilobites. Because the Taylors Falls fauna is slightly younger, it might have been colonized by migrants from Antarctica. Regardless of what configuration of continental margins and ocean basins is inferred from palaeomagnetic data for the Late Cambrian, these two areas were quite widely separated (Fig. 1).

The biogeographic inference that may be drawn from the similarity between these two molluscan assemblages is that these molluscs, that is the monoplacophorans and the gastropods, had a long planktonic larval stage with a potential for transportation over great distances by ocean currents. Partial support of our interpretation comes from investigations of present-day planktonic larvae. Scheltema (1986, p.307) notes '...transport for thousands of kilometers,

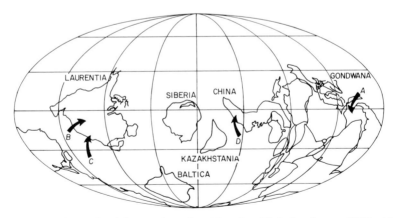

Fig. 1. Reconstruction of Late Cambrian continental positions (modified after Scotese 1987) with arrows showing the approximate locations of the Ellsworth Mountains molluscan fauna and other faunas from China and North America. A, Ellsworth Mountains Fauna, Gondwanaland (Antarctica), Late Dresbachian; B, Taylors Falls Fauna, Laurentia (North America), Early Franconian; C, Hoyt Limestone Fauna, Laurentia (North America), Trempealeauan; D, Wanwankou Limestone Fauna, North China, Trempealeauan.

either to the east or to the west is possible...'
Studies of the distribution of Cretaceous gastropods (Jablonski 1982) further support the view that long-ranging larvae have occurred in the past.

Rostroconch molluscs are present in the Ellsworth fauna but extremely rare; they are not known from the Cambrian of Minnesota. The most likely explanation to us seems to be an ecological one, perhaps based on the differences between a quartz sand environment in Minnesota and a lime-mud bottom in western Antarctica.

Trempealeauan molluscs

In connection with early Late Cambrian palaeobiogeography, it is worthwhile to discuss briefly the younger Late Cambrian molluscs. The Hoyt Limestone (Fig. 1, locality C) of eastern New York is of Trempealeauan age (Fisher 1977). The fauna is similar at the generic level to that of the Taylors Falls biota, though the similarities are not nearly as strong as they are between the Taylors Falls and the Ellsworth Mountains faunas. On the other hand the Hoyt fauna (Walcott 1912) is remarkably similar to that of the Wanwankou Formation (Kobayashi 1933) of the Liaoning Province in northeastern China (Fig. 1, locality D). There may be species in common in these two faunas (Yochelson, unpublished data). As opposed to the Ellsworth and the Taylors Falls fauna, high-coned monoplacophorans do not characterize these two faunas. The lack of modern day restudy of the latest Cambrian gastropod and monoplacophoran faunas masks the similarity, but even half a century ago, Kobayashi (1933) noted the close relationships to the Hoyt fauna. Both are associated with *Cryptozoon* stromatolites in shallow water.

In terms of present day geography, New York and NE China are also separated by approximately one-third of the circumference of the earth, and form an axis at approximately right angles to the Ellsworth Mountains/east-central Minnesota faunas. As with the previous pair of faunas, no existing reconstruction of Late Cambrian geography places these areas close together. Nevertheless, even assuming that one of the two pairs (i.e. either Minnesota and western Antarctica, or New York and northern China) of areas were adjacent in either Franconian or Trempealeauan time, it seems extremely implausible that movement of plates could separate the other pair in the short span of about 10 Ma between these two ages.

All four faunas are of high diversity, suggesting warm-water environments. This consideration is supported by plate reconstruction for the Late Cambrian (Fig. 1) showing widely-separated equatorial positions of the four faunas. The palaeogeography further reinforces the view that the molluscs spread by larval means (biological) rather than by plate movement (geological).

Biogeographical and palaeoecological summary

In younger Palaeozoic faunas, gastropods and monoplacophorans seldom form as large a proportion of the biota. They have tended to be either ignored or slighted by many investigators. However, our data suggest that the commonly worldwide distribution of these molluscs might be exceedingly useful for intercontinental correlation.

Various factors of ecology and biogeography are involved in any discussion of biostratigraphy. As is well known, many of the fossil forms such as arthropods and brachiopods, show provinciality during various times of the Palaeozoic. In marked contrast, the Late Cambrian gastropods and monoplacophorans are cosmopolitan. We believe this difference in distribution is explained by these molluscs having long-lived planktotrophic larva, as described by Jablonski (1986).

We are puzzled as to why the fossil gastropods and monoplacophorans, which are characteristic of extremely shallow, nearshore environments, are so widespread. One might theoretically expect that the forms subject to greater environmental stress would be more restricted in distribution than those of the slightly deeper water habitat offshore. Our observations on the Late Cambrian forms and on the Palaeozoic gastropods in general, refute this expectation. In spite of extensive study by many researchers (Bretsky 1969; Bretsky & Lorenz 1970; Sepkoski 1987; Tevesz & McCall 1983; Valentine 1973, 1985; and others) this problem remains unsolved.

Occurrence of the earliest cephalopods

A century ago Schmidt (1888) described the Early Cambrian *Volborthella* as the oldest cephalopod; much of the earlier literature on the origin of cephalopods considered the position of the 'orthoconic' *Volborthella* within the class. It is appropriate to note this centennial, even though that genus is no longer considered to be a mollusc, let alone a cephalopod.

For many years now, it has been generally accepted that *Plectronoceras* (see Ulrich *et al.* 1944) from the Trempealeauan limestones of NE China is the oldest cephalopod. This form was first described by Walcott (1905; see also Walcott 1913) as *Cyrtoceras cambria*. Three-quarters of a century has brought no older contenders for the title of the oldest authentic cephalopod. However, a second species of *Plectronoceras*, *P. liaotungense* (Kobayashi 1935) was described from coeval strata.

Recently, Cambrian cephalopods have been the subject of considerable research within China. In a summary of these investigations, Chen & Teichert (1983) have masterfully monographed the plectronocerids and their younger Cambrian descendants. In NE China *Plectronoceras* is exceedingly rare in the oldest beds where it occurs. Chen (pers. comm. 1981) notes that a team of five palaeontologists who visited the type locality of *P. cambria* found one additional specimen and a fragment of a second during a week of collecting. In marked contrast, specimens of this genus are relatively abundant in slightly younger beds of similar lithology (Chen & Teichert 1983).

The *Knightoconus* hypothesis

In proposing the '*Knightoconus* hypothesis,' Yochelson *et al.* (1973) suggested that this monoplacophoran genus had the necessary features to be the precursor of the cephalopods. They assumed that *Plectronoceras* was the oldest cephalopod; we reaffirm that assumption based on the work of Chen & Teichert (1983) and Chen *et al.* (1983). A fundamental fact of the palaeontological record is the stratigraphic occurrence of a fossil. At the time *Knightoconus* was described, it was older than the oldest known cephalopod. In spite of the intense field examination and study given to *Plectronoceras* and its allies, no older cephalopods have been found. It is still a valid point that the youngest *Knightoconus*, from Antarctica, is slightly older than *Plectronoceras*, from China.

Knightoconus antarcticus Yochelson, Flower and Webers is a relatively large species, some specimens attaining a maximum dimension of 7 cm. The shell is a curved cone which expands at a uniform rate; the apex does not overhang the apertural margin. The aperture is broadly oval, the narrower part of the oval being on the concave side. The convex side is taken to be anterior.

In addition to its generally cyrtoconic shell, the principal morphological feature of *Knightoconus* of concern here is the presence of multiple internal septa. For such septation to form, certain conditions appear to be necessary. At intervals, soft parts in the apical area should be unattached within the shell, the epithelial tissues in the area ought to have the ability to secrete calcium carbonate, and the body should have the ability to move forward within the shell so as to free the apical area of the soft parts and thereby allow formation of another septum.

Whether the forward internal movement in *Knightoconus*, as now represented by septal formation, was periodic cannot be conclusively determined. One cap-shaped specimen only 2 mm (Fig. 2 c, d) in length, the smallest stage known of a larger specimen, is already terminated by a septum. Many broken specimens of *Knightoconus* show regularly spaced septa. By reconstruction, we are convinced that more than one-third of the length of the cyrtoconic shell is occupied by septa. The presence of a septum in the 2 mm fragment combined with observations on several hundred specimens of various sizes, would argue for ten or more septa in a mature adult. The spacing of septa that we have observed is not erratic and there is a suggestion of uniform or near uniform spacing in the specimens of *Knightoconus*.

The hypothesis suggests that the formation of a siphuncle was fortuitous and was an inadvertant consequence of the septal formation. In one animal or in a small number of specimens, when the soft parts were initially moved forward in a small-sized shell, the epithelial covering was not completely detached from the interior of the apex. Thus, when the first septum was deposited, there could be no deposition in the area occupied by the strand of stretched tissue still attached to this apex. As each succeeding septum was deposited, it also 'continued' non-deposition in the area occupied by the strand of tissue.

Calcium carbonate was also deposited in a continuum along the strand at right angles to the main plane of the septum, forming the septal neck. Eventually, the septal neck was reduced in length. Once the chambers were interconnected by mantle tissue, the principal step in formation of a new class had been completed. This inner tissue was modified so that eventually it could control the composition and amount of fluid in each chamber. In some younger cephalopods, secondary deposits were produced in the chambers or in the siphuncle. Clearly this process involved much physiological change, but it certainly required no more change

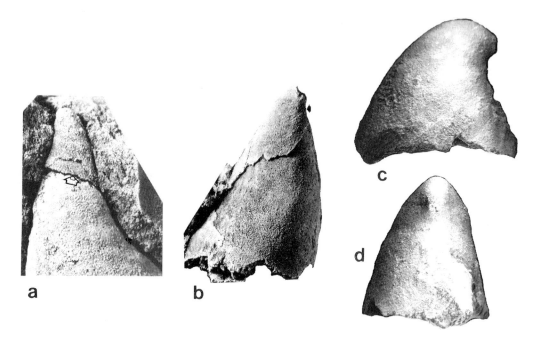

Fig. 2. *Knightoconus antarcticus* Yochelson, Flower and Webers. (**a**) Left side view of the upper part of a specimen showing recurvature at the upper broken apex of the specimen, arrow shows location of a septum, × 6.6 (**b**) Right side view of another specimen showing recurvature of the broken apical area. Small arrow shows concave portion. × 5.3; (**c & d**) Right side and anterior views of early growth stages of small specimen truncated by a septum, × 18. Photographs from Webers *et al.* in press.

than, for example, the modification of the fore-fin of a fish to the wing of an eagle.

New data on the ontogeny of *Knightoconus*

Additional material from the Minaret Formation containing numerous specimens of *Knightoconus* was collected in 1979–1980 (Webers *et al.* in press). Examination of these specimens has supported our view of the origin of the cephalopods through *Knightoconus*, or a very similar form.

Yochelson *et al.* (1973) made the suggestion that the concave surface of the cyrtoconic shell in *Knightoconus* should be considered posterior and the convex surface, anterior. This was apparently the reverse of the curvature in the shell of spoon-shaped Monoplacophora. The evidence from muscle scars in spoon-shaped monoplacophorans suggests that the concave surface is anterior. However, the oval shape of the aperture is similar in both cyrtoconic and low spoon-shaped forms. In the absense of muscle scars in the cyrtoconic forms, interpreting the widest part of the oval as posterior is consistent with the apertural shape of the low spoon-shaped group.

Admittedly, no muscle scars were known on *Knightoconus* in 1973, and in spite of many specimens examined, none have been found. Yochelson *et al.* (1973) concluded that the curvature of *Knightoconus* and other high-coned monoplacophorans did not necessarily indicate anterior and posterior, but that in all probability, the convex side of a cyrtoconic shell was anterior. This uncertainty as to the curvature in low spoon-shaped shells being a guide to orientation of cyrtoconic forms has been borne out by studies of new specimens of *Knightoconus*.

Early growth stages of *Knightoconus* as outlined by growth lines show a small cap-shaped stage which has a curvature that is the reverse of the mature shell (Fig. 2 a, b). Assuming the concave portion of the spoon-shaped stage to be anterior, the recurvature of the shell produces the same anterior/posterior orientation in *Knightoconus* as in *Plectronoceras*.

Plectronoceras cambria (Walcott)

As is well-known, the cyrtoconic *Plectronoceras cambria* (Walcott) has a curved shell, completing less than one-quarter of an open-coiled whorl. Like almost all cephalopods, the shell lies in one plane, that is, it is bilaterally symmetrical. Because of the curvature of the shell, the two sides of a longitudinal section cut along the plane of bilateral symmetry are of unequal length, the 'outer' or convex edge being longer than the 'inner' or concave side. This is the same general morphology found in *Knightoconus*.

Within the shell, more than half of the total length is occupied by closely spaced partitions, thereby forming camerae, or chambers; Chen & Teichert (1983, p. 32) report 7 septa in 2 cm of shell length. When seen in cross-section, the septa, which are much thinner than the shell wall, are gently and uniformly curved. This curvature is interrupted by a large opening that in the type lot is about 10–15% of the total length of the septum were it not interrupted; Chen & Teichert (1983; p. 32) have arrived at about the same proportion of septal diameter to chamber opening. The opening is formed by each septum bending abruptly at nearly a right angle and proceeding straight backward toward the apex, almost touching the earlier septum behind it.

On one septum in the type lot it is possible to observe the cameral opening in plan view (Fig. 3). The septal opening is essentially U-shaped, with the ends of the 'U' intersecting the curved shell wall. This opening is not circular or oval-shaped within the septum. In younger *Plectronoceras* and Early Ordovician cephalopod genera, the siphuncle is not in contact with the wall (Flower 1964, p. 28–29), so that a cross-section shows septal necks on both sides of the perforated septum. This migration away from the wall seems to us another fundamental step in development of the regulatory function of the siphuncle.

Investigators of cephalopods commonly describe the position of an eccentric siphuncle as either endogastric or exogastric; *Plectronoceras* is endogastric. In our view, it is better to discuss the morphology of the cyrtoconic shell of *Plectronoceras* in an objective manner, without reference to the hypothetical position of soft parts. Crick (1988) notes the confusion which the use of 'endogastric' and 'exogastric' have introduced into the literature on cyrtoconic cephalopods. Thus, the siphuncle of *Plectronoceras* is on the concave side of the shell. We are satisfied that mature *Plectronoceras* and mature

Knightoconus had the same orientation. During life, in our view, the longer convex surface was forward (or anterior) and the shorter concave surface was posterior.

This species is not circular but is slightly oval. The narrower part of the oval is on the concave (posterior or endogastric) side. In our experience with high conical monoplacophorans, the aperture is invariably oval with the narrower part on the concave side.

The presumed life habit of *Plectronoceras* is not directly germane to the argument, but is of interest; in the absence of any data on this issue, one can only speculate. If this cephalopod had the chambers filled with sea water ('water ballast' of Crick), it probably crawled like *Knightoconus*. Younger species of *Plectronoceras*, and presumably other Cambrian cephalopods, modified the siphuncle so that buoyancy almost immediately became a factor in the life style of this new class. There is no evidence as to whether these fossils were carnivorous or herbivorous. We believe that *Plectronoceras* was a herbivore.

Comments on the *Knightoconus* hypothesis

In the decade and a half since the Yochelson, Flower, and Webers concept has been in the literature, several workers on molluscs have considered it. We shall discuss their ideas in alphabetical order; there may be comments by others of which we are not aware. We have not discussed papers by Holland (1979, 1987), for he seemingly has accepted and understood the *Knightoconus* hypothesis as it was originally presented.

Bandel

In two allied works, Bandel (1982, 1983) in general supports the notion that a *Knightoconus*-like ancestor gave rise to *Plectronoceras*. The author adds several modifications to the hypothesis which are best illustrated by a drawing (Bandel 1982, p. 84, fig. 58B). In his reconstruction, the septal necks are inclined, the only detail of fact which can be demonstrated to be in error; as noted, the 'necks' in *P. cambria* bend abruptly back, forming essentially a right angle. We also doubt the presence of connecting rings as illustrated by Bandel and find no convincing evidence for their presence in the earliest representatives of *Plectronoceras*. Bandel reconstructs the animal with a long foot, on the back of which an operculum rests; we doubt that this structure was present but cannot argue against it. He also

a   b

Fig. 3. *Plectronoceras cambria* (Walcott). Photomicrographs of fragmented specimen truncated at a septum. The inverted U-shaped opening for the siphuncle is in contact with the outer surface of the steinkern. Paratype USNM 57821. (a) × 52. SEM Laboratory, National Museum of Natural History.

adds a bulbous larval shell to the apex for which there is no evidence.

Within the shell, Bandel modifies the musculature we presented earlier, and again, we cannot argue against his modification. There is some confusion in the literature concerning the musculature of *Hypseloconus* and the hypseloconids. Berkey (1898) reported muscle scars in his type lot, but Knight (1941) re-examined the material and refuted these presumed scars. Years later, Yochelson (1958) reported a closed ring-shaped welt on an Early Ordovician steinkern assigned to *Hypseloconus*; in all likelihood, this has nothing to do with retractor muscles, but is the 'pallial line' of this form where the mantle was attached to the shell. More recently, Stinchcomb (1980) found discrete paired muscle scars in a Late Cambrian *Hypseloconus* species. Muscle scars remain unknown in any septate hypseloconids. Thus, precisely where retractor muscles were located in *Knightoconus* and how many were present remain matters of uncertainty.

Bandel (1982, p. 84) indicates that Yochelson *et al.* (1973) suggested that there was tissue within the chambers. However, this is a misreading of our wording, for that point was not mentioned; the long septal necks of *Plectronoceras cambria* make it a virtual impossibility. We are satisfied that in the various species of *Plectronoceras*, the chambers contained only liquid, either sea water or body fluids. In that

sense, it was still nothing more than a *Knightoconus* with incomplete chambers. Bandel suggests that *Plectronoceras* had a carnivorous mode and indicates that jaws might have been present. We have no evidence to argue this point one way or the other, but think that this interpretation is quite unlikely. We have already stated our interpretation of the life habits of this genus. Although we quibble somewhat with several of the Bandel modifications, we are pleased with his general acceptance of the '*Knightoconus*' model.

Chen & Teichert

Chen and Teichert (1983, p. 40–41) concur that it is a reasonable assumption to derive the Cephalopoda from the Monoplacophora. However, they raise objections to the *Knightoconus* hypothesis.

> ...Absence of a hole would prove absence, presence of a hole, presence of a siphuncle. Once a phragmocone is built having one or more imperforate septa, there is no way in which a siphuncle could form in such a shell and the septa become perforate in later evolutionary stages.

> Since the siphuncle is an 'all-or-nothing phenomenon', its presence must be a primary characteristic of cephalopod organization. Only after it was preformed in the cephalopod body could construction of septa have begun,

and already the first septum must have been perforated to allow passage for the rudimentary siphonal strand. Presence of a septal foramen thus is a fundamental property of the cephalopodean septum which distinguishes it radically from the septum of monoplacophorans and gastropods. In other words, a monoplacophoran or gastropodan septum could never have developed in a cephalopodean septum, simply because Monoplacophora and Gastropoda have no siphon. The presence of a 'presiphuncle' in cephalopods (called 'Protosipho' by Hengsbach 1981) was already suggested by Yochelson *et al.* (1973), but the wider implications of this concept seem to have escaped these authors.

We are unconvinced by suggestions to derive the cephalopods from the Monoplacophora by way of such forms as *Yochelcionella* as proposed by Runnegar & Jell (1976) and Pojeta (1980). We find it difficult to conceive of an evolutionary mechanism by which such a highly specialized feature as the 'snorkel' of *Yochelcionella* could have developed into a specialized feature such as the siphuncle of cephalopods.

In a later work Teichert (1988, p. 22) has briefly repeated his objection cited above, but no alternative is proposed.

We concur that a hole is an all or nothing phenomenon. However, Chen and Teichert seemingly suggest that both siphuncle and septa developed simultaneously in a shell. To us this seems a remarkable coincidence. It is more parsimonious to suggest an already septate shell was modified by the formation of the siphuncle.

Our first presentation of the '*Knightoconus* hypothesis' may not have been entirely clear as to our notion concerning forward movement of the soft parts, relative to the interior of the shell apex. If all of the body mass moves forward, a fluid-filled cavity remains between the apex and the tissue. The epithelial tissue on the part of the body mass enclosed by the shell would then secrete a septum; eventually the body mass would move forward again. We concur with Chen & Teichert (1983) that the siphuncle probably could not have developed after several complete non-pierced septa were formed. It is for this reason we suggested the siphuncle developed within the first septum of a *Knightoconus* descendant.

What we proposed was that at the first forward movement of the body mass, a strand of tissue remained attached near the apex. Thus, the earliest septum could not be deposited where this strand was attached; all succeeding septa

would also contain an opening in the same position. It is this initial forward movement of the body which is critical. In that regard, we judge both the position and shape of the siphuncle to be significant.

On the convex side of the shell, musculature must be a little longer and presumably stronger than on the concave side. Anyone who has attempted to loosen a moistened postage stamp from one free corner is aware that, as the size of the loosened part increases, the resistance also increases so that commonly the last part of the stamp will not pull free from the underlying paper. The same principle applies here; of necessity, there was less of a pull by muscles on the concave side. The shape of the siphuncle of *Plectronoceras cambria* described earlier is the shape that a stretched piece of tissue would have taken rather than a structure formed completely separately. In summary, we repeat our suggestion that one time a small organism was unable to pull all of its interior epithelial tissue free of the apical shell on the concave side; this was the origin of the siphuncle, and of the Cephalopoda.

Dzik

A fundamentally different interpretation of the origin of the cephalopods is given by Dzik, a worker who has proposed radically different interpretations for a number of diverse groups. The essence of his hypothesis is given in his abstract (Dzik 1981, p. 191). This is that cephalopods were derived from '...planktic monoplacophorans possibly related to the circothecid hyoliths.'

Dzik places considerable reliance on a bulbous protoconch as an indicator of relationship, even though this same simple unornamented structure occurs in the number of unrelated organisms. For example, the *Cyrtocycloceras* sp. illustrated by him is probably a dacryoconarid, and is not a mollusc at all. Likewise, systematic position of the circothecids is quite uncertain, but they are not related to Hyolitha.

Quite apart from that, Dzik hypothesizes that straight larval shells developed a low density water bubble apically to aid flotation. Because the soft parts of fossil larvae are not known, one may speculate as one wishes. Among living planktic forms, it is known that oil vacuoles within the protoplasm seem to be the preferred mode of flotation. Indeed, from what is known of the vertical movement of many naked plankton, probably some sort of buoyancy control was developed before a shell was developed.

We cannot find any good rationale for such an apical bubble. Even if it were present, we still cannot see how the formation of several chambers which are completely closed off, and fill the apex except on one side, could have lead to a siphuncle.

There is no evidence to support Dzik's view that straight larval shells lead to cephalopods, for the orthoconic cephalopods were a later development. There is no evidence for his speculation that the tri-part Silurian aptychus gave rise to jaws. There is no evidence for the homologies that Dzik draws among muscle patterns of various groups of molluscs; his ancestral form may well be an operculum of a worm. In short, we cannot see any reason to abandon our hypothesis on the basis of his paper.

Jell, Pojeta & Runnegar

These authors in various joint works and alone have repeatedly discussed the early evolution of the molluscs. We will touch on only those papers in which cephalopods and the *Knightoconus* hypothesis was considered. Several different views have been propounded; they will be treated in chronological order. In following the chronology, readers should remember that *Yochelsonella* and *Yochelcionella* are fundamentally different taxa. The first is a smooth, compressed large form not too dissimilar to *Hypseloconus*. The second is a tiny rugose shell whose profile is interrupted by the development of a tube or 'snorkel'. Runnegar & Pojeta (1974) stated '...the Late Cambrian-Early Ordovician genera *Hypseloconus* Berkey, *Yochelsonella* Flower, and *Knightoconus* Yochelson, Flower and Webers are tall, laterally compressed shells that appear to have been curved away from the head. Such shells first appear in the early Late Cambrian. *Knightoconus* has apical septa, and Yochelson *et al.* have suggested that such septate monoplacophorans became primitive cephalopods when they developed a siphuncle...', and '... We believe that the hypseloconids were derived from cone-shaped shells like *Tannuella*...'. Their acceptance of the *Knightoconus* hypothesis at this point is indicated by their illustration (Runnegar & Pojeta 1974, p. 314, fig. 4) which shows *Knightoconus* as ancestral to *Plectronoceras*. Nearly the same remarks are repeated in another work (Pojeta & Runnegar 1976, p. 32).

Runnegar & Jell (1976) also accept the origin of the cephalopods from a *Knightoconus*-like ancestor. They state (Runnegar & Jell 1976, p. 125) '...Perhaps *Hypseloconus* is a tall cone because it is endogastrically curved. If so, endogastric curvature must have been the most important single character for the production of the Cephalopoda...'. Additionally, they suggest that the development of endogastric shells might have developed through a morphological transition from the exogastrically curved *Yochelcionella cyrano* and *Y. daleki*, to the endogastrically curved *Y. ostentata*.

Jell (1976) suggested the origin of the cephalopods from yochelcionellids. *Yochelcionella* is a small helcionellid with a large 'snorkel' projecting from the shell near the apex. The snorkel was interpreted as an inhalent siphon. In some forms the snorkel is on the concave side of the apex, and in others, it is on the convex side. In every species the side of the shell bearing the snorkel is considered by all these authors to be anterior; where the snorkel is on the convex side of the apex, the shell is considered to be recurved. Jell (1976, p. 270) states, 'The arrival and retention of this reverse coiling in tall yochelcionellids may have been followed in turn by the loss of the snorkel, perhaps due to a change in life habit from benthonic to nektobenthonic; the development of apical septa, as the first step in developing a buoyancy mechanism; and finally, the development of a siphuncle, a regulator for the buoyancy mechanism, as the final step in the progression to the nektonic Cephalopoda'.

This model has become known as the 'snorkel hypothesis'. We judge the concept implausible, and we consider *Yochelcionella* quite unlike *Plectronoceras*. Recurvature is the only feature demonstrated by the model, and then only if the snorkel had the inhalent function assumed for it. For cephalopods to evolve, first the snorkel must disappear; then the critical features of a cephalopod (septation and a siphuncle) must somehow develop.

Runnegar (1980) and Pojeta (1980) accept the snorkel hypothesis for the origin of the Cephalopoda. Pojeta stresses the importance of the development of a siphuncle relative to the development of septa and derives the siphuncle from the snorkel. Concerning *Yochelcionella ostentata* he states (Pojeta 1980, p. 71) '...Such a shell could have been ancestral to the cephalopods, the snorkel becoming a siphuncle after its end was sealed...' It is not stated how the end of the snorkel became sealed. Further, if the snorkel was sealed during life, it could not have functioned as an inhalent siphon.

Runnegar (1983, p. 129) states '...It is also clear that the ancestry of the Cephalopoda lies in tall, backwardly-curving (endogastric) monoplacophorans such as *Knightoconus*...However, the direct ancestor is not yet known, nor is the mechanism clear...' He further comments that 'Knightoconus is not the only tall, septate Cambrian monoplacophoran...and it is still not known how the siphuncle might have evolved'. The snorkel hypothesis is not discussed.

In another joint paper Runnegar & Pojeta (1985) devote two pages to the cephalopods. Part of this space is occupied by a diagram of a cross-section of *Plectronoceras cambria* (Runnegar & Pojeta 1985, fig. 19). The authors illustrate a large number of septa: they show several septa within the siphuncle, and make the point that these block off access to the apex. Septal 'necks' are drawn with reverse curvature. The authors further show within the siphuncle near the body chamber half a dozen tusk-like structures attached to the concave wall and indenting the siphuncular neck; these structures are referred to as connecting rings. However, Runnegar (pers. comm. 1988) indicated that the siphuncular deposits and connecting rings were based on structures that are present in younger (Lower Ordovician) cephalopod genera such as *Ellesmeroceras* (e.g. as illustrated by Flower 1964); it was assumed by these authors that the illustrated structures might have been present in *Plectronoceras cambria*, but were not preserved. We view this drawing as being based on unfounded assumptions and are satisfied that no connecting rings or siphuncular structures are known in *Cyrtoceras cambria*, the type species of *Plectronoceras*.

It is guardedly accepted by Runnegar & Pojeta (1985, p. 37) that a tall endogastrically curved shell such as *Knightoconus* was ancestral to the cephalopods. 'However, *Knightoconus* is not the only tall, septate Cambrian monoplacophoran — a tall septate monoplacophoran occurs in earliest Middle Cambrian strata in China — and *Knightoconus* is considerably larger than *Plectronoceras*. Furthermore, there is no entirely adequate explanation for the origin of the siphuncle.' The snorkel hypothesis is not mentioned.

By way of reply, it is sufficient to note that there may well have been conical, tall, septate monoplacophorans older than *Knightoconus*; because this Chinese form is not named, it is impossible to investigate this remark and find out whether it is multiseptate. To the best of our knowledge, *Knightoconus* and *Shelbyoceras* are the only tall, multiseptate, conical monoplacophorans which have been described. Even if this enigmatic Chinese form was multiseptate, the important point (to repeat) is not how old is the earliest ancestor, but how young is the latest

relative of the oldest cephalopod.

It is freely granted that some *Knightoconus* are larger than *Plectronoceras*; *Plectronoceras* is a more slender shell, as is well known. The comment on size has nothing to do with the basic issue, namely development of the siphuncle. The *Knightoconus* hypothesis was concerned with the origin and original positioning of the siphuncle, not with its subsequent modification for interacting with the fluid in the chambers. The imaginative addition of connecting rings and siphuncular structures to the oldest cephalopod complicates a straightforward interpretation.

Finally, Pojeta (1987, figs 14, 15) in a sketch of a proposed phylogeny of the Mollusca, illustrates two species of *Yochelcionella* as directly ancestral to *Plectronoceras*, repeating without amplification the view propounded in 1980.

Kobayashi

In a recent short work, Kobayashi (1987) points out, quite correctly, that he deserves credit for first noting multiple septation in Cambrian Mollusca and for considering this as fundamental to the origin of the class. He does not elaborate on the details of siphuncle formation.

Several points in his note deserve comment. Kobayashi mentions septation in Early and Middle Cambrian species assigned to *Helcionella*. For present purposes, as noted, the issue is not the oldest forms with septation, but rather the youngest prior to the appearance of the oldest cephalopod. In this regard, *Knightoconus* still seems particularly appropriate. Further, there is confusion as regards the systematic position of *Helcionella*. One of us (E. L. Y.) is firmly of the opinion that *Helcionella* and its close allies, which are laterally compressed, are so different from the broad monoplacophorans as to constitute a separate class. Some of the species which various authors have assigned to *Helcionella* are more properly placed within the hypseloconic monoplacophorans.

Kobayashi remarks on the changing angle of the septal necks with growth and on the presence of one connecting ring in *P. liaotungense* Kobayashi. Neither of these features occurs in the type species *Cyrtoceras cambria*. In passing, it should be mentioned that Kobayashi makes comparisons of the curved shell of *Plectronoceras* to the similar-appearing Middle Ordovician *Pollicina*. Although *Pollicina* was placed within the Monoplacophora, it is now known to be an open-coiled, very slightly asymmetric gastropod (Yochelson, unpublished data).

The angle of the septal necks in Kobayashi's species of *Plectronoceras* is indeed inclined relative to the septum, not at right angles to it. However, the presumed connecting ring is seen in one cross-section at one place, on one side, not both sides, of the siphuncle. Re-examination of Kobayashi's illustration convinces us that the 'connecting ring' is actually a fossil fragment, perhaps part of a trilobite. In our view, connecting rings were a later development of cephalopods. The septal necks first should have been reduced in length, allowing the tissue to be exposed to the fluid within the chamber. Only at that stage would connecting rings and ability to control buoyancy be developed. We surmise that this stage may have come shortly after the first development of the siphuncle.

Kobayashi (1988) has reviewed the history of study of the older cephalopods from Asia, but adds no new data. He views his species as living with the apex forward and perhaps proceeding by a series of jumping movements, thereby allowing it greater ability to capture animals. We disagree with the orientation and find no basis for the locomotion or indeed the general concept that *Plectronoceras* was necessarily a carnivore.

Wade

The most recent comment on the *Knightoconus* hypothesis to come to our attention is that of Wade (1988, p. 18), as part of a review of the phylogeny of the nautiloids. She states 'Chen and Teichert (1983: 40−41) pointed out, once again, that either a septum has a hole in or it has not. It follows that to seek an ancestor among Monoplacophora with more than one imperforate septum (Yochelson, *et al.* 1973; Holland 1979, 1987) is of doubtful value; the more septa the more doubt. The siphuncle had to be useful in an incipient form, indeed throughout development, to develop at all...'. In the remaining portion of a long, detailed paragraph she considers the 'snorkel' hypothesis and rejects it. She also finds one positive and one negative point in Dzik's hypothesis. Wade concludes this paragraph with 'The primary function of a siphuncle is the removal of liquid from newly made float chambers, and this it has had — since basic need and basic structure have not been changed — since cephalopods' first evolved.'

Her discussion of other hypotheses helps clarify several aspects of the objections to the *Knightoconus* hypothesis. The positive point for Dzik's notion is that he used the concept of larval muscles to pull the shell forward. This is also what we imply. Muscles move by contraction and for the soft parts to be pulled forward, musculature must be near the aperture. There was no indication in our original work of musculature in the apical portion of the soft parts. In any shelled mollusc, the 'visceral hump' must for practical purposes be in contact with the shell at a very early growth stage. In larval forms, although there is differentiation of mantle and visceral tissue, the geometric distance separating them is inconsequential. In a young animal, the visceral tissue is adjacent to, and in all probability is in contact with, the inner surface of the shell. All that we have stated is that a tiny portion of the tissue did not detach from the interior of the shell when the soft parts were pulled forward.

One of the several objections to the 'snorkel' hypothesis is that the tissue of siphons and siphuncles may be different in some features such as musculature and enervation. The tissue strand that we consider as having been left behind would come from the epithelial covering of the visceral hump and would not be the precursor of a siphon formed from a fold of the mantle. Thus, another of Wade's objections to the *Knightoconus* hypothesis seems to be satisfied.

There is one point of disagreement and this is with her statement that the siphuncle must have been functional from the beginning. We find the concept of simultaneous laying down of septa, development of a hole through the septa, and formation of a fully functional siphuncle as stretching the laws of probability to breaking point. Viewing these as three sequential steps, as we have done, seems more parsimonious.

Discussion

Crick (1988) has produced the most detailed study to date of cephalopod evolution from the standpoint of buoyancy regulation. He notes that the geometry of the relatively long body chamber in the curved shell of Cambrian forms may have been sufficient to regulate buoyancy, combined with 'water ballast' and the weight of numerous closely spaced septa. To restate this, there is no evidence in the earliest cephalopods of any particular mechanism to decrease weight and increase the buoyancy. It is because these animals are so little different from the hypseloconid monoplacophorans that we assume they were crawling herbivores or omnivores.

The development of connecting rings, the only hard-part morphological evidence related to the nature of the soft tissue of the siphuncle, may be a key point. Flower (1964, p. 30, fig. 5) discusses and illustrates them in *Paleoceras*, a late Trempealeauan form. He notes that 'fragility of the connecting ring seems to be a common feature of the Plectronoceratidae; indeed destruction of the ring seems far commoner than its preservation' (Flower 1964, p. 29). If connecting rings are fragile in the younger Plectronoceridae, and are strong in the younger Ellesmeroceratidae, it seems reasonable to us that these structures may not have been present in the oldest *Plectronoceras*. Chen & Teichert (1983, p. 32) report that connecting rings 'have been destroyed' in *Plectronoceras*, the only one preserved being the suspect ring of *P. liaotungense* Kobayashi. In the absence of any other evidence, we wonder if such rings were present.

In restudying the thin-section in the type lot of *Plectronoceras cambria* (Walcott), we are struck by the remarkably long extension of the septal neck toward the apex. As noted, each septum bends abruptly at nearly a right angle and proceeds adaperturally until it almost touches the next older septum. It seems fairly obvious that the ability of the apical portion of the soft parts to secrete calcium carbonate, which formed the septa, continued in the trapped and stretched strand of tissue still attached to the apex.

To move from the organization of *Plectronoceras cambria* to that of younger cephalopods, it is necessary to shift the siphuncle from its position in contact with the shell wall to a position within the septum, and to shorten septal necks. Which of these steps took place first, or whether they occurred together, we do not know. We believe both steps are associated with the development, or the increase in efficiency, of the removal of cameral fluids by the siphuncle. We suspect that the siphuncle moved away from the shell wall first. This would give more flexibility to the siphuncle and provide greater surface area. That would then be followed by the shortening of the septal necks, which would also provide greater surface area for the siphuncle.

We suspect that connecting rings could not have developed before these internal modifications were made. Connecting rings are not reported until the evolution of younger plectronocerids, such as *Lunanoceras* and *Eodiaphragmoceras* (Chen & Teichert 1983, p. 51–52). From this point onward, physiological modification of fluid regulation within the chambers was the key to development of the cephalopods.

We cannot say that *Knightoconus* is precisely the direct ancestor of *Plectronoceras*. Nevertheless, we see no reason to change our basic point that development of multiple septation in a shell was a precursor to the Cephalopoda. In our view, the formation of a siphuncle was an accident in the development of septation, one that led to unexpected pathways of animal evolution. Even if our model of the origin of the siphuncle should prove to be in error in detail, we suggest that *Knightoconus*, or a very closely related form, is an appropriate ancestor to the cephalopods on the basis of both its stratigraphic position, and its multisepate, 'endogastric' curved shell.

'Ontogeny recapitulates phylogeny' is one of the great half-truths in biology and palaeontology. Living *Nautilus* has been shaped by 500 Ma of evolution. The development of its embryonic shell and early septation are only just now being revealed (Arnold *et al.* 1987). In spite of the gross oversimplification of Haeckel's law, and all its pitfalls, we are impressed with the similarity between these dramatic new discoveries and our suggestions as to developments within a shell of *Knightoconus*.

This paper is dedicated to the memory of our co-worker Rousseau H. Flower.

References

ARNOLD, J. M., LANDMAN, N. H., & MUTVEI, H. 1987. Development of the embryonic shell of *Nautilus*. *In:* SAUNDERS, W. B., & LANDMAN, N. H. (eds), *Nautilus: The biology and palaeobiology of a living fossil*, Plenum Press, New York & London. 373–400.

BANDEL, K. 1982. Morphologie und Buildung der frühtogenetischen Gehäuse des conichiferen Mollusken. *Facies*, 7.

—— 1983. Wondel der Vorstelungen von Frühevolution der Mollusken, besonders der Gastropoda und Cephalopoda. *Palaeontologische Zeitschrift*, **57**, 271–84.

BERKEY, C. P. 1898. The Geology of the St. Croix

Dalles, pt. 3, Paleontolgy. *American Geologist*, **21**, 270–274.

BRETSKY, P. W., 1969. Evolution of Paleozoic benthic marine invertebrate communities, *Palaeogeography, Palaeoclimatology, Palaeoecology*. **6**, 45–49.

—— & LORENZ, D. M. 1970. Adaptive response to environmental stability: a unifying concept in paleoecology. *Proceedings of the North American Paleontological Convention*, **I**, Allen Press, Lawrence, Kansas. 522–520

CAVALERI, M. E., MOSSLER, J. H., & WEBERS, G. F. 1987. The Geology of the St. Croix River Valley. *In: Field Trip Guidebook for the Upper*

Mississippi Valley, Minnesota, Iowa, and Wisconsin, Minnesota Geological Survey, Guidebook Series 15, 23–43

CHEN, Jun-yuan, & TEICHERT, C. 1983. Cambrian Cephalopoda of China. *Palaeontographica*, Bd. **181**, Abt. A, 1–102.

——, TEICHERT, C., ZHOU, Zhi-yi, LIN, Yao–kin, WANG, Zhi-hao, & XU, Jun-tao 1983. Faunal sequence across the Cambrian-Ordovician boundary in northern China and its international correlation. *Geologica et Palaeontologica*, **17**, 1–15.

CRICK, R. E. 1988. Buoyancy regulation and macroevolution in nautiloid cephalopods. *Senckenbergiana Lethaea*, **69**, 13–42.

DZIK, J. 1981. Origin of the Cephalopoda, *Acta Paleontologica Polonica*, **26**. 161–191.

FISHER, D. W. 1977. Plate 2, Hadrynian, Cambrian and Ordovician rocks. New York State Museum, Map and Chart Series, **25**.

FLOWER, R. H. 1964. The Nautiloid Order Ellsemeroceratida (Cephalopoda). *New Mexico Institute of Mining and Technology, State Bureau of Mines and Mineral Resources, Memoir*, **12**, 1–234.

GORDON, M., Jr. & YOCHELSON, E. L. 1987. Late Mississippian Gastropods of the Chainman Shale, West-Central Utah. *Professional Paper. United States Geological Survey*, **1368**.

HOLLAND, C. H. 1979. Early Cephalopoda. *In:* HOUSE, M. R., (ed). *The origin of major invertebrate groups*. Systematics Association, Special Publication, **12**, 367–378.

—— 1987. The nautiloid cephalopods: a strange success. *Journal of the Geological Society London*, **144**, 1–15.

JABLONSKI, D. 1982. Evolutionary rates and modes in Late Cretaceous gastropods: role of larval ecology. *Proceedings of the North American Paleontological Convention*, **3** (Vol. I), 257–67.

—— 1986. Larval ecology and macroevolution in marine invertebrates. *Bulletin of Marine Science*. **36**, 565–87.

JELL, P. A. 1976. Mollusca. *In: McGraw-Hill Encyclopedia of Science and Technology, Yearbook 1976*, 269–71.

KOBAYASHI, T. 1933. Faunal study of the Wanwanian (Basal Ordovician) Series with special notes on the Ribeiridae and the Ellesmeroceroids. *Journal of the Faculty of Science, Imperial University of Tokyo, Section II, Geology, Mineralogy, Geography, Seismology*, **3**, (7), 249–238.

—— 1935. On the phylogeny of the primitive nautiloids, with descriptions of *Plectronoceras liaotungense*, new species and *Iddingsia* (?) *shantungensis*, new species, *Japanese Journal of Geology* and Geography, **12**, (1–2), 17–26.

—— 1987. The ancestry of the Cephalopoda from Helcionellacea to *Plectronoceras. Proceedings of the Japan Academy, Series B*, 135–138.

—— 1988. The early Palaeozoic cephalopods of eastern Asia, Senckenbergiana Lethaea, in press.

Knight, J.B. 1941. Paleozoic gastropod genotypes. *Special Papers. Geological Society of America*,

32, 1–510.

POJETA, J. Jr. 1980. Molluscan Phylogeny. *Tulane Studies in Geology and Paleontology*. **16**, 55–80.

—— 1987. Phylum Overview. In: BOARDMAN, R. S., CHEETHAM, A. H., & ROWELL, A. J. (eds) *Fossil Invertebrates*. Blackwell Scientific Publications, Oxford, 270–293.

—— & RUNNEGAR, B. 1976. The paleontology of rostroconch mollusks and the early history of the phylum Mollusca. *Professional Paper. United States Geological Survey*, **968**, 1–88.

RUNNEGAR, B. 1980. Mollusca: the first hundred million years. *Journal of the Malacological Society of Australia*, **4**, 223–224.

—— 1983. Molluscan phylogeny revisited. *In:* ROBERTS. J. & JELL, P. A. (eds). *Memoir. Association of Australasian Palaeontologists*, **1**, 121–144

—— & JELL, P. A. 1976. Australian Middle Cambrian molluscs and their bearing on early molluscan evolution. *Alcheringa*, **1**, 109–138.

—— & POJETA, J. Jr. 1974. Molluscan phylogeny: The paleontological viewpoint. *Science*, **186**, 311–316.

—— & —— 1985. Origin and diversification of the *Mollusca. In:* TRUEMAN, E. R. & CLARKE, M. R. (eds). *The Mollusca. 10. Evolution*, Academic Press, London. 1–57

SCHELTEMA, R. S. 1986. On dispersal and planktonic larvae of benthonic invertebrates, an eclectic overview and summary of problems. *Bulletin of Marine Science*, **39**, 290–322.

SCHMIDT, F. 1888. Uber ein neuentdeckte unter-cambrische Fauna, *Mémoires del Académie Impériale des Sciences de St. Petersbourg, 7th series*, **36**, 1–29.

SCOTESE, C. R. 1987. Phanerozoic Plate Reconstructions, *University of Texas Technical Report*, **90**.

SEPKOSKI, J. J. 1987. Environmental trends in extinction during the Paleozoic. *Science*, **23**, 64–66.

SHERGOLD, J., & WEBERS, G. F. *in press*. Late Cambrian Trilobites from the Ellsworth Mountains, West Antarctica. *Memoirs. Geological Society of America*, **170**.

STINCHCOMB, B. L. 1980. New Information on Late Cambrian Monoplacophora *Hypseloconus* and *Shelbyoceras* (Mollusca), *Journal of Paleontology*, **54**, 45–49.

TEICHERT, C., 1988, Main features of cephalopod evolution. *In:* CLARKE, M. R. & TRUEMAN, E. R. (eds.). *The Mollusca. 12. Palaeontology and Neontology of Cephalopods* Academic Press, London. 11–75

TEVESZ, M. J., & McCALL, P. L., (eds) 1983. *Biotic Interactions in Recent and Fossil Benthic Communities*. Plenum Press, New York & London.

ULRICH, E. O., FOERSTE, A. F., MILLER, A. K. & UNKLESBAY, A. G. 1944. Ozarkian and Canadian cephalopods. *Special Papers. Geological Society of America*, **58**, 1–217.

VALENTINE, J. W. 1973. *Evolutionary paleoecology of the marine biosphere*. Prentice-Hall, Englewood Cliffs, N.J.

—— (ed) 1985, *Phanerozoic diversity patterns, profiles*

in macroevolution. Princeton University Press, Princeton, N.J.

WADE, M. 1988. Nautiloids and their descendants: cephalopods classification in 1986. *New Mexico Bureau of Mines and Mineral Resources, Memoir*, **44**, 15–25.

WALCOTT, C. D. 1905. Cambrian faunas of China. *Proceedings of the United States National Museum*, **29**, 1–106.

— — 1912 New York Potsdam-Hoyt Fauna, Cambrian Geology and Paleontology, II, *Smithsonian Miscellaneous Collections*, **57**, 251–304.

— — 1913, The Cambrian faunas in China. *Publications Carnegie Institution of Washington*, **54**.

WEBERS, G. F., POJETA, J. Jr & YOCHELSON, E. L. *in press*. Molluscan faunas from the Minaret Formation, Ellsworth Mountains, West Antarctica, *Memoirs. Geological Society of America*, 170.

YOCHELSON, E. L. 1958. Some Lower Ordovician monoplacorphoran mollusks from Missouri, *Journal of the Washington Academy of Science* **4**, 8–14.

— —, FLOWER, R. H., & WEBERS, G. F. 1973. The bearing of the new Late Cambrian mono-placophoran genus *Knightoconus* upon the origin of the Cephalopoda. *Lethaia*, **6**, 275–310.

The Aztec fish fauna (Devonian) of Southern Victoria Land: Evolutionary and biogeographic significance

G. C. YOUNG

Division of Continental Geology, Bureau of Mineral Resources, PO Box 378, Canberra, ACT, 2601, Australia

Abstract: The Aztec fauna is one of the most diverse fossil vertebrate faunas from Antarctica, with over 30 taxa belonging to the four major groups of jawed fishes (acanthodians, placoderms, osteichthyans, chondrichthyans), and one group of agnathans (thelodontids). Biogeographically the strongest affinity is with eastern Australia, as part of the East Gondwana Province. Biostratigraphically the Aztec succession includes probably the youngest known non-marine thelodont agnathans at the base, and the oldest known phyllolepids at the top. This transition, and an association of species of *Turinia* and *Bothriolepis*, is not seen in any other sequence. Most or all of the Aztec sequence is regarded as Givetian (late Middle Devonian) in age. The Aztec fauna contains older and/ or more primitive representatives of several groups typical of younger (late Palaeozoic) strata in Euramerica, suggesting a biotic dispersal episode between Gondwana and Euramerica in the Late Devonian. Centres of origin and dispersal hypotheses in Palaeozoic biogeography can be related to disappearance of faunal barriers as manifested in different stratigraphic ranges in different areas. The change for Devonian vertebrates from the endemic patterns seen in the Early Devonian to the close similarity of Famennian faunas between East Gondwana and Euramerica suggests disappearance of a major faunal barrier in the Late Devonian, presumably the ocean between Gondwana and Euramerica.

History of discovery

Fish remains of Devonian age were the first vertebrates to be discovered as fossils on the Antarctic continent. The fauna from the Aztec Siltstone of the Beacon Supergroup in southern Victoria Land (Fig. 1), first collected from moraine of the Mackay Glacier near Granite Harbour (Taylor 1913), is now the best known Palaeozoic vertebrate assemblage from Antarctica (White 1968; Ritchie 1975; Young 1982, 1988). A slightly older Early Devonian fish fauna occurs in the Horlick and Ellsworth Mountains (Doumani *et al.* 1965; Bradshaw & McCartan 1983; V. T. Young 1986; Young in press) but is still poorly known. In this paper I summarise current knowledge of the Aztec fish fauna, and consider its evolutionary and bio-geographic significance in the context of early vertebrate history generally, and Devonian palaeogeography.

Material of the Aztec fauna was first collected by the Western Party of Scott's 'Terra Nova' expedition (1910–1913), and was described by Woodward (1921), who recognised eight taxa of Devonian fishes, including three new species of European genera (Table 1). This was the first time that typical 'Old Red Sandstone' type Devonian fishes were identified from the Southern Hemisphere. The first in situ remains of the Aztec fauna were collected during the 1955–58 Trans-Antarctic Expedition, from the region of the Skelton Névé in southern Victoria Land (Gunn & Warren 1962). These specimens were described by White (1968), who revised Woodward's determinations, and added five new species (in four new genera) to the faunal list (Table 1). Other fish remains from the same general area were recorded by Matz & Hayes (1966) and Helby & McElroy (1969). The most significant collections were made by the Victoria University of Wellington Antarctic Expeditions of 1968–69 (McKelvey *et al.* 1972), and 1970–71 (Ritchie 1971*a*, *b*). The present account is based mainly on these collections, which are under study at the Australian Museum, Sydney, and the Bureau of Mineral Resources, Canberra. A current faunal list is given in Table 1; several important groups remain to be described, and some of these are illustrated here for the first time.

Occurrence

The Aztec Siltstone is the uppermost formation of the Taylor Group (age: Devonian), which together with the overlying Victoria Group (Permo-Triassic) comprises the Beacon Supergroup (McKelvey *et al.* 1972). The Aztec Silt-stone crops out over an area some 150 km long, and about 50 km wide, in the Skelton Névé-Mackay Glacier region (Fig. 2). It is inter-

From Crame, J. A. (ed.), 1989, *Origins and Evolution of the Antarctic Biota,*
Geological Society Special Publication No. 47, pp. 43–62.

43

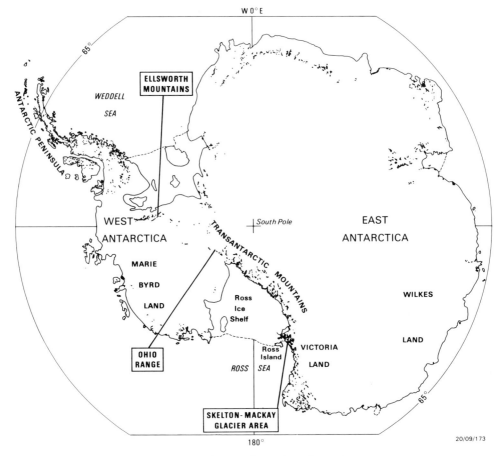

Fig. 1. The three known localities for Devonian vertebrates from Antarctica.

preted as an alluvial plain deposit, with point bar, backswamp and lacustrine lithologies (McPherson 1978). Associated conchostracans (Tasch 1987), plant macro- and micro-fossils (Plumstead 1962; Helby & McElroy 1969), various trace fossils, and other evidence (Barrett & Kohn 1975; Barrett 1980) are consistent with a non-marine depositional environment. The vertebrate assemblage is similar to those occurring elsewhere in fluviatile red bed sequences (for example the Old Red Sandstone fish faunas of Europe). The large size of some of the fishes, particularly in the upper parts of the formation, indicates substantial bodies of permanent water. However much of the sequence also includes numerous soil and caliche horizons, indicating subaerial exposure (McPherson 1979).

Currently the Aztec fish fauna is recorded from some 24 localities (Fig. 2), of which Mount Fleming, Mount Crean (Lashly Range), Portal Mountain, Mount Metschel, Alligator Peak,

and Mount Ritchie (southern Warren Range) have yielded the most important specimens. The original derivation of the 'Terra Nova' collection from moraine at Granite Harbour (2, Fig. 2) remains unknown, but there is little doubt that most of it came from the lower part of the Aztec Siltstone, and not from lower formations of the Taylor Group as suggested by Matz *et al.* (1972; their 'Windy Gully Formation', beneath the Beacon Heights Orthoquartzite). However two fish spines are recorded as coming from the uppermost beds of the Beacon Heights orthoquartzite, which is conformably overlain by the Aztec Siltstone.

The Aztec fish fauna

Early-middle Palaeozoic gnathostome (jawed) fishes belong to four major groups. The placoderms (armoured fishes) and acanthodians (spiny-finned fishes) are extinct Palaeozoic

Table 1. *Faunal lists for the Aztec fish fauna, after Woodward (1921), White (1968), and current work*

	Woodward (1921)	White (1968)	Current faunal list
Agnathans	?primitive ostracoderms	—	*Turinia* sp. nov.
Acanthodians	*Cheiracanthus*	*Cosmacanthus* *Gyracanthides warreni* *Antarctonchus glacialis* *Byssacanthoides debenhami*	*Gyracanthides warreni* White *Antarctonchus glacialis* White *Byssacanthoides debenhami* Woodward *Culmacanthus antarctica* Young
Elasmobranchs	dermal tubercles	*Mcmurdodus featherensis*	*Mcmurdodus featherensis* White *Antarctilamna prisca* Young xenacanthid sp. nov.
Placoderms	*Bothriolepis antarctica* *Byssacanthoides debenhami*	*Bothriolepis antarctica* antiarch indet. *Antarctaspis mcmurdoensis* *Antarctolepis gunni* arthrodires indet.	*Bothriolepis antarctica* Woodward *Bothriolepis alexi* Young *Bothriolepis askinae* Young *Bothriolepis barretti* Young *Bothriolepis karawaka* Young *Bothriolepis kohni* Young *Bothriolepis macphersoni* Young *Bothriolepis mawsoni* Young *Bothriolepis portalensis* Young *Bothriolepis vuwae* Young *Bothriolepis* sp. indet. 1–13 *Pambulaspis antarctica* Young *Antarctolepis gunni* White *Groenlandaspis antarcticus* Ritchie *Groenlandaspis* spp. phlyctaeniid arthrodires indet. phyllolepid *Antarctaspis mcmurdoensis* White
Crossopterygians	*Holoptychius antarcticus* osteolepid scales	*Gyroptychius? antarcticus* ?rhizodontid	*Gyroptychius? antarcticus* (Woodward) osteolepiform sp. 1 osteolepiform sp. 2 rhizodontiform gen. nov.
Actinopterygians	palaeoniscid scales	palaeoniscid scales, types I, II, III.	palaeoniscoid gen. nov. ?palaeoniscoid indet.
Dipnoans	—	—	?ctenodontid

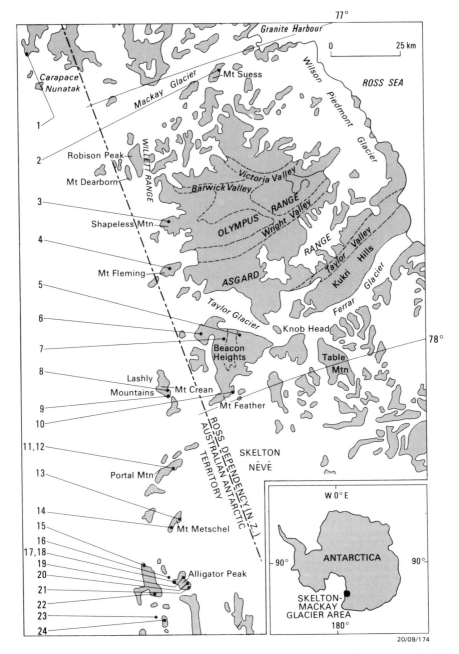

Fig. 2. Twenty-four recorded localities for the Aztec Siltstone fish fauna (Middle–Late Devonian) in the Mawson–Mulock Glacier region of southern Victoria Land. 1, Allan Hills (breccia); 2, Gondola Ridge, Mount Suess (moraine); 3, Shapeless Mountain; 4, Mount Fleming; 5, Beacon Heights; 6, Kennar Valley; 7, Aztec Mountain; 8, Mount Crean; 9, Lashly Mountains; 10, Mount Feather; 11, 12, Portal Mountain; 13, 14, Mount Metschel; 15, northern Warren Range; 16, northern Boomerang Range; 17, 18, Alligator Peak; 19, Alligator Ridge; 20, 21, eastern and southeastern spurs, Alligator Peak; 22, Mount Warren; 23, southern Warren Range; 24, Mount Ritchie.

groups (Fig. 6); the two extant groups (Fig. 3) are the chondrichthyans (sharks and their relatives) and osteichthyans (bony fishes), which make up the fish fauna of today. The Aztec fish fauna contains representatives of all four gnathostome groups, together with small scales belonging to one group of agnathans or primitive jawless vertebrates (Fig. 6B). Noteworthy is the absence from the fauna of various groups of armoured agnathans (osteostracans, heterostracans) which are abundant and diverse in Devonian fish faunas of Europe and North America.

Osteichthyans

The bony fishes of the Devonian Period include three major subgroups: the lungfishes (Dipnoi), various lobe-finned 'crossopterygian' fishes (rhipidistians, coelacanths, onychodontids), and the ray-finned palaeoniscoids (Actinopterygii). Coelacanths and onychodontids have not yet been identified in the Aztec fauna.

Lungfishes were rare, being represented only by a few isolated toothplates. Ritchie (1972) tentatively referred a specimen from Mount Fleming to the family Dipteridae, and Long & Campbell (1985) noted resemblances to the genus *Delatitia* from the Early Carboniferous of Victoria. Campbell & Barwick (1987, fig. 2) have illustrated a specimen from Mount Crean.

Much more common are remains of rhipidistian crossopterygians (Fig. 3B). Woodward (1921) referred a few imperfect scales to a new species of the porolepiform *Holoptychius*, and also identified osteolepid scales and an incomplete cheek plate. The Osteolepiformes and Porolepiformes are two major groups within

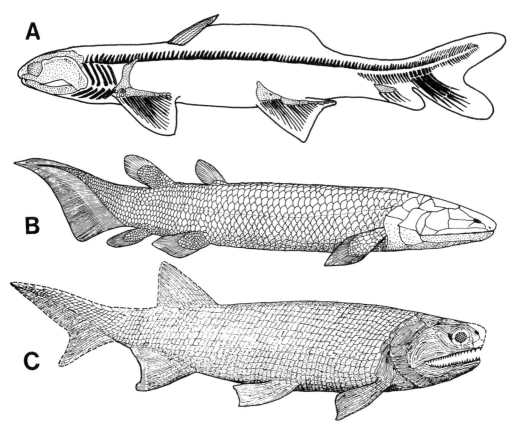

Fig. 3. Restorations based on fossils from the Aztec fauna of representatives of the two major living groups of jawed fishes (chondrichthyans and osteichthyans). (**A**)the chondrichthyan *Antarctilamna prisca* from Mount Crean (locality 8, Fig. 2; actual length about 400 mm); (**B**) a crossopterygian osteichthyan, also from Mount Crean (a new genus of rhipidistian; actual length about 1 m); (**C**) an actinopterygian osteichthyan, belonging to a new genus of palaeoniscoid fish from Portal Mountain (locality 12, Fig. 2; actual length about 200 mm). (**A**) modified after Dick 1981; (**B**) modified after Thomson 1973 and Long 1985.

the Rhipidistia. White (1968) found that some of Woodward's supposed '*Holoptychius*' scales were composed of the typical osteolepid tissue cosmine, and that the specimen identified by Woodward as a clavicle of '*Holoptychius*' was a pectoral fin element belonging to an antiarchan placoderm. White (1968) referred all the rhipidistian remains except for a possible 'rhizodont' tooth from the Boomerang Range to the osteolepid, which he renamed *Gyroptychius? antarcticus* (Table 1).

New material includes several articulated or partly associated specimens, confirming the presence of three types of rhipidistians in the fauna, belonging to at least four genera (Young, Long & Ritchie in prep.). A rare porolepiform and several cosmine-covered osteolepiforms occur in the lower part of the sequence, although cosmoid scales occur right through to the top of the sequence. In the upper part of the sequence is a large rhizodontiform rhipidistian with fangs in the jaw over 45 mm long. This is the largest predator in the fauna, and attained a size of over 2.5 m. The Rhizodontiformes are a third major group within the Rhipidistia, which typically occur in Carboniferous and Permian strata in Europe and North America. Members of the group are characterised by their large size and several special features of the shoulder girdle (Andrews & Westoll 1970).

The third major group within the Osteichthyes is the Actinopterygii (Fig. 3C), also first recognised by Woodward (1921) on the basis of a small number of palaeoniscoid scales. Woodward compared these scales to those of a Late Devonian North American form (*Rhadinichthys devonicus*), but White (1968) noted that similar scales occur in other genera, for example *Elonichthys* and *Acrolepis* from the Carboniferous. The two other palaeoniscoid scale types of large size identified by White (1968) are equally unreliable, and only articulated remains provide sufficient characters for assessment. A reconstruction based on several articulated specimens from Mount Crean and Portal Mountain (Fig. 3C) shows some similarity to the Australian Late Devonian form *Howqualepis* (Long 1988) in its cranial morphology.

Chondrichthyans

These are represented in the modern fauna by the sharks and rays (Elasmobranchii), and the ghost sharks and chimaeras (Holocephali). The fossil record of chondrichthyans is poor because their cartilaginous skeleton is not readily preserved. Fossil remains are mainly isolated teeth, spines, and scales, and scales were recognised in the Aztec fauna by Woodward (1921) on the evidence of their characteristic microstructure. White (1968) described a single shark tooth as a new genus and species, *Mcmurdodus featherensis* White, placed in a new family Mcmurdodontidae, and a new species of this genus has recently been discovered in Australia (Turner & Young 1987). Diplodont teeth resembling those of *Xenacanthus* were identified by Ritchie (1972), and described by Young (1982). These have two prominent lateral cusps projecting from a well defined base (Fig. 4D). *Xenacanthus* is a form well represented in late Palaeozoic deposits of Europe and North America. Another form also referred to the xenacanths by Young (1982) is *Antarctilamna prisca* (Fig. 3A), in which a fin-spine, scales of various types, and diplodont teeth are preserved in association (Fig. 4E). Similar remains referred to the same species have been found in Devonian strata on the south coast of New South Wales, Australia (Fergusson *et al.* 1979; Young 1982).

The teeth of *Mcmurdodus* resemble those of modern hexanchoid sharks, as first noted by White (1968), although this group is only known otherwise from Jurassic and younger strata. The presence of two types of xenacanth sharks in the Aztec fauna is also of interest because they typically occur in late Palaeozoic freshwater deposits. The earliest representatives previously known are a few isolated teeth and spines from marine early Upper Devonian rocks in North America, and possible diplodont teeth from the Early Devonian of Spain (Mader 1986). The oldest articulated remains from the Northern Hemisphere (*Diplodoselache* from the Early Carboniferous of Scotland; Dick 1981) supplements the evidence of *Antarctilamna*, which

Fig. 4. Specimens from the Aztec fauna of southern Victoria Land (housed in the Bureau of Mineral Resources, Canberra, and the Australian Museum, Sydney). (**A**) median dorsal plate of a phyllolepid from Alligator Ridge (CPC 26408; × 2); (**B**) head-shield of the antiarch *Bothriolepis askinae** from Mount Crean (holotype, CPC 25985; × 1); (**C**) turiniid thelodont and acanthodian scales from Mount Fleming (AF 168; × 20); (**D**) xenacanth shark tooth from Portal Mountain (AMF 54330; × 3); (**E**) part of the squamation and dorsal fin-spine from the holotype of *Antarctilamna prisca* Young from the Lashly Range (CPC 21187; × 5). [* specific name amended from Young 1988 in accordance with I.C.Z.N. Art. 31]

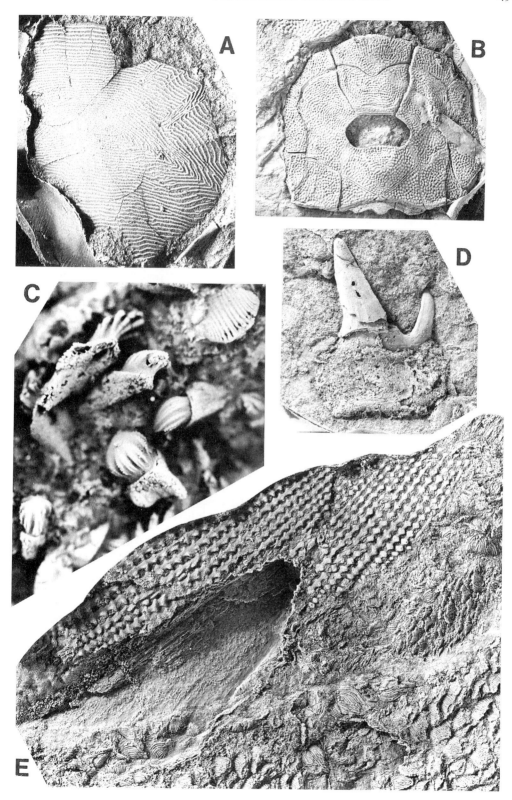

is the earliest known xenacanth represented by semi-articulated remains. Both forms indicate that primitively there was a well developed squamation, although other xenacanth sharks had a naked skin. *Antarctilamna* had a similar spine structure and attachment to that of another major group of Palaeozoic chondrichthyans, the ctenacanth sharks (Fig. 3A). Thus the characteristic single dorsal spine attached to the back of the skull or the shoulder-girdle of typical late Palaeozoic xenacanths must have developed within the group by modification of a dorsal fin spine. Of interest is the record of Devonian and Early Carboniferous xenacanths from South Africa (Oelofsen 1981), South

America (Zangerl 1981; Goujet *et al.* 1984), and Iran (Blieck *et al.* 1980). From Seripona in Bolivia Gagnier *et al.* (1988) have recently described a new species, *Antarctilamna seriponensis.*

The isolated teeth referred to *Xenacanthus* sp. by Young (1982) are considerably larger than those of *Antarctilamna*. They differ in having two cusps of different length which are twisted in different planes, in lacking the smaller central cusps, and in having a deeper more bulbous base (Fig. 4D). They are readily distinguished from all previously described teeth of xenacanth sharks, and probably belong to a new species.

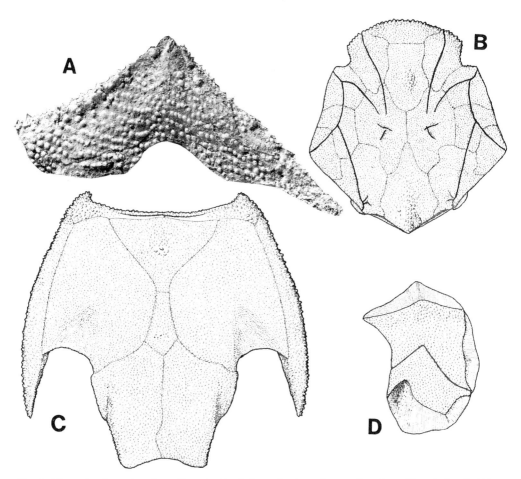

Fig. 5. The arthrodiran placoderm *Groenlandaspis* from the Aztec fauna of southern Victoria Land. (**A**) a median dorsal plate of the trunk-shield in right lateral view, from Portal Mountain (CPC 26409; × 1.8); (**B–D**) *Groenlandaspis antarcticus* Ritchie, restorations of the skull-roof in dorsal view (**B**), and trunk-shield (**C**) in ventral view (both × 0.45), and posterior dorsolateral plate from the trunk-shield (**D**) in right lateral view (× 0.675). ((**B–D**) modified after Ritchie 1975).

Acanthodians

The acanthodians (Early Silurian–Early Permian) were generally small fusiform fishes, characterised by their minute scales, small bones covering the skull, cheek, and shoulder girdle, and the prominent spines supporting the leading edge of the paired and unpaired fins (Fig. 6D). Fossil remains of acanthodians are mainly isolated scales and spines.

Woodward (1921) identified isolated acanthodian scales in the Granite Harbour samples as very close to the Middle Devonian *Cheiracanthus murchisoni* from Scotland. White (1968) described spines of several distinctive types (Table 1), including a new species (*G. warreni*) of the genus *Gyracanthides* Woodward from the upper fossiliferous horizon in the Boomerang Range. Similar isolated spines have since been found at Mount Ritchie (Warren Range), Mount Crean (Lashly Range), Mount Metschel, and north of Alligator Peak (Boomerang Range). The Antarctic species may have attained a similar size to the type species, *Gyracanthides murrayi* from eastern Victoria (Woodward 1906). *Gyracanthides* was apparently restricted to the upper horizons of the Aztec Siltstone.

A second acanthodian (*Antarctonchus glacialis* White 1968) is represented by long slender spines with numerous strong longitudinal ribs. Much smaller but otherwise similar spine fragments (*Byssacanthoides debenhami* Woodward), originally described as an antiarch, were shown by White to be acanthodian spines. Such spines occur at most localities where the Aztec fauna has been collected, although they are never common. Partly articulated material from the lower part of the Aztec Siltstone at Portal Mountain, with finwebs, spines, and much of the scale cover preserved, resembles the Middle Devonian Scottish form *Diplacanthus*. A third acanthodian genus (*Culmacanthus* Long) is represented by a single distinctive cheek plate from Mount Crean, which is placed by Young (1989) in a new species, *C. antarctica*. The type species from Victoria (Fig. 6D) has spines similar to those of *Diplacanthus* (Long 1983). Thus at least three acanthodian taxa are present in the Aztec fauna, but only one is represented by well preserved remains.

Placoderms

This group (the armoured fishes) is characterised by a well developed armour of large bony plates covering the skull, cheek, and shoulder girdle, which are robust and readily preserved. Placoderms are the most abundant and diverse group of Devonian vertebrates, and this is exemplified in the Aztec fauna, where placoderm remains belonging to the arthrodire *Groenlandaspis* (Fig. 5), and the antiarch *Bothriolepis* (Fig. 6C) are the most commonly preserved vertebrate remains.

Woodward (1921) recognised both of these placoderm groups in the Granite Harbour material, and he correctly identified the antiarch *Bothriolepis antarctica*. Young (1988) described nine new species of *Bothriolepis* from the Aztec fauna (Fig. 4B), and a new species of the asterolepid antiarch *Pambulaspis* Young 1983. White (1968) confirmed the presence of arthrodiran remains with the description of two new genera (*Antarctaspis mcmurdoensis* White, *Antarctolepis gunni* White), and other plates of uncertain affinity. Arthrodire remains belonging to *Groenlandaspis* spp. (Fig. 5) are, after *Bothriolepis*, the most common elements in the fauna. *Groenlandaspis* was previously known only by the type species, *G. mirabilis* from the Late Devonian of eastern Greenland (Ritchie 1975).

The third group of placoderms represented in the Aztec fauna is the Phyllolepida, first noted by Ritchie (1972) in material from the Boomerang Range, and also recorded from scree at Mount Ritchie. The genus *Phyllolepis* is a distinctively ornamented placoderm of uncertain affinity which is widely distributed in the Upper Old Red Sandstone of Europe, where it is a vertebrate index fossil for Famennian strata (Fig. 7). Phyllolepids have long been known from the Late Devonian of eastern Australia (e.g. Hills 1929; Long 1984; Ritchie 1984), but they were not recognized in the Antarctic collections studied by Woodward (1921) and White (1986), which were derived mainly from the lower part of the Aztec Siltstone. In the biostratigraphic scheme of Young (1988) the Antarctic phyllolepid (Fig. 4A) is used to define the highest zone in the sequence.

Agnathans

Woodward (1921) studied various minute dermal denticles in the Granite Harbour material, which he attributed 'either to primitive Ostracoderms or to Elasmobranchs'. The former is a now little-used collective term for the various armoured agnathan (jawless) vertebrates of the early and middle Palaeozoic. Two denticle types were described by Woodward, and those which in thin section are solid, with histological structure described as 'typically Elasmobranch' (Woodward 1921, p. 57), probably belong to one of the chondrichthyans in the fauna (White

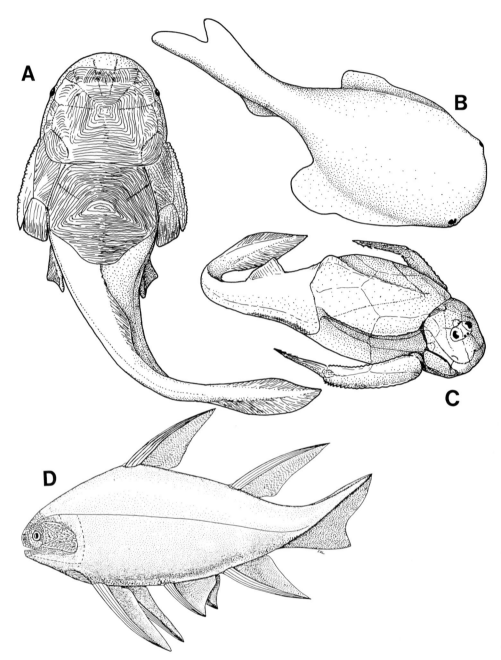

Fig. 6. Restorations of the major extinct groups of jawless agnathans (thelodontids) and jawed fishes (acanthodians and placoderms) represented in the Aztec fauna. (**A**) a phyllolepid placoderm in dorsal view (actual length about 220 mm); (**B**) a thelodontid agnathan in dorsolateral view (actual length about 340 mm); (**C**) the placoderm *Bothriolepis* in dorsolateral view (actual length about 250 mm); (**D**) the acanthodian *Culmacanthus* in lateral view (actual length about 150 mm). ((**A**) modified after long 1984 and Ritchie 1984; (**B**) based on Turner 1982*b*; (**D**) modified after Long 1983).

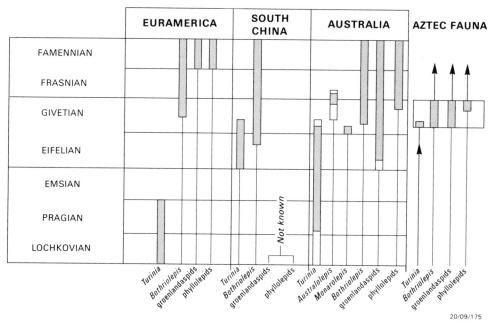

Fig. 7. Stratigraphic ranges of various taxa of Devonian fishes from Euramerica, China, and Australia compared with the Aztec fish fauna of southern Victoria Land.

1968; Young 1982). Other scales, with a conspicuous central pulp cavity, were regarded by Gross (1950) as possibly belonging to psammosteid heterostracans, a view reiterated by Tarlo (1964), because at that time psammosteids were the only heterostracans known to persist into the Upper Devonian.

New examples of these scales collected in 1970–71 have now confirmed that thelodontids are present in the Aztec fauna (Turner & Young, in prep.). The thelodontids (Fig. 6B) are a poorly known agnathan group, with a dermal cover of minute denticles, each containing a single pulp cavity. These are well represented in microvertebrate assemblages of Silurian and Early Devonian age, but in recent years they have been recovered in strata of Middle or Late Devonian age in Australia (Turner & Dring 1981; Young & Gorter 1981). Examples from the Aztec fauna exhibit the typical fluted crown, constricted neck lacking pulp openings, and thick annular base of the genus *Turinia* (Fig. 4C). In the biostratigraphic scheme for the Aztec Siltstone of Young (1988), turiniid thelodont scales are used to define the two lowest zones (associated with *Bothriolepis askinae* and *B. kohni*).

Discussion

Age of the Aztec fauna

Some consideration of the biostratigraphy and age of the Aztec fauna is necessary to place it in a time perspective. Woodward (1921) first suggested a Late Devonian age for the Granite Harbour assemblage, but with more extensive material White (1968) noted the mixture of Early and Late Devonian elements, and concluded that the fauna was more likely Middle Devonian in age. This view was contradicted by Ritchie (1975), who judged the evidence of the placoderms *Groenlandaspis* and cf. *Phyllolepis* to indicate a latest Devonian (Famennian) age. White (1968) also noted the possibility that faunas of different ages might be represented at the various localities, which are spread over distances of up to 200 km.

With detailed systematic study of the extensive collections of 1969–71 these problems have been largely resolved. In a comprehensive account of the placoderm genus *Bothriolepis*, nine new species have been described and named, with remaining *Bothriolepis* material organized into 13 indeterminate species (Young 1988).

Bothriolepis is the most common fossil in the Aztec fauna, and occurs at most localities and horizons, including the oldest and youngest. The association of various *Bothriolepis* species with other elements in the fauna forms the basis for a zonal scheme for the Aztec sequence, with the lowest *Bothriolepis* zones characterised by a thelodont association, and the uppermost by the presence of phyllolepid placoderms (Young 1988).

The vertebrate faunal successions documented from other regions, including the European sequences on which the earlier assessments of age were based, are compared with the Aztec sequence in Fig. 7. It is now clear that the Aztec fauna includes some unique faunal associations, which were the source of previous confusion about age. In Europe, for example, turiniid thelodont scales range up only into the middle part of the Lower Devonian (Karatajute-Talimaa 1978). An association of *Bothriolepis* with *Phyllolepis* and *Groenlandaspis* is typical of the latest Devonian (Famennian), although *Bothriolepis* first appears in the latter part of the Middle Devonian (*varcus* zone based on conodonts of the Russian Platform sequence; Young 1988). In China the ranges of *Turinia* and *Bothriolepis* apparently overlap, but they are not known to be associated in any sequence. Middle Devonian turiniids occur in the Eifelian-Givetian of West Yunnan (Wang *et al.* 1986), whereas the earliest known *Bothriolepis* (*B. tungseni*) occurs in the late Eifelian of east Yunnan (e.g. Pan 1981), a region in which Middle Devonian turiniids have not been found. In southeastern Australia the species '*Bothriolepis*' *verrucosa* Young & Gorter 1981, which is associated with Middle Devonian turiniids, has been referred by Young (1988) to the new bothriolepid genus *Monarolepis*.

On present knowledge the Aztec fauna is unique in containing the only known association of species of *Turinia* and species of *Bothriolepis*. At the top of the Aztec sequence different *Bothriolepis* species are associated with phyllolepid placoderms. Thus the Aztec fauna seems to be a transitional succession, with overlapping ranges of what may be the youngest non-marine occurrences of *Turinia* with a very early species of *Bothriolepis* at the base, and the first appearance of phyllolepid placoderms recorded at the top. In southeastern Australia phyllolepids occur at least as early as early Frasnian (see below), which suggests that most or all of the Aztec sequence should be placed in the Givetian (Fig. 7).

Biogeography of the Aztec fauna

The Aztec fauna as currently known contains at least six endemic genera and over 20 endemic species, which might be taken to indicate the existence of a distinctive Antarctic vertebrate fauna at that time. However, I judge this apparent endemism to reflect the fact that the Aztec fauna is slightly older than comparable assemblages from other areas, as just discussed. On the other hand there are several genera (*Culmacanthus*, *Mcmurdodus*, *Pambulaspis*) and species (*Antarctilamna prisca*) which otherwise occur only in eastern Australia. This faunal resemblance is consistent with standard Gondwana reconstructions for the Palaeozoic, which place Australia and East Antarctica together (Fig. 8), and also with the identification of a distinctive East Gondwana vertebrate faunal province, previously defined for the Early Devonian by Young (1981*a*). Endemics shared between the Aztec fauna and eastern Australia indicate that this faunal province persisted at least until the late Middle Devonian. In this context, and consistent with the differences in the vertebrate successions of different areas discussed above, it is noteworthy that Euramerica belonged to a quite distinctive faunal province, characterised by such major agnathan groups as cephalaspids and psammosteid heterostracans (Young 1981*a*). These groups are completely unknown in the Devonian of the East Gondwana Province.

Origin of the early vertebrate fauna of Antarctica

As noted above an older (Early Devonian) vertebrate fauna from Antarctica is known from the Ohio Range and Ellsworth Mountains (Fig. 1). This is a marine assemblage, and it shows that by Early Devonian time two groups of early fishes (placoderms and acanthodians) inhabited shallow epicontinental seas of Antarctica. This fauna may be a vertebrate equivalent of the Malvinokaffric Province defined on invertebrates of Early Devonian age (Young in press). The younger Aztec fauna is however much better known, and represents the oldest significant evidence of a continental biota in Antarctica. What can this fauna tell us about the origin and evolution of early vertebrates in Antarctica?

Antarctica today is a geographic entity with special climatic features, but in the Palaeozoic it was part of the Gondwana supercontinent, and for this reason it cannot necessarily be expected

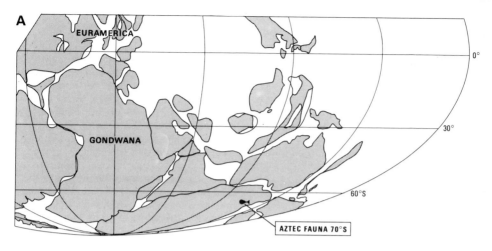

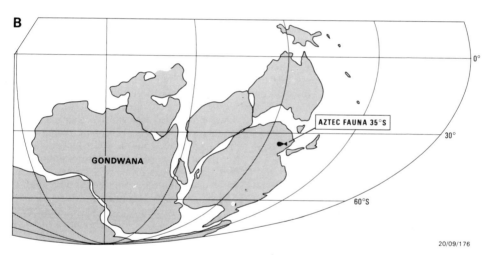

20/09/176

Fig. 8. Alternative reconstructions of Gondwana for the Late Devonian. (**A**) after the model of Scotese (1986), in which the ocean between Gondwana and Euramerica is closed by clockwise rotation of Gondwana during the Devonian, to permit Late Devonian fresh-water and terrestrial faunal exchange between the two regions, but with an unlikely position of about 70 degrees south palaeolatitude for the Aztec fauna; (**B**) based on an alternative APWP for Gondwana, using new palaeomagnetic data from Australia (Klootwijk & Giddings 1988).

that fossils from the Palaeozoic of Antarctica would be characterizable as a distinct fauna. As noted above, the Aztec fauna shares a significant number of taxa with the faunas of eastern Australia, placing it in the 'East Gondwana Province' of Young (1981a).

Regarding the origin of this early vertebrate fauna, it is possible that the Antarctic region, as part of the major landmass of the middle Palaeozoic, was the site of the initial radiation of jawed vertebrates, evidence of which remains completely unknown because of the poor fossil record and the very limited exposures on the Antarctic continent. However other evidence shows this to be most unlikely, because the major vertebrate groups in the Aztec fauna were already widespread by Early Devonian time. Elsewhere in Gondwana Early Devonian acanthodians, turiniids, and placoderms are known at least from South America and Australia, and the latter also has osteichthyans. Even so, an East Gondwanan origin for these groups is not supported by the fossil record as currently known. Thus, in eastern Australia, there are no significant vertebrate faunas (except for acanthodian and turiniid micro-vertebrate remains) before the Emsian (late Early Devonian), whereas much older (Loch-kovian) faunas are well known from Europe, and the oldest representatives of many groups, including some very primitive taxa, occur in the Late Silurian and Early Devonian of South China. If evolutionary centres are to be sought, then on current evidence areas such as South China or Europe are much better candidates than Antarctica or eastern Australia.

Although all the major groups represented in the Aztec fauna evidently had attained a cosmopolitan distribution well before the late Middle Devonian, at lower taxonomic levels evolution within the province can be assumed. The acanthodian *Culmacanthus* probably orig-inated in the East Gondwana Province, and the origin of the chondrichthyan *Antarctilamna* may be placed somewhere in Gondwana. In both cases, possible evolutionary centres can be assumed to lie within the known geographic range of the taxon concerned. Finally, there is no evidence, nor any reason, to doubt that the various endemic species in the Aztec fauna evolved in place.

Subsequent evolution of the early vertebrate fauna of Antarctica

In previous publications (Young 1981*a*, 1984, 1987*b*) I have proposed that some early vertebrate taxa may have originated in the Gondwana region, and subsequently dispersed into other areas, in particular the Euramerican region. The Aztec fauna provides important evidence supporting this hypothesis, and I summarize it here in terms of the empirical observations which the hypothesis attempts to explain, and the theoretical basis for inter-preting those observations. The theoretical question is important because the search for evolutionary centres, and dispersal explanations in biogeography generally, have been severely criticised in recent literature (e.g. Croizat *et al.* 1974). Furthermore the validity of fossil evi-dence in biogeographic analysis has been challenged (Platnick & Nelson 1978; Patterson 1981*a*, *b*). These theoretical issues are discussed in the next section.

In recent years there has been a great increase in our knowledge of Devonian vertebrate faunas from poorly explored regions such as Australia, South China, the Middle East, South America,

and Africa. Using this newly acquired global data base, several broad features of Devonian vertebrate distributions in time and space are now evident. The first is that a major change can be seen in the pattern of vertebrate distribution between the Early and the Late Devonian. As noted above, Early to Middle Devonian fish faunas of the 'East Gondwana Province' (Young 1981*a*) differ markedly from those of Europe and North America in the complete absence of some major groups, such as the osteostracan and heterostracan agna-thans. The only agnathans known from the East Gondwana region are thelodontids. Amongst the diverse placoderm faunas, both marine (e.g. Young 1980, 1981*b*; Long & Young 1988) and non-marine (e.g. Ritchie 1973) faunas contain major taxa not present in the well studied faunas of comparable age in Europe.

On the other hand, if we compare the vertebrate faunas of Euramerica and East Gondwana in the latest Devonian, there are some remarkable similarities in the freshwater vertebrate assemblages of the two regions (Table 2). This resemblance is due in part to the presence in both regions of various placo-derms such as phyllolepids, *Bothriolepis*, and *Groenlandaspis*, which were already established in the upper Aztec sequence in the late Middle Devonian. Also noteworthy is the distribution of early tetrapods. The ichthyostegids and acanthostegids from the latest Devonian of east Greenland are well known (e.g. Jarvik 1980), but tetrapod footprints and other remains are now documented from the Middle and Late Devonian of the Gondwana areas of eastern Australia and South America (Young 1987*a*), and possibly in the Early Devonian of Victoria (Warren *et al.* 1986). This evidence shows that early tetrapods were widespread in Gondwana during the Devonian, and they were also present in Euramerica during the Late Devonian.

A second general feature well illustrated by taxa from the Aztec fauna is a pattern of older and/or more primitive occurrences in East Gondwana than in Europe. This is seen in several groups (xenacanth sharks, phyllolepid and bothriolepid placoderms etc.), and could be considered to indicate a Gondwana origin, and subsequent dispersal to produce the charac-teristic Late Devonian and Early Carbon-iferous fish faunas of Euramerica. These East Gondwana taxa fall into two groups:

(a) Those for which there is biostratigraphic evidence of older occurrence in Gondwana for closely related taxa, but the detailed inter-relationship of taxa between the two areas has not been worked out. For these groups, appli-

Table 2. *Comparison of the Late Devonian (Famennian) vertebrate faunas of East Greenland and eastern Australia*

	East Greenland	Eastern Australia
Placoderms	*Bothriolepis*	*Bothriolepis*
	Remigolepis	*Remigolepis*
	Phyllolepis	phyllolepids
	Groenlandaspis	*Groenlandaspis*
Lungfishes	*Soederberghia*	*Soederberghia*
	Nielsenia	
	Oervigia	
	Jarvikia	
Rhipidistians	*Holoptychius*	*Holoptychius*
	Eusthenodon	*Canowindra*
Tetrapods	*Ichthyostega*	*Metaxygnathus*
	Ichthyostegopsis	
	Acanthostega	

cation of 'Matthew's rule of thumb', that the site of the oldest fossils is the centre of origin (Patterson 1981*a*, p. 461), would indicate a Gondwana origin, with subsequent dispersal into Euramerica at some time before they appear in the Euramerican fossil record.

(b) Groups which not only have an older biostratigraphic occurrence, but are represented by primitive taxa in the Gondwana assemblages. For these groups, 'Hennig's progression rule', that the distribution of the most primitive members of a group indicates its centre of origin, gives the same result as Matthew's Rule, namely a Gondwana origin, and subsequent dispersal into Euramerica.

Examples of the first group from the Aztec fauna (based on its assumed Givetian–early Frasnian age) include rhizodontid crossopterygians (earliest record in Euramerica *Sauripterus* in the Famennian of eastern North America), and the arthrodire *Groenlandaspis* (earliest record in the late Famennian of east Greenland). Of the second group there are several examples. The primitive xenacanth shark *Antarctilamna* lacks specializations (loss of squamation, dorsal spine attached to skull or shoulder girdle) which characterize all late Palaeozoic representatives of the group in Euramerica. In a cladistic analysis Young (1982) and Maisey (1984) placed *Antarctilamna* as the most primitive known xenacanth. For placoderms the phyllolepids are an example for which cladistic relationships have been partly worked out. Whether *Antarctaspis* (Denison 1978) or *Wuttagoonaspis* (Miles 1971) is the sister group to typical phyllolepids, the genera *Austrophyllolepis* and *Placolepis* (Long 1984; Ritchie

1984) represent a paraphyletic stem group in the East Gondwana region (Young 1987*b*, fig. 5).

A more detailed cladistic analysis of bothriolepid antiarchs (Young 1988) shows a more complex pattern than suggested by biostratigraphic evidence alone. It seems (e.g. Pan 1981) that the oldest known species of *Bothriolepis* occur in the Eifelian of South China, where there are also several endemic bothriolepids, suggesting a south Chinese origin for the group (Young 1984). However the most primitive species of *Bothriolepis* in the cladogram of Young (1988) occurs in Antarctica, so application of the progression rule would seem to resolve East Gondwana, or East Gondwana plus South China, as the centre of origin, depending on palaeogeographic interpretations of the relative position of these regions in the middle Palaeozoic. The fact that within several clades a pattern of areas involving East Gondwana is repeated may indicate a vicariance event dividing several sympatric taxa, as in the example given by Nelson & Platnick (1981, fig. 7.7).

The complexity of this example is accentuated by the fact that *Bothriolepis* attained a much wider distribution in the Late Devonian than in the Middle Devonian, so a cladogram of all well known species covers more than one palaeogeographic setting, and probably several different distribution patterns at different times for the group.

In summary, the hypothesis of Gondwana origin and dispersal during the Devonian is based primarily on application of Matthews rule and progression rule biogeography to the occurrence of older and/or more primitive representatives of various groups in the East Gondwana region. It is this 'dispersalist' approach that has been severely criticised in the extensive recent literature on biogeographic theory (see Nelson & Rosen 1981; Nelson & Platnick 1981, and references cited therein), but the relevance of such criticisms in the context of Palaeozoic palaeogeography may be questioned.

Fossil data in biogeography and palaeogeography

The traditional 'dispersalist' approach uses biostratigraphic and phylogenetic evidence to deduce a centre of origin and dispersal history to explain the observed distribution of individual groups. Some of the main criticisms of this approach are summarised below:

(1) The fossil record is notoriously incomplete, yet both Matthew's rule and progression

rule biogeography assume a complete fossil record (Patterson 1981*a*, *b*). The search for ancestral distributions is thus analogous to the search for ancestors in phylogeny reconstruction, and is to be rejected for the same reasons (e.g. Patterson 1983).

(2) Geological and palaeogeographic ideas should play no part in biogeographic analysis, since there can be no guarantee that a particular view of the geographic history of an area or areas is correct (Patterson 1981*a*).

(3) Dispersal explanations (of organisms crossing pre-existing barriers) are inferior to vicariance explanations (involving geographic subdivision of an ancestral biota), because the former applies to individual groups, whilst the latter affects whole biotas, and individual groups can be tested against a general pattern.

(4) Widespread taxa are biogeographically uninformative (Platnick & Nelson 1978; Nelson & Platnick 1981).

(5) Biogeographic relationship between two areas can have no meaning, because 'relationship' is a comparative concept, requiring a minimum of three areas.

A general point to be made first about such criticisms is that they have arisen in the context of analysing the distribution of the modern biota to elucidate a coherent historical explanation. In that context they may have some validity, but for an entirely fossil group, distributed in the Palaeozoic, when global palaeogeography is very poorly understood, a different perspective is necessary.

Regarding point 1 above, the recognition of ancestor−descendant relationships in phylogeny reconstruction assumes a complete fossil record, because to assert that one particular fossil species is the ancestor of another implies that undiscovered potential ancestors do not exist elsewhere. In contrast, in biogeographic analysis, assumptions about unknown fossils are much more specific, involving specific taxa in specific areas (Young 1984). Whilst the absence of taxa in a particular area is negative evidence which can never be definitely established, such absences might be judged as biogeographically significant in cases where fossil faunas of the right age are well known in the area concerned. The empirical observation that different taxa of the same age occur in different areas permits the biogeographic hypothesis that a faunal barrier separated those areas at that time.

Application of the progression rule depends on the presence of a paraphyletic stem group in a particular area, which documents the occurrence of early speciation events of that group in

the area concerned. Involved here is an assumption that unknown taxa placed cladistically within that paraphyletic group do not exist in other areas. In the case of phyllolepid placoderms discussed above, the negative biostratigraphic evidence that they did not occur in pre-Famennian strata in Euramerica rests on many empirical observations about the composition of Euramerican vertebrate faunas of that age. Yet the hypothesis that the paraphyletic stem group was restricted to the East Gondwana region might be refuted by new discoveries in any part of the world currently poorly known with respect to its Devonian vertebrates. For example, discovery of a primitive phyllolepid in the Middle Devonian of the Kolyma block (northeastern Siberia) could either imply a cosmopolitan ancestral distribution for the group, or (as in the case regarding known early phyllolepid taxa in southeastern Australia and southern Victoria Land), it could be hypothesised that the Kolyma region is a displaced portion of middle Palaeozoic Gondwana. On the other hand, discovery of pre-Famennian phyllolepids in South America (where, apart from tetrapod footprints, Late Devonian fossil vertebrate faunas are not yet known), would extend the ancestral range of the group within the hypothesized Palaeozoic Gondwana, and thus not affect dispersal explanations of phyllolepids outside this palaeogeographic entity.

This illustrates the point already made above in discussing the South China−East Gondwana relationships of *Bothriolepis*: palaeogeographic hypotheses must be considered at the outset to decipher fossil distributions, for which modern geography is largely irrelevant (point 2 above). Furthermore, although the distribution of extant forms can theoretically be established unequivocally by empirical observation, modern geography is also the result of historical geological processes, which therefore must impinge on historical explanations of the observed distributions. Any palaeogeographic hypothesis can be subjected to further tests using other fossil groups, or geological and geophysical data, so in the case of biotic dispersal (in contrast to 'chance' dispersal explanations for individual groups), a general pattern is implied for which new empirical data represent valid tests (point 3 above).

Under the vicariance paradigm (Croizat *et al.* 1974) the appearance of geographic barriers is an accepted historical explanation of disjunct distributions. But the disappearance of barriers, to cause biotic dispersal events, is also a phenomenon relevant to historical biogeography. Disappearance of a barrier would manifest itself in fossil sequences in the increased similarity of faunas from two areas through time. Extinctions of some groups in one or both areas might be expected to accentuate the faunal change. The timing of the change could be estimated from biostratigraphic studies, but the cause of the change (e.g. alterations in palaeogeography, palaeoclimate etc.) could only be determined by non-biological data.

Regarding point 4, there is an obvious distinction to be made between living and fossil species. A living species which is widespread with respect to two or more areas recognised as biogeographic units on the basis of their endemic biotas is biogeographically uninformative with respect to the history of those areas, because its distribution defines a biogeographic unit at a higher hierarchical level, for which a different set of biogeographic problems applies. But a fossil species 'widespread' in relation to modern geography, may indicate that areas now far apart may have been together in the past. As just noted, the distribution of living species can be established empirically, but for fossil species distribution must be analysed in the context of palaeogeographic hypotheses.

On the final point (5), the examples discussed above concern two middle Palaeozoic vertebrate provinces, but we are concerned with differences of pattern at different times. The empirical observation that vertebrate faunas in the two areas are much more similar in the Late Devonian than in the Early Devonian is a significant difference in pattern which calls for an explanation.

Conclusion and summary

The Aztec fish fauna of southern Victoria Land is one of the most diverse fossil vertebrate faunas from Antarctica. It is interpreted as the youngest known example of the East Gondwana vertebrate Province of Young (1981a). It contains several taxa which are more primitive and/ or occur earlier in the fossil record than in Euramerican sequences. This pattern is apparently repeated in several groups; it may be explained under a model of biotic dispersal, which may have resulted from major palaeogeographic change during the Late Devonian, apparently at or near the Frasnian–Famennian boundary. This hypothesis explains the following five observations about the Devonian fish faunas of the Australian–Antarctic region compared to those of Europe and North America:

(i) marked differences in composition of the non-marine vertebrate faunas of the two regions in the Early and Middle Devonian;

(ii) close similarities in the non-marine vertebrate faunas of the two regions in the Famennian;

(iii) earlier stratigraphic occurrence in Australian–Antarctic sequences compared to those of Euramerica, a pattern repeated in several groups, including phyllolepid and groenlandaspid placoderms, rhizodontid crossopterygians, and xenacanth sharks;

(iv) occurrence of primitive taxa, or paraphyletic stem groups, in the Australian–Antarctic sequences, a pattern repeated in several groups (phyllolepid placoderms, xenacanth sharks, bothriolepid antiarchs);

(v) sudden extinction of a major Euramerican group, the psammosteid heterostracans, at the end of the Frasnian, to be replaced by adaptively similar phyllolepids (Young 1987b).

I am indebted to the Trans-Antarctic Association for a travel grant permitting my attendance at the London meeting in May 1988. For advice on the fauna I am grateful to S. M. Andrews, K. S. W. Campbell, P. L. Forey, B. G. Gardiner, D. Goujet, J. A. Long, A. Ritchie, and S. Turner. P. J. Barrett, M. A. Bradshaw, C. T. McElroy, and J. G. McPherson provided information on fossil localities, stratigraphic occurrence, and palaeoenvironment, and M. A. Bradshaw and A. Ritchie made material available for study. Assistance and logistic support during field work under the New Zealand Antarctic Research Program is gratefully acknowledged. R. W. Brown, P. W. Davis, and H. M. Doyle assisted with fossil preparation and photography. Published with the permission of the Director, Bureau of Mineral Resources, Canberra.

References

ANDREWS, S. M. & WESTOLL, T. S. 1970. The postcranial skeleton of rhipidistian fishes excluding *Eusthenopteron. Transactions of the Royal Society of Edinburgh*, **68**, 391–489.

BARRETT, P. J. 1980. The non-marine character of the Devonian Taylor Group (Beacon Supergroup) in south Victoria Land, Antarctica. *In*: LASKOR, B. (ed.). *4th International Gondwana Symposium*, **2**, 478–480.

—— & KOHN, B. P. 1975. Changing sediment transport directions from Devonian to Triassic in the Beacon Super Group of south Victoria

Land, Antarctica. *In*: CAMPBELL, K. S. W. (ed.) *Gondwana Geology*. ANU Press, Canberra, 15–35.

BLIECK, A., GOLSHANI, F., GOUJET, D., HAMDI, A., JANVIER, P., MARK-KURIK, E. & MARTIN, M. 1980. A new vertebrate locality in the Eifelian of the Khush-Yeilagh Formation, eastern Alborz, Iran. *Palaeovertebrata*, **9–V**, 133–154.

BRADSHAW, M. A. & McCARTAN, L. 1983. The depositional evironment of the Lower Devonian Horlick Formation, Ohio Range. *In*: OLIVER, R. L., JAMES, P. R. & JAGO, J. B. (eds) *Antarctic Earth Science*. Australian Academy of Science, Canberra, 238–241.

CAMPBELL, K. S. W. & BARWICK, R. E. 1987. Paleozoic lungfishes – a review. *Journal of Morphology, Supplement 1 (1986)*, 93–131.

CROIZAT, L., NELSON, G. J. & ROSEN, D. E. 1974. Centers of origin and related concepts. *Systematic Zoology* **23**, 265–287.

DENISON, R. H., 1978. Placodermi. *Handbook of Paleoichthyology, volume 2* (SCHULTZE, H-P. ed.). Gustav Fisher Verlag, Stuttgart.

DICK, J. R. F. 1981. *Diplodoselache woodi* gen. et sp. nov., an early Carboniferous shark from the Midland Valley of Scotland. *Transactions of the Royal Society of Edinburgh: Earth Sciences*, **72**, 99–113.

DOUMANI, G. A., BOARDMAN, R. S., ROWELL, A. J., BOUCOT, A. J., JOHNSON, J. G., McALESTER, A. L., SAUL, J., FISHER, D. W. & MILES, R. S. 1965. Lower Devonian fauna of the Horlick Formation, Ohio Range, Antarctica. *American Geophysical Union, Antarctic Research Series*, **6**, 241–281.

FERGUSSON, C. L., CAS, R. A., COLLINS, W. J., CRAIG, G. Y., CROOK, K. A. W., POWELL, C. McA., SCOTT, P. A. & YOUNG, G. C. 1979. The Late Devonian Boyd Volcanic Complex, Eden, N. S. W. *Journal of the Geological Society of Australia*, **26**, 87–105.

GAGNIER, P. Y., TURNER, S., FRIMAN, L., SUAREZ-RIGLOS, M. & JANVIER, P. 1988. The Devonian vertebrate and mollusc fauna from Seripona (Dept. of Chuquisaca, Bolivia). *Neues Jahrbuch für Geologie und Paläontologie, Abhandlungen*, **176**, 269–297.

GOUJET, D., JANVIER, P. & SUAREZ-RIGLOS, M. 1984. Devonian vertebrates from South America. *Nature*, **312**, 311.

GROSS, W. 1950. Die palaeontologische und stratigraphische Bedeutung der Wirbeltierfaunen des Old Reds und der marinen altpalaozoischen Schichten. *Abhandlungen der Deutschen Akademie der Wissenschaften zu Berlin. Mathematisch-naturwissenschaftliche Klasse*, **1949**, 1–130.

GUNN, B. M. & WARREN, G. 1962. Geology of Victoria Land between the Mawson and Mulock Glaciers, Antarctica. *Trans-Antarctic Expedition 1955–1958. Scientific Reports, Geology*, **11**, 1–157.

HELBY, R. J. & McELROY, C. T. 1969. Microfloras from the Devonian and Triassic of the Beacon Group, Antarctica. *New Zealand Journal of Geology & Geophysics* **12**, 376–382.

HILLS, E. S. 1929. The geology and palaeontography of the Cathedral Range and Blue Hills in north western Gippsland. *Proceedings of the Royal Society of Victoria*, **41**, 176–201.

JARVIK, E. 1980. *Basic structure and evolution of vertebrates, Vol. 1 & 2*. Academic Press, London.

KARATAJUTE-TALIMAA, V. N. 1978. *Silurian and Devonian thelodonts of the USSR and Spitsbergen*. Vilnius, Mokslas. (in Russian).

KLOOTWIJK, C. & GIDDINGS, J. 1988. An alternative APWP for the Middle to Late Palaeozoic of Australia – Implications for terrane movements in the Tasman Fold Belt. *Ninth Australian Geological Convention, Abstracts*, **21**, 219–220.

LONG, J.A. 1983. A new diplacanthoid acanthodian from the Late Devonian of Victoria. *Memoir of the Association of Australasian Palaeontologists*, **1**, 51–65.

—— 1984. New phyllolepids from Victoria and the relationships of the Group. *Proceedings of the Linnean Society of New South Wales* **107**, 263–308.

—— 1985. New information on the head and shoulder girdle of *Canowindra grossi* Thomson, from the Upper Devonian Mandagery Sandstone, New South Wales. *Records of the Australian Museum*, **37**, 91–99.

—— 1988. New palaeoniscoid fishes from the Late Devonian and Early Carboniferous of Victoria. *Memoirs of the Association of Australasian Palaeontologists*, **7**, 1–64.

—— & CAMPBELL, K. S. W. 1985. A new lungfish from the Lower Carboniferous of Victoria, Australia. *Proceedings of the Royal Society of Victoria*, **97**, 87–93.

—— & YOUNG, G. C. 1988. Acanthothoracid remains from the Early Devonian of New South Wales, including a complete sclerotic capsule and pelvic girdle. *Memoirs of the Australasian Association of Palaeontologists*, **7**, 65–80.

MADER, H. 1986. Schuppen und Zahne von Acanthodiern und Elasmobrachiern aus dem Unter-Devon Spaniens (Pisces). *Gottinger Arbeiten zur Geologie und Paläontologie* **28**, 1–59.

MAISEY, J. G. 1984. Chondrichthyan phylogeny: a look at the evidence. *Journal of Vertebrate Paleontology*, **4**, 359–371.

MATZ, D. B. & HAYES, M. O. 1966. Sedimentary petrology of Beacon sediments. *Antarctic Journal of the United States*, **1**, 134–135.

—— PINET, P. R. & HAYES, M. O. 1972. Stratigraphy and petrology of the Beacon Supergroup, southern Victoria Land. *In*: ADIE, R. J. (ed.) *Antarctic Geology and Geophysics*. Universitetsforlaget, Oslo, 353–358.

McKELVEY, B. C., WEBB, P. N., GORTON, M. P. & KOHN, B. P. 1972. Stratigraphy of the Beacon Supergroup between the Olympus and Boomerang Ranges, Victoria Land. *In*: ADIE, R. J. (ed.) *Antarctic Geology and Geophysics*. Universitetsforlaget, Oslo, 345–352.

McPHERSON J. G. 1978. Stratigraphy and sedimentology of the Upper Devonian Aztec Siltstone, southern Victoria Land, Antarctica. *New Zealand Journal of Geology & Geophysics* **21**, 667—683.

—— 1979. Calcrete (Caliche) palaeosols in fluvial redbeds of the Aztec Siltstone (Upper Devonian), Southern Victoria Land, Antarctica. *Sedimentary Geology*, **22**, 267—285.

MILES, R. S., 1971. *Palaeozoic fishes*. 2nd edn. Chapman & Hall, London.

NELSON, G. & PLATNICK, N. 1981, *Systematics and biogeography: cladistics and vicariance*. Columbia University Press, New York.

—— & ROSEN, D. E. 1981. *Vicariance biogeography. A critique*. Columbia University Press, New York.

OELOFSEN, B. W. 1981. The fossil record of the class Chondrichthyes in southern Africa. *Palaeontologica Africana*, **24**, 11—13.

PAN, J. 1981. Devonian antiarch biostratigraphy of China. *Geological Magazine*, **118**, 69—75.

PATTERSON, C. 1981a. Methods of palaeobiogeography. *In*: NELSON. G. & ROSEN, D. E. (eds) *Vicariance biogeography. A critique*. Columbia University Press, New York, 446—489.

—— 1981b. The development of the North American fish fauna — a problem of historical biogeography. *In*: FOREY, P. L. (ed.) *The Evolving Biosphere*. Cambridge University Press, Cambridge, 265—281.

—— 1983. Aims and methods in biogeography. *In*: SIMS, R. W., PRICE, J. H. & WHALLEY, P. E. S. (eds) *Evolution, Time, Space: the Emergence of the biosphere*. Systematics Association Special Volume, Academic Press, London, **23**, 1—28.

PLATNICK, N. I. & NELSON, G. J. 1978. A method of analysis for historical biogeography. *Systematic Zoology*, **27**, 1—16.

PLUMSTEAD, E. P. 1962. Fossil floras of Antarctica. *Trans-Antarctic Expedition 1955—1958. Scientific Reports 9, Geology*. 154 pp.

RITCHIE, A. 1971a. Ancient animals of Antarctica — Part 2. *Hemisphere*, **15** (12), 12—17.

—— 1971b. Fossil fish discoveries in Antarctica. *Australian Natural History*, **17**, 65—71.

—— 1972. Appendix. Devonian fish. *In*: McKELVEY, B. C., WEBB, P. N., GORTON, M. P. & KOHN, B. P. 1972. Stratigraphy of the Beacon Supergroup between the Olympus and Boomerang Ranges, Victoria Land. *In*: ADIE, R. J. (ed.) *Antarctic Geology and Geophysics*. Universitetsforlaget, Oslo, 345—352.

—— 1973. *Wuttagoonaspis* gen. nov., an unusual arthrodire from the Devonian of Western New South Wales, Australia. *Palaeontographica*, **143A**, 58—72.

—— 1975. *Groenlandaspis* in Antarctica, Australia and Europe. *Nature*, **254**, 569—573.

—— 1984. A new placoderm, *Placolepis* gen. nov. (Phyllolepidae) from the Late Devonian of New South Wales, Australia. *Proceedings of the Linnean Society of New South Wales*, **107**, 321—353.

SCOTESE, C. R. 1986. Phanerozoic reconstructions: a new look at the assembly of Asia. *University of Texas Institute for Geophysics Technical Report*, **66**.

TARLO, L. B. H. 1964. Psammosteiformes (Agnatha). A review with descriptions of new material from the Lower Devonian of Poland. *Palaeontologica Polonica*, **13**, 1—135.

TASCH, P. 1987. Fossil Conchostraca of the Southern Hemisphere and continental drift. Paleontology, biostratigraphy, and dispersal. *Memoir, Geological Society of America*, **165**, 1—290.

TAYLOR, T. G. 1913. The geological expedition to Granite Harbour. *In*: SCOTT, R. F., *Scott's Last Expedition*, **2**, 222—290.

THOMSON, K. S. 1973. Observations on a new rhipidistian fish from the Upper Devonian of Australia. *Palaeontographica*, **143A**, 209—220.

TURNER, S. 1982. A new articulated thelodont (Agnatha) from the Early Devonian of Britain. *Palaeontology*, **25**, 879—889.

—— & DRING, R. S. 1981. Late Devonian thelodonts (Agnatha) from the Gneudna Formation, Carnarvon Basin, Western Australia. *Alcheringa* **5**, 39—48.

—— & YOUNG, G. C. 1987. Shark teeth from the Early—Middle Devonian Cravens Peak Beds, Georgina Basin, Queensland. *Alcheringa*, **11**, 233—244.

—— & —— in prep. Middle-Late Devonian thelodont scales from Antarctica.

WANG, S.-T., DONG, Z-Z, & TURNER, S. 1986. Discovery of Middle Devonian Turiniidae (Thelodonti: Agnatha) from western Yunnan, China. *Alcheringa*, **10**, 183—186.

WHITE, E. I. 1968. Devonian fishes of the Mawson-Mulock area, Victoria Land, Antarctica. *Trans-Antarctic Expedition 1955—1958. Scientific Reports, Geology*. **16**, 1—26.

WOODWARD, A. S. 1906. On a Carboniferous fish fauna from the Mansfield district, Victoria. *Memoirs of the National Museum, Melbourne*, **1**, 1—32.

—— 1921. Fish-remains from the Upper Old Red Sandstone of Granite Harbour, Antarctica. *British Antarctic ('Terra Nova') Expedition, 1910. Natural History Report (Geology)*, **1**, 51—62.

YOUNG, G. C. 1980. A new Early Devonian placoderm from New South Wales, Australia, with a discussion of placoderm phylogeny. *Palaeontographica*, **167A**, 10—76.

—— 1981a. Biogeography of Devonian vertebrates. *Alcheringa*, **5**, 225—243.

—— 1981b. New Early Devonian brachythoracids (placoderm fishes) from the Taemas—Wee Jasper region of New South Wales. *Alcheringa* **5**, 247—271.

—— 1982. Devonian sharks from south-eastern Australia and Antarctica. *Palaeontology*, **25**, 817—843.

—— 1983. A new antiarchan fish (Placodermi) from the Late Devonian of southeastern Australia. *BMR Journal of Australian Geology & Geophysics*, **8**, 71—81.

—— 1984. Comments on the phylogeny and biogeography of antiarchs (Devonian placoderm fishes),

62   G. C. YOUNG

and the use of fossils in biogeography. *Proceedings of the Linnean Society of N.S.W.* **107**, 443–473.

—— 1987*a*. Devonian vertebrates of Gondwana. *In*: McKenzie, G. D. (ed), *Gondwana Six. Stratigraphy, Sedimentology, and Paleontology*. Geophysical Monograph, American Geophysical Union, **41**, 41–50.

—— 1987*b*. Devonian palaeontological data and the Armorica problem. *Palaeogeography, Palaeoclimatology, Palaeoecology*, **60**, 283–304

—— 1988. Antiarchs (placoderm fishes) from the Devonian Aztec Siltstone, southern Victoria Land, Antarctica. *Palaeontographica*, **A202**, 1–125.

—— 1989. New occurrences of culmacanthid acanthodians (Pisces, Devonian) from Antarctica and southeastern Australia. *Proceedings of the Linnean Society of New South Wales* **111**, 12–25.

—— in press. Description of the fish spine. *In*:

Craddock, C., Splettstoesser, J. F. & Webers, G. F. (eds). *Geology and Paleontology of the Ellsworth Mountains, Antarctica*. Memoir. Geological Society of America, **170**.

—— & Gorter, J. D. 1981. A new fish fauna of Middle Devonian age from the Taemas/Wee Jasper region of New South Wales. *Bureau of Mineral Resources Geology & Geophysics, Bulletin*, **209**, 83–147.

——, Long, J. A., & Ritchie, A. in prep. Rhipidistian fishes from the Devonian of Antarctica: systematics, relationships and biogeographic significance. *Records of the Australian Museum*.

Young, V. T. 1986. Early Devonian fish material from the Horlick Formation, Ohio Range Antarctica. *Alcheringa*, **10**, 35–44.

Zangerl, R. 1981. *Chondrichthyes I. Paleozoic Elasmobranchii*. Handbook of Paleoichthyology, Volume 3a. (Schultze, H-P. ed.), Gustav Fischer Verlag, Stuttgart.

Patterns of evolution in the Dicynodontia, with special reference to austral taxa

SHERRI L. DEFAUW

Department of Biology, Berry College, Mt Berry Station, GA 30149, USA

Abstract: Dicynodonts were the dominant primary consumers of the latest Palaeozoic and early Mesozoic ecosystems. Of the 66 currently accepted genera of dicynodonts, three (possibly four) of these occur in the Fremouw Formation of Antarctica (i.e. *Kingoria*, *Lystrosaurus*, *Myosaurus* and ?*Kannemeyeria*). Modifications to the basic shearing mechanism of the jaws as well as postcranial morphology presumably reflect differences in the physical nature of the food consumed and foraging techniques used by dicynodonts. To date, five dicynodont feeding types have been recognized: invertebrate specialists, grubbers, browsers, forest litter foragers and flexible foragers. Three out of the five 'central' dicynodont feeding types are represented among the members of the early Mesozoic Antarctic palaeofauna. In order to more accurately assess systematic and ecomorphic relationships among dicynodonts, comprehensive generic diagnoses (that incorporate postcranial characters) are warranted; suggested revisions to the generic diagnoses of Antarctic dicynodonts are provided. Dicynodont distributions are evaluated in light of ecomorphic considerations, tectonic-orogenic events, sea-level fluctuations, palaeoclimates and floral changes. Available evidence, pertaining to austral taxa, suggests that: (1) Antarctica served as a refugium for kingoriamorphs in the Early Triassic, and (2) the informal members of the Fremouw Formation are correlated with more than one African biozone.

The dicynodonts were a widespread, morphologically diverse group of mammal-like reptiles prevalent during the Permian and Triassic periods (approximately 240–190 Ma). Considered to be the most successful taxon of all therapsids, these 'two-tusked', principally herbivorous forms achieved nearly worldwide distribution and are typically the dominant tetrapods in the fossil assemblages collected from Late Permian continental deposits. The most extensive record of the group, however, occurs in depositional basins restricted to the southern half of the African continent. From the late Middle Triassic onward, the fossil record reveals a steady decline in the abundance and diversity of these forms. By the close of the Triassic period, the dicynodonts became extinct.

Members of the suborder Dicynodontia were described as early as 1845 (Owen 1845*a* and *b*), yet a pattern of the overall relationships and evolution within this taxon is just beginning to emerge. Recent work focusing on palatal and lower jaw construction (Cluver & Hotton 1981; Cluver & King 1983; Hotton 1986) has served to clarify generic diagnoses, re-align some evolutionary relationships, and group the dicynodonts into several functional types. Most recently, King (1988) has reassessed the relationships within the Dicynodontia. King recognizes 66 valid genera; 18 genera have been relegated to *incertae sedis* status. Two peaks in taxonomic diversity are now evident within the

Dicynodontia (Fig. 1.) A listing of the temporal and spatial distribution of dicynodont genera has been printed separately and has been deposited in the Geological Society Library and with the British Library at Boston Spa, W. Yorkshire, UK, as Supplementary Publication No. SUP 18056 (also available from the author on request). The first occurred during the Tatarian stage of the Late Permian (32 genera in total, 25 of these are found exclusively in southern Africa). The second diversification took place in the early Middle Triassic (17 genera have been recorded from the Anisian, with the largest fraction of taxa coming from the Donguz Group of the USSR). The mercurial state of dicynodont taxonomy has long hampered the recognition of macroevolutionary patterns in the group. The purpose of this paper is to summarize various lines of evidence (systematic and ecomorphological) derived from the current body of knowledge on dicynodonts (with special reference to austral forms), and integrate this information with data on extrinsic geological events, palaeoclimates, and floral changes in order to refine our views concerning the evolution of this group.

Dicynodonts have long been considered the principal herbivores or primary consumers of the latest Palaeozoic and early Mesozoic ecosystems. Their success has been largely attributed to their unique bicondylic masticatory apparatus, which permitted extensive antero–

From Crame, J. A. (ed.), 1989, *Origins and Evolution of the Antarctic Biota*, Geological Society Special Publication No. 47, pp. 63–84.

63

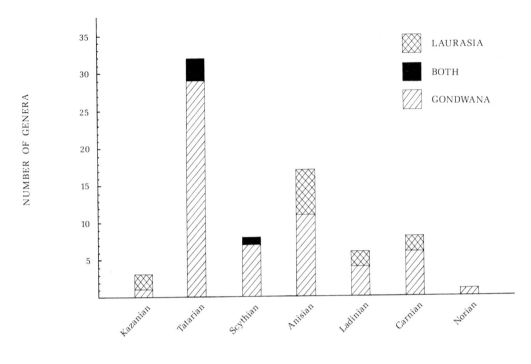

CHRONOSTRATIGRAPHIC UNITS

Fig. 1. Taxonomic diversity and coarse spatial distribution of dicynodont genera.

posterior movements of the lower jaw (Crompton & Hotton 1967). A number of authors have commented on the types of vegetation dicynodonts may have consumed, but little in the way of rigorous functional analyses have been conducted (8 out of 66 genera have been analysed). Variations on the basic quadrate articular sliding mechanism have been described by some investigators (Crompton & Hotton 1967; Cluver 1971, 1974a and b; King 1981; Hotton 1986) and it has been suggested that some dicynodonts may have consumed fruits, fungi, invertebrates, and carrion (Kemp 1982; Hotton 1986; Cox 1972; Sushkin 1926). Thus, some taxa are more appropriately considered as secondary consumers.

To date, five feeding types have been distinguished among the dicynodont genera. Hotton (1986) in his review of 'central' dicynodonts (namely, *Kingoria*, *Dicynodon*, *Oudenodon*, and *Diictodon*) recognized four types based on modifications to the basic shearing mechanism of the jaws. He also considered: (1) the habitual orientation of the head (as discerned by reference to the horizontal semicircular canal), with emphasis on the position of the plane of the palate; (2) snout configuration

(i.e. shape of the caniniform blade, position of the external nares relative to the plane of the palate, and the presence or absence of maxillary tusks); (3) wear patterns on the tusks; (4) lateral flaring of the zygomatic arches (which would enhance the transverse component of muscular force) and (5) foot structure (i.e. the relative length of the ungual phalanges). His proposed feeding types are: forest litter forager (e.g. *Kingoria*); flexible forager (e.g. *Dicynodon* and *Lystrosaurus*); browser (e.g. *Oudenodon*); and grubber (e.g. *Diictodon*). A fifth type, referred to here as invertebrate-specialist, was originally proposed by Cox (1972) to describe the feeding habits of *Kawingasaurus* (a small fossorial form restricted to East Africa); Cluver (1974a) applied this feeding type to the genus *Myosaurus*. The feeding regimens of the Antarctic dicynodont genera (i.e. forest litter foragers, flexible foragers, and invertebrate-specialists) are evaluated in detail here (refer to Hotton 1986 for a discussion of 'grubbers' and 'browsers'). In addition to summarizing and discussing aspects of functional anatomy and palaeoecological interpretations, this paper will also review the palaeobiogeography as well as the bio- and lithostratigraphic ranges of Early Triassic austral dicynodonts (i.e. *Kingoria, Lystrosaurus, Myosaurus,* and ?*Kannemeyeria*). The attributes of these four genera are briefly compared with those of other rather widespread dicynodonts (i.e. genera known to occur on two or more cratons, including *Diictodon, Oudenodon, Endothiodon, Dicynodon,* and possibly *Pristerodon* and *Cistecephalus* — see Fig. 2).

The major groups of dicynodonts

Cluver & King (1983) recognized six major clades within the suborder Dicynodontia, namely the infraorders Venjukoviamorpha, Eodicynodontia, Endothiodontia, Pristerodontia, Diictodontia, and Kingoriamorpha. These divisions are based primarily on features of the premaxilla (fused or unfused), maxilla (embayment anterior to caniniform process present or absent), palatine (small or large with a smooth or 'sculptured' ventral surface) and dentition (premaxillary teeth present or absent; postcanine teeth present or absent). Carroll (1988) includes the infraorder Dromasauria (a group of small, agile herbivores known from only four specimens recovered from the Upper Permian of southern Africa) in the suborder Dicynodontia; this taxon was not addressed in Cluver and King's assessment. King's (1988) revision of the Dicynodontia encompasses seven clades (including dromasaurs); however, the ranks of many taxa have been reduced. As this paper does not directly address systematic issues, the conservative taxonomic scheme of Carroll (1988) is used here.

Although the search for patterns (systematic or ecomorphological) is a relatively straightforward approach, Radinsky (1985, p. 13) has cautioned that, 'characters that are highly correlated and are necessary integrated parts of a single biomechanical system should not be treated as independent characters in formulating hypotheses of phylogenetic relationship, for they represent different aspects of the same feature.' Radinsky resurrected van der Klaauw's concept of cranial functional components and suggested that investigators engaged in systematic and economorphological analyses recognize the inherent biases of correlated character complexes, and temper their interpretations accordingly.

The initial attempt by Cluver & King (1983) to update dicynodont generic diagnoses and erect a more coherent classification scheme focused mainly on Upper Permian forms. Following Toerien's lead (1953) in revising phylogenetic relationships, palatal characters have

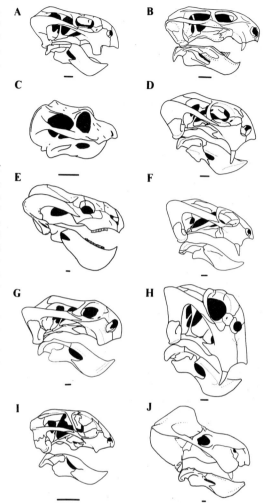

Fig. 2. Late Permian and Early Triassic dicynodont genera that reportedly occur on two or more cratons. (**A**) *Diictodon* (after Cluver & Hotton 1981). (**B**) *Pristerodon* (illustrated with lower jaw fully retracted, after Crompton & Hotton 1967). (**C**) *Cistecephalus* (USNM 22942). (**D**) *Dicynodon* (after Cluver & Hotton 1981). (**E**) *Endothiodon* (after Broili & Schroeder 1936). (**F**) *Kingoria* (after Cluver & Hotton 1981). (**G**) *Oudenodon* (after Cluver & Hotton 1981). (**H**) *Lystrosaurus* (after Cluver 1971). I, *Myosaurus* (after Cluver 1974a). (**J**) *Kannemeyeria* (after Pearson 1924a). Bar scale beneath each skull equals 1 cm.

been heavily emphasized. Prior to this, dicynodont classification was based on the relationships of the dermal roof elements, overall size and shape of the skull and dentition. Palates, because of more rigorous functional constraints, do not exhibit the plasticity evident in the dermal roof elements of the skull. Variations in palatal construction presumably reflect structural modifications related to the differential forces encountered during comminution of various food items. Although testing the fidelity of the current systematic synthesis is beyond the scope of this paper, refinements to the approach taken by Cluver & King (1983) are offered here. As Radinsky has urged, future systematic analyses should be based on recognized functional complexes. Variations in the basic organization of the dicynodont palate and jaw margins have been scrutinized; however, morphological modifications evident in the components of the jaw joint, auditory system, body and ramus of the lower jaw, facial region, neurocranium, girdles, propodials and podials also warrant investigation. Such diversity requires more rigorous comparison and quantification before accurate functional and phylogenetic interpretations can be formulated. Aspects of the jaw joint, jaw ramus, girdles and propodials are addressed in this paper. An analysis of the form and function of dicynodont stapes is currently in progress (Hotton, DeFauw & Colbert, in preparation).

The Antarctic dicynodonts include representatives from three of the six infraorders erected by Cluver & King (1983), namely Kingoriamorpha, Pristerodontia and Diictodontia. At present, the most speciose of these is the Pristerodontia which encompasses all of the large, Late Permian as well as Triassic forms. The Early Triassic members of four dicynodont families are investigated here: Kingoriidae, Emydopidae, Lystrosauridae, and Kannemeyeriidae. These families encompass a wide size range of specimens (collectively ranging from 39–500 mm) and purportedly represent forms with different foraging strategies (Crompton & Hotton 1967; Hotton 1986). Therefore, an examination of cranial and postcranial characteristics and scaling relationships of Antarctic dicynodonts represents a convenient point of departure for discerning ecomorphic patterns of diversity within the suborder Dicynodontia.

Review of Antarctic dicynodont genera

The Fremouw Formation of Antarctica has yielded three species of Lystrosaurus (L. murrayi, L. curvatus, and L. maccaigi), a small rare dicynodont, Myosaurus gracilis, two specimens pertaining to the genus Kingoria (Hotton & Colbert, in preparation), and a new fossil assemblage containing an undescribed kannemeyeriid, provisionally identified as belonging to the genus Kannemeyeria (Hammer et al.

1986, 1987). The following sections provide brief historical, morphological and systematic sketches of each genus. The diagnoses of three austral forms, Kingoria, Lystrosaurus and Kannemeyeria have been expanded here to include postcranial characters. Although incorporating postcranial characters at the generic level is atypical for most vertebrates, at present, this practice is appropriate for dicynodonts (see Cox 1959, 1969, 1972; DeFauw 1986). The postcranial morphology of Myosaurus is currently unknown, therefore its generic diagnosis remains unrevised.

The genus Kingoria Cox, 1959

Of the dicynodont genera considered in detail here, the taxon Kingoria is historically the youngest (Figs 2F & 3A). Cox (1959) derived the name from the village of Kingori (Tanzania) which is close to the site where most of the material was collected. Fossils pertaining to this genus have been recovered from the Aulacephalodon–Cistecephalus and Dicynodon–Whaitsia assemblage zones (as defined by Keyser & Smith 1979) of southern Africa and correlated horizons of Tanzania. These palaeofaunas occur in deposits of the Middleton, Teekloof and Balfour Formations of South Africa, and the Kawinga Formation of Tanzania. Two partial skulls (one with associated lower jaw) recovered from the lower member of the Fremouw Formation (Shenk Peak locality) of Antarctica (during the austral field season of 1970–71) have been referred, by Dr J. W. Kitching, to the genus Kingoria (Colbert 1982; Hotton & Colbert, in preparation). The occurrence of this genus raises questions concerning the relative age of the Fremouw Formation. Heretofore, this Antarctic unit has been considered earliest Triassic in age (Early Scythian) because roughly 85% of the fossils retrieved pertain to the genus Lystrosaurus (Colbert 1974, 1982). The presence of Kingoria in the Fremouw Formation may be interpreted as either the latest occurrence of this taxon, or construed as evidence that the Fremouw Formation is correlated with more than one African biozone. Of these options, the former is more suitable in this case, as the Antarctic Kingoria material was collected within 50 m of the formational base (Lystrosaurus specimens were also recovered from horizons within this stratigraphic interval). In addition, the Antarctic Kingoria specimens exhibit several distinct morphological differences compared to the Late Permian African forms (N. Hotton, pers comm).

Table 1. Comparison of functionally significant features of Antarctic dicynodont crania

CHARACTER	Kingoria	Myosaurus	Lystrosaurus	Kannemeyeria
1 Canine tusks	present or absent (wear facets lacking)	absent	present (terminal & medial wear facets)	present (terminal & medial wear facets)
2 Paired, anterior palatal ridges on the premaxilla	absent	absent	present	present
3 Low, longitudinal ridge medial to the premaxilla-maxilla contact	present	absent	absent	absent
4 Jaw margins (premaxilla & maxilla)	rounded	sharp-edged	sharp-edged	sharp-edged
5 Caniniform process	present	present, but small, sharp posterior edge	present, sharp anterior edge	present, relatively large
6 Maxilla with ventral keel	present	absent	absent	absent
7 Palatal exposure of palatine	small, smooth-surfaced	small, smooth-surfaced	relatively large, heavily pitted	relatively small, pitted anteriorly
8 Interorbital v. intertemporal breadth	IO >IT	IO <IT	IO >IT	IO >IT
9 Anterior tip of dentaries	blunt, tapering 'break'	fairly sharp, shovel-shaped tip	sharp, squared-off tip	sharp, squared-off tip
10 Dentary tables	absent	absent	present	present
11 Dentary shelf	present (widely expanded)	present (widely expanded)	present (variable development)	present
12 Mandibular fenestra	reduced in size or absent	present	present	present
13 Articular condyles	subequal in size	subequal in size	lateral condyle 1.5 × larger than medial articular condyle	subequal in size
14 Quadrates	immovable	immovable	movable	immovable

Members of the genus *Kingoria* are characterized as small to medium-sized dicynodonts with basal skull lengths ranging from 100–150 mm (Hotton 1986). The taxon currently contains seven species (listed in Cluver & Hotton 1981). So far, all have been transferred from the genus *Dicynodon* to the genus *Kingoria*, and to date, their validity has not been challenged. The Antarctic form from Shenk Peak represents another new species (N. Hotton, pers comm).

The most comprehensive generic diagnosis for *Kingoria* was produced by Cox (1959). Since that time, revised diagnoses for the genus have been submitted by Cluver & Hotton (1981) and Cluver & King (1983). Combined features of the skull which distinguish *Kingoria* from other dicynodont genera are listed in Table 1.

Palaeoecological interpretations based on cranial remains have been put forward by Kemp (1982) and Hotton (1986). Comments on feeding habits were prompted by the rounded nature of the upper jaw margin and mandibular symphysis as well as the unusual 'inflated' appearance of the lower jaw. Earlier, Cox (1959) had suggested that the jaws appeared to be more suitable for crushing than for cutting. For this genus, Kemp (1982) proposed a diet of fruits rather than leaves, stems, and roots. Hotton (1986) re-evaluated the masticatory apparatus in more detail, and affirmed that the shearing component of the cycle was of little significance compared to the other central dicynodonts of the Late Permian (i.e. *Diictodon*, *Oudenodon* and *Dicynodon*) and the Early Triassic genus *Lystrosaurus*. Hotton concluded (p. 286), 'The configuration of the symphysis suggests a crushing function, but the palate seems to have lacked the horn covering of other dicynodonts...whatever the food of *Kingoria*, it must have been of small size, soft and abundant.' He dubbed this form a 'forest litter forager'.

Compared to the other Antarctic dicynodont genera, *Kingoria* displays the most dramatic foreshortening of the lower jaws; a 'beak' bite is biomechanically impossible (Fig. 3). The symphyseal region is long and rather narrow (however not as lengthy as the symphysis in *Kannemeyeria*) which immediately suggests that crushing may be the primary means of processing food prior to swallowing, but the lower jaws lack dentary tables and a horny covering on the premaxillary and palatines is presumed absent. Hand manipulation of a specimen and paper models indicate that some ental–ectal movements (i.e. transverse jaw movements) may have been possible, but the principal jaw movements are propalinal. It is conceivable that some shearing could occur between the sharpened dorsal edge of the mandible and the keeled region posterior to the caniniform process. A lengthy mandibular symphysis would serve to distribute the bite forces generated in this region of the jaw more effectively. Crushing or mashing actions (depending on the hardness of the food item) were presumably confined to the premaxilla, and were applied with increasing force toward the posterior portion of the palate (the alignment of adductor musculature becomes more favourable as the jaw is retracted). The apparent lack of food rendering capabilities in the 'precanine' region of the mouth would place an initial constraint on the size of the food item selected. The limited extent of facial and buccal horny sheaths, combined with the near terminal position of the external nares, lack of wear facets on the canine tusks, gracile limbs, and sharp, conical ungual phalanges suggest that foraging activities may have been highly selective; superficially buried items were probably removed using the forelimbs rather than snout grubbing.

King (1985) examined *Kingoria*'s ecological role from a postcranial perspective. Based on fragmentary material, she concluded that the humerus was typical, and indistinguishable from those of other Permian dicynodonts. Morphological modifications evident on the femur, however, indicated a more upright stance had been achieved in the hindquarters. According to King's functional scenario, the principal advantage of an 'in-turned' hindlimb is improved manoeuvrability. The description of more complete propodial material (DeFauw 1986) refines King's functional assessment of *Kingoria* (these skeletal attributes are elaborated on in the next few paragraphs).

It is proposed that the following postcranial characters be appended to the revised generic diagnosis outlined in Cluver & King (1983, p. 252–3): (1) a procoracoid notch is present on the shoulder girdle instead of the typical procoracoid foramen; (2) the humeral shaft is unusually long (approximately 25% of the overall humeral length); (3) a bulbous trochlea protrudes from the distal end of the humerus; (4) the femoral head is well ossified and almost hemispherical, and appears to be separated from the greater trochanter by an incipiently developed 'neck'; and (5) the unfinished edge of the greater trochanter extends distally, close to one half the length of the femur. Characters 2 and 4 are autapomorphic for the genus. Cox's original diagnosis (1959) included a dozen postcranial characters; of these only one, the configuration of the sacrum (the pubis and ischium are located posterior to the ilium), serves to distinguish *Kingoria* from all other dicynodonts.

Kingoria, the putative 'forest litter forager' (Hotton 1986), is strikingly different in its appendicular morphology from the other genera reviewed. Features such as well ossified, strongly convex articular surfaces, and lengthening of the humeral and femoral shafts suggest that this form was more agile than its predecessors as well as its contemporaries (DeFauw 1986). Other aspects of its osteology, however, are similar to those expressed in the burrowing cistecephalids (e.g. well developed dorsal flexure of both humeral and femoral heads; foreshortened deltopectoral crest; distinct capitellum and trochlea; trochlea protrudes from distal end of humerus; relatively high degree of scapulocoracoid 'twist'). A peculiar flange on the anterolateral aspect of the fibula was observed by King (1985), and she proposed that it served as an attachment site for the fibulotibialis inferior. An examination of lacertilians with climbing capabilities (e.g. *Sceloporus*) reveals that the insertion of the iliofibularis muscle is more distally located than in non-climbers (Snyder 1954); perhaps the iliofibularis inserted on this feature. Based on the evidence at hand, *Kingoria* possesses a blending of characters that suggest moderate digging as well as climbing abilities (DeFauw 1986). Its overall gracile appearance coupled with an evident shift in fore- and hindlimb posture may have resulted in a significant improvement in its locomotor stamina (see discussion in Carrier 1987). The rarity of this form is the faunas of the *Aulacephalodon–Cistecephalus* and *Dicynodon–Whaitsia* assemblage zones may be the result of its preference for more 'upland' (riparian?) habitats.

Also included in the infraorder Kingoriamorpha is the genus *Kombuisia* (Hotton 1974), a small, tuskless, relatively rare member of the Early Triassic *Kannemeyeria–Diademodon* assemblage zone (as defined by Keyser & Smith 1979; this replaces the designation '*Cynognathus* zone') of southern Africa. Hotton (1974) provides a detailed discussion of the possible relationships between these two genera. The absence of kingoriamorphs from the *Lystrosaurus–Thrinaxodon* assemblage zone of southern Africa suggests that portions of Antarctica may have served as a refugium for these forms during this interval (i.e. Early Scythian age). Kingoriamorphs reappear in South African deposits of Late Scythian–Anisian age.

The genus Myosaurus *Haughton, 1917*

Members of the genus *Myosaurus* (Figs 2I and 3B) occur in the Lower Triassic *Lystrosaurus–Thrinaxodon* assemblage zone of southern Africa (Keyser & Smith 1979), and its biostratigraphic equivalent (the '*Lystrosaurus* zone', roughly the lower 200 m of the Fremouw Formation) in Antarctica (Cosgriff & Hammer 1979; Hammer & Cosgriff 1981). Keyser (1973b) lists *Myosaurus* in the '*Daptocephalus* zone' (currently referred to as the *Dicynodon–Whaitsia* assemblage zone, Keyser & Smith 1979), however, his biozone assignment is questionable (see Cluver 1974a and Kitching 1977). To date, less than a dozen specimens have been recovered from exposures of the Katberg Formation of Harrismith, Orange Free State; the Fremouw Formation (Collinson Ridge locality) has yielded one skull, a partial lower jaw, and an unprepared block presumably containing postcrania.

Myosaurus gracilis is a small dicynodont, with an average skull length of 40 mm (Cluver 1974a). The genus is unique in that it represents the smallest form to survive the Permo-Triassic transition with a distribution that extends beyond the limits of the South African Beaufort group. The most recent diagnosis of the genus *Myosaurus* was provided by Cluver (1974a); Hammer & Cosgriff (1981) supplied a differential diagnosis distinguishing this form from small Late Permian dicynodonts. Cluver & King (1983) place *Myosaurus* and many other small forms in the infraorder Diictodontia (based on the possession of a notched maxilla, reduced palatal exposure of the palatine, and the loss of marginal postcanine teeth); Diictodontia is the sister group of Kingoriamorpha. Combined features which serve to characterize the genus *Myosaurus* are listed in Table 1.

The feeding habits of *Myosaurus* have been discussed by Cluver (1974a) and Kemp (1982). Cluver (p. 48) remarked on the distinct 'lack of well-developed crushing and shearing areas in the mouth'. The presumed lack of a substantial horny sheath developed on the palatal surface (as 'evidenced' by the absence of bone surfaces pitted with nutrient foramina) and limited extent of horn on the external surface of the muzzle combined with the absence of dentary tables and the narrow, yet smoothly rounded protracted edges of the dentary probably prompted this interpretation. On the basis of the sharply angled surfaces of the medial and lateral quadrate condyles as well as the relatively close fit between the modest palatal rim and symphyseal region, Cluver (1974a, p. 48) concluded that 'quadrate-articular rotation was highly unlikely' and the jaw excursion was limited to 'strictly antero-posterior sliding movements'. Cluver also concluded (p. 48) that the putative diet of *Myosaurus* was 'roughly similar' to the invertebrate-rich diet suggested for the genus *Kawingasaurus* (Cox 1972). Kemp (1982, p. 138) placed *Myosaurus* in the 'second [distinct] group of Permian dicynodonts' based on a reduction of the palatine bone (similar to that observed in *Kingoria*) and the development of a notch in the maxillary margin. These modifications presumably indicated that the anterior aspect of the feeding apparatus was more important than the posterior part; the maxillary notch may have further facilitated the processing of slender, but tough aerial stems, rhizomes and roots (Kemp 1982).

Comparison of *Myosaurus* with proportionately scaled drawings of the remaining three antarctic dicynodonts reveals the following (Fig. 3):

(1) Premaxilla moderately, roughly one-third basal skull length. The premaxilla of *Kingoria* and *Kannemeyeria* are slightly longer, whereas the premaxilla of *Lystrosaurus* was significantly shorter.

(2) Dentary symphysis of *Myosaurus* rather 'weak' relative to overall jaw length. In dorsal view, this measurement is apparently the shortest in the genus *Lystrosaurus*, with *Kingoria* and *Kannemeyeria* exhibiting significant increases in symphyseal length.

(3) Dentary shelves are considerably larger than those evident in *Kingoria*, *Lystrosaurus* and *Kannemeyeria*. This feature gives the jaws of *Myosaurus* a more inflated appearance in dorsal view.

(4) Reflected lamina better developed and more extensive, relative to overall mandible length and basal skull length, than in the other three genera reviewed.

The size of the reflected lamina may facilitate the detection of vibrations, and thus assist or complement the conventional auditory system. King (1981) has suggested that the size and complexity of the reflected lamina may be related to a well-developed tongue. The function of the reflected lamina is currently being evaluated in conjunction with the form and function of the dicynodont stapes (Hotton, DeFauw & Colbert, in preparation). The features set forth above, combined with the functional attributes outlined in the preceding paragraph suggest that this diminutive dicynodont was capable of generating and sustaining a powerful bite force over the entire functional jaw edge (rather than concentrating

it in the anterior or posterior parts of the dentary's dorsal surface). It is proposed that precise cutting, perhaps more appropriately termed nipping, occurred between the 'fairly sharp' symphyseal margin and the premaxilla and anterior and medialmost portions of the maxilla. Small items could conceivably be crushed or ground between the narrow, round-edged dentaries and the palate proper. It is also feasible that food was actively triturated during both fore and aft jaw movements.

An actual specimen of *Myosaurus* (with associated lower jaws) was unavailable for direct manipulation, however, the following suite of features suggests that some transverse (i.e. ental−ectal) jaw movement may have been possible: (1) anterior palatal ridges absent; (2) relatively weak caniniform process with fairly sharp posterior edge (similar in overall construction to most *Oudenodon* specimens); (3) lateral and medial articular condyles subequal in size (similar to the condition exhibited in *Cistecephalus*, which has ental−ectal capabilities, Cluver 1974a and b); (4) relatively wide intertemporal region (also evident in the genus *Cistecephalus*); (5) flared out zygomatic arches (similar to the genus *Oudenodon*, which Hotton (1986) considers a browsing form, with well-developed ental−ectal capabilities). The placement of external nares in close proximity to the upper jaw margin combined with the limited extent of facial horn and small caniniform processes suggest that this form did not intensively engage in snout grubbing for food (as another relatively small wide-spread genus, *Diictodon*, presumably was capable of doing). The masticatory apparatus of *Myosaurus* appears best equipped to process small, rather soft food items and this form is therefore provisionally considered an invertebrate-specialist.

The South African locality (at Harrismith) which yielded the majority of specimens has also yielded much wood pertaining to *Dadoxylon* sp.; perhaps, *Myosaurus* was a forest-edge species (which would explain, in part, its rarity in the fossil record). A functional analysis of available postcranial material will greatly facilitate the refinement of *Myosaurus*'s foraging habits.

The genus Lystrosaurus *Cope, 1870*

Members of the genus *Lystrosaurus* (Figs 2H and 3C) derive from the Lower Triassic *Lystrosaurus−Thrinaxodon* assemblage zone (Keyser & Smith 1979) of southern Africa and many of its stratigraphic equivalents throughout the world. In addition to the African forms,

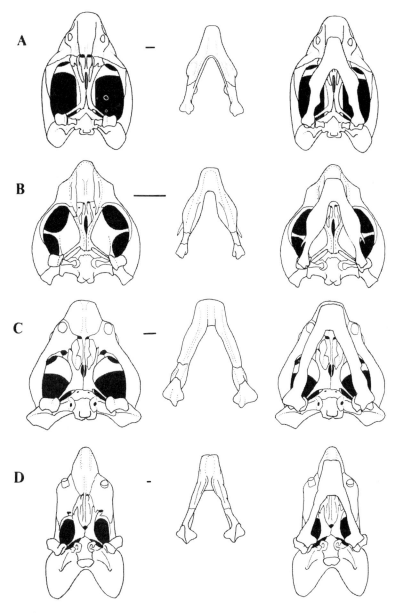

Fig. 3. Palates and lower jaws of the Antarctic dicynodont genera. (**A**) *Kingoria* (after Cluver & Hotton 1981). (**B**) *Myosaurus* (after Cluver 1974*a*). (**C**) *Lystrosaurus* (after Cluver 1971). (**D**) *Kannemeyeria* (after Pearson 1924*a*). The skulls have been drawn to the same unit length. The lower jaws (in the middle column) have also been proportionately scaled, and are depicted in dorsal view. Bar scales to the right of each palatal view are equal to 1 cm.

representatives of the genus have been described from Antarctica (Fremouw Formation, Elliot *et al.* 1970; Kitching *et al.* 1972; Colbert 1974; Cosgriff *et al.* 1978, 1982; Cosgriff & Hammer 1979, 1981), Asia (Luang Prabang, Laos, Repelin 1923; Jiucaiyuan Formation, China, Yuan & Young 1934, Young 1935, 1939 and Sun 1964, 1973*a*), the European regions of the USSR (Ryabinin horizon, Kalandadze 1975), India (Panchet Formation, Huxley 1865; Lydekker 1887; Tripathi 1962; Tripathi & Satsangi 1963) and probably Australia (Arcadia

Formation, Thulborn 1983; King 1983). Combining the abundance of fossil material referrable to this genus and the palaeobiogeographical distribution of the taxon, *Lystrosaurus* is considered to be the most successful mammal-like reptile of all time (Colbert 1974).

The probable Permian ancestor of the genus *Lystrosaurus* has eluded investigators for decades. Proposed ancestral forms have included: '*Dicynodon*' [*Diictodon*] *testudirostris* (Haughton 1917); *Dicynodon gilli* (Broom 1932); and *Dicynodon annae* (Toerien 1954; Camp 1956). Cluver (1971, p. 260) remarked that '*Daptocephalus* (now a junior synonym of *Dicynodon*) stands near the ancestral stock of the Triassic dicynodonts', but it is 'an unlikely direct ancestor for *Lystrosaurus*'. Most of these authors cautiously hint at a non-South African region of origin. Recently, King (1981) commented that *Dicynodon trigonocephalus* (a Late Permian form from Zambia) may be the closest to *Lystrosaurus*. Hotton (pers comm, 1988) contends that *Lystrosaurus* is most closely allied to *Dicynodon lacerticeps*.

Members of the genus are characterized as small to large, tusked dicynodonts with an average skull length of about 120 mm (Cluver 1971; skull lengths range from 49–390 mm, Kitching 1968). First described by Huxley in 1859 (*a* and *b*), the group has since had a very diverse taxonomic history. Considering for a moment only the African forms, at one time, as many as 22 species were assigned to the genus (Broom 1932; Tripathi & Satsangi 1963). If all African species ever described are tallied, the total number of species climbs to 28. The contents of the genus have been reworked and revised by Broom (1932, recognized 5 species), Brink (1951, 12 species), Kitching (1968, 6 species), Cluver (1971, 9 species), Colbert (1974, concurred with Kitching) and Cosgriff *et al.* (1982, advocated 3–6 species). Aside from the African forms, a total of 11 species of *Lystrosaurus* have been described from other parts of the world. Two of these species (*L. orientalis* and *L. incicivum*) are considered *nomina dubia* (Colbert 1974). The validity of the remaining non-African species has not been challenged. The most comprehensive diagnosis of the genus *Lystrosaurus* was produced by Cluver (1971). Cranial features which characterize the genus are included in Table 1.

Crompton & Hotton (1967) produced the first detailed analysis of the dicynodont masticatory apparatus; the chief model for their characterization was the genus *Lystrosaurus*. Reconstructive refinements were proposed by Cluver (1971) and Hotton (1986). Briefly stated,

'the masticatory cycle of *Lystrosaurus* is...a combination of slicing, which takes place in front of the caniniform processes, and crushing, which takes place in the mouth at the level of the palatal horn layers on the palatine and adjoining maxilla' (Cluver 1971, p. 214). This cycle is quite different from the jaw movements proposed for either *Kingoria* or *Myosaurus* (this paper). Hotton (1986) suggests that the feeding mechanism of *Lystrosaurus* (as well as *Dicynodon*) reflects a greater flexibility in foraging ability than the other dicynodont genera considered in his review (i.e. *Diictodon*, *Oudenodon*, *Kingoria*, and *Dicynodon*). As reconstructed, the grubbing mode of food procurement in *Lystrosaurus* was presumably augmented by the foliage-cropping facility of the caniniform blade, hence the designation 'flexible forager'.

An inspection of scaling relationships among Antarctic dicynodonts reveals that the internal choanae of *Lystrosaurus* are situated farther forward than in other genera (Fig. 3). This observation is in sharp contrast with one of the evolutionary trends perceived in Triassic dicynodonts, i.e. lengthening of the premaxilla (Toerien 1953; Keyser 1974). The dentary symphysis is deeper (dorso–ventrally) than the other forms examined, but the symphysis is not as extensive (antero–posteriorly) as in *Kingoria* and *Kannemeyeria*. From manipulation of skulls and models, it is apparent that jaw movements were largely constrained by palatal 'guides' (a pair of anterior palatal ridges and a central median ridge) and the close fit between upper and lower mandibles. Food was comminuted using propalinal and just plain orthal movements. *Lystrosaurus* also exhibits very robust quadrates (these are more heavily built than *Kingoria*, *Myosaurus* or *Kannemeyeria*). Overall, the lower jaw is of sturdy construction and much more massive in appearance than the other Antarctic dicynodonts.

Modifications evident in the lystrosaurid snout have often been interpreted as adaptations to feeding in an aquatic environment (Watson 1912; Broom 1932; Romer 1966; Colbert 1969; Cluver 1971). Watson (1912, p. 293) noted that the sheer massiveness of lystrosaurid jaws is 'rather difficult to reconcile with the softness of most aquatic plants', and he went on to suggest that *Lystrosaurus* may have consumed molluscs. More recently, lystrosaurids have been characterized as reptilian analogues of hippopotami (Cox 1965; Kemp 1982). The ecological impact of these mammalian herbivores on their environs yields some intriguing comparative information. Contrary to popular belief, the principal fodder of hippopotami is not aquatic vegetation. These ungulates contribute to the formation of 'lawns' in forested or thicketed areas which, in turn, enhances species diversity, because a variety of diminutive herbivores are attracted to the patchy habitats created and sustained by the foraging habits of 'megaherbivores' (Owen-Smith 1987). According to Owen-Smith (p. 354), 'Megaherbivores are supreme generalists not only in habitat tolerance, but also in their dietary acceptance of a wide range of plant species and parts...they can support themselves on nutrient-poor tissues...due to their low specific metabolic rates, a direct consequence of large body size'. It is indeed a curious coincidence that large dicynodonts such as lystrosaurids and kannemeyeriids exhibit the widest palaeobiogeographic ranges and are 'accompanied' in the faunal assemblages by diminutive forms (i.e. *Myosaurus* and *Kombuisia*). In addition, *Lystrosaurus* possesses long dentary tables (horizontally expanded surfaces located just posterior to the dentary symphysis) on the lower jaw as well as wear facets on the tusks. The presence of these features is suggestive of feeding on plants with an abundance of sclereids (i.e. 'stone cells' or phytolithic tissues), which lends support to the proposition that lystrosaurids foraged primarily on land (see Hotton 1986).

Various aspects of the postcranial skeleton of *Lystrosaurus* have been described by Huxley (1859b), Lydekker (1887), Watson (1912, 1913), Van Hoepen (1915), Haughton (1917), Young (1935, 1939) and Li (1983). The functional anatomy of the shoulder girdle and forelimb was evaluated by DeFauw (1981), who also compared both girdle and propodial elements with the appendicular skeletons of other common African dicynodont genera (DeFauw 1986). Combinations of postcranial characters that serve to characterize the genus *Lystrosaurus* include: (1) glenoid facet present on the procoracoid; (2) procoracoid notch present in the shoulder girdle; (3) prominent interclavicular peg present; (4) 3–5 discrete emarginations are evident on the iliac blade (in some specimens a foramen is formed at the position of the second emargination); (5) 6 sacral ribs adjoin the medial surface of the iliac blade; (6) an incipient 'ischial keel' is present on the posterodorsal margin of the ischium. It is proposed that these features be appended to Cluver's diagnosis of the genus (Cluver 1971, p. 237).

Lystrosaurus has been characterized as a 'flexible forager' (Hotton 1986). It is most similar in appendicular morphology to the genus *Dicynodon* (DeFauw 1986), which Hotton

(1986) also dubbed as a flexible forager. For decades, *Lystrosaurus* was portrayed as an aquatic reptile that foraged in shallow water. Features used to substantiate this conclusion included: (1) elevated external nares (the steep-sided construction suggested the presence of a muscular flap or valve, Watson 1912); (2) ventral displacement of the feeding apparatus; (3) reduced size of the clavicles and interclavicle compared to other 'anomodonts' (Broom 1903); (4) short, flattened radius and ulna (reminiscent of the forearm of a seal, Watson 1912); (5) shallow pelvis and posteriorly extended ischium (viewed as adaptations to swimming, Watson 1912); and (6) lack of ossification of the distal tarsal elements (Watson 1913). Some of these (items 3, 4 and 5) are inaccurate assessments based on scrappy or incomplete specimens, coupled with a paucity of comparative material. For example, the elements of the forearm are not as short and flat (relatively speaking) as Watson indicates. Propodial-epipodial indices for the genus average out to 1.3; the range for the common dicynodonts measured by DeFauw (1986) varied from 1.2 to 1.6. When the antebrachial elements are articulated properly, the bones are not flattened in the same plane as Watson (1912, 1913) has restored them (DeFauw 1981). The clavicles are more gracile in appearances than in earlier therapsids (this observation does not always hold true when compared to other dicynodont genera), and this may be due to a reduction in the amount of torque generated at the glenohumeral articulation as a result of the construction and orientation of the scapulocoracoid (DeFauw 1981). On the basis of apparent morphological modifications of the appendicular skeleton (i.e. dorsal expansions of the scapular blade and ilium, strongly convex and rather well-ossified humeral and femoral heads, limb ratios, robust digits and spade-shaped ungual phalanges) and taphonomic evidence, the wide-ranging *Lystrosaurus* is considered a semi-aquatic form with rather well-developed digging capabilities (DeFauw 1981, 1986; Hotton 1986).

The genus Kannemeyeria *Seeley, 1908*

The first Triassic dicynodont to be described, '*Dicynodon*' *simocephalus* (Weithofer 1888), was later referred to the genus *Kannemeyeria* (Broom 1932). Members of the genus occur in the Lower and Middle Triassic deposits of southern Africa, South America, India and possibly Antarctica. According to some investigators, the Puesto Viejo Formation (Late Scythian) of Argentina yields the earliest kannemeyeriids, which may have been contemporaneous with *Lystrosaurus* (Bonaparte 1966). Additional lithostratigraphic units in which *Kannemeyeria* occurs include: the Burgersdorp (Late Scythian, South Africa, Kitching 1977), N'tawere (Early to Middle Anisian, Zambia, Keyser & Cruickshank 1979; Crozier 1970), Omingonde (Early to Middle Anisian, Namibia, Keyser 1973a) and Manda (Middle Anisian, Tanzania, Keyser & Cruickshank 1979; Cruickshank 1965) Formations of Africa; the Yerrapalli Formation (Anisian) of India (Roy Chowdhury 1970); and possibly the Fremouw Formation (Scythian) of Antarctica (Hammer *et al.* 1987). A dicynodont quadrate, recently reported from the Arcadia Formation (Scythian) of southeast Queensland, Australia (Thulborn 1983), was provisionally identified as a dicynodont closely related (or perhaps even identical) to *Kannemeyeria*. The fragment is now assigned to the genus *Lystrosaurus* (King 1983). Kannemeyeriids are also known from China (the Scythian-Anisian age Er-Ma-Ying Formation of North China) and the USSR (the Anisian–Ladinian age Donguz and Bukobay Groups). Keyser & Cruickshank (1979) have remarked that some of the Chinese and Russian forms may be members of the genus *Kannemeyeria*. At any rate, kannemeyeriids (like lystrosaurids) were extremely successful members of the infraorder Pristerodontia, and they achieved wide geographical distributions.

Kannemeyeria was presumably derived from the *Dicynodon* lineage (Camp 1956; Keyser & Cruickshank 1979). The genus has been referred to as the 'typical' kannemeyeriid (Keyser & Cruickshank 1979); however, considering the range of diversity evident within the family, this is not an accurate assessment. Camp (1956) suggested that the ancestral kannemeyeriid was broad-headed, and thus similar to some of the Russian Permian species of '*Dicynodon*', but more distantly related to '*Daptocephalus*' (= *Dicynodon*) and *Lystrosaurus*; these authors proposed a northern origin (Laurasian) for the group. On the basis of similarities in the premaxillary portion of the snout and configuration of the intertemporal region, Keyser & Cruickshank (1979) allied *Kannemeyeria* with the species *Dinanomodon rubidgei* (considered a member of the genus *Dicynodon* by Cluver & King 1983), but these two taxa are separated by a considerable time gap.

Kannemeyeriids were large dicynodonts, with basal skull lengths typically ranging from 240–500 mm (Keyser & Cruickshank 1979). Camp (1956) provided the most comprehensive definition of the genus *Kannemeyeria*, which currently contains four species (see Walter 1985, pp. 19–22). Functionally significant cranial features for this genus are listed in Table 1.

Palaeoecological interpretations of kannemeyeriids (*sensu* Keyser & Cruickshank 1979) have been advanced by Pearson (1924a and b), Cox (1965), Cruickshank (1978), Walter (1985) and Hotton (1986). Pearson postulated that *Kannemeyeria* grubbed for its food, on the basis of tusk wear and capabilities for digital flexion. Cox (1965) and Cruickshank (1978) offered more detailed characterizations of the kannemeyeriid niche, though these authors had divergent opinions on browsing and grazing dicynodonts. The controversy concerned the two main Triassic taxa, *Kannemeyeria* and *Stahleckeria* (Cox considered these representatives of two families; Cruickshank differentiated them at the level of subfamily), and serves to highlight the pitfalls of circumstantial analogies. Cox (1965) proposed that the Black and White rhinos (*Diceros bicornis* and *Ceratotherium simum*, respectively) were suitable analogues for the dicynodont families Kannemeyeriidae (browsers) and Stahleckeriidae (grazers), respectively. In brief, Cox characterized kannemeyeriids as large dicynodonts with pointed snouts and high occiputs. Stahleckeriids, on the other hand, possessed blunt snouts and wide occiputs. Cruickshank (1978) disagreed with Cox's analogy, contending that these dicynodont taxa were closer to pigs in size, and refuted Cox's perceived variations in occipital proportions. Instead, he proposed that the Warthog (*Phacochoerus aethiopicus*) and the African Bush Pig (*Potamochoerus porcus*) were more suitable analogues. He noted an obvious difference in the cranial morphology of these suids, namely, the angle of the occiput with respect to the plane of the palate. When this feature was compared with the occipital orientation of the two dicynodont groups, the kannemeyeriids (*sensu* Cox) were interpreted as selective 'grazers' and the stahleckeriids (*sensu* Cox) were 'browsers'. Walter (1985) suggested that the term 'grazer' and 'browser' should be discarded. She also pointed out that plant resources were most likely partitioned by maximum head height among the kannemeyeriid dicynodonts (*sensu* Keyser & Cruickshank 1979). Most recently, Hotton (1986) has proposed that kannemeyeriids were 'flexible foragers' (he also included *Dicynodon* and *Lystrosaurus* in this feeding regimen).

Comparing the scaling relationships of cranial features of Antarctic genera reveals: (1) the premaxilla is most extensively developed in *Kannemeyeria*; (2) the system of three palatal ridges reaches its greatest development in

Kannemeyeria; (3) the dentary symphysis is very elongate and robust (more so than the other Antarctic forms); and (4) a foreshortening of the lower jaw is evident. A true 'beak' bite (i.e. occlusion of the anteriormost jaw margins) does not appear to be possible in the forms with pointed snouts (e.g. *Kannemeyeria*). The presence of facial ridges and nutrient foramina, as well as evident tusk wear, suggest that snout grubbing played an important role in food procurement (Hotton 1986). From manipulation of models, propalinal and orthal jaw movements dominate the masticatory cycle. Shredding and shearing occurred anterior to the canines (the caniniform process is sharp anteriorly), whereas crushing occurred posterior to the canines. The lengthy mandibular symphysis coupled with the foreshortening of the lower jaw would serve to more effectively distribute the powerful bite forces generated as the jaw is retracted.

The limb adaptations in kannemeyeriid dicynodonts were recently evaluated in detail by Walter (1985). It is proposed that the following autapomorphic postcranial characters, discerned by her, be appended to Cruickshank's (1970, p. 50) diagnosis for the genus *Kannemeyeria*: (1) femoral head inclined anteriorly by approximately 50 degrees; (2) a flattened medial distal femoral condyle is present; (3) the posterior margin of the ischium is slightly inflected.

Although *Kannemeyeria* has been characterized as a 'flexible forager' (Hotton 1986), morphological modifications evident in the facial, palatal and intertemporal regions of other kannemeyeriids indicate the need for future interpretive refinements of the roles of Middle and Late Triassic forms. Based on available sedimentological data, Walter (1985, p. 237) has observed that 'kannemeyeriid fossils are always found in shallow to very shallow lacustrine sediments', and the abundant plant resources of these lake margins presumably were partitioned among the different kannemeyeriid species (or among ontogenetic stages) by maximum head height. These ponderous, slow moving herbivores were the largest primary consumers of the communities in which they occurred (Walter 1985, 1986). Kannemeyeriids persisted through the early Norian stage of the Late Triassic.

Palaeobiogeographic distributions and associations of dicynodont feeding types

Forty two dicynodont genera occur in Late Permian and Early Triassic age deposits. Eight to possibly eleven genera are known to occur on two or more cratons during this interval

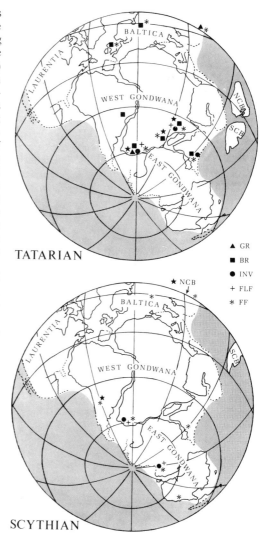

Fig. 4. A preliminary examination of the palaeogeographic distributions of the five recognized dicynodont feeding types during the Tatarian and Scythian stages (maps modified from Irving (1983) and Lin *et al.* (1985)). Abbreviations: BR, Browser; FF, Flexible forager; FLF, Forest litter forager; GR, Grubber; INV, Invertebrate specialist; NCB, North China block; SCB, South China block. Starred localities indicate the presence of dicynodont genera whose feeding types have not been characterized. Symbols located outside of the global perimeters (see upper right quadrant of each map) denote dicynodont feeding types present on the Tarim craton.

(Figs 2 and 4). These widespread dicynodont genera include: *Diictodon, Dicynodon, Endothiodon, Kingoria, Oudenodon, Lystrosaurus, Myosaurus,* and possibly *Pristerodon, Cistecephalus* and *Geikia* (see Keyser (1981)

regarding the first two genera of question-able distribution, and Rowe (1980) and Cruickshank & Keyser (1984) regarding *Geikia*). Currently, *Kannemeyeria* is the only genus that persists on at least two cratons from the Early through the Middle Triassic.

Various authors have pointed out the prob-lems encountered in extracting reliable in-terpretations from the fossil record (Cox 1973; Bakker 1977; Benton 1979). Factors that ob-scure the actual distributions of extinct organ-isms include: (1) the distribution, accessibility and extent of fossil horizons; (2) the loss of fossil horizons due to erosion, deep burial, sub-duction, metamorphism, and diagenesis; (3) the distribution of palaeontologists and the con-comitant extent of exploration and study; (4) facies control; (5) the relative dating of fossi-liferous horizons; and (6) systematic and taxonomic refinements. The distributions of dicynodont genera were plotted on recently revised Permo-Triassic palaeogeographic re-constructions. Integration of these data with information on sea level fluctuations, tectonic-orogenic events, and floral changes provides some interesting insights on dicynodont evolution.

Late Permian dicynodont-bearing beds exhib-it a bizonal distribution throughout Pangaea; these localities are clustered between approxi-mately 15−40° North latitude and 32−62° South latitude (see Fig. 4, modified from the palaeo-geographic reconstructions of Irving 1983). In the Northern Hemisphere, the Permian genera *Diictodon* and *Dicynodon* occupied the Tarim block of the Asian tectonic collage (also see maps in Lin *et al.* 1985). *Oudenodon*, *Geikia* and *Dicynodon* have been reported from components of the Baltica assemblage (Amalitzky & Karpinski 1922; Rowe 1980; Keyser 1981; see maps in Irving (1983) and Lin *et al.* (1985)). In the Southern Hemisphere, most of the dicynodont genera are found in southern Africa. India yields four genera, *Endothiodon*, *Oudenodon*, ?*Pristerodon* and ?*Cistecephalus* (Kutty 1972; Keyser 1981). One genus, *Endothiodon*, occurs in South America (Barbarena *et al.* 1985).

From the evidence available, *Diictodon* and *Dicynodon* were the widest ranging (perhaps most environmentally tolerant) dicynodonts of the Late Permian. Apparent sites of origin can be discerned for some of the widespread Permian forms. *Pristerodon* and *Diictodon* first appear in South African deposits. During the Late Permian, *Pristerodon* and cistecephalids spread to Tanzania and, possibly, India; *Diictodon* dispersed to China. *Kingoria* occurs in the Late Permian deposits of both South and East Africa, and a relict form is found in the Early Triassic of Antarctica. The infor-mation on hand for *Oudenodon*, *Endothiodon*, *Dicynodon*, and *Geikia*, however, does not provide unambiguous answers as to their geographic origins.

As mentioned previously, the first of two peaks in dicynodont taxonomic diversity occurred during the Tatarian stage (Upper Permian). In total, 32 genera have been re-corded from this interval, with 30 of these occurring in southern Africa. Is this view reasonable or is it an artifact of taxonomic splitting, facies control, and over 140 years of collecting from accessible exposures? At this point, the query is left unresolved.

A review of Pangaean topographic diversity during the Late Permian through Early Triassic reveals that the supercontinent exhibited little internal relief; most mountain ranges were con-fined to continental peripheries (Ziegler *et al.* 1982). Therefore, at that time, there were few zoogeographical barriers to impede the dispersal of tetrapods, and a comparatively cosmopolitan fauna developed (Colbert 1973; Cox 1973). An inspection of cycles of sea-level fluctuations indicates that a steady long-term fall in sea-level culminated in the Late Permian (Haq *et al.* 1987). This long-term regression was punctuated by two dramatic short-term declines in sea level, as compared with the present day mondial mean sea level (see Haq *et al.* 1987, fig. 5). Earlier, Schopf (1974) had demonstrated reductions in the areal extent of shallow marine seas (as well as significant reductions in marine invertebrate diversity) throughout the Permian.

The plant megafossil succession throughout the Permian of Gondwana is characterized by an increase in the dominance of *Glossopteris* and a decline of *Gangamopteris*. The diversifi-cation of dicynodonts presumably coincided with the establishment of a *Glossopteris* flora. In contrast to the rather homogeneous *Glossopteris-Gangamopteris* flora of the Permian of Gondwana, three floral provinces have been traditionally recognized in the Northern Hemisphere: (1) the Cathaysian; (2) the Angaran; and (3) the Euramerican (Chaloner & Meyen 1973). Recently, Wang (1985) has summarized the Permo-Triassic palaeophytogeography of North China (former-ly a component of the Cathaysian Province). He assigns these floral assemblages to the Euramerican Province rather than to the Cathaysian Province. Earlier, Bock (1969) had concluded that the Euramerican Province is the closest to the *Glossopteris* flora in composition.

Thus, the floristic barriers that King (1988) recognizes, as influencing the dispersal of dicynodonts in the Northern Hemisphere, are based on out-dated interpretations. The distributions of *Diictodon* and *Geikia*, in particular, are not problematic (King states that they are) for the floral assemblages presumably associated with these dicynodonts were quite similar.

The Scythian stage marks a 'bottleneck' in the evolutionary history of the Dicynodontia, and also coincides with a 'peak' in the areal extent of shallow marine seas (Schopf 1974). By the earliest Triassic, new plant associations were established (a noticeable change in floral composition, highlighted in the microfossil record, occurred in the latest Gondwanan Permian assemblages), and the glossopterids were reduced to relicts (Schopf & Askin 1980). The widespread occurrence of low-diversity coastal plant associations (i.e. lycopods and gymnosperms) coincided with the extensive marine transgression at the base of the Triassic.

Eight genera of dicynodonts have been reported from Early Triassic age deposits worldwide. These taxa are located on components of both East and West Gondwana (i.e. Antarctica & Australia and South America & Africa, respectively), India, the 'near equatorial' North China and Indochina–Malaya blocks, and the Baltica block (Fig. 4). Thus, during the Scythian, dicynodont-bearing localities ranged from approximately 70° South latitude to 35° North latitude; a bizonal distribution of sites is, once again, apparent.

Lystrosaurus and *Kannemeyeria* are clearly the most widespread Early Triassic dicynodont taxa. The rather sudden appearance of *Lystrosaurus* in the Early Triassic of South Africa, coupled with the difficulty in selecting a suitable ancestor, resulted in a number of investigators concluding that the 'frequent lizard' was an immigrant (Camp 1956; Cluver 1971). Recently, King (1981) has proposed that *Dicynodon trigonocephalus*, an East African form, is 'closest to *Lystrosaurus*'. *Kannemeyeria* has also been identified as a form that may have immigrated from the north. Kannemeyeriid ancestors may have occupied the North China block. Based on our current knowledge of the assembly of Asia, it is conceivable that kannemeyeriids (*sensu* Keyser & Cruickshank 1979) evolved, in isolation, on the North China block (which traversed the Tethys and joined with the Siberian–Kazhakstan–Tarim cratonic assemblage during the Late Permian, see Lin *et al.* 1985). These forms may have then spread through the European block, to North and South America, Africa, Antarctica and India.

However, these controversial points remain to be more firmly resolved.

The second peak in dicynodont taxonomic diversity occurred in the Anisian (early Middle Triassic, see Fig. 1). Seventeen genera have been recorded from this interval, with the largest fraction of taxa coming from the Donguz Group of the USSR. In contrast to the Tatarian radiation, this taxonomic diversification occurred during a long term transgression (which was initiated during the Late Tatarian and completed at the beginning of the Norian, Haq *et al.* 1987, fig. 5). The two apparent radiations are similar in that both occur over time intervals that encompass two significant short-term declines in sea level (see fig. 5 in Haq *et al.* 1987). Orogenic activity intensified during the long-term marine transgression of the Triassic (Dingle 1978; Lefeld 1978; Lin *et al.* 1985), and Pangaea drifted northwards about 30 degrees (Smith *et al.* 1981). These events presumably precipitated the breakup of the relatively cool, humid climate that prevailed during the Early Triassic (Frakes 1979). Johnson (1971) presented compelling evidence linking peaks in orogenic intensity with maximum transgression of shallow seas. This phenomenon, termed the Haug Effect, may be the mechanism that controls land diversity cycles (Bakker 1977). The dicynodont fossil record, on its own, provides ambiguous support for the Haug Effect and its impact on land diversity cycles (a comprehensive treatment of Permo-Triassic tetrapod distributions is in preparation). The Late Permian through Middle Jurassic temnospondyl record lends support to Bakker's hypothesis (DeFauw 1989).

A 'transitional flora' was evident in Gondwana during the latest Permian and earliest Triassic. By the end of the Early Triassic, a rather uniform *Dicroidium* flora was dominant. Floral provincialism developed in Gondwana during the Middle and Late Triassic (Schopf & Askin 1980). In contrast, the provinciality apparent in the northern floras of the late Palaeozoic declined throughout the Triassic, but two major zones can still be recognized in the Late Triassic (i.e. the 'Siberian' floristic region and a 'Eurasian' floristic region, Ash 1986).

The pervasive floral changes throughout the latest Permian and Triassic may have influenced the spatial and temporal distributions of dicynodonts; however, plant fossils are rarely associated with vertebrate faunas (thereby diminishing the significance of the perceived correlations). Presumably, the majority of Permian genera were associated with the *Glossopteris* flora, but *Diictodon*, *Dicynodon*,

Oudenodon and *Geikia* were also associated with a mixture of components from the Late Permian Euramerican and Angaran floras. *Lystrosaurus*, *Myosaurus*, *Kingoria* and *Kannemeyeria* foraged, for the most part, on components of the 'transitional flora' of the Southern Hemisphere . The taxa of North China were presumably associated with the Eurasian Arid 'subprovince' (Wang 1985) of the Euramerican floral province. Near the end of the Early Triassic, herbivorous components of the *Kannemeyeria* assemblage subsisted, for the most part, on the *Dicroidium* flora that became widely established. Plant material was found in association with the new '*Cynognathus* zone' faunal assemblage discovered during the 1985−86 austral field season (Hammer *et al.* 1986, 1987), and may provide information on the progress of floral change in the southernmost reaches of Gondwana.

To date, five basic dicynodont feeding types have been recognized. Although Hotton (1986) has cautioned that some feeding types may have been overlooked, the types currently recognized provide a basis for exploring spatial and temporal patterns of ecomorphic diversity (also see DeFauw 1987, 1989). The feeding habits of the seven widespread dicynodont genera not discussed in detail in the previous section on Antarctic forms (i.e. *Diictodon*, *Dicynodon*, *Endothiodon*, *Oudenodon*, *Geikia*, *Pristerodon* and *Cistecephalus*) are briefly reviewed here. Three of these forms have already been characterized by Hotton (1986). *Diictodon* has been depicted as a grubber; the recent discoveries of helical burrows containing articulated skeletons demonstrate its rather prodigious digging capabilities (Smith 1987). *Oudenodon* is considered to be the reptilian analogue of a mammalian browser, and *Dicynodon* is classified as a flexible forager. Cruickshank & Keyser (1984) portray *Geikia* as a selective browser. Provisional feeding types are proposed for the remaining three genera (i.e. *Pristerodon*, *Endothiodon* and *Cistecephalus*). *Pristerodon* is considered a small, flexible forager. This tentative assignment is supported by: the sharp-edged jaw margins, beak bite capabilities, effective 'shearing and shredding' batteries of postcanine teeth, and morphological modifications evident in the appendicular skeleton that are comparable to those observed in other flexible foragers (DeFauw 1986). *Endothiodon* is tentatively characterized here as a browser. It possesses a massive dentary symphysis with a pointed, upturned tip, and Cox (1964, p. 18) suggested that 'the lower jaw may have been used for grubbing vegetable matter out of the ground'. In addition, *Endothiodon* displays large moveable quadrates

with medial and lateral condyles roughly equal in size (comparable to the condition observed in *Oudenodon*), and aspects of its appendicular skeleton are similar to those observed in *Oudenodon* (DeFauw 1986). *Cistecephalus* possesses a unique mosaic of features (discussed in detail by Cluver 1974*a* and *b*) including: a deep, robust dentary symphysis that terminates in a square, shovel-like tip; well-developed dentary tables (for crushing food items); relatively flat quadrate and articular articulation surfaces (suggesting that transverse jaw movements may have been part of the mastication cycle); and a postcranial skeleton presumably specialized for fossorial activities (Cluver 1978). It is tentatively considered to be an invertebrate-specialist.

A preliminary examination of the palaeogeographic distributions of the five recognized dicynodont feeding types from the Late Permian reveals the following pattern (Fig. 4). Representatives of all five types are present in South Africa during the Late Permian. Three different foraging types occur in India (i.e. browsers, flexible foragers and invertebrate specialists). The Chinese forms include members of two out of the five feeding regimes (i.e. grubbers and flexible foragers). Dicynodont browsers and flexible foragers were recovered from the Russian Platform and the United Kingdom. A browsing form has recently been reported from the Late Permian of South America.

Shifts in feeding diversity are apparent from the Late Permian through the Early Triassic. The rather small grubbing forms as well as small to medium-sized browsing forms (*sensu* Hotton 1986) apparently did not survive the Permo-Triassic transition. However, invertebrate specialists, flexible foragers and forest litter foragers persisted on the African and Antarctic continents. To date, only flexible foragers have been recovered from the remaining Scythian age localities in South America, India, the European regions of the USSR, China, and most recently, Australia. As previously mentioned, application of the term 'flexible forager' to all kannemeyeriids (*sensu* Keyser & Cruickshank 1979) may not be appropriate. A detailed analysis of the morphological modifications evident in the masticatory apparatuses (and appendicular skeletons) of these forms is necessary to refine their roles.

Discussion

The external 'settings' of the two diversifications evident in the dicynodont fossil record differ in a few ways, but also exhibit some similarities.

The initial 'Gondwanan' radiation of the Tatarian (Late Permian) has been associated with the absence of competing herbivore groups (i.e. the dinocephalian and pareiasaur reptiles, Kemp 1982). Other associated factors included: a long-term decline in sea level; a general warming of the climate throughout the interval; the confinement of tectonic-orogenic activity to the periphery of Pangaea (for the most part); and the development of a rather homogeneous *Glossopteris* flora in the Southern Hemisphere. In contrast, the second diversification was associated with an increase in 'intracontinental' orogenic activity, and a long-term rise in sea level. The second diversification (which took place in the Anisian) apparently occurred in the Northern Hemisphere, but this may be the result of taxonomic splitting (as some of the six Russian genera are based on very fragmentary remains). With the appearance of five 'new' genera in Africa, two in both China and India, and one in South America, perhaps it is more appropriate to frame the second diversification on a 'Pangaean scale'. Enhanced orogenic activity alters habitat diversity, as do transgressions. Mountain building increases habitat heterogeneity, whereas increases in sea level restrict the areal extent of peripheral lowlands. The tetrapod 'generalists' (those that exhibit wide environmental tolerances) would clearly have the advantage under these conditions. The list of similarities exhibited during the two apparent radiation events include: (1) both diversifications coincide with the establishment of rather homogeneous floras in Gondwana (as a corollary, floral provinciality declined throughout the Triassic in the northern hemisphere); (2) both encompass two significant short-term declines in sea level; (3) both occur during a general warming phase (as evidenced by the distribution of red beds and evaporites). Each diversification (and extinction event, for that matter) is the result of a coincidence-of-events; perhaps the identification of similarities in the 'staging' of these events is not as relevant as the differences which contribute to their unique 'signature' in the fossil record.

Obviously, between the two 'peaks' in the fossil record, there exists a 'trough'; a sharp decline in dicynodont diversity occurred during the latest Permian through the basal Triassic. It is recognized that dicynodont diversity may be 'artificially' constrained by facies control. However, a decline in the diversity of coastal plant associations coupled with the widespread occurrence of these relatively depauperate assemblages and floral transitions in the Pangaean interior (Schopf & Askin 1980) would effectively curb the diversity of primary consumers. In

these situations, once again, the herbivore generalist would have an advantage. During this time, the glossopterids were reduced to relicts, but it is conceivable that isolated 'pockets' of the *Glossopteris* flora persisted in Gondwana in the basal Triassic. These would represent potential refugia for specialized forms. Based on the persistence of kingoriamorphs in the Fremouw Formation, it is suggested that Antarctica may have served as a refugium. The nature of the flora, at this time, is under investigation.

Lastly, some comments on the apparent extinction of dicynodonts in the late Triassic are in order here. During the Triassic–Jurassic transition, faunal dominance shifts from therapsids to archosaurs. Walter (1985, p. 250) attributes the extinction of kannemeyeriids to a 'more global mechanism than predation and competition pressure from archosaurs', but does not elaborate further. Here I shall present a scenario, which may represent the chain of events that brought about the 'wholesale' collapse of therapsid-based communities (or at least those communities that sustained dicynodonts).

What we are examining, in the Late Triassic, is the collapse of patchy (perhaps relict) ecosystems; the key issue being the minimal areal extent necessary to achieve a self-sustaining, self-perpetuating biotic community. It is now known that mammalian megaherbivores (i.e. elephants, rhinos, hippopotami) whose natural ranges are restricted by the confines of reserves have the ability to cause dramatic changes in floral composition in a short period of time (see examples cited in Owen-Smith 1987). Typically, these animals cannot survive in restricted areas under the conditions that they have created by virtue of their foraging habits. In addition, local or small-scale climatic perturbations can conceivably 'push' an already stressed ecosystem beyond a critical threshold, and bring about its decline and, if unchecked, its eventual collapse.

Walter (1985) has observed that kannemeyeriid dicynodonts ranged widely, but their remains are always found in shallow to very shallow lacustrine sediments. Their large body size rendered them immune to predation from most Triassic carnivores, with the exception of some 'thecodont' taxa. As a result of their relative invulnerability to predation, they could attain saturation densities. In addition, Walter noted that although kannemeyeriids were contemporaneous with prosauropods, they did not coexist in the same communities.

Owen-Smith (1987) has recently proposed that megaherbivores play a pivotal role in perpetuating habitat diversity in regions with marked seasonal differences in precipitation throughout the year. Late Triassic continental sediments in North America, South America, Africa and Europe indicate semi-arid to arid conditions with periods of high rainfall alternating with dry spells (see Tucker & Benton 1982). According to Owen-Smith, the foraging activities of megaherbivores produce local pockets of nutrient-rich, spatially diverse vegetation that attracts a host of smaller herbivores. The foraging activities of kannemeyeriids may have had a similar impact on Late Triassic vegetation. If communities (containing dicynodonts or rhynchosaurs) were isolated, and of limited areal extent, then it is feasible that the foraging activities of these large, Late Triassic herbivores coupled with local to regional climatic perturbations may have been sufficient to upset the biotic balance (e.g. dramatic short-term oscillations in sea level occurred in the Carnian and Norian, see Haq *et al.* 1987, fig. 5) and bring about the collapse of these relict ecosystems.

If this scenario is accurate, then we should find the following sequence of occurrences in the fossil record: (1) indications of floral change and, perhaps, contemporaneous shifts in local to regional climates (as indicated by the distribution of climatically sensitive lithologies); (2) local extinctions of large herbivores; (3) significant declines in the abundance and diversity of small to medium-sized herbivores, with concomitant changes in the structure of higher-order consumers. The most likely places to search for evidence that will either support or refute this scenario are Late Triassic to Early Jurassic sequences that yield fine-scaled documentation of the biotic transition, such as the Newark Supergroup of eastern North America (see Olsen & Sues 1986).

Summary

(1) A selective examination of aspects of the skull and appendicular skeleton of austral dicynodonts, as well as an inspection of cranial scaling relationships, highlight the morphological mosaicism evident in this group. Apparently a great deal of experimentation was occurring in the Dicynodontia, and some of the morphological modifications observed may represent multiple solutions to the same functional adaptation. Only rigorous biomechanical analyses can resolve this issue.

(2) The diagnoses of three austral dicynodont genera (*Kingoria*, *Lystrosaurus* and *Kannemeyeria*) have been expanded here to include distinctive suites of postcranial characters.

(3) The Shenk Peak vertebrate faunule of the Fremouw Formation contains relicts from the Late Permian of southern Africa (i.e. *Kingoria*), which suggests that Antarctica may have served as a refugium during the Permo-Triassic transition.

(4) Vertebrate fossils collected from the Gordon Valley of Antarctica (during the 1985–86 austral field season) demonstrate that the informal members of the Fremouw Formation are correlated with more than one African biozone. Undescribed cranial fragments of a dicynodont, provisionally identified as belonging to the genus *Kannemeyeria*, were recovered along with remains pertaining to cynodonts and various temnospondyls.

(5) Two peaks in taxonomic diversity are evident within the Dicynodontia. The first peak occurred during the Tatarian (Late Permian) and the second diversification took place during the Anisian (Middle Triassic). Endemism increased throughout the Triassic. The intervening Scythian stage marks a 'bottleneck' in the evolutionary history of the Dicynodontia.

(6) Of the five ecomorphic types recognized to date, the invertebrate-specialists, forest litter foragers and flexible foragers survive the Permo-Triassic transition. The small grubbers and small to medium-sized browsers apparently do not persist, and may have been ecologically replaced by herbivorous procolophonoids, therocephalians (bauriids) and cynodonts (diademodontids and traversodontids).

(7) Pervasive climatic changes (triggered by tectonic-orogenic events and sea-level fluctuations) throughout the Late Permian and Triassic resulted in a series of biotic shifts (i.e. the disassembly of previously 'stable' communities and the formation of new species associations). Aspects of Owen-Smith's 'keystone megaherbivore' hypothesis may be applied to explain the 'collapse' of Late Triassic communities containing dicynodonts.

I am most grateful to the late J. W. Cosgriff, Jr. for provocative discussions on the paleobiology of Triassic vertebrates, critical appraisals of portions of my dissertation (aspects of which are presented in this paper), and the opportunity to study Triassic Antarctic vertebrates. An earlier draft of the manuscript was reviewed by N. Hotton, G. M. King and an anonymous reviewer; their comments and suggestions were very helpful. Special thanks are extended to L. Folder (Memorial Library, Berry College) for the acquisition of pertinent articles, and to P. English for typing portions of the manuscript and producing the text figures. This endeavor was supported, in large measure, by a faculty development award from Berry College. Travel funds to present this paper were provided by the Division of Polar Programs, National Science Foundation; the efforts of W. Zinsmeister and J. A. Crame are also gratefully acknowledged.

References

AMALITZKY, V. & KARPINSKI, A. 1922. Diagnoses of the new forms of vertebrates and plants from the upper Dvina. *Bulletin of the Russian Academy of Science*, **1922**, 1–12.

ASH, S. 1986. Fossil plants and the Triassic–Jurassic boundary. *In*: PADIAN, K. (ed.) *The beginning of the age of Dinosaurs, Faunal change across the Triassic–Jurassic boundary*. Cambridge University Press, Cambridge, 12–30.

BAKKER, R. T. 1977. Tetrapod mass extinctions — a model of the regulation of speciation rates and immigration by cycles of topographic diversity. *In*: HALLAM, A. (ed.) *Patterns of evolution, as illustrated by the fossil record*. Elsevier Scientific Publishing Company, Amsterdam, 439–468.

BARBARENA, M. C., ARAUJO, D. C. & LAVINA E. L. 1985. Late Permian and Triassic Tetrapods of Southern Brazil. *National Geographic Research*, **1**, 5–20.

BENTON, M. J. 1979. Ecological succession among Late Palaeozoic and Mesozoic tetrapods. *Palaeogeography, Palaeoclimatology and Palaeoecology*, **26**, 127–150.

BOCK, W. 1969. The American Triassic flora and global distribution. *In*: BOCK, W., (ed.) *Triassic Global Plant Distribution*. **4**. Geological Center, North Wales, Pennsylvania, 363–401.

BONAPARTE, J. F. 1966. Una nueva "fauna" Triasica de Argentina (Therapsida: Cynodontia, Dicynodontia), consideraciones filogeneticas y paleobiogeograficas. *Ameghiniana*, **4**, 243–296.

BRINK, A. S. 1951. On the genus *Lystrosaurus* Cope. *Transactions of the Royal Society of South Africa*, **33**, 107–120

BROILI, F. & SCHROEDER, J. 1936. Beobachtungen an Wirbeltieren der Karrooformation. XVI. Beobachtungen am Schadel von *Emydochampsa* Broom. *Sitzungsberichte der Bayerischen Akademie der Wissenschaften*, **1936**, 45–60.

BROOM, R. 1903. On the structure of the shoulder girdle in *Lystrosaurus*. *Annals of the South African Museum*, **4**, 139–141.

—— 1932. *The mammal-like reptiles of South Africa*. H. F. & G. Witherby, London.

CAMP, C. L. 1956. Triassic dicynodont reptiles. Part 2. Triassic dicynodonts compared. *Memoirs of the University of California*, **13**, 305–348.

CARRIER, D. R. 1987. The evolution of locomotor stamina in tetrapods: circumventing a mechanical constraint. *Paleobiology*, **13**, 326–341.

CARROLL, R. L. 1988. *Vertebrate Paleontology and Evolution*. W. H. Freeman and Company, New York.

CHALONER, W. G. & MEYEN, S. V. 1973. Carboniferous and Permian floras of the northern continents *In*: HALLAM, A. (ed.) *Atlas of Palaeobiogeography*. Elsevier Scientific Publishing Company, Amsterdam, 169–186.

CLUVER, M. A. 1971. The cranial morphology of the dicynodont genus *Lystrosaurus*. Annals of the South African Museum, **56**, 155–274.

—— 1974a. The cranial morphology of the Lower Triassic dicynodont *Myosaurus gracilis*. Annals of the South African Museum, **64**, 35–54.

—— 1974b. The skull and mandible of a new cistecephalid dicynodont. *Annals of the South African Museum*, **64**, 138–155.

—— 1978. The skeleton of the mammal-like reptile *Cistecephalus* with evidence for a fossorial mode of life. *Annals of the South African Museum*, **76**, 213–246.

—— & HOTTON, N. 1981. The genera *Dicynodon* and *Diictodon* and their bearing on the classification of the Dicynodontia (Reptilia, Therapsida). *Annals of the South African Museum*, **83**, 99–146.

—— & KING, G. M. 1983. A reassessment of the relationship of Permian Dicynodontia (Reptilia, Therapsida) and a new classification of dicynodonts. *Annals of the South African Museum*, **91**, 195–273.

COLBERT, E. H. 1969. *Evolution of the Vertebrates*, 2nd Ed. John Wiley & Sons, New York.

—— 1973. Continental drift and the distribution of fossil reptiles. *In*: TARLING, D. H. & RUNCORN, S. K. (eds) *Implications of continental drift to the earth sciences*. Academic Press, New York, 395–412.

—— 1974. *Lystrosaurus* from Antarctica. *American Museum Novitates*, **2535**, 1–44.

—— 1982. Triassic vertebrates in the Transantarctic Mountains. *In*: M. TURNER & J. SPLETTSTOESSER (eds) *Geology of the Central Transantarctic Mountains*. Antarctic Research Series, **36**, 11–35.

COPE, E. D. 1870. On the skull of dicynodont Reptilia. *Lystrosaurus frontosus* from Cape Colony. *Proceedings of the American Philosophical Society*, **11**, 419.

COSGRIFF, J. W. & HAMMER, W. R. 1979. New species of Dicynodontia from the Fremouw Formation. *Antarctica Journal of the United States*, **14**, 30–32.

—— & —— 1981. New skull of *Lystrosaurus curvatus* from the Fremouw Formation. *Antarctica Journal of the United States*, **16**, 52–53.

——, —— W. J. RYAN, 1982. The Pangaean reptile *Lystrosaurus maccaigi*, in the lower Triassic of Antarctica. *Journal of Paleontology*, **56**, 371–385.

——, ——, ZAWISKIE, J. M. & KEMP, N. R. 1978. New Triassic vertebrates from the Fremouw Formation of the Queen Maud Mountains. *Antarctic Journal of the United States*, **13**, 23–24.

COX, C. B. 1959. On the anatomy of a new dicynodont genus with evidence of the position of the tympanum. *Proceedings of the Zoological Society of London*, **132**, 321–366.

—— 1964. On the palate, dentition, and classification of the fossil reptile *Endothiodon* and related genera. *American Museum Novitates*, **2171**, 1–25.

—— 1965. New Triassic dicynodonts from South America, their origins and relationships. *Philo-*

sophical *Transactions of the Royal Society of London*, **248B**, 457–516.

— — 1969. Two New Dicynodonts from the Triassic Ntawere Formation, Zambia. *Bulletin of the British Museum (Natural History) Geology*, **17**, 257–294.

— — 1972. A new digging dicynodont from the Upper Permian of Tanzania. *In*: Joysey K. A. & Kemp T. S. (eds) *Studies in Vertebrate Evolution*. Oliver & Boyd, Edinburgh, 173–189.

— — 1973. Triassic Tetrapods. *In*: Hallam, A. (ed.) *Atlas of Palaeobiogeography*. Elsevier Scientific Publishing Company, Amsterdam, 213–223.

Crompton, A. W. & Hotton, N. 1967. Functional morphology of the masticatory apparatus of two dicynodonts (Reptilia, Therapsida). *Postilla*, **109**, 1–51.

Crozier, E. A. 1970. Preliminary report on two Triassic dicynodonts from Zambia. *Palaeontologia Africana*, **13**, 39–45.

Cruickshank, A. R. I. 1965. On a specimen of the anomodont reptile *Kannemeyeria latifrons* (Broom) from the Manda Formation of Tanganyika, Tanzania. *Proceedings of the Linnean Society of London*, **176**, 149–157.

— — 1970. The taxonomy of the Triassic anomodont genus *Kannemeyeria* Seeley 1908. *Palaeontologia Africana*, **13**, 47–55.

— — 1978. Feeding adaptations in Triassic dicynodonts. *Palaeontologia Africana*, **21**, 121–132.

— — & Keyser, A. W. 1984. Remarks on the genus *Geikia* Newton, 1983, and its relationships with other dicynodonts: (Reptilia: Therapsida). *Transactions of the Geological Society of South Africa*, **87**, 35–39.

DeFauw, S. L. 1981. *The pectoral girdle, sternum and forelimb of the dicynodont genus* Lystrosaurus. Master's Thesis. Wayne State University.

— — 1986. *The Appendicular Skeleton of African Dicynodonts*. Doctoral Dissertation. Wayne State University.

— — 1987. A new perspective on the last phase in the evolution of the Temnospondyli (Amphibia, Labyrinthodontia). *In*: Currie, P. J. & Koster, E. H. (eds) *Fourth Symposium on Mesozoic Terrestrial Ecosystems, Short Papers*. Tyrrell Museum of Palaeontology, Drumheller, 61–69.

— — 1989. Temnospondyl amphibians: a new perspective on the last phases in the evolution of the Labyrinthodontia. *Michigan Academician*, **22**, 7–32.

Dingle, R. V. 1978. South Africa. *In*: Moullade, M. & Nair, A. E. M. (eds) *The Phanerozoic Geology of the World II, The Mesozoic, A*. Elsevier Scientific Publishing Company, Amsterdam, 401–434.

Elliot, D. H., Colbert, E. H., Breed, W. J., Jensen, J. A. & Powell, J. S. 1970. Triassic tetrapods from Antarctica: evidence for continental drift. *Science* **169**, 1197–1201.

Frakes, L. A. 1979. *Climates throughout geologic time*. Elsevier Scientific Publishing Company, Amsterdam.

Hammer, W. R. & Cosgriff, J. W. 1981. *Myosaurus*

gracilis, an anomodont reptile from the Lower Triassic of Antarctica and South Africa. *Journal of Paleontology*, **55**, 410–424.

— — DeFauw, S. L., Ryan, W. J. Tamplin, J. W. 1986. New vertebrates from the Fremouw Formation (Triassic), Beardmore Glacier Region, Antarctica. *Antarctica Journal of the United States*, **21**, 26–28.

— — Ryan, W. J. & DeFauw, S. L. 1987. Comments on the vertebrate fauna from the Fremouw Formation (Triassic), Beardmore Glacier Region, Antarctica. *Antarctic Journal of the United States*, **22**, 32–33

Haq, B. U., Hardenbol, J. & Vail, P. R. 1987. Chronology of fluctuating sea levels since the Triassic. *Science*, **235**, 1156–1167.

Haughton, S. H. 1917. Investigations in South African reptiles and Amphibia, Part 10. Descriptive catalogue of the Anomodontia, with special reference to the examples in the South African Museum. *Annals of the South African Museum*, **12**, 127–174.

Hotton, N. 1974. A new dicynodont (Reptilia, Therapsida) from *Cynognathus* zone deposits of South Africa. *Annals of the South African Museum*, **64**, 157–165.

— — 1986. Dicynodonts and their role as primary consumers. *In*: Hotton, N., MacLean, P. D., Roth, J. J. & Roth, E.C. (eds), *The Ecology and Biology of Mammal-like Reptiles*. Smithsonian Institution Press, Baltimore.

Huxley, T. H. 1859a. On a new species of dicynodont (*D. murrayi*) from near Colesburg, South Africa; and on the structure of the skull in dicynodonts. *Quarterly Journal of the Geological Society of London*, **15**, 555–556.

— — 1859b. On some amphibian and reptilian remains from South Africa and Australia. *Quarterly Journal of the Geological Society of London*, **15**, 642–658.

— — 1865. On a collection of vertebrate fossils from the Panchet rocks, Raniganj Coalfield. *Paleontologia Indica (Series 4)*, **1**, 2–24.

Irving, E. 1983. Fragmentation and assembly of the continents, mid-Carboniferous to present. *Geophysical Surveys*, **5**, 299–333.

Johnson, J. G., 1971. Timing and coordination of orogenic, epeirogenic and eustatic events. *Bulletin of the Geological Society of America*, **5**, 67–82.

Kalandadze, N. N. 1975. The first discovery of *Lystrosaurus* in the European regions of the USSR. *Paleontologisheskii Zhurnal*, **1975**, 140–142 (Russian).

Kemp, T. S. 1982. *Mammal-like reptiles and the origin of mammals*. Academic Press, London.

Keyser, A. W. 1973a. A new Triassic vertebrate fauna from southwest Africa. *Palaeontologia Africana*, **16**, 1–15.

— — 1973b. A preliminary study of the type area of the *Cistecephalus* zone of Beaufort series and a revision of the anomodont family Cistecephalidae. *Memoirs of the Geological Survey of South Africa*, **62**, 1–17.

—— 1974. Evolutionary trends in Triassic Dicynodontia. *Palaeontologia Africana*, **17**, 57–68.

—— 1981. The stratigraphic distribution of the Dicynodontia of Africa reviewed in a Gondwana context. *In*: CRESSWELL, M. M. & VELLA, P. (eds) *Gondwana 5 — Selected papers and abstracts of papers presented at the Fifth International Gondwana Symposium, Wellington, New Zealand, 1980*. A. A. BALKEMA, Rotterdam, 61–63.

—— & CRUICKSHANK, A. R. I 1979. The origins and classification of Triassic dicynodonts. *Transactions of the Geological Society of South Africa*, **82**, 81–108.

—— & SMITH, R. M. H. 1979. Vertebrate biozonation of the Beaufort Group with special reference to the western Karoo basin. *Annals of the Geological Survey of South Africa*, **12**, 1–36.

KING, G. M. 1981. The functional anatomy of a Permian dicynodont. *Philosophical Transactions of the Royal Society*, **291B**, 243–322.

—— 1983. First mammal-like reptile from Australia [reply to Thulborn 1983]. *Nature*, **306**, 209.

—— 1985. The postcranial skeleton of *Kingoria nowacki* (von Huene) (Therapsida: Dicynodontia). *Zoological Journal of the Linnean Society*, **84**, 263–289.

—— 1988. Anomodontia. *In*: WELLNHOFER, P. (ed.) *Handbuch der Palaoherpetologie, Teil 17 C*. Gustav Fischer Verlag, Stuttgart, 1–174.

KITCHING, J. W. 1968. On the *Lystrosaurus* zone and its fauna with special reference to some immature Lystrosauridae. *Palaeontologia Africana*, **11**, 61–76.

—— 1970. A short review of the Beaufort zoning in South Africa. *In*: HAUGHTON, S. H. (ed.) *Second Symposium on Gondwana Stratigraphic and Palaeontology, Cape Town and Johannesburg, 1970*. C.S.I.R., Pretoria, 309–312.

—— 1977. The distribution of the Karroo vertebrate fauna. *Bernard Price Institute for Palaeontological Research, Memoir*, **1**, 1–131.

—— COLLINSON, J. W., ELLIOT, D. H. & COLBERT, E. H. 1972. *Lystrosaurus* zone fauna from Antarctica. *Science*, **175**, 524–526.

KUTTY, T. S. 1972. Permian reptilian fauna from India. *Nature*, **237**, 462–463.

LEFELD, J. 1978. Mongolia. *In*: MOULLADE, M. & NAIRN, A. E. M. (eds) *The Phanerozoic Geology of the World II, The Mesozoic, A*. Elsevier Scientific Publishing Company, Amsterdam, 55–78.

LI, Y., 1983. Restoration of the pelvic muscles of *Lystrosaurus*. *Vertebrata Palasiatica*, **21**, 328–339.

LIN, J., FULLER, M. & ZHANG, W. 1985. Preliminary Phanerozoic polar wander paths for the North and South China blocks. *Nature*, **313**, 444–449.

LYDEKKER, R. 1887. The fossil Vertebrata of India. *Records of the Geological Survey of India*, **20**, 51–80.

OOLSEN, P. E. & SUES, H-D. 1986. Correlation of continental Late Triassic and Early Jurassic sediments, and patterns of the Triassic–Jurassic tetrapod transition. *In*: PADIAN, K. (ed.) *The Beginning of the Age of Dinosaurs*. Cambridge University Press, Cambridge, 321–351.

OWEN, R. 1845a. Description of certain fossil crania, discovered by A. G. BAIN, Esq., in sandstone rocks at the south-eastern extremity of Africa, referable to different species of an extinct genus of Reptilia (*Dicynodon*), and indicative of a new tribe or sub-order of Sauria. *Proceedings of the Geological Society of London*, **4**, 500–504.

—— 1845b. III. Report on the reptilian fossils of South Africa. Part I. Description of certain fossil crania, discovered by A. G. BAIN, Esq., in the sandstone rocks at the south-eastern extremity of Africa, referable to different species of an extinct fenus of Reptilia (*Dicynodon*) and indicative of a new tribe or sub-order of Sauria. *Transactions of the Geological Society London (Second Series)*, **7**, 59–84.

OWEN-SMITH, N. 1987. Pleistocene extinctions: the pivotal role of megaherbivores. *Paleobiology*, **13**, 351–362.

PEARSON, H. G. 1924a. The skull of the dicynodont reptile *Kannemeyeria*. *Proceedings of the Zoological Society of London*, **1924**, 793–826.

—— 1924b. A dicynodont reptile reconstructed. *Proceedings of the Zoological Society of London*, **1924**, 827–855.

RADINSKY, L. B. 1985. Approaches in Evolutionary Morphology: A Search for Patterns. *Annual Review of Ecology and Systematics*, **16**, 1–14.

REPELIN, J. 1923. Sur un fragment de crane de *Dicynodon*, recueilli par H. Cuouillon dans les environs de Luang-Prabang (Haut-Laos). *Bulletin du Service Geologique de L'Indochine*, **12**, 1–7.

ROMER, A. S. 1966. *Vertebrate Paleontology*, Third Edition. The University of Chicago Press, Chicago and London.

ROWE, T. 1980. The morphology, affinities, and age of the dicynodont reptile *Geikia elginensis*. *In*: JACOBS, L. L. (ed.) *Aspects of vertebrate history*. Museum of Northern Arizona Press, 269–294.

ROY CHOWDHURY, T. 1970. Two new dicynodonts from the Triassic Yerrapalli Formation of central India. *Palaeontology*, **13**, 132–144.

SCHOPF, T. J. M. 1974. Permo-Triassic extinctions: relation to sea-floor spreading. *Journal of Geology*, **82**, 129–143.

—— & ASKIN, R. A. 1980. Permian and Triassic Flora Biostratigraphic Zones of Southern Land Masses. *In*: DILCHER, D. L. & TAYLOR, T. N. (eds) *Biostratigraphy of Fossil Plants, Successional and Paleoecological Analyses*. Dowden, Hutchinson and Ross, Inc., Stroudsburg, 119–152.

SEELEY, H. G. 1908. On a fossil reptile with a trunk from upper Karroo rocks of Cape Colony. *Report to the British Association (Dublin)*, **78**, 713.

SMITH, A. G., HURLEY, A. M. & BRIDEN, J. C. 1981. *Phanerozoic paleocontinental world maps*. Cambridge University Press, Cambridge.

SMITH, R. M. H. 1987. Helical burrow cast of therapsid origin from the Beaufort Group (Permian) of South Africa. *Palaeogeography, Palaeoclima-*

tology, Palaeoecology, **60**, 155–170.

SNYDER, R. C. 1954. The anatomy and function of the pelvic girdle and hindlimb in lizard locomotion. *American Journal of Anatomy*, **95**, 1–45.

SUN, A-L., 1964. Preliminary report on a new species of *Lystrosaurus* of Sinkiang. *Vertebrata Palasiatica*, **8**, 217.

— — 1973*a*. Dicynodont fossils from the Turfan. Academia Sinica, *Memoirs of the Institute of Vertebrate Palaeontology Palaeoanthropology*, **10**, 53–68.

— — 1973*b*. A new species of the genus *Dicynodon* from the northern piedmont of the Tian Shan. *Vertebrata Palasiatica*, **11**, 52–58.

SUSHKIN, P. P. 1926. Notes on the pre-Jurassic Tetrapoda from Russia. 2. Contributions to the morphology and ethology of the Anomodontia. *Palaeontologia Hungarica*, **1**, 328.

THULBORN, R. A. 1983. A mammal-like reptile from Australia. *Nature*, **303**, 330–331.

TOERIEN, M. J. 1953. The evolution of the palate in South African Anomodontia and its classificatory significance. *Palaeontologia Africana*, **1**, 49–117.

— — 1954. *Lystrosaurus primitivus*, sp. nov. and the orgin of the genus *Lystrosaurus*. *Annals and Magazine of Natural History*, **12**, 934–938.

TRIPATHI, C. 1962. On the *Lystrosaurus* remains from the Panchets of Raniganj Coalfields. *Records of the Geological Survey of India*, **89**, 407–426.

— — & SATSANGI, P. P., 1963. *Lystrosaurus* fauna of the Panchet Series of the Raniganj Coalfield. *Memoirs of the Geological Survey of India, Palaeontologia Indica (New Series)*, **37**, 1–66.

TUCKER, M. E. & BENTON, M. J. 1982. Triassic environments, climates and reptile evolution. *Palaeogeography, Palaeoclimatology, Palaeoecology*, **40**, 361–379.

VAN HOEPEN, E. C. N., 1915. Contributions to the knowledge of the reptiles of the Karroo Formation. 3. The skull and other remains of *Lystrosaurus putterilli* n. sp. *Annals of the Transvaal Museum*, **5**, 70–82.

WALTER, L. R. 1985. *Limb adaptations in kannemeyeriid dicynodonts.* PhD Dissertation. Yale University.

— — 1986. The limb posture of kannemeyeriid dicynodonts: functional and ecological considerations. In: PADIAN, K. (ed.) *The Beginning of the Age of Dinosaurs: Faunla change across the Triassic–Jurassic boundary.* Cambridge University Press, Cambridge, 89–97.

WANG, Z. 1985. Palaeovegetation and plate tectonics: palaeophytogeography of North China during the Permian and Triassic times. *Palaeogeography, Palaeoclimatology, Palaeoecology*, **49**, 25–45.

WATSON, D. M. S. 1912. The skeleton of *Lystrosaurus*. *Records of the Albany Museum*, **2**, 287–295.

— — 1913. The limbs of *Lystrosaurus*. *Geological Magazine*, **5**, 256–258.

WEITHOFER, A. 1888. Uber einen neuen Dicynodonten (*Dicynodon simocephalus*) aus der Karrooformation, Sudafrika. *Annalen aus der Naturhaftigen Museum*, **3**, 1–6.

YOUNG, C. C. 1935. On two skeletons of Dicynodontia from Sinkiang. *Bulletin of the Geological Society of China*, **14**, 483–517.

— — 1939. Additional Dicynodontia from Sinkiang. *Bulletin of the Geological Society of China*, **19**, 111–146.

YUAN, P. L. & YOUNG, C. C. 1934. On the occurrence of *Lystrosaurus* in Sinkiang. *Bulletin of the Geological Society of China*, **13**, 575–580.

ZIEGLER, A. M., SCOTESE, C. R. & BARRETT, S. F. 1982. Mesozoic and Cenozoic paleogeographic maps. *In*: BROSCHE, F. & SUNDERMANN, J. (eds) *Tidal Friction and Earth's Rotation II.* Springer-Verlag, Berlin.

The phenomenon of forest growth in Antarctica: a review

W. G. CHALONER & G. T. CREBER

Department of Biology, Royal Holloway and Bedford New College, Egham Hill, Egham, Surrey TW20 OEX, UK

Abstract: The existence of a temperate Antarctic flora in the Permian, Mesozoic and early Tertiary poses a number of problems that are not soluble by reference to any environmental situation obtaining at the present day (i.e. in which plants live in a regime where a warm summer is followed by a winter without sunlight).

Certain characteristics of the polar climate during a number of geological periods can be deduced from the remains of forests which grew at very high latitudes where tree growth cannot take place at the present day. The demands of tree growth of the order of 3 to 4 mm increase in trunk radius each year are such that an input of light energy of at least $3000 \text{ MJ m}^{-2} \text{ a}^{-1}$ would be necessary.

Light energy inputs of this magnitude have been measured at very high latitudes in both North and South polar regions and it would appear that polar forest growth would be possible today if the ambient temperatures were sufficiently raised above their present levels. It is also significant that the forest productivity of about 10 t ha^{-1} as estimated for the Lower Cretaceous forest on Alexander Island (Palaeolat. 70°S), is a terrestrial productivity that is only matched at the present day in a variety of temperate and subtropical regions of the world.

The evolution of photoperiodic ecotypes in tree species must have facilitated the colonization of polar regions by tree varieties adapted to make all of their annual growth during the long summer light period. There is evidence also that these photoperiodic ecotypes have a high rate of cambial activity concentrating the more rapid growth of wood into the polar summer.

The realization that fossil plants occur abundantly in areas that are now ice desert (Greenland, Spitzbergen, Antarctica) goes back to the earliest days of polar exploration (Seward 1892). The recognition of 'continental drift' and 'polar wandering' (now incorporated in the concept of plate tectonics) raised the possibility that such plants had grown under at least a temperate climate, and that they had moved into their present inhospitable climatic regions. However, as the palaeopositions of polar land masses have become established, it is clear that plants did indeed grow within what were then (and are now) the polar circles. Fossil wood, especially if in the form of in situ 'fossil forest', has the special merit that it can give a direct measure of bioproductivity. Also, as distinct from other plant fossils, certain rather precise climatic parameters apply to the growth of forests.

Ever since the discovery of the Permian, Mesozoic and early Tertiary polar fossil forests (Halle 1913; Nathorst 1914; Seward 1914) there has been speculation as to the required conditions that would permit tree growth in very high latitudes where trees do not grow at the present day. Various questions may be posed: (1) Would the annual input of solar energy be adequate to enable trees to produce the substantial growth rings seen in the fossil wood? (2) Would the annual polar solar cycle, with its dark winter, permit trees to grow? (3) Would the low angle of sunlight cause excessive mutual shading of trees when aggregated into forests? It is our intention in this paper to show that these three questions can be answered without invoking any special circumstances unsupported by observational evidence. To do so we have adopted techniques used by foresters at the present day to estimate tree productivity.

Estimation of the productivity of fossil forests

If an area of forest floor becomes fossilized with the stumps in situ, it is possible to count the number of trees in a given area and to measure their diameters and ring-width sequences. The contemporaneity of the trees may be established by examination of the stumps for indications of the state of the wood and some cross-dating may also be attempted (Jefferson 1982). From the data, as in modern forestry practice, the volume of wood produced annually per unit area of forest floor can be calculated (Creber &

From Crame, J. A. (ed.), 1989, *Origins and Evolution of the Antarctic Biota,*
Geological Society Special Publication No. 47, pp. 85–88.

85

Chaloner 1984a, b, 1985; Creber 1986; Creber & Francis 1987; Hamilton & Christie 1971). Since the calorific value of conifer wood is about $21\,000\,\mathrm{J\,g^{-1}}$(Miller 1955) there must be a relationship between the annual solar energy input to each tree and the amount of wood it produces each year. Only about 0.4% of the Sun's energy intercepted by the tree is eventually bound up in the trunk wood (Kozlowski 1962; Ovington 1961). Hence the amount of wood produced annually by an area of forest sets a lower limit to the annual solar energy input.

In cases where there are no fossilized stumps in situ, data from individual pieces of fossil wood may be used. The method is due to Assmann (1970) who has used it on extant trees. He calculates the cross-section area added by each successive growth ring to the existing cross-section area of the trunk before the ring was formed. This additional area he calls the area increment and it is of course directly proportional to the amount of wood produced by the whole trunk during the year. The advantage of the method is that the area increments do not depend on the ring widths alone but are very much influenced by the diameters of the cross-sections of the trunks on which they accrue. The larger the initial trunk diameter, the larger will be the area increment for a given ring width.

The area increment added by a ring of width x mm to the existing area of trunk cross-section can be found by using the following formula:

$$\pi(r + x)^2 - \pi r^2$$

where r is the radius of the trunk before the new ring is added. If it is supposed that a tree has a sequence of ring widths of 1 mm per year it is possible to calculate the area increments due to the addition of 1 mm to the existing radius annually. A graph of these increments plotted against years is a straight line of a certain slope. The same exercise can be carried out with ring widths of 3 and 5 mm to produce lines of greater slopes. By reference to Hamilton & Christie (1971), these lines can be seen to represent tree growths of high (5 mm), medium (3 mm) and low (1 mm) annual productivity. Fossil ring-width series can be plotted on the same graph as the sequences for the 1, 3 and 5 mm ring widths and an assessment can then be made of the productivity represented by the fossil series (Creber & Francis 1987).

Frequently, specimens of fossil wood may lack the central pith region. In these cases the original trunk diameter can be estimated from the radius of curvature of the rings. If the original trunk had been very large, the outer rings may show virtually no curvature. However, it is still possible to estimate the productivity by plotting the sequence of area increments and comparing the slope with the 1, 3 and 5 mm ones as described above. Clearly it is better if these calculations are carried out on a large number of wood specimens from the same site, avoiding branch wood as far as possible, so that the average productivity can be calculated. The wood specimens should also be examined to check for compression effects.

Estimation of land productivity

A knowledge of the trunk wood (bole wood) productivity enables an estimate to be formed of the total productivity of the entire forest. A number of authors (Forrest & Ovington 1970; Jordan 1971; Kanninen et al. 1982; Ovington 1961; Whittaker 1966) have studied extant forests composed of a wide variety of tree species and in many parts of the world. From their published data it is evident that trunk wood productivity is about 40–50% of total forest productivity. Bazilevich et al. (1971) have produced world maps of forest productivity and interesting comparisons may be made between fossil forest productivity at the palaeolatitudes of growth and that of extant forests in their present day latitudes.

The Permian Period

There are many records of Lower Permian fossil wood from the Transantarctic Mountains in Antarctica at a palaeolatitude of about 80°S. (Creber & Chaloner 1987; Doumani & Long 1962; Maheshwari 1972). An examination of a number of specimens of this wood collected by the late James Schopf and presently held in the Botany Department of Ohio State University, Columbus, Ohio reveals that maximum ring widths of about 10 mm were formed by the trees and the average ring width was about 2.2 mm. The latter would indicate a productivity approaching the medium level (3 mm) indicated above. This level of productivity supports the evidence from other palaeobotanical specimens that the palaeoclimate relatively close to the South Pole at the time of growth was very different from that at the present day.

The Mesozoic and Early Tertiary Periods

It is in these periods that the largest amount of fossil wood is found in polar latitudes (Creber & Chaloner 1984a, b, 1985, 1987). Some is located in the same regions of Antarctica as those in which the Lower Permian specimens are seen. Examination of in situ Triassic material also held in the Botany Department at Columbus has yielded an average ring width of 2.3 mm, with a maximum of 10 mm.

However, the discovery of the remains of the in situ stumps of a fossil forest in the Lower Cretaceous of Alexander Island, Antarctic Peninsula, at a palaeolatitude of about 70°S (Jefferson 1982), has enabled a detailed estimate to be made of its productivity (Creber & Chaloner 1984a, b, 1985). This forest appears to have been overwhelmed by a single volcanic event and the trees show indications of contemporaneity through the uniformity of trunk diameter, lack of dead and rotted wood and cross-dating (Jefferson 1982). The forest was apparently producing about 5 t ha^{-1}(0.5 kg m^{-2}) of trunk wood with a mean ring width of 2.5 mm. Using the principle explained above this would have been equivalent to a total forest productivity of about 10 t ha^{-1}. Such a forest productivity at the present day occurs only in much lower latitudes (Creber & Francis 1987). As in the Triassic, the mean ring width indicates also a medium productivity close to the 3 mm level. A large number of other specimens of fossil wood of Cretaceous and early Tertiary age has been described from the Antarctic Peninsula region (Francis 1986). These also showed a medium level of productivity with a mean ring width of about 2.3 mm, an indication of a productivity very similar to the forest on Alexander Island.

Discussion

The annual solar input at the present day (Farman & Hamilton 1978; LaGrange 1963) in the region of Alexander Island is about 3200 MJ m^{-2}. Since the energy content of the 0.5 kg m^{-2} of wood produced by the forest would have represented about 0.3% of the total solar energy input there would appear to have been an adequate light supply. The rather wide observed spacing of the fossil trees (563 trees ha^{-1}) would have tended to minimize mutual shading and was no doubt due to the predominantly low angle insolation. In an attempt to indicate the likely incidence of shadows in the forest, a reconstruction was made (Creber & Chaloner 1984b) which showed the shadows cast by a single 'model' coniferous tree at three times during a midsummer day at 70°S. It was clear that no trees would have been severely affected. It must be borne in mind also that in a polar summer the sun makes a complete circuit of the sky so that one tree will not shade another for much of the day as in temperate latitudes.

Even on a cloudless day when hard shadows are cast there may still be as much as 25% of the incident radiation in the form of diffuse light (Monteith 1973) which could be intercepted by the tree crown from all directions.

It has been demonstrated by Vaartaja (1959, 1962) that certain modern tree species have photoperiodic ecotypes which grow near to the northern tree line. These ecotypes are found to concentrate their annual growth into the short polar summer and to grow very poorly if transplanted to lower latitudes. Similarly Gregory & Wilson (1968) showed that the production of xylem cells by the cambium of *Picea glauca* (white spruce) trees growing near College, Alaska (65°N) was at a much higher rate than in trees of the same species growing at Petersham, Massachusetts (42°N). This facility enables the high latitude members of this species to complete a growth ring in a much shorter time during the short polar summer. Another characteristic of trees near the northern timber line is their tall conical shape. Jahnke & Lawrence (1965) have shown that this form of crown is ideally suited to intercepting the maximum amount of light energy from sunlight at low angles to the horizon. Were the high latitude trees of past Antarctic floras to have adopted this tall habit, it would have been a further adaptation enabling them to grow successfully in forests.

It would appear, therefore, that the only necessary factor for trees to grow at very high latitudes at the present day is an increase in the average ambient temperature. All of the other factors required for the evolution of polar forest trees are within the scope of adaptations that are known to be available to modern trees, such as photoperiodic ecotypes, higher rates of cambial activity during the growing season and an appropriate crown structure. Additionally the trees could have availed themselves of the facility to adopt the deciduous habit as an adaptation to the long winter darkness, although there is evidence that the evergreen habit might have been sustainable in a dark winter. Larson (1964) showed that conifer leaves may be covered in black bags for up to 140 days and still retain the capacity to regenerate their pigments and recommence photosynthesis. Furthermore, if the winters had been relatively mild the trees might have used the adaptation shown by the Bristlecone pines studied by Mooney & Brayton (1966). They brought the pines down from the White Mountains of California to sea level and found that the trees drastically reduced their respiration rate in the higher ambient temperature.

References

ASSMANN, E. 1970. *The principles of forest yield study*. Pergamon Press, Oxford.

BAZILEVICH, N. I., DROZDOV, A. V. & RODIN, L. E. 1971. World forest productivity. *In*: DUVIGNEAUD, P. (ed.) *Productivity of forest ecosystems*. UNESCO, Paris, 345–353.

CREBER, G. T. 1986. Tree growth at very high latitudes in the Permian and Mesozoic. Colloque Internationale sur l'Arbre. *Naturalia Monspeliensa*, 487–493.

— — & CHALONER, W. G. 1984a. Climatic indications from growth rings in fossil woods. *In*: BRENCHLEY, P. J. (ed.) *Fossils and climate*. Wiley, Chichester, 49–74.

— — & — — 1984b. Influence of environmental factors on the wood structure of living and fossil trees. *Botanical Review*, **50**, 357–448.

— — & — — 1985. Tree growth in the Mesozoic and early Tertiary and the reconstruction of palaeoclimates. *Palaeogeography Palaeoclimatology, Palaeoecology*, **52**, 35–60.

— — & — — 1987. The contribution of growth rings to the reconstruction of past climates. *In*: WARD, R. G. (ed.) *Application of tree-ring studies: current research in dendrochronology and related areas*. British Archaeological Reports, International Series, **333**, 37–67.

— — & FRANCIS, J. E. 1987. Productivity in fossil forests. *In*: JACOBY, G. C. (ed.) *Proceedings of the International Symposium on Ecological Aspects of Tree-ring analysis*. U.S. Dept. of Energy, Carbon Dioxide Research Division, Washington, DC, 319–26.

DOUMANI, G. A. & LONG, W. E. 1962. The ancient life of the antarctic. *Scientific American*, **207(3)**, 168–84.

FARMAN, J. C. & HAMILTON, R. A. 1978. Measurements of radiation at the Argentine Islands and Halley Bay, 1963–1972. *British Antarctic Survey Scientific Report*, **99**.

FORREST, W. G. & OVINGTON, J. D. 1970. Organic matter changes in an age series *Pinus radiata* plantations. *Journal of Applied Ecology*, **7**, 177–186.

FRANCIS, J. E. 1986. Growth rings in Cretaceous and Tertiary wood from Antarctica and their palaeoclimatic implications. *Palaeontology*, **29**, 665–684.

GREGORY, R. A. & WILSON, B. F. 1968. A comparison of cambial activity of white spruce in Alaska and New England. *Canadian Journal of Botany*, **58**, 687–692.

HALLE, T. G. 1913. The Mesozoic flora of Graham Land. *Wissenschaftliche Ergebnisse der Schwedischen Südpolar-Expedition 1901–03*. Bd. III, Lief. 14, 123 pp., Taf. 1–9.

HAMILTON, G. J. & CHRISTIE, J. M. 1971. *Forest management tables (metric)*. Forestry Commission, HMSO.

JAHNKE, L. S. & LAWRENCE, D. B. 1965. Influence of photosynthetic crown structure on potential productivity of vegetation, based primarily on mathematical models. *Ecology*, **46**, 319–326.

JEFFERSON, T. H. 1982. Fossil forests from the Lower Cretaceous of Alexander Island, Antarctica. *Palaeontology*, **25**, 681–708.

JORDAN, C. F. 1971. Productivity of a tropical forest and its relation to a world pattern of energy storage. *Journal of Ecology*, **59**, 127–143.

KANNINEN, M., HARI, P. & KELLOMAKI, S. 1982. A dynamic model for above-ground growth and dry matter production in a forest community. *Journal of Applied Ecology*, **19**, 465–476.

KOZLOWSKI, T. T. 1962. Photosynthesis, climate and tree growth. *In*: KOZLOWSKI, T. T. (ed.) *Tree growth*, Ronald Press, New York. 149–164.

LaGRANGE, J. J. 1963. Meteorology I. Shackleton, Southice and the journey across Antarctica. *Trans-Antarctic Expedition 1955–58 Scientific Report*, **13**.

LARSON, P. R. 1964. Contribution of different aged needles to growth and wood formation of young red pines. *Forest Science*, **10**, 224–238.

MAHESHWARI, H. K. 1972. Permian wood from Antarctica and revision of some Lower Gondwana woods. *Palaeontographica B*, **138**, 1–43.

MILLER, D. H. 1955. Snow cover and climate in the Sierra Nevada, California. *University of California Publications in Geography*, **11**, 1–218.

MONTEITH, J. L. 1973. *Principles of environmental physics*. Edward Arnold, London.

MOONEY, H. A. & BRAYTON, R. 1966. Field measurements of the metabolic responses of bristlecone pine and big sagebrush in the White Mountains. *Botanical Gazette*, **127**, 105–113.

NATHORST, A. G. 1914. Nachträge zur Paläozoischen Flora Spitzbergens. *Zur fossilen Flora der Polarländer*. Teil I. Lief. IV. Stockholm.

OVINGTON, J. D. 1961. Some aspects of energy flow in plantations of *Pinus sylvestris*. *Annals of Botany*, **25**, 12–20.

SEWARD, A. C. 1892. *Fossil plants as tests of climate*. Clay & Sons, London.

— — 1914. Antarctic fossil plants. British Antarctic (Terra Nova) Expedition 1910. *Natural History Reports, Geology*. **1(1)**. London.

VAARTAJA, O. 1959. Evidence of photoperiodic ecotypes in trees. *Ecological Monographs*, **29**, 91–111.

— — 1962. Ecotypic variation in photoperiodism of trees with special reference to *Pinus resinosa* and *Thuja occidentalis*. *Canadian Journal of Botany*, **40**, 849–856.

WHITTAKER, R. H. 1966. Forest dimensions and productivity in the Great Smoky Mountains. *Ecology*, **47**, 103–121.

Antarctica: Cretaceous cradle of austral temperate rainforests?

MARY E. DETTMANN

Department of Botany, University of Queensland, St Lucia, Qld 4067, Australia

Abstract: Although it is well known that certain Gondwanic elements of present-day austral temperate rainforests occurred on Antarctica during latest Cretaceous to early Tertiary times, there has been insufficient factual evidence for pinpointing the cradle of these forests. Fossil evidence from Antarctica and closely associated regions in the Creataceous southern Gondwanan assembly confirms that Antarctica was a Cretaceous origination and dispersal region of certain elements of today's southern hemispheric humid and perhumid forests. Antarctic origins are indicated for the fern *Lophosoria*, the podocarp gymnosperms *Lagarostrobus* and *Dacrydium*, *Nothofagus*, *Ilex*, and several lineages of the Proteaceae; migration to their present regions of distribution was probably step-wise. Antarctica also served as a Cretaceous dispersal corridor for other angiosperms represented today in mid to low latitude austral regions. These include *Ascarina* (or its stock), Myrtaceae, Gunneraceae, and Winteraceae, all of which had earlier histories in northern Gondwana or southern Laurasia. Origination and dispersal appears to be related to changing environmental circumstances associated with fragmentation of Gondwana and opening and enlargement of the southern oceans.

The concept of Antarctica as the source for plants now widely distributed in southern regions is not new. Both Darwin (1859) and Hooker (1847) suggested that Antarctica must have, in the past, supported a flora containing elements now restricted to southern perhumid forests. Many of these elements are now confined to southern South America, New Zealand, and the frontal arc from eastern Australia to the Indo-Malaysian region (Fig. 1). Although extending into the tropics, the plants there occur in upland rainforests and fringing communities and have mesothermal, not megathermal, requirements (e.g. Webb & Tracey 1981). They include southern conifers (the podocarps and araucarians), *Nothofagus*, rainforest to sclerophyllous members of the Proteaceae, and Winteraceae, *Ilex*, *Gunnera*, Myrtaceae, and several cryptogams.

Knowledge of the Cretaceous vegetation of Antarctica is mainly from the Antarctic Peninsula and nearby islands where a near-complete Cretaceous sequence is exposed (Ineson *et al.* 1986; Farquharson 1982); diverse plant megafossil and spore–pollen assemblages reported from these sequences provide a broad perspective of the contemporaneous floral succession (Truswell in press, and references cited therein). Evidence of the Cretaceous vegetation in the vicinity of the Ross and Weddell seas comes from recycled spores and pollen believed to have been sourced from subglacial Cretaceous sequences and redeposited in Recent marine muds (Truswell 1983; Truswell & Drewry 1984).

Recent marine muds from a range of localities off the coast of George V Land and Wilkes Land, Eastern Antarctica, have also yielded data on the Antarctic Cretaceous flora (Truswell 1983). The recycled palynomorphs were probably sourced from Antarctic counterparts of Cretaceous sequences in southern marginal basins of Australia. Remnants of one of those sequences, near the coast of George V Land, have yielded palynofloras comparable to those of Early Cretaceous age in the Otway and Gippsland basins, southeastern Australia (Domack *et al.* 1980). Plant fossil evidence from the Cretaceous of southern Australia, New Zealand, and southern South America is also relevant as these areas were associated with Antarctica for most of the Cretaceous. India and southern Africa were also closely aligned with Antarctica at the close of the Jurassic and their Cretaceous floras may be expected to reveal evidence of floral exchange and migration pathways between northern and southern Gondwana.

Cretaceous vegetation of southern high latitude regions: a perspective

From the latest Jurassic to the close of the Cretaceous, Antarctica's land vegetation comprised a series of evergreen coniferous rain-forests (Jefferson 1983; Truswell 1983, in press; Dettmann 1986*a*; Dettmann & Thomson 1987; Askin 1988). Similar forests occurred in high latitude regions of adjacent land masses of the southern Gondwana assembly, but from the Antarctic Peninsula northwards to Patagonia, floral zonation was steep with an interfingering of austral podocarp and northern Gondwanan

From Crame, J. A. (ed.), 1989, *Origins and Evolution of the Antarctic Biota*, Geological Society Special Publication No. 47, pp. 89–105.

89

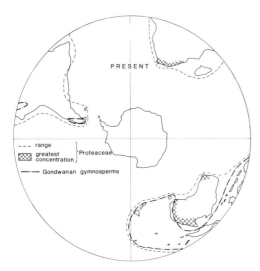

Fig. 1. Present distribution ranges of Proteaceae and Gondwanan gymnosperms including the Podocarpaceae. Map is South Polar Lambert equal-area (adapted from Smith *et al.* 1981).

cheirolepidacean communities (Figs 2 and 3). The geographical limits of the Antarctic-type vegetation fluctuated during opening and enlargement of the South Atlantic, Indian, and Southern oceans. With progressive opening of these oceans austral planktonic dinocyst floras (Helby *et al.* 1987) achieved circumpolar distribution in mid- to high latitude regions (Fig. 2).

The Antarctic Early Cretaceous forests were rich in podocarps and araucarians. Associated with them were ginkgos, taeniopterids, and bennettitaleans together with understorey and ground communities of ferns, lycopods, and bryophytes; angiosperms, of chloranthaceous stock, appeared no later than the early Albian and may have been represented by Barremian–Aptian times (Dettmann 1986*b*). There was marked regionalism in understorey associations between the Antarctic Peninsula and the eastern Antarctic–Australasian area (Dettmann & Thomson 1987). Exchange of floral elements between these and neighbouring areas of southern Gondwana was probably step-wise and mostly west to east (Fig. 2a, b).

The open-canopied forests of the northern Antarctic Peninsula region (for full locality map see Francis 1986, text. fig. 1) were of high productivity and climates were equable, wet, and temperate (Creber & Chaloner 1985; Francis 1986). To the south, on Alexander Island, growth rings in wood suggest marked seasonality or trees growing at the limits of their ecological range (Jefferson 1982). The abrupt

zonation in vegetation from the Antarctic Peninsula to southern South America and the Falkland Plateau implies a steep climatic gradient (Dettmann 1986*a*) which appears to have decreased during enlargment of the South Atlantic Ocean. The vegetation fringing this ocean on the Falkland Plateau and in southern South Africa retained its cheirolepidacean character for much of the Early Cretaceous (Figs 2b and 3).

The Early Cretaceous climate of the eastern Antarctic–southeastern Australian region is also considered to have been cool to warm temperate, (Dettmann 1981; Douglas 1986) and was favourable for forest and peat swamp growth. Evidence for ice rafting in sediments of Early Cretaceous age in South Australia has been advanced by Frakes & Francis (1988), but they note that summer temperatures were probably warm.

During the Late Cretaceous, Antarctica and southern Australasia were vegetated by podocarp-rich coniferous forests that extended into southern South America and over the Falkland Plateau (during Campanian–Maastrichtian times; Fig. 2d, e). In the Turonian, *Lagarostrobus*, early Proteaceae, and *Ilex* were introduced into these forests, and in the Campanian *Dacrydium* and *Nothofagus* were established. As for the Early Cretaceous, there was marked regionalism involving both cryptogams and angiosperms (Dettman & Thomson 1987). Angiosperm pollen is neither common nor diverse in Cenomanian–Turonian palynofloras, reflecting a flora rich in cryptogams; however, during the Campanian–Maastrichtian there was considerable turnover of taxa with the loss of many of the cryptogam elements and the introduction of angiosperms. Amongst the latter were Myrtaceae, Gunneraceae, probable Epacridaceae, Winteraceae, Trimeniaceae, and an array of Proteaceae including probable *Macadamia*, *Gevuina*/*Hicksbeachia*, *Knightia*, *Xylomelum*, and *Beauprea* (Dettmann & Jarzen in press). Some were emigrants from northern Gondwana, but others almost certainly evolved in Antarctica (see later discussion). Migration between eastern and western Antarctica may have been bidirectional and there was exchange with northern Gondwana, possibly via South America.

Late Cretaceous climates are thought to have been cool to warm temperate with little or no seasonality and high moisture levels (Dettmann 1986*a*; Askin 1988), and are therefore similar to those of the Early Cretaceous.

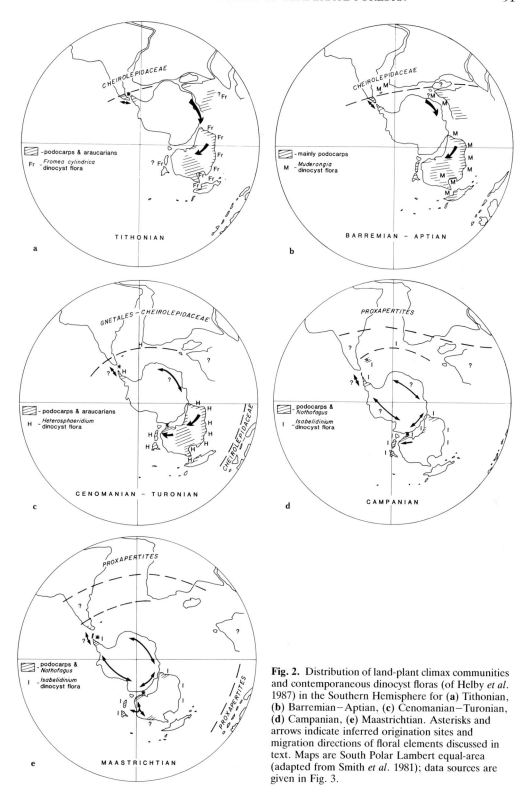

Fig. 2. Distribution of land-plant climax communities and contemporaneous dinocyst floras (of Helby *et al.* 1987) in the Southern Hemisphere for (**a**) Tithonian, (**b**) Barremian–Aptian, (**c**) Cenomanian–Turonian, (**d**) Campanian, (**e**) Maastrichtian. Asterisks and arrows indicate inferred origination sites and migration directions of floral elements discussed in text. Maps are South Polar Lambert equal-area (adapted from Smith *et al.* 1981); data sources are given in Fig. 3.

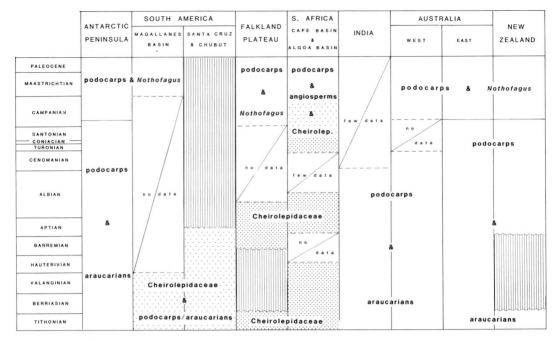

Fig. 3. Chronostratigraphic and geographic distributions of latest Jurassic to earliest Tertiary land-plant climax communities of southern Gondwanic regions. Vertical hatching denotes hiatuses within sedimentary sequences from regions specified. Principal sources of data are: Antarctic Peninsula, Askin (1983, 1989), Dettmann & Thomson (1987); Magallanes Basin, Baldoni & Archangelsky (1984), Archangelsky *et al.* (1984); Santa Cruz and Chubut, Archangelsky *et al.* (1981, 1984), Volkheimer *et al.* (1977); Falkland Plateau, Harris (1977), Hedlund & Beju (1977), Kotova (1983); South Africa, McLachlan & Pieterse (1978); India, Varma *et al.* (1984), Venkatachala (1974), Venktachala *et al.* (1980); Australia, Dettmann (1981), Helby *et al.* (1987 and references cited therein); New Zealand, Couper (1953, 1960), Raine *et al.* (1981), Raine (1984).

Selected floristic elements in the Cretaceous of Antarctica

Although many are preliminary accounts, palaeobotanical and palynological studies have provided a wealth of information on the character and composition of Antarctica Cretaceous floras. Represented in the vegetation were affiliates of extant cryptogams and phanerogams, many of which are important in today's austral floras.

Selected taxa identified amongst Antarctic Cretaceous plant fossils are summarized in Table 1 and discussed below with reference to their early histories in and beyond southern Gondwana.

Bryophyta

Anthocerotaceae, *Phaeoceros/Nothylas*-type (Figs 4, 5a; Table 1): Both *Foraminisporis dailyi* (Cookson & Dettmann) Dettmann 1963 and *Foraminisporis wonthaggiensis* (Cookson & Dettmann) Dettmann 1963 have been allied to

the hornworts genera *Phaeoceros* and *Nothylas* (Dettmann 1963, 1986b; Jarzen 1979). *F. dailyi* occurs in Berriasian–Cenomanian sediments on the Antarctic Peninsula, and although unreported in situ from eastern Antarctic sequences, has a Berriasian–Campanian range in the Otway Basin. Morphologically similar spores of *Nevesisporites* attest to a history dating back to the Early Jurassic in Australia and South America. *F. wonthaggiensis* has Valanginian first appearances in the Otway Basin and occurs in probable remnants of that sequence in eastern Antarctica (Domack *et al.* 1980). However, it is not known from the Antarctic Peninsula region.

Ricciaceae/Riellaceae (Fig. 4; Table 1): Described species of *Triporoletes* are believed to represent spores borne by aquatic and terrestrial members of these families. A ricciaceous affinity for *T. reticulatus* (Pocock) Playford 1971 is strengthened by its association with the plant macrofossil *Hepaticites discoides* Douglas 1973 in Early Cretaceous sediments of the Otway Basin. Here *T. reticulatus* appears in the

Table 1. *Cretaceous range and appearance in the Antarctic region of selected plant groups as indicated by spore and pollen evidence.*

Plant family/genus	Spore/pollen taxa	Antarctic Peninsula	Cretaceous range New Zealand/ Campbell Plat.	E. Antarctica S. Australia	Appearances older (younger) elsewhere	Present distribution
MUSCI Sphagnaceae	Stereisporites spp.	ranges	throughout	Cretaceous		cosmopolitan
HEPATICEAE Anthocerotaceae: Phaeoceros	Foraminisporis dailyi (Cookson & Dettmann)	Berrias.–Cenoman.	Albian–Cenoman.	Berrias.–Campan.	Berrias.; Queensland	cosmopolitan
Nothylas	F. wonthaggiensis (Cookson & Dettmann)	—	—	Valangin.–Maastr.	Valangin.; Queensland	cosmopolitan
Ricciaceae/Riellaceae: Riccia, Riella	Triporoletes reticulatus (Pocock)	Albian–Maastr.	Albian–Maastr.	Barrem.–Maastr.	Valangin.; Queensland	cosmopolitan
LYCOPODIOPSIDA Lycopodium (Lycopodium)	Retitriletes spp.	ranges	throughout	Cretaceous		cosmopolitan
FILICOPSIDA Cyatheaceae/Dicksoniaceae:	Cyathidites spp.	ranges	throughout	Cretaceous		tropical & temperate regions
Dicksoniaceae:	Dictyophyllidites spp.	ranges	throughout	Cretaceous		mainly topical
	Trilites tuberculiformis Cookson	Cenom.–Maastr.	—	Albian–Maastr.	Valangin.; Queensland	
Gleicheniaceae: Gleichenia, Dicranopteris	Gleicheniidites spp.	ranges	throughout	Cretaceous	Albian; Queensland	tropical & subtropical regions
	Clavifera spp.	l. Maastr.	Campan.–Maastr.	Turon.–Maastr.	Albian; Queensland	
Lophosoriaceae: Lophosoria	Cyatheacidies annulatus Cookson	Berrias.–Campan.	—	Cenom. Santon.	Albian; Queensland	South & central America
Marsilaceae (?):	Crybelosporites striatus (Cookson & Dettmann)	late Albian	Albian–Cenom.	early Albian–Santon.	late Aptian; Falkland Plat.	tropical & temperate regions
Osmundaceae:	Baculatisporites spp. Osmundacidites spp.	ranges	throughout	Cretaceous		tropical & temperate regions
Parkeriaceae: Ceratopteris	Cicatricosisporites australiensis (Cookson)	Berrias.–Albian	Aptian–Cenom.	Berrias.–Campan.	Tithonian; India	tropical & subtropical regions

Table 1. *cont.*

Plant family/genus	Spore/pollen taxa	Antarctic Peninsula	Cretaceous range New Zealand/ Campbell Plat.	E. Antarctica S. Australia	Appearances older (younger) elsewhere	Present distribution
Schizaeaceae:						
Anemia (Anemirhiza), *Mohria*	*Cicatricosporites hughesii* Dettmann	early Albian	–	–	Berrias.; Queensland	tropical America, Africa
Anemia (Anemia)	*Nodosisporites* spp.	middle Albian	–	–	Barrem.–Apt; Patagonia	mainly tropical America
Anemia (Coptophyllum)	*Appendicisporites* spp.	middle Albian	–	late Albian	Barrem.; Queensland, Patagonia	tropical America, Africa, Madagascar, India.
GYMNOSPERMAE						
Araucariaceae:						
Araucaria	*Araucariacites australis* Cookson	ranges	throughout	Cretaceous		mainly Southern Hemisphere
Brachyphyllum irregulare	*Balmeiopsis limbata* (Balme)	Albian–Cenom.	–	Barrem.–Cenom.	Late Jurassic; N. Africa	extinct
Podocarpaceae:						
Microcachrys	*Microcachryidites antarcticus* Cookson	Berrias.–Maastr.	Apt.–Maastr.	Berrias.–Maastr.	Late Jur.; India N. Aust.	Tasmania
Trisacocladus tigrensis	*Trichotomosulcites subgranulatus* Couper	Alb.–Maastr.	Alb.–Maastr.	Barrem.–Maastr.	Berrias.–Apt.; S. America	extinct
Lagarostrobus	*Phyllocladidites mawsonii* Cookson	Tur./Con.–Maastr.	Tur.–Maastr.	Tur.–Maastr.	(Campan./ Maastr.); Falkland Plat.	Tasmania
Dacrydium	*Lygistepollenites* spp.	Santon.–Maastr.	Camp.–Maastr.	Santon.–Maastr.	(Mioc.); New Guinea	South America to Malaysia; not in Australia
Dacrycarpus	*Dacrycarpites australiensis* Cookson & Pike	Maastr.	(Eocene)	Maastr.	(Eocene); S. America	New Zealand, E. Australia, New Caledonia, New Guinea, E. Asia
Podocarpus	*Podocarpidites ellipticus* Cookson	ranges	throughout	Cretaceous		as above
ANGIOSPERMAE						
Aquifoliaceae: *Ilex*	*Ilexpollenites* spp.	Maastr.	Camp.–Maastr.	Turon.–Maastr.	(Senon.); Borneo	widespread, not in New Zealand
Chloranthaceae: *Ascarina*	*Clavatipollenites hughesii* Couper	e. Albian–Campan.	l. Albian–Cenom.	Barrem.–Santon.	Aptian; Falkland Plat.	New Zealand, Malaysia, Pacific Isl., Madagascar

Taxon	Pollen type					Southern Hemisphere
Gunneraceae: *Gunnera*	*Tricolpites reticulatus* Cookson	Maastr.	Campan.–Maastr.	Campan.–Maastr.	Turon.; Peru	extinct form
Fagaceae: *Nothofagus* Ancestral group	*Nothofagites senectus* Dettmann & Playford	Campan.–Maastr.	Campan.–Maastr.	Campan.–Maastr.	(Maastr.); S. Amer.	S. Amer., New Zealand, Tasmania
N. fusca group	*N. fusca* type	Maastr.	Maastr.	Maastr.	Maastr.; S. Amer.	S. Amer., New Zealand, E. Aust.
N. menziesii group	*N. menziesii* type	Maastr.	(Paleocene)	(Eocene)	Maastr.; S. Amer.	
N. brassii group	*N. brassii* type	Maastr.	(Eocene)	Maastr.	Maastr.; S. Amer.	New Caledonia, New Guinea
Myrtaceae:	*Myrtaceidites* spp.	e. Campan.–Maastr.	(Paleocene)	(Paleocene)	Senon.; Borneo, N. Africa	Australia, Polynesia to Malaya, South America
Proteaceae: *Beauprea*	*Beaupreaidites* spp.	l. Maastr.	l. Campan.–Maastr.	e. Campan.–Maastr.	(Eocene); N. Aust.	New Caledonia
Macadamia (?)	*Propylipollis* cf. *scaboratus* (Couper)	Campan.–Maastr.	Campan.–Maastr.	Campan.–Maastr.	–	Australia to E. Malaysia
Gevuina/Hicksbeachia	*Propylipollis reticuloscabratus* (Harris)	Maastr.	–	Campan.–Maastr.	(Oligoc.); Queensland	N. Australia, S. Amer., and tropical Pacific.
Xylomelum (?)	*Propylipollis annularis* (Cookson)	Maastr.	(Paleocene)	(Paleocene)	–	Australia
Winteraceae: *Bubbia, Belliolum*	*Pseudowinterapollis wahooensis* (Stover)	–	Maastr.	Campan.–Maastr.	Similar forms in Apt.–Alb. Israel	New Guinea, E. Australia, New Caledonia

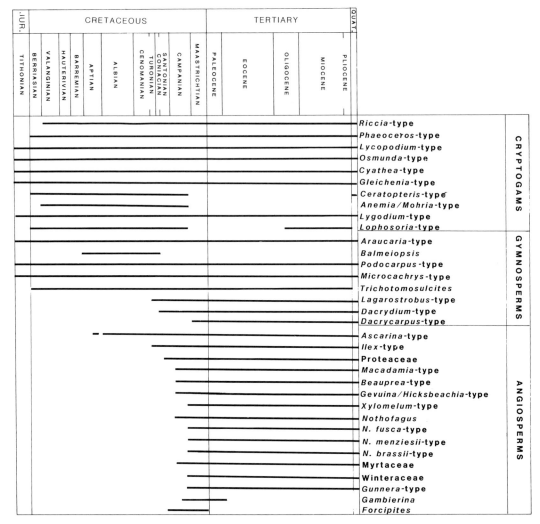

Fig. 4. Known stratigraphic ranges of selected floral elements in Antarctica and Australasia (including New Zealand, New Caledonia, Papua New Guinea, and intervening islands). Based on published data referred to under descriptions of the various groups.

Valanginian and is a common component of low gradient fluviatile/estuarine sediments of Albian–Turonian age. In contrast, the species occurs infrequently in Albian–Maastrichtian fan slope sediments in the James Ross Island region, Antarctic Peninsula (Dettmann & Thomson 1987). It seems likely that the parental source of *T. reticulatus* migrated to Antarctica during the Cretaceous, as older, Late Jurassic, records are known from Laurasia.

Lycopodiopsida

Lycopodiaceae, *Lycopodium*-type (Figs 4, 5b; Table 1): *Retitriletes*, which has cosmopolitan Late Mesozoic–Recent distribution and records that extend into the Triassic, replicates spores of *Lycopodium* (*Lycopodium*). Extant members of this subgenus are mainly terrestrial, occurring on disturbed sites adjacent to rainforests and on moorlands (Øllgard 1979).

Retitriletes species are common and diverse in Early Cretaceous palynofloras of Australia and New Zealand, but less so in coeval sediments of the Antarctic Peninsula region.

Filicopsida

Gleicheniaceae, *Gleichenia/Dicranopteris*-type (Fig. 4; Table 1): Dispersed spores included in *Gleicheniidites* verify the widespread distribution of the family during Jurassic−Cenozoic times. *Clavifera* incorporates fossil spores that probably were borne by an extinct alliance of the family which evolved during the Late Jurassic or Early Cretaceous. As discussed by Askin (1989), oldest appearances of *Clavifera triplex* (Bolkhovitina) Bolkhovitina 1966 imply earlier representation in eastern Australia (Albian) than in the Peninsula region of Antarctica (Late Cretaceous).

Lophosoriaceae, *Lophosoria*-type (Figs 4, 5; Table 1): This monotypic genus of ferns now restricted to South and central America had wide austral distribution during the Cretaceous and Tertiary as revealed by its unique spores, *Cyatheacidites annulatus* Cookson ex Potonié 1956 (Dettmann 1986a). It evolved in the Antarctic Peninsula−southern South America region during earliest Cretaceous times and the spore record suggests further diversification there of allied forms that are now extinct. Migration eastwards to Australia and northwards in South America coincided with environmental changes associated with Early Cretaceous opening of the South Atlantic Ocean. Dispersal across Australia was from west to east and occurred during Albian−Cenomanian phases of rifting between Australia and Antarctica prior to opening of the Southern Ocean.

Marsiliaceae (Fig. 5c; Table 1): *Crybelosporites striatus* (Cookson & Dettmann) Dettmann 1963 includes microspores of hydropteridean, possibly marsiliaceous, affinity. The Albian incoming of *C. striatus* in Australia and Antarctica is later than appearances on the Falkland Plateau (Aptian) and North America (earliest Cretaceous). During Albian−Cenomanian times these aquatic ferns apparently thrived in and about low relief depositional areas of Australia. The rare occurrence of spores in Antarctic Peninsula Cretaceous sediments may be related to the steep terrain on which suitable habitats of standing water may have been lacking (Dettmann 1986a).

Parkeriaceae, *Ceratopteris*-type (Fig. 4; Table 1): Species of *Cicatricosisporites* in which the mural sets are in discrete bundles in the equa-torial radial regions (Dettmann 1986b) are believed to be allied to *Ceratopteris*, an aquatic fern of tropical regions. Earliest records of *Ceratopteris*-like spores are from the Middle Jurassic in the Caribbean and northern African regions, but in Australia and Antarctica it is not recorded until latest Jurassic time. The *Ceratopteris*-like spores represented include *Cicatricosisporites australiensis* (Cookson) Potonié 1956 and *C. ludbrookiae* Dettmann 1963, both of which were widely distributed throughout austral regions during Early−mid Cretaceous times.

Schizaeaceae, *Mohria/Anemia*-type (Figs 4, 5d, e; Table 1): Species of *Cicatricosisporites*, *Nodosisporites*, and *Appendicisporites* having mural sets that coalesce in the equatorial radial regions can be allied to the Schizaeaceae. The African genus *Mohria*, and the circum-Caribbean subgenus *Anemia* (*Anemirhiza*) have spores comparable to *Cicatricosisporites hughesii* Dettmann 1963 and *C. pseudotripartitus* (Bolkhovitina) Dettmann 1963. Both occur in Cretaceous sediments of Australia and Antarctica and elsewhere in southern high latitude regions. *C. hughesii* ranges from Barremian−Campanian in the Otway Basin and is known from the early Albian on the Antarctic Peninsula. Older records are unknown from the peninsula although Aptian sediments have yet to be studied. Other forms reminiscent of *Mohria*- and *Anemirhiza*-type spores are known from the Middle Jurassic in northern Gondwana and southern Laurasia, a region that may have been an early radiation centre for these alliances of the Schizaeaceae. Migration into austral areas probably involved India. Available records suggest that introduction in western and northern regions of Australia was earlier (earliest Cretaceous) than in the southeast (Barremian).

Nodosisporites (Fig. 5e) and *Appendicisporites* (Fig. 5d) are morphologically comparable to spores of the subgenera *Anemia* and *Coptophyllum*, respectively, of *Anemia*. Extant members are concentrated in tropical America with ranges that extend to Africa, Madagascar, and India. However, the fossil spore record implies a wider Cretaceous distribution in both the Northern and Southern Hemispheres. Both *Appendicisporites* and *Nodosisporites* are represented in middle Albian−Cenomanian sections of the Antarctic Peninsula sequence, but the latter is unknown from Australia. Earliest appearances of *Appendicisporites* in Australia are markedly heterochronous, with older, Barremian, occurrences in northern and western sequences in contrast to late Albian introduc-

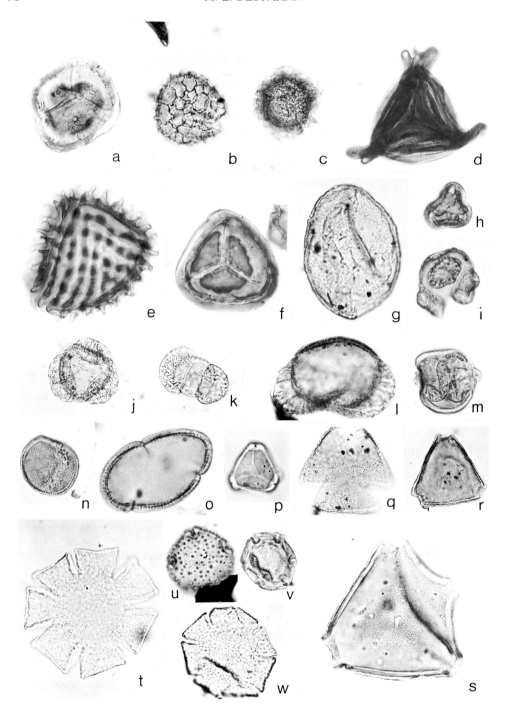

tions in the southeast. Late Jurassic records of both genera from the Northern Hemisphere confirm a lag in migration to Australasia and Antarctica.

Gymnospermae

Araucariaceae (Figs 4, 5g; Table 1): This family, which has a history that extends into the Triassic, evidently achieved global or near-global distribution during the Cretaceous. *Araucariacites australis* Cookson 1947 probably includes pollen of araucarian origin and is a persistent and sometimes common element of Jurassic–Cretaceous palynofloras of Australasia and Antarctica. The more distinctive *Balmeiopsis limbatus* (Balme) Archangelsky 1979 is less certainly of araucarian derivation. It occurs in pollen sacs of *Brachyphyllum irregulare* Archangelsky 1966 and its first appearances imply radiation from northern Gondwana, southwards to Antarctica and Australasia during the Early Cretaceous. On the Antarctic Peninsula it is known from early Albian–Cenomanian sediments and in Australia has progressively younger incomings from Valanginian in western depositional basins (Backhouse 1988) to Barremian and Albian in central and southeastern areas. Step-wise eastwards migration seems likely.

Podocarpaceae (Figs 4, 5h–m; Table 1): Antarctic Cretaceous palynofloras contain common and diverse podocarpaceous pollen, several morphotypes of which can be allied to extant genera.

The most frequently occurring pollen are the bisaccate *Podocarpidites ellipticus* Cookson 1947 (Fig. 5k), which may represent pollen of *Podocarpus* and certain *Dacrydium* (Pocknall 1981), and trisaccate *Microcachrys*-like *Microcachryidites antarcticus* Cookson 1947 (Fig. 5i). *Microcachrys*, now relict in Tasmania, had widespread southern Gondwana distribution from latest Jurassic–Cretaceous times, but an earlier history in Europe is suggested by the occurrence of *Microcachrys*-like cones in the Middle Jurassic of Poland (Reymanova 1987).

The trisaccate pollen of *Dacrycarpites australiensis* Cookson & Pike 1953 (Fig. 5j) replicates that of *Dacrycarpus*. It has earliest occurrences in Campanian–Maastrichtian sediments of Antarctica and southern Australia, but, thus far, is known only from Eocene and younger sequences in South America and New Zealand (Mildenhall 1980). *Dacrycarpus*-like megafossils from the Jurassic of India (Florin 1940) suggest an earlier history of the genus or its stock outside the Antarctic/Australian region.

Antarctic Cretaceous sequences contain a further distinctive podocarpaceous trisaccate pollen type, *Trichotomosulcites subgranulatus* Couper 1953 (Fig. 5h), which is known from cones of extinct *Trisacocladus tigrensis* Archangelsky 1966. Oldest reported occurrences of the pollen are from the earliest Cretaceous of southern South America. A spread to Antarctica, Australia, and New Zealand was achieved by the Albian and the pollen persists into early Tertiary sediments throughout the Cretaceous distribution area.

Fig. 5. Photomicrographs of fossil cryptogam spores (a–f) and pollen of gymnosperms (g–m) and angiosperms (n–w); magnifications × 375 (j), × 562 (a–i, k–m), and × 750 (n–w). All specimens, except o, r, u, and w, are figured in Dettmann & Thomson (1987) and Dettmann & Jarzen (in press) where locality and repository details are given; specimens in o and w are from sample D.3122.3, Vega Island, and those in r and u from sample 8540, James Ross Island as designated by Dettmann & Thomson. (**a**) *Foraminisporis dailyi* (Cookson & Dettmann) Dettmann 1963 (*Phaeoceros*-type); (**b**) *Retitriletes austroclavatidites* (Cookson) Döring *et al.* 1963 (*Lycopodium*-type); (**c**) *Crybelosporites striatus* (Cookson & Dettmann) Dettmann 1963 (Marsiliaceae); (**d**) *Appendicisporites* cf. *insignis* (Markova) Chlonova 1976 (*Coptophyllum*-type); (**e**) *Nodosisporites* cf. *crenimurus* (Srivastava) Davies 1986 (*Anemia*-type); (**f**) *Cyatheacidites annulatus* Cookson ex Potonié 1956 (*Lophosoria*-type); (**g**) *Araucariacites australis* Cookson 1947 (Araucariaceae); (**h**) *Trichotomosulcites subgranulatus* Couper 1953 (Podocarpaceae); (**i**) *Microcachryidites antarcticus* Cookson 1947 (*Microcachrys*-type); (**j**) *Dacrycarpites australiensis* Cookson & Pike 1953 (*Dacrycarpus*-type); (**k**) *Podocarpidites ellipticus* Cookson 1947 (*Podocarpus*-type); *Lygistepollenites florinii* (Cookson & Pike) Stover & Evans 1973 (*Dacrydium balansae*-type); (**m**) *Phyllocladidites mawsonii* Cookson ex Couper 1953 (*Lagarostrobus*-type); (**n**) *Clavatipollenites hughesii* Couper 1958 (*Ascarina*-type); (**o**) *Tricolpites reticulatus* Cookson 1947 (*Gunnera*-type); (**p**) *Myrtaceidites eugeniioides* Cookson & Pike 1953 (Myrtaceae); (**q**) *Peninsulapollis truswelliae* Dettmann & Jarzen 1988 (?Proteaceae); *Propylipollis* cf. *crassimarginus* Dudgeon 1983 (*Macadamia*-type); (**s**) *Propylipollis* cf. *annularis* (Cookson) Martin & Harris 1974 (*Xylomelum*-type); (**t**) *Nothofagidites asperus* (Cookson) Romero 1973 (*Nothofagus menziesii*-type); (**u**) *Nothofagidites senectus* Dettmann & Playford 1968 (*Nothofagus* ancestral-type); (**v**) *Nothofagidites lachlaniae* (Couper) Pocknall & Mildenhall 1984 (*Nothofagus fusca*-type); (**w**) *Nothofagidites dorotensis* Romero 1973 (*Nothofagus brassii*-type).

Lagarostrobus, now confined to Tasmania and with bisaccate pollen identical with *Phyllocladidites mawsonii* Cookson 1947 (Fig. 5m), had an Australasian−Antarctic−southern South American distribution during the Late Cretaceous and Tertiary (Playford & Dettmann 1978). Origination was no later than the Turonian, but the southern high latitude radiation centre has not yet been pinpointed from available pollen evidence.

The history of the *Dacrydium balansae/ D. bidwillii* alliance, which has a distribution range encompassing South America, New Zealand, and Pacific islands to Malaysia, can be traced by the bisaccate pollen genus *Lygistepollenites* (Fig. 5l). Oldest records of the pollen are from Coniacian−Santonian sediments in southeastern Australia and the Antarctic Peninsula (Dettmann & Jarzen in press). The latest Cretaceous−mid Tertiary distribution of the pollen is similar to that of *Phyllocladidites mawsonii* (*Lagarostrobus*); retraction and northwards migration to its present range occurred during and after the Miocene.

Angiospermae

Aquifoliaceae, *Ilex* (Fig. 4; Table 1): *Ilex*, which has cosmopolitan distribution, has a history that extends back to the Late Cretaceous. Oldest indubitable occurrences of *Ilex*-like pollen, *Ilexpollenites*, are in the Turonian of the Otway Basin (Martin 1977). Late Cretaceous migration, with routes involving Antarctica, is implied by respective Campanian and Maastrichtian first occurrences of *Ilexpollenites* in New Zealand (Raine 1984) and on the Antarctic Peninsula. Radiation to Borneo and Africa may have been earlier, in the Senonian (Muller 1981).

Chloranthaceae, *Ascarina* (Figs 4, 5n; Table 1): The oldest angiospermous pollen recorded from Antarctic and Australasian sequences is *Clavatipollenites hughesii* Couper 1958 which is comparable to pollen of *Ascarina*, a shrub now confined to New Zealand, Pacific islands, Malaysia, and Madagascar. Cretaceous records of the pollen suggest origination in northern Gondwana followed by Aptian−Albian radiation worldwide. Australian incomings are not certainly dated but may be as old as Barremian−Aptian (Dettmann 1986b) and predate appearances (late Albian) in New Zealand (Raine et al. 1981). Elucidation of the timing of introduction in Antarctica awaits palynological analyses of Barremian−Aptian sediments that underlie the early Albian sections from which C. hughesii

has been reported (Dettmann & Thomson 1987).

Fagaceae, *Nothofagus* (Figs 4, 5t−w; Table 1): A southern high latitude evolution of *Nothofagus* during the Late Cretaceous is supported by pollen and leaf evidence (Tanai 1986; Dettmann et al. in press). Early Campanian inception of ancestral types, such as *Nothofagidites senectus* (Dettmann & Playford 1968; Fig. 5u) in the Australian/Antarctic region was followed by Maastrichtian diversification and evolution of *brassii* (Fig. 5w), *fusca* (Fig. 5v), and *menziesii* (Fig. 5t) pollen producers on the Antarctic Peninsula and in Tierra del Fuego, South America. The later, staggered appearances of the extant pollen types in New Zealand and Australia imply step-wise migration with routes involving Antarctica (Fig. 6a).

Gunneraceae, *Gunnera* (Figs 4, 5o; Table 1): As indicated by the pollen record, *Gunnera* was established by the Turonian in northern Gondwana (Jarzen 1980) and radiated northwards and southwards during the Late Cretaceous. Its introduction into Antarctica and Australasia during the Late Cretaceous is confirmed by the presence of *Tricolpites reticulatus* Cookson 1947 in Campanian−Maastrichtian sediments of southern Australia, New Zealand, and the Antarctic Peninsula (Dettmann & Jarzen in press).

Myrtaceae (Figs 4, 5p; Table 1): Myrtaceous pollen similar to that of *Szygium* and *Eugenia* of the Myrtoideae occur in early Campanian and Maastrichtian sediments on the Antarctic Peninsula. These records predate earliest occurrences (Paleocene) in Australia and New Zealand, but are younger than the lower Senonian report from Borneo (Muller 1968) and the Santonian of Gabon (Boltenhagen 1976). The pollen evidence is sufficiently persuasive for a Late Cretaceous evolution of the family outside Australia which is now its centre of diversity. Introduction there may have been via Antarctica (Fig. 6b), although migration from Borneo cannot be discounted until Late Cretaceous sequences from intervening regions, including Papua New Guinea, are thoroughly investigated.

Proteaceae (Figs 4, 5r,s; Table 1): Proteaceous pollen represented in Campanian−Maastrichtian sediments of the Antarctica Peninsula confirm the former presence of the Proteoideae and Grevilleoideae (Dettmann & Jarzen in press). Amongst the Proteoideae is the New Caledonian endemic, *Beauprea*, which, as indicated by *Beaupreaidites*, probably evolved adjacent to the seaway between eastern Antarctica and southern Australasia (Pocknall

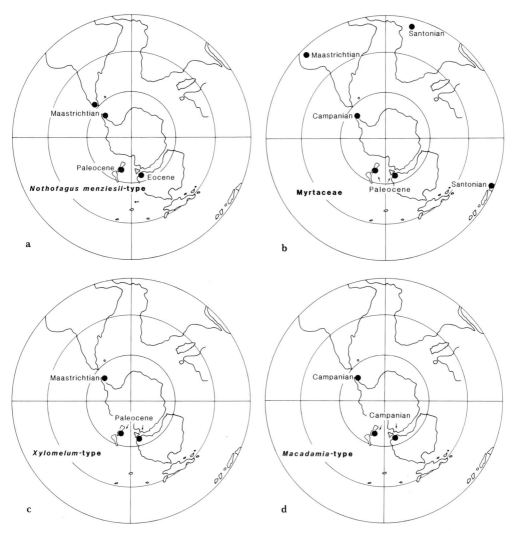

Fig. 6. Time of first appearance of pollen of (**a**) *Nothofagus menziesii*-type, (**b**) Myrtaceae, (**c**) *Xylomelum*-type, (**d**) *Macadamia*-type. Map is South Polar Lambert equal-area for the Santonian (after Smith *et al*. 1981).

& Crosbie 1988; Dettmann & Jarzen 1988). Although *Beaupreaidites* occurs infrequently in the Antarctic Peninsula region, a possible affiliate, *Peninsulapollis* (Fig. 5q), is both common and diverse.

Propylipollis, species of which have been associated with *Gevuina/Hicksbeachia*, *Knightia*, *Macadamia*, and *Xylomelum*, occur in Campanian–Maastrichtian sequences or in sediments sourced from those sequences in Antarctica. Pollen of the first two alliances appear in early Campanian sediments in southeastern Australia; this area may have been the site of evolution of the New Caledonian/

New Zealand *Knightia* and the *Gevuina/Hicksbeachia* alliance (Dettmann & Jarzen in press) which today has a disjunct distribution in northeastern Australia and Chile. But an earlier incoming of *Propylipollis annularis* (Cookson) Martin & Harris 1974 (Fig. 5s) on the Antarctic Peninsula than in southern Australia suggests that possible affiliates of the Australian sclerophyllous *Xylomelum* may have migrated in the reverse direction from western Antarctica to southern Australia (Fig. 6c). The timing and place of origin of *Macadamia*, a northeastern Australian–Malaysian rainforest component, has yet to be ascertained precisely, but pollen evidence suggests introduction in Antarctica and in southern Australasia by the Campanian (Fig. 6d; Dettmann & Jarzen, in press).

Winteraceae (Fig. 4; Table 1): Winteraceous pollen similar to *Belliolum* and *Bubbia* occur in Upper Cretaceous sediments of southeastern Australia and New Zealand, but are as yet unreported from Antarctica. Records of *Drimys* leaves from the early Tertiary on Seymour Island (Dusén 1908) suggest widespread austral distribution of the family during the Tertiary. However, further investigations are required to establish radiation routes from a possible Aptian–Albian source area in Israel (Walker *et al*. 1983).

Phytogeographic implications

Of the elements discussed above, many now have disjunct distributions in the mid- to low latitudes of southern regions. A large number of the phanerogams and certain of the crypto-gams have similar or overlapping distribution ranges and are concentrated in rainforests of South America, eastern Australasia, and islands northwards to Malaysia. Some also occur in Africa and/or Madagascar. Not all are Gondwanan in origin as had been postulated from biogeographical studies. The Winteraceae may have had an earlier history in southern Laurasia than in Gondwana, and the same may be true for certain early (Jurassic) Podocar-paceae, including *Microcachrys* (Reymanova 1987), and for many of the cryptogams. Radi-ation into the Antarctic region was not synchronous and may have involved separate routes through northern Gondwana.

Another group of immigrant taxa probably had origins in northern Gondwana. Amongst these are early chloranthaceous angiosperms, Gunneraceae, Proteaceae, and Myrtaceae. The last-mentioned seems to have occurred in Antarctica prior to introduction in Australia; migration from a northern distribution centre may well have been via a South American–Antarctic route. Floristic evidence for a north-ern Gondwanan origin of early Proteaceae (Johnson & Briggs 1975) is supported by pollen evidence. This large family is now concentrated in South Africa and Australia and with a distri-bution range that encompasses South America, and the southwestern Pacific region to Malaysia. Several of its alliances probably evolved in areas fringing the Late Cretaceous seaway between eastern Antarctica and southern Australasia. Amongst these are *Beauprea*, *Gevuina/ Hicksbeachia*, and *Knightia*, all of which date

from the Campanian. *Ilex* also probably orig-inated in the region, but at an earlier, Turonian, date.

Antarctica and closely associated areas of southern South America and southern Australasia was almost certainly the site of origin of other floral elements including *Lagarostrobus*, certain *Dacrydium*, *Nothofagus*, and the fern *Lophosoria*. The radiation centre of the last-mentioned was in the vicinity of the Antarctic Peninsula. Evolution occurred in earliest Cretaceous times (Dettmann 1986a) during initial opening of the South Atlantic Ocean. The same general region was the site of Maastrichtian diversification of *Nothofagus* (Dettmann *et al*. in press). Pollen evidence also indicates evolution of early *Nothofagus* in south-ern high latitudes, but the precise centre in Antarctica or contiguous regions has not been delineated as its pollen, which may have been widely disseminated by wind, had synchron-ous early Campanian appearances in western Antarctica, New Zealand, and southern Australia. The same holds true for the Turonian inception of *Lagarostrobus* and the *Dacrydium* allies that shed *Lygistepollenites*.

Many of the taxa discussed above now occur in mid- to low latitudes and are there concen-trated in upland perhumid rainforests or on oceanic islands. They mostly prefer habitats in-fluenced by equable, humid climatic conditions that can be classed as temperate (Webb & Tracey 1981). Some, including *Lagarostrobus*, *Microcachrys*, and the *fusca* and *menziesii* pol-len producers of *Nothofagus*, occur in perhumid temperate forests of South America and/or southern Australasia. None of the evidence conflicts with a temperate humid climate.

Implications for origination and dispersal of plants

From the above discussion it is evident that Antarctica was most likely a source area and served as a dispersal corridor during the Cre-taceous for certain floral elements now with disjunct distribution in austral mid to low lati-tude regions. Several phases of floral inception and migration are delineable from the fossil record (Figs 2, 6). They are contemporaneous with opening and enlargement of oceans and fragmentation of the southern Gondwana as-sembly. Causative factors may well be related to environmental disturbances and climatic changes resulting from volcanic and tectonic activity concurrent with rifting, drifting, and opening of oceans.

Western Antarctica was most likely the source region of *Lophosoria* and extant groups of *Nothofagus*. Inception of the former in earliest Cretaceous times was on a magmatic arc during a phase of considerable volcanic and tectonic activity associated with early opening of the South Atlantic Ocean (e.g. Farquharson 1983). Migration northwards to South America and the Falkland Plateau occurred as this ocean opened progressively from south to north; floral changes recorded there are believed to reflect alterations to terranes and climatic moderation (Dettmann 1986*a*). Migration eastwards to and within Australia coincided with phases of rifting and early drifting between Antarctica and Australia, and with depocentre displacements in intracratonic basins of Australia. Dispersal patterns of other floral elements indicate eastwards migration from western Antarctica and southern South America to Australia during the Early Cretaceous (Fig. 2a, b).

Near the close of the Cretaceous there was renewed volcanism in the Peninsula region (Farquharson 1983) and resultant environmental disturbances may have triggered evolution of primordial *Nothofagus* groups. Dispersal of these groups in a westwards direction to New Zealand and Australia must have involved trans-Antarctic routes (Fig. 6a). The same routes may have been utilized by the Myrtaceae and *Gunnera* after the radiation from northern Gondwana. But other floral elements appear to have migrated in the opposite direction during the Late Cretaceous (Fig. 2c–e). Certain Proteaceae have later inceptions in western Antarctica than in southern Australasia and probably evolved in the embryonic Southern Ocean region (Dettmann & Jarzen, in press). Campanian evolution was contemporaneous with rifting, early drifting, and volcanic activity (Boddard *et al.* 1986; Thompson 1986). Environmental stresses related to these events appear to be mirrored in the spore–pollen assemblages that record turnover of taxa and an expansion of angiosperm communities.

The Transantarctic Association is thanked for facilitating my attendance at the London meeting.

References

ARCHANGELSKY, S., BALDONI, A., GAMERRO, J. C., PALAMARCZUK, S. & SEILER, S. 1981. Palinológia estratigráfica del Cretácico de Argentina Austral. Diagrams del grupos polinicos de Sudroeste de Chubut y Noroeste de Santa Cruz. *Actas VIII Congreso Geologico Argentino, San Luis 1981,* 719–742.

——, ——, —— & SEILER, S. 1984. Palinológia estratigráfica del Cretácico de Argentina Austral. III. Distribucion de las especies y conclusiones. *Ameghiniana,* **21,** 15–33.

ASKIN, R. A. 1983. Tithonian (uppermost Jurassic)–Barremian (Lower Cretaceous) spores, pollen and microplankton from the South Shetland Islands, Antarctica. *In:* OLIVER, R. L., JAMES, P. R. & JAGO, J. B. (eds) *Antarctic earth science.* Australian Academy of Science, Canberra, and Cambridge University Press, Cambridge, 295–297.

—— 1988. Campanian to Paleocene palynological succession of Seymour and adjacent islands, northeastern Antarctic Peninsula. *In:* FELDMAN, R. M. & WOODBURNE, M. O. (eds) *Geology and palaeontology of Seymour Island, Antarctic Peninsula.* Geological Society of America Memoir, **169,** 131–53.

—— 1989. Endemism and heterochroneity in the Seymour Island Campanian to Paleocene palynofloras: implications for origins, dispersal and palaeoclimates of southern floras. *In:* CRAME, J. A. (ed.) *Origins and evolution of the Antarctic Biota,* Geological Society, London, Special Publication, **47,** 107–120.

BACKHOUSE, J. 1988. Late Jurassic and Early Cretaceous palynology of the Perth Basin, Western Australia. *Bulletin of the Geological Survey of Western Australia,* **135.**

BALDONI, A. & ARCHANGELSKY, S. 1983. Palinológia de la Formación Springhill (Cretácico Inferior), subsuelo de Argentina y Chile Austral. *Revista Espanola de Micropaleontologia,* **15,** 47–101.

BODARD, J. M., WALL, V. J. & KANEN, R. A. 1986. Lithostratigraphic architecture of the Latrobe Group, offshore Gippsland Basin. *In:* GLENIE, R. C. (ed.) *Second south-eastern Australian Oil Exploration Symposium, Melbourne, 1985.* Petroleum Exploration Society of Australia, 113–136.

BOLTENHAGEN, E. 1976. Pollen et spores Senoniennes du Gabon. *Revue de Micropaléontologie,* **18,** 191–199.

COUPER, R. A. 1953. Upper Mesozoic and Cainozoic spores and pollen grains from New Zealand. *Paleontological Bulletin New Zealand Geological Survey,* 22.

—— 1960. New Zealand Mesozoic and Cainozoic plant microfossils. *Paleontological Bulletin New Zealand Geological Survey,* 32.

CREBER, G. T. & CHALONER, W. G. 1985. Tree growth in the Mesozoic and early Tertiary and the reconstruction of palaeoclimates. *Palaeogeography, Palaeoclimatology, Palaeoecology,* **52,** 35–60.

DARWIN, C. 1859. *On the origin of species by means of natural selection.* Murray, London.

DETTMANN, M. E. 1963. Upper Mesozoic microfloras from southeastern Australia. *Proceedings of the*

Royal Society of Victoria, **77**, 1–148.

— — 1981. The Cretaceous flora. *In*: KEAST, A. (ed.) *Ecological biogeography of Australia*. Junk, The Hague, 357–375.

— — 1986a. Significance of the Cretaceous-Tertiary spore genus *Cyatheacidites* in tracing the origin and migration of *Lophosoria* (Filicopsida). *Special Papers in Palaeontology*, **35**, 63–94.

— — 1986b. Early Cretaceous palynoflora of subsurface strata correlative with the Koonwarra Fossil Bed, Victoria. *Memoirs of the Association of Australasian Palaeontologists*, **3**, 79–110.

— — & JARZEN, D. M. 1988. Angiosperm pollen from uppermost Cretaceous strata of southeastern Australia and the Antarctic Peninsula. *Memoirs of the Association of Australasian Palaeontologists*, **5**, 217–237.

— — & — — In press. The Antarctic/Australian rift valley: Late Cretaceous cradle of northeastern Australasian relicts? *In*: TRUSWELL, E. M., BALME, B. E., DETTMANN, M. E. & WILSON, G. J. (eds) *Proceedings of the 7th International Palynological Congress, Brisbane 1988*. Australian Academy of Science, Canberra.

— — & THOMSON, M. R. A. 1987. Cretaceous palynomorphs from the James Ross Island area — a pilot study. *British Antarctic Survey Bulletin*, **77**, 13–59.

— —, POCKNALL, D. T., ROMERO, E. J. & ZAMALOA, M. In press. *Nothofagidites* Erdtman ex Potonié 1960; a catalogue of species with notes on the palaeogeographic distribution of *Nothofagus* B1. *Paleontological Bulletin New Zealand Geological Survey*.

DOMACK, E. W., FAIRCHILD, W. W. & ANDERSON, J. B. 1980. Lower Cretaceous sediment from the East Antarctic continental shelf. *Nature*, **287**, 625–626.

DOUGLAS, J. G. 1986. The Cretaceous vegetation, and palaeoenvironments of Otway Group sediments. *In*: GLENIE, R. C. (ed.) *Second southeastern Australia Oil Exploration Symposium, Melbourne*, 1985. Petroleum Exploration Society of Australia, 233–240.

DUSÉN, P. 1908. Uber die Tertiäre Flora der Seymour Insel. *Wissenschaftliche Ergebnisse der Scwedishen Südpolar-expedition 1901–1903*, **3**, 1–127.

FARQUHARSON, G. W. 1982. Late Mesozoic sedimentation in the northern Antarctic Peninsula and its relationship to the southern Andes. *Journal of the Geological Society, London*, **139**, 721–728.

— — 1983. Evolution of Late Mesozoic sedimentary basins in the northern Antarctic Peninsula. *In*: OLIVER, R. L., JAMES, P. R. & JAGO, J. B. (eds) *Antarctic earth science*, Australian Academy of Science, Canberra, and Cambridge University Press, Cambridge, 323–327.

FLORIN, R. 1940. The Tertiary fossil conifers in South Chile and their phytogeographical significance, with a review of the fossil conifers of southern lands. *Kungliga Svenska vetenskapsakademiens handlingar*, **19**, 1–107.

FRAKES, L. A. & FRANCIS, J. E. 1988. Australian

Cretaceous ice. *9th Australian Geological Convention, Brisbane 1988, Abstracts*, 144–145.

FRANCIS, J. E. 1986. Growth rings in Cretaceous and Tertiary wood from Antarctica and their palaeoclimatic implications, *Palaeontology*, **29**, 665–684.

HARRIS, W. K. 1977. Palynology of cores from Deep Sea Drilling Sites 327, 328, and 330, South Atlantic Ocean. *In*: BARKER, P. F., DALZIEL, I. W. D. *et al. Initial Reports Deep Sea Drilling Project*, **36**. United States Government Printing Office, Washington, 761–815.

HEDLUND, R. & BEJU, D. 1977. Stratigraphic palynology of selected Mesozoic samples, DSDP Hole 327A and Site 330. *In*: BARKER, P. F., DALZIEL, I. W. D. *et al. Initial Reports Deep Sea Drilling Project*, **36**. United States Government Printing Office, Washington, 817–827.

HELBY, R. J., MORGAN, R. & PARTRIDGE, A. D. 1987. A palynological zonation of the Australian Mesozoic. *Memoirs of the Association Australasian Palaeontologists*, **4**, 1–93.

HOOKER, J. D. 1847. *The botany of the Antarctic voyage of H. M. discovery ships Erebus and Terror in the years 1839–1843. I Flora Antarctica*. Reeve Bros., London. Reprinted 1963, J. Cramer, Weinheim.

INESON, J. R., CRAME, J. A. & THOMSON, M. R. A. 1986. Lithostratigraphy of the Cretaceous strata of west James Ross Island, Antarctica. *Cretaceous Research*, 7,141–159.

JARZEN, D. M. 1979. Spore morphology of some Anthocerotaceae and the occurrence of *Phaeoceros* spores in the Cretaceous of North America. *Pollen et Spores*, **21**, 211–231.

— — 1980. The occurrence of *Gunnera* pollen in the fossil record. *Biotropica*, **12**, 117–123.

JEFFERSON, T. H. 1982. The Early Cretaceous fossil forests of Alexander Island, Antarctica. *Palaeontology*, **25**, 681–708.

— — 1983. Palaeoclimatic significance of some Mesozoic Antarctic fossil forests. *In*: OLIVER, R. L., JAMES, P. R. & JAGO, J. B. (eds) *Antarctic Earth Science*, Australian Academy of Science, Canberra, and Cambridge University Press, Cambridge, 593–598.

JOHNSON, L. A. S. & BRIGGS, B. G. 1975. On the Proteaceae — the evolution and classification of a southern family. *Botanical Journal of the Linnean Society*, **70**, 83–185.

KOTOVA, I. Z. 1983. Palynological study of Upper Jurassic and Lower Cretaceous sediments, Site 511, Deep Sea Drilling Project Leg 71 (Falkland Plateau). *In*: LUDWIG, W. J., KRASHENINNIKOV, V. A., *et al. Initial Reports of the Deep Sea Drilling Project*, **71**, United States Government Printing Office, Washington, 879–906.

MARTIN, H. A. 1977. The history of *Ilex* (Aquifoliaceae) with special reference to Australia: evidence from pollen. *Australian Journal of Botany*, **25**, 655–673.

MCLACHLAN, I. R. & PIETERSE, R. E. 1978. Preliminary palynological results. Site 361 Leg 40, Deep Sea Drilling Project. *In*: BOLLI, H., RYAN,

W. B. F. *et al. Initial Reports of the Deep Sea Drilling Project* **40**, United States Government Printing Office, Washington, 857–881.

MILDENHALL, D. C. 1980. New Zealand Late Cretaceous and Cenozoic plant biogeography: a contribution. *Palaeogeography, Palaeoclimatology, Palaeoecology*, **31**, 197–233.

MULLER, J. 1968. Palynology of the Pedawan and Plateau Sandstone Formations (Cretaceous–Eocene) in Sarawak, Malaysia. *Micropaleontology*, **14**, 1–37.

—— 1981. Fossil pollen records of extant angiosperms. *The Botanical Review*, **47**, 1–142.

ØLLGARD, B. 1977. *Lycopodium* in Ecuador — habits and habitats. *In*: LARSEN, L. & HOLM–NIELSEN, P. (eds) *Tropical Botany*. Academic Press, London, New York, 381–395.

PLAYFORD, G. & DETTMANN, M. E. 1978. Pollen of *Dacrydium franklinii* Hook.f. and comparable early Tertiary microfossils. *Pollen et Spores*, **20**, 513–534.

POCKNALL, D. T. 1981. Pollen morphology of the New Zealand species of *Dacrydium* Solander, *Podocarpus* L'Heritier, and *Dacrycarpus* Endlicher (Podocarpaceae). *New Zealand Journal of Botany*, **19**, 67–95.

—— & CROSBIE, Y. M. 1988. Pollen morphology of *Beauprea* (Proteaceae): modern and fossil. *Review of Palaeobotany and Palynology*, **53**, 305–327.

RAINE, J. I. 1984. Outline of a palynological zonation of Cretaceous to Paleogene terrestrial sediments in west coast region, South Island, New Zealand. *Reports of the New Zealand Geological Survey*, **109**, 1–82.

——, SPEDEN, I. G. & STRONG, C. P. 1981. New Zealand *In*: REYMENT, R. A. & BENGTSEN, P. (eds) *Aspects of mid-Cretaceous regional geology*. Academic Press, London, 221–267.

REYMANOVA, M. 1987. A Jurassic podocarp from Poland. *Review of Palaeobotany and Palynology*, **51**, 133–143.

SMITH, A. G., HURLEY, A. M. & BRIDEN, J. C. 1981. *Phanerozoic paleocontinental maps*. Cambridge University Press, Cambridge.

TANAI, T. 1986. Phytogeographic and phylogenetic history of the genus *Nothofagus* B1. (Fagaceae in the Southern Hemisphere). *Journal of the Faculty of Science Hokkaido University Series 4*, **21**, 505–582.

THOMPSON, B. R. 1986. The Gippsland Basin — development and stratigraphy. *In*: GLENIE, R. C. (ed.) *Second southeastern Australia Oil Exploration Symposium, Melbourne 1985*. Petroleum Exploration Society of Australia, 57–64.

TRUSWELL, E. M. 1983. Recycled Cretaceous and Tertiary pollen and spores in Antarctic marine sediments: a catalogue. *Palaeontographica*, **186B**, 121–174.

—— in press. Antarctica: a history of terrestrial vegetation. *In*: TINGEY, B. (ed) *Geology of Antarctica*, Oxford University Press.

—— & DREWRY, D. J. 1984. Distribution and provenance of recycled palynomorphs in surficial sediments of the Ross Sea, Antarctica. *Marine Geology*, **59**, 187–214.

VARMA, Y., RAO, N. & RAMANUJAM, G. G. K. 1984. Palynology of some Upper Gondwana deposits of the Palar Basin, Tamil Nadu, India. *Palaeontographica*, **190B**, 37–86.

VENTKATACHALA, B. S. 1974. Palynological zonation of the Mesozoic and Tertiary subsurface sediments in the Cauvery Basin. *In*: SURANGE, *et al.* (eds) *Aspects and appraisal of Indian palaeobotany*. Birbal Sahni Institute of Palaeobotany, Lucknow, 476–495.

—— RAWAT, M. S., SHARMA, K. D. & JAIN, A. K. 1980. Aptian-Albian spore and pollen assemblages from the Cauvery Basin, India. *Proceedings of the IV International Palynological Conference, Lucknow 1976*, **2**, 396–402.

VOLKHEIMER, W., CACCAVARI DE FILICE, M. A. & SEPULVEDA, E. 1977. Datos palinologicos de la Formacion Ortiz (Grupo la Amarga). Cretacico inferior de la cuenca nequina (Republica Argentina). *Ameghiniana*, **14**, 59–74.

WALKER, J. W., BRENNER, G. J. & WALKER, A. G. 1983. Winteraceous pollen in the Lower Cretaceous of Israel: early evidence of a magnolialean angiosperm family. *Science*, **220**, 1273–1275.

WEBB, L. J. & TRACEY, J. G. 1981. Australian rainforest: patterns and change. *In*: KEAST, A. (ed.) *Ecological biogeography of Australia*. Junk, The Hague, 605–694.

Endemism and heterochroneity in the Late Cretaceous (Campanian) to Paleocene palynofloras of Seymour Island, Antarctica: implications for origins, dispersal and palaeoclimates of southern floras

ROSEMARY A. ASKIN

Department of Earth Sciences, University of California, Riverside, California 92521, USA

Abstract: Late Cretaceous (Campanian) to Paleocene nearshore marine-coastal/deltaic sediments (López de Bertodano and Sobral Formations) from Seymour Island, Antarctica, contain abundant marine and terrestrial palynomorphs. Marine dinoflagellate cysts, and spores and pollen of land plants from 510 samples exhibit varying levels of provincialism: (1) Seymour Island-James Ross Island basin endemic species; (2) Antarctic endemic species; (3) Weddellian Province (southern South America—Western Antarctica—New Zealand—southeastern Australia) species; (4) Austral species; and (5) Cosmopolitan species.
A significant proportion of the Antarctic Cretaceous—Tertiary terrestrial vegetation evolved in southern polar latitudes. Almost half of the late Maastrichtian angiosperm pollen flora is apparently endemic to the Seymour Island—James Ross Island basin or Antarctica. The late Campanian—Maastrichtian dinoflagellate cyst assemblages also include a significant Antarctic or Weddellian provincial component. Climatic conditions in the Late Cretaceous were moist and equable. Many species that evolved in these polar latitudes became isolated due to various geographical, climatic or biological barriers. Other species dispersed northward, presumably following routes of similar climate. Some Seymour Island latest Cretaceous angiosperm species which subsequently appeared in the Paleocene or Eocene in New Zealand and southeastern Australia, have modern relatives living in mid- to low latitudes.

Abundant and often exquisitely preserved late Cretaceous—early Tertiary fossils on Seymour Island led Zinsmeister (1986) to refer to the island as a palaeontological 'Rosetta stone' for understanding the evolution of Southern Hemisphere life. Among this multitude of fossils are ubiquitous palynomorphs, providing information on the origins of southern floras.

Antarctic Cretaceous—Tertiary marine and nonmarine palynomorph assemblages are highly endemic or provincial. They contain numerous previously undescribed species, and taxa of limited areal distribution. Moreover, for the species that also occur outside Antarctica, their first (and last) occurrences are often disparate, or 'heterochronous' (Hickey *et al.* 1983) when their high latitude stratigraphic ranges are compared with those in lower latitudes. Some species have their earliest occurrences in Antarctica, suggesting that the region acted as a site of origin or evolutionary 'cradle' for these taxa.

The idea of Antarctica as a cradle for plant taxa is not new. Hooker (1847; and in Skottsberg 1953) and Darwin (1859) noted the coherence of the southern flora, and its probable origin in the southern high latitudes. More recently, Cranwell (1964), for example, discussed an Antarctic origin for southern beeches (*Nothofagus*). Many fossil plant and animal species first appear in older rocks in Antarctica (or the high southern latitudes) than in lower latitudes (e.g. Zinsmeister & Feldmann 1984; Dettmann & Thomson 1987; Huber 1988; Dettmann 1989; Feldmann & Tshudy 1989). The long held view, put forth, for example, as a query by Dobzhansky (1950), that the tropics serve as the principal evolutionary cradle and dispersal centre no longer seems viable for some groups. Although much of the Antarctic Cretaceous—Tertiary endemic flora remained isolated because of geographical, climatic or biological barriers, many plants spread northwards, introducing major taxa to later mid-latitude floras, and other taxa to mid- to low latitude floras.

It could be argued that, because the fossil record on Seymour Island may be more complete and better preserved than elsewhere, the numerous 'oldest' first occurrences are more readily identified on Seymour Island than at other locations. However, fossil palynomorph successions have been sufficiently studied in some other areas, e.g. Paleocene—Eocene of southeastern Australia (M. E. Dettmann, pers. comm. 1988), so that reported palynomorph

From Crame, J. A. (ed.), 1989, *Origins and Evolution of the Antarctic Biota,*
Geological Society Special Publication No. 47, pp. 107–119.

107

species absences probably do reflect their absence from the fossil record. Nevertheless,
examples are presented here with the caveat
that interpretations are tentative and await the
acquisition of more data from Antarctica and
from other southern continents.

Regional setting

Seymour Island is a small (about 20 km long),
seasonally ice-free island located in the James
Ross Island basin at the tip of the Antarctic
Peninsula (Fig. 1). Although Mesozoic tectonic
reconstructions of the Antarctic Peninsula—
South America region are not universally accepted, it is generally agreed that the James
Ross Island basin contains sedimentary and
volcanic strata deposited in a back-arc terrain
(e.g. Elliot 1988). Cretaceous—Paleogene sediments on Seymour Island have an Antarctic
Peninsula provenance, and are poorly-
consolidated, well-exposed, relatively un-
weathered and richly fossiliferous. They include
upper Campanian to Paleocene sediments of
the López de Bertodano and overlying Sobral
Formations (Fig. 2), which are mainly fine-
grained (muddy to sandy silts) and become
sandier in the upper Sobral Formation. Depositional environments for most of the López de
Bertodano Formation are shallow shelf, mainly
nearshore marine (Askin 1988; Harwood 1988;
Huber 1988; Macellari 1988). A regression
is evident high in the formation and coastal/
estuarine conditions prevailed in the uppermost
(Danian) part. Nearshore marine environments
in the basal Sobral Formation were followed
by deltaic conditions for the remainder of the
Sobral. The overlying Paleocene Cross Valley
and Eocene La Meseta Formations include
coarse to fine sands representing channel fill
and deltaic deposits.

These uppermost Cretaceous and Paleogene
sediments on Seymour Island provide the only
near-continuous, outcropping record of the fossil succession in Antarctica during the last stages
of Gondwana breakup. There is ample evidence
from terrestrial biotas, particularly certain vertebrates and southern beeches (*Nothofagus*),
which were unlikely to have dispersed via
'island-hopping', that a South America—
Antarctica land connection survived through
the Paleocene and into the Eocene (e.g. Woodburne & Zinsmeister 1984; Case *et al.* 1987;
Case 1988). Most Paleocene and Eocene reconstructions (e.g. Smith *et al.* 1981; Grunow *et al.*
1987), however, depict a severed land-
connection between South America and the
Antarctic Peninsula.

Biostratigraphic framework

The Upper Cretaceous—Paleogene sediments
of Seymour Island contain fossils of most major
floral and faunal groups. The ranges of these
fossil biotas are becoming known, despite the
problems that endemism and heterochroneity
present to biostratigraphy.

Palynomorph species discussed here are from
510 outcrop samples from the upper Campanian
to Paleocene López de Bertodano Formation
and lower Paleocene Sobral Formation from
the southern part of Seymour Island (Fig. 1).
Age control for these sections, summarized in
Fig. 2, is based on evidence from dinoflagellate
cysts (Askin 1988) and other fossil groups.

General aspect of palynofloras

Most samples contain both marine (dinoflagellate cysts, acritarchs and other algae) and non-
marine palynomorphs (spores and pollen from
land plants, fungal spores, freshwater algae).
Palynomorph assemblages from the Cretaceous
part of the succession are dominated by marine dinoflagellate cysts of the *Manumiella/
Isabelidinium* complex (Askin 1988). More variable Paleocene assemblages include species of
Spinidinium, *Deflandrea/Cerodinium*, and local
concentrations of *Palaeoperidinium pyro-
phorum*. The land-derived palynomorphs
are dominated by podocarpaceous conifer and
relatively diverse angiosperm pollen, with a
minor cryptogam spore component. The terrestrial palynomorphs were transported into the
depositional basin from various habitats,
including coastal lowlands fringing the depositional basin, the alluvial plain, and hilly
to possibly mountainous terrain along the
Antarctic Peninsula.

Provincialism

Some Seymour Island Cretaceous—Tertiary
terrestrial and marine palynomorph species
were apparently areally restricted, whereas
others were more widespread (Fig. 3). Five
levels of provincialism are recognized.

(1) Seymour Island—James Ross Island basin
endemic species.

(2) Antarctic endemic species. These occur
on Seymour Island, and have been found recycled in Recent sediments from coastal East
Antarctica, and the Weddell and Ross Seas
(Truswell 1983; Truswell & Drewry 1984).

(3) Weddellian Province species. The
southern high latitude Weddellian Province
(Zinsmeister 1979, 1982) comprised southern

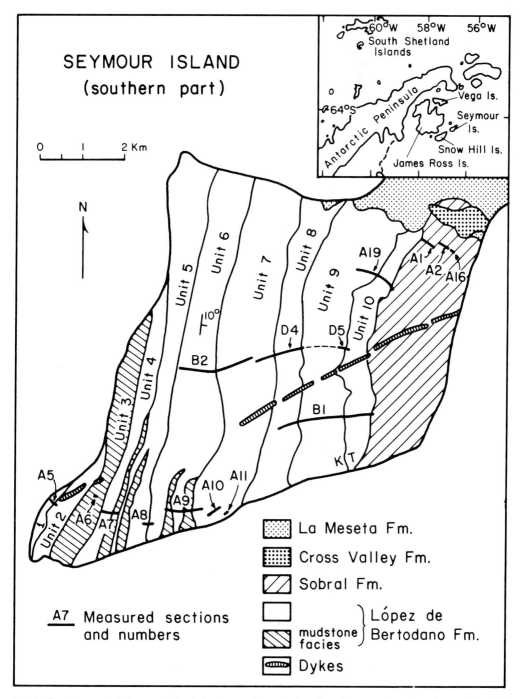

Fig. 1. Southern part of Seymour Island showing geology and location of measured sections. Some of the 510 samples were collected from additional locations in upper unit 9 and unit 10. Lithological units of López de Bertodano Formation after Macellari (1988).

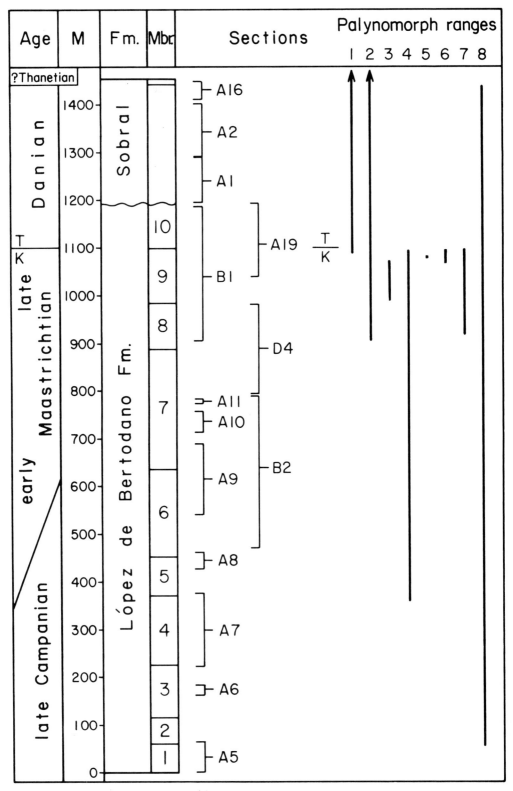

Fig. 2. Diagram to show age and stratigraphic position of sections, and stratigraphic ranges of eight selected palynomorph taxa. 1, *Clavifera triplex*; 2, *Azolla* spp.; 3 *Bombacacidites* spp.; 4, *Cranwellia striata*; 5, *Anacolosidites sectus*; 6, *Beaupreaidites* spp.; 7, *Cupanieidites orthoteichus*; 8, *Polycolpites langstonii*.

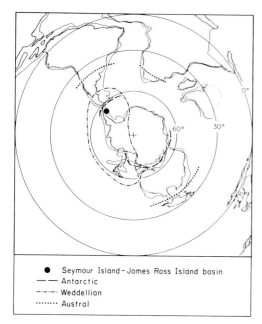

●   Seymour Island–James Ross Island basin
– –   Antarctic
–·–·–   Weddellian
········   Austral

Fig. 3. Approximate limits of Antarctic, Weddellian and Austral Provinces for Campanian to Paleogene palynofloras. Reconstruction after Smith *et al.* (1981) for 80 Ma.

South America (including the Falkland Plateau), West Antarctica, New Zealand and southeastern Australia.

(4) Austral species. By the Late Cretaceous, the 'reduced Gondwana' Austral province embraced the southern part of South America (including Chubut, Argentina), the southernmost tip of Africa, and Australia (and areas between). Many authors have discussed the character of this province. It is equivalent to, for example, the '*Proteacidites-Nothofagidites* paleophytogeoprovince' of Srivastava (1978). It is characterized by the southern podocarpaceous conifers, and/or *Nothofagus*, and/or other typically austral angiosperms (e.g. Proteaceae, Myrtaceae). During the Neogene, some of these taxa spread northward into New Guinea–New Caledonia–Indomalaysia, and farther into Africa.

(5) Cosmopolitan species. As might be expected for primarily planktonic marine organisms, only a small proportion (<20%) of the dinoflagellate cyst flora was endemic to the James Ross Island basin during the Late Cretaceous. Increasing numbers of endemic species and species with Weddellian affinities appear through the Paleogene (see also Wrenn & Hart 1988), possibly due to changing

palaeogeography, ocean currents and greater isolation of the Antarctic region.

Among terrestrial palynomorphs, most gymnosperm species have Weddellian to Austral distribution patterns. Cryptogam spores show Weddellian to cosmopolitan distributions, although some simple spore morphotypes are produced by several different plant taxa (including different families). The cosmopolitan distribution of such spores thus may not reflect a cosmopolitan distribution for their parent plants.

The angiosperm record is provincial with mainly endemic Seymour Island–James Ross Island and Weddellian species. Weddellian species prevailed during the late Campanian, but endemic species became predominant as the continents drifted apart toward the end of the Cretaceous, and through the Paleocene–Eocene. Angiosperm species diversity increased through the latest Cretaceous, from approximately 33 pollen species in the oldest late Campanian samples on Seymour Island, to 44 species near the end of the Campanian, 50 species in the early Maastrichtian, and 59 species in the late Maastrichtian (Askin, unpublished data). At the end of the Cretaceous, almost 50% of the angiosperm pollen species found on Seymour Island were apparently endemic Seymour Island–James Ross Island or Antarctic species. Relative abundances of these endemic species increased through the Paleogene, with a corresponding decrease of Weddellian species.

Observations on the distribution of selected palynomorph taxa

Dinoflagellate cysts

First (and last) occurrences of many dinoflagellate cyst species differ (Dettmann & Thomson 1987; Askin 1988; Pirrie & Riding 1988) among the James Ross Island basin, New Zealand, and southeastern Australia, although these areas were relatively close geographically, and lay at approximately the same palaeolatitude in the Late Cretaceous (Fig. 3). Furthermore, there was some differentiation in composition of the dinoflagellate cyst floras between the James Ross Island basin and Australasian sectors of the Weddellian Province. This is exemplified by related, but consistently distinguishable, dinoflagellate cyst species in upper Campanian–Maastrichtian rocks, such as *Manumiella* 'sp. 3' (Askin 1988) on Seymour Island and '*Isabelidinium*' *haumuriense* in New Zealand

112   R. A. ASKIN

(Wilson 1984). Another related, though bizarre form, *Satyrodinium bengalense*, was described from Campanian sediments in the Bay of Bengal (Lentin & Manum 1986), then much closer to the East Antarctic coast.

Reasons for these range and compositional differences are not clear, but must lie in the configuration of West Antarctic palaeogeography, oceanic circulation patterns and facies differences. Zinsmeister (1979, 1982) depicted the James Ross Island basin in free oceanic contact with the distal part of the Weddellian Province. In his reconstruction, West Antarctica consisted of a series of islands surrounded by shallow

interconnected seaways. Case (1988) and Case *et al.* (1987), however, proposed continuous land through the Antarctic Peninsula−West Antarctica for at least parts of the latest Cretaceous to the Eocene. This would constitute a geographical barrier to marine forms.

Cryptogam spores

The spores of cryptogams (mosses, liverworts, lycopods, ferns) occur throughout the Seymour Island succession. Heterochronous first appearance trends (Figs 4(1) and 4(2)) for certain fern taxa, like *Clavifera triplex* (Bolkhovitina)

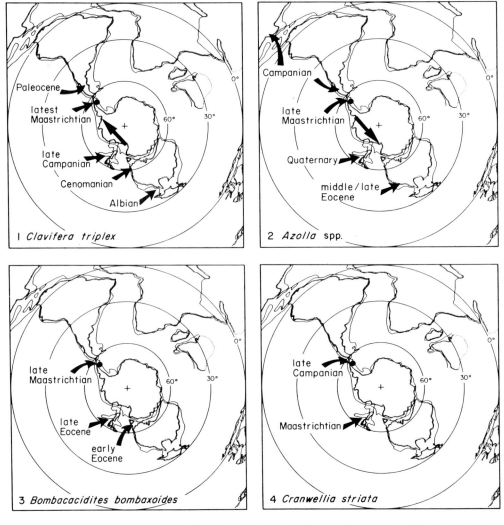

Fig. 4. Time of first occurrences of (**1**) *Clavifera triplex*; (**2**) *Azolla* spp.; (**3**) *Bombacacidites bombaxoides*; (**4**) *Cranwellia striata*. Areas indicated on diagrams are generalized areas rather than precise locations, and additional younger occurrences are not noted. These data are based on published sources cited in the text.

Bolkhovitina spores from a gleicheniaceous fern (Fig. 5(2)) and massulae of *Azolla*-type salviniaceous ferns (Fig. 5(1)), provide evidence that the Antarctic Peninsula acted as a dispersal corridor in both directions between South America and Australasia (Askin, in prep.). Dettmann (1986, 1989) discussed the origin and dispersal of other older Cretaceous ferns.

Gymnosperm pollen

Pollen of podocarpaceous conifers that originated in high southern latitudes, and were well-established there by the Late Cretaceous (Dettmann 1989), are abundant throughout the late Campanian−Paleocene of Seymour Island. They include *Phyllocladidites mawsonii* Cookson ex Couper (the extant plant, *Lagarostrobus franklinii*, grows in the western Tasmanian rainforest) and other '*Phyllocladidites*' types, plus pollen related to extant *Dacrydium*, *Dacrycarpus*, *Podocarpus* and *Microcachrys* species.

Other gymnosperm pollen species found on Seymour Island include *Cycadopites* spp. (Cycadales or Ginkgoales) and *Araucariacites australis* Cookson, which are long-ranging geographically and stratigraphically, and *Ephedra*

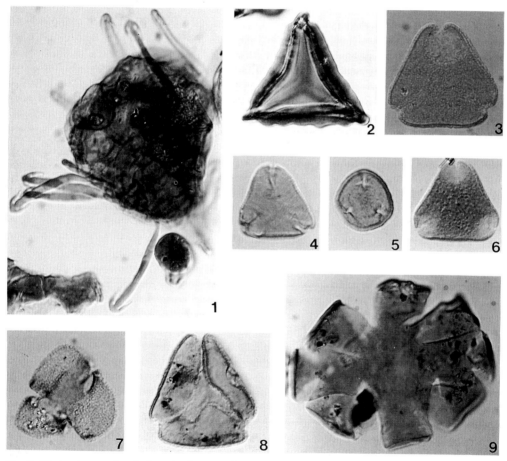

Fig. 5. Photomicrographs of selected palynomorphs. Magnification approximately × 1000. Sample and slide number, National Museum of Natural History (Washington, DC) collection number (prefix USNM), and England Finder references are provided for each specimen. (**1**) *Azolla* sp. massula, B1−92/1, USNM 435353, D39; (**2**) *Clavifera triplex* (Bolkhovitina) Bolkhovitina, A19−12/2, USNM 435354, N32/1; (**3**) *Beaupreaidites elegansiformis* Cookson, B1−76/1, USNM 435355, Q37; (**4**) *Cranwellia striata* (Couper) Srivastava, A19−02/1, USNM 435356, H17/2; (**5**) *Anacolosidites sectus* Partridge, B1−76/1, USNM 435357, N48; (**6**) *Beaupreaidites verrucosus* Cookson, A19−12/1, USNM 435358, K26; (**7**) *Bombacacidites bombaxoides* Couper, B1−66/1, USNM 435359, G25/2; (**8**) *Cupanieidites orthoteichus* Cookson & Pike, B1−103/1, USNM 435360, V18; (**9**) *Polycolpites langstonii* Stover, A19−13/1, USNM 435361, J49/4.

notensis Cookson, which first appeared during the Maastrichtian throughout the Weddellian Province.

Angiosperm pollen

Many angiosperm pollen species display heterochronous first appearance trends. Some have modern counterparts restricted to mid- to low latitudes, as outlined below. Additional taxa represented, but not discussed here, in the upper Campanian to Paleocene pollen record of Seymour Island include species of Liliaceae, Aquifoliaceae, Casuarinaceae, Ericales, Gunneraceae and Myrtaceae.

Bombacaceae. A small ($28\,\mu m$) specimen of *Bombacacidites bombaxoides* Couper (Fig. 5(7)) and similar specimens (*Bombacacidites* sp., more rounded form with finer reticulum) occur in the Seymour Island upper Maastrichtian (sections A19 and B1). First occurrences of *B. bombaxoides* (Fig. 4(3)) are early Eocene in southeastern Australia (Stover & Partridge 1973) and late Eocene in New Zealand (Mildenhall 1980). In North America and northern South America other species of *Bombacacidites* appeared in the Paleocene (e.g. Anderson 1960; Frederickson 1979; Muller *et al.* 1987). Wolfe (1975) discussed occurrences of *Bombax*-type pollen from the early Maastrichtian, and a possible early form (small, unsculptured) from the Campanian of eastern USA. He suggested an eastern North American origin for the Bombacaceae. *Bombax*-type plants grow today in tropical America, Africa, and Indomalaysia to northern Australia.

Fagaceae. Diverse southern beech (*Nothofagus*) pollen occur throughout the Campanian to Paleocene succession of Seymour Island. Despite the potential over-representation of *Nothofagus* species in the fossil record (e.g. Hill & MacPhail 1983), they are usually not abundant in these units. They are more abundant in the Eocene palynofloras of Seymour Island. All three groups of southern beeches (*brassii*, *fusca* and *menziesii*) were established by the late Campanian. Their origin in the southern high latitudes and dispersal are discussed by Dettmann (1989). *Nothofagus* spp. are widespread today throughout the southern South America–New Zealand–southeast/east Australia–New Guinea–New Caledonia region.

Loranthaceae. *Cranwellia striata* (Couper) Srivastava pollen (Fig. 5(4)) appear in latest Campanian sediments (section A7) on Seymour Island. Reported first occurrence of this species in New Zealand (Fig. 4(4)) is in the Maastrichtian (Mildenhall 1980). *Cranwellia striata* pollen are not direct ancestors of late Eocene-modern New Zealand *Elytranthe* (Couper 1960; Mildenhall 1978, 1980), and North American and Russian specimens assigned to this species are unrelated to both (Mildenhall 1978). Modern southern loranthaceous plants include parasitic to hemiparasitic creepers in Australasia (south to cool temperate Stewart Island, N. Z.), South America (to southern Chile), the tropical Pacific and Indomalaysia.

Olacaceae. A single specimen of *Anacolosidites sectus* Partridge (Fig. 5(5)) was found in latest Maastrichtian (B1) sediments, pre-dating (Fig. 6(1)) the Eocene occurrences of this distinctive species in southeastern Australia (Stover & Partridge 1973), southwestern Australia (Stover & Partridge 1982) and Queensland (Foster 1982). Other species of *Anacolosidites* are found in widely-separated Maastrichtian and Paleocene rocks, with the Maastrichtian records restricted to northern localities (summarized in Muller 1981) far removed from Seymour Island. Where this family originated remains in question. The North American Maastrichtian pollen *Libopollis* was suggested as a possible ancestral type to some *Anacolosidites* species (Farabee 1986). Extant *Anacolosa* grows in Indomalaysia, tropical Pacific and Africa-Madagascar. These occurrences were interpreted by Truswell *et al.* (1987) as relict from a formerly widespread distribution.

Proteaceae. Various proteaceous taxa were well-established in the Antarctic Peninsula area by the late Campanian (e.g. Askin 1983; Dettmann & Thomson 1987; Dettmann & Jarzen 1988; Dettmann 1989), including species of *Proteacidites*, *Propylipollis* and *Cranwellipollis*. A small specimen of *Beaupreaidites verrucosus* Cookson (22 μm, with poorly developed verrucae, Fig. 5(6)), and rare *B. elegansiformis* Cookson (Fig. 5(3)) occur in the latest Maastrichtian (A19, B1) of Seymour Island, and the latter species also occurs in the Seymour Island Eocene. *B. verrucosus* (Fig. 6(2)) first occurs in the Eocene in New Zealand (Raine 1984) and southern Australia (Stover & Partridge 1973, 1982). *B. elegansiformis* appeared earlier, in the late Campanian and Maastrichtian of the New Zealand–southeastern Australia region (Wilson 1975; Pocknall & Crosbie 1988; Dettmann & Jarzen

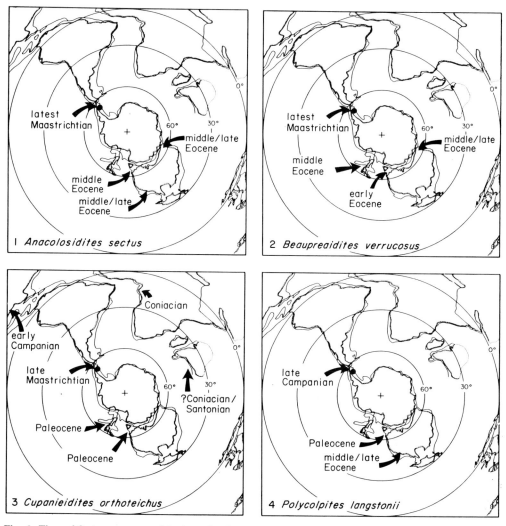

Fig. 6. Time of first occurrences of 1. *Anacolosidites sectus*; 2. *Beaupreaidites verrucosus*; 3. *Cupanieidites orthoteichus*; 4. *Polycolpites langstonii*. Sources cited in text.

1988). *Beauprea* grows today in New Caledonia. Species of *Peninsulapollis*, common throughout the Seymour Island Campanian–Paleocene, are also possibly related to *Beauprea* (Dettmann & Jarzen 1988). The place of origin of *Beauprea*-type proteaceous plants during the Campanian was suggested as the Australasian–Antarctic region (Pocknall & Crosbie 1988; Dettmann & Jarzen 1988).

Sapindaceae (Cupanieae tribe). Pollen assignable to *Cupanieidites orthoteichus* Cookson & Pike (Fig. 5(8)) occur in the Seymour Island late Maastrichtian (B1). This pollen type, which includes *C. major* Cookson & Pike and *C. reticularis* Cookson & Pike, appeared earlier in

the Cretaceous (Coniacian) in Africa and India (summarized in Muller 1981), ranging into North America in the early Campanian and Maastrichtian (e.g. Drugg 1967; Nichols *et al.* 1982). First appearances (Fig. 6(3)) of *C. orthoteichus* in southeastern Australia (Harris 1965) and New Zealand (Raine 1984) were in the Paleocene. Extant *Cupania* and related types grow throughout tropical America, Africa, Indomalaysia to Australia and tropical Pacific.

Polycolpate pollen, (?Pedaliaceae). Pollen of *Polycolpites langstonii* Stover (Fig. 5(9)) occur in the Seymour Island upper Campanian (A5) to Paleocene, pre-dating (Fig. 6(4)) their Paleocene first occurrence in southeastern

Australia (Stover & Partridge 1973). This may be the same pollen type referred to Pedaliaceae (similar to *Josephinia* of Australia and Kenya) by Cranwell (1969). Cranwell's drawing depicts more colpi (11) than are typical (8–9) for *P. langstonii* noted here.

Palaeoclimatic inferences

Some of the above examples are heterochronous taxa whose extant relatives (*Bombax, Anacolosa, Beauprea, Cupania*) live in low to mid-latitudes (in the tropics and subtropics). Their rarity in the well-studied New Zealand Cenozoic fossil record, in essentially cool- to warm-temperate assemblages, led Mildenhall (1980) to caution that the New Zealand fossil floras should not be regarded as tropical. He suggested that the floras perhaps reflect high humidity, but not necessarily high temperatures. The same is probably true for the latest Cretaceous floras of Seymour Island, where ample moisture has been documented (e.g. Francis 1986; Askin in prep.) and modelled (e.g. Parrish *et al.* 1982), and where 'tropical' and 'cool temperate' species occur together. These transported assemblages represent plants from a variety of habitats, from wetlands to more arid areas, and low to high altitude. Furthermore, some extant, low latitude forms (for example, *Nothofagus brassii* group, some *Cupania, Anacolosa, Beauprea*) grow in mid- to high altitudes, indicating warm to mild rather than high temperatures.

The presence of these 'tropical' species on Seymour Island suggests, however, that the climate may have been warmer in the late Maastrichtian than the 'cool temperate' conditions often ascribed to the Antarctic Peninsula fossil rainforests. The apparent disappearance of most of these 'tropical' forms from the Seymour Island Paleocene vegetation is consistent with cooling temperatures at the end of the Cretaceous or in the earliest Tertiary (Fig. 2, *N.B.* Some of these forms are extremely rare and their ranges on Fig. 2 are unlikely to represent true ranges of the parent plants). No significant decrease (or increase) in moisture is yet discernible for the Seymour Cretaceous/Tertiary flora. Cranwell (1969) suggested warm temperate conditions for the Seymour Island– Snow Hill Island Late Cretaceous. Seasonally warm and dry conditions were suggested for the latest Cretaceous of the nearby South Shetland Islands (summarized in Birkenmajer 1985), with cooling somewhat later in the Tertiary. Francis (1986) suggested warm conditions ('warm/cool-temperate') for the Cretaceous of Seymour

Island–James Ross Island basin, with possibly deteriorating conditions in the Paleocene. Recently obtained palaeotemperature data from the nearby marine record (ODP site 690, Maud Rise; Stott & Kennett (& ODP Leg 113 Scientific Party) 1988) indicate a warming trend for surface waters through the latest Maastrichtian with a cooling event immediately preceding the Cretaceous/Tertiary boundary. Paleocene surface water temperatures were cooler than in the Late Cretaceous.

Comparisons with climatic preferences of extant plants must be made with caution. It is probable that plant taxa dispersed along routes of similar climate. In some cases, however, extinct ancestors may have lived in somewhat different conditions, and their modern counterparts may represent relatively recent adaptations, or relict occurrences. Global climate patterns today are very different from, and globally less equable than, in the Late Cretaceous, when temperature gradients from poles to equator were much lower. The modern restricted distribution of *Anacolosa*, for example, differs from the widespread occurrences of *Anacolosidites* species in the Maastrichtian–Paleocene (*c.* 65°N through the palaeotropics to *c.* 65°S palaeolatitude). These fossil occurrences may not reflect widely varying temperature (or moisture) norms or extremes. Furthermore, different light regimes seem not to adversely affect plant growth and fertility, given adequate warmth and moisture (Axelrod 1984; Creber & Chaloner 1985). Both the Maastrichtian Seymour Island plants and their living low latitude relatives seem to have moist, equable climates in common.

Presence of 'tropical' plant species, and a moist, warm, equable climate in the Antarctic Peninsula area during the latest Cretaceous, imply that Antarctica may have served as a dispersal pathway more frequently than previously thought. There is no longer a need to invoke a northern 'subtropical–tropical route' (Raven & Axelrod 1974) for certain plant taxa, and a southern route might solve some of the problems associated with timing of migration and palaeogeography.

Besides serving as a dispersal corridor, Antarctica was also a cradle for new plant taxa. Unusual climatic conditions (e.g. polar winter/summer light cycles, low light angle, relatively mild climate, ample moisture) in the high southern latitudes of the Late Cretaceous–Paleogene apparently provided the 'stress impetus' for species diversification in a wide variety of plants and animals. Important plant taxa apparently evolved in the Antarctic region, and now form significant parts of the southern vegetation outside of Antarctica. Although families such as Bombacaceae and Olacaceae may have originated elsewhere, it seems that some branches diversified in the southern high latitudes, with the subsequent dispersal of species northward to lower latitudes. Species that grew on the Antarctic Peninsula at the end of the Cretaceous occurred during the Eocene in warm mid-latitudes in Australia and New Zealand, after these Gondwana fragments had drifted away from Antarctica. Modern low latitude representatives of these plants may provide us with further insight into Late Cretaceous vegetation and climate.

This manuscript was improved by critical reading by the reviewers and S. R. Jacobson, and comments by R. W. J. M. van der Ham. I also thank J. A. Case, M. E. Dettmann and E. M. Truswell for helpful discussion. Support for this project was provided by National Science Foundation grants DPP 8314186, 8719840 and 8716484.

References

ANDERSON, R. Y. 1960. Cretaceous–Tertiary palynology, eastern side of the San Juan Basin, New Mexico. *New Mexico Institute of Mining and Technology, Memoir*, **6**, 1–58.

ASKIN, R. A. 1983. Campanian palynomorphs from James Ross and Vega Island, Antarctic Peninsula. *Antarctic Journal of the United States*, **18**(5), 63–64.

—— 1988. Campanian to Paleocene palynological succession of Seymour and adjacent islands, northeastern Antarctic Peninsula. *In*: FELDMANN, R. M. & WOODBURNE, M. O. (eds) *Geology and Paleontology of Seymour Island, Antarctic Peninsula*. Geological Society of America Memoir, **169**, 131–153.

—— In prep. Cryptogam spores from the upper Campanian and Maastrichtian of Seymour Island, Antarctica.

AXELROD, D. I. 1984. An interpretation of Cretaceous and Tertiary biota in polar regions. *Palaeogeography, Palaeoclimatology, Palaeoecology*, **45**, 105–147

BIRKENMAJER, K. 1985. Onset of Tertiary continental glaciation in the Antarctic Peninsula sector (West Antarctica). *Acta Geologica Polonica*, **35**, 1–31.

CASE, J. A. 1988. Paleogene floras from Seymour Island, Antarctic Peninsula. *In*: FELDMANN, R. M. & WOODBURNE, M. O. (eds) *Geology and Paleontology of Seymour Island, Antarctic Peninsula*. Geological Society of America Memoir, **169**, 523–530.

——, WOODBURNE, M. O & CHANEY, D. S. 1987. A gigantic phororhacoid(?) bird from Antarctica. *Journal of Paleontology*, **61**, 1280–1284.

COUPER, R. A. 1960. New Zealand Mesozoic and Cainozoic plant microfossils. *New Zealand Geological Survey, Paleontological Bulletin* **32**, 1–87.

CRANWELL, L. M. 1964. Antarctica: cradle or grave for its *Nothofagus*. *In*: CRANWELL, L. M. (ed.) *Ancient Pacific Floras, the Pollen Story*. University of Hawaii Press, Honolulu, 87–93.

—— 1969. Palynological intimations of some pre-Oligocene Antarctic climates. *In*: VAN ZINDEREN BAKKER, E. M. (ed.), *Palaeoecology of Africa*. S. Balkema, Cape Town, 1–19.

CREBER, G. T. & CHALONER, W. G. 1985. Tree growth in the Mesozoic and Early Tertiary and the reconstruction of palaeoclimates. *Palaeogeography, Palaeoclimatology, Palaeoecology*, **52**, 35–60.

DARWIN, C. 1859. *On the origin of species by means of natural selection*. Murray, London.

DETTMANN, M. E. 1986. Significance of the Cretaceous-Tertiary spore genus *Cyatheacidites* in tracing the origin and migration of *Lophosoria* (Filicopsida). *Special Papers in Palaeontology*, **35**, 63–94.

—— 1989. Antarctica: Cretaceous cradle of austral temperate rainforests? *In*: CRAME, J. A. (ed.) *Origins and Evolution of the Antarctic Biota*: Geological Society, London, Special Publication, **47**, 89–105.

—— & JARZEN, D. M. 1988. Angiosperm pollen from uppermost Cretaceous strata of south-eastern Australia and the Antarctic Peninsula. *Association of Australasian Palaeontologists Memoir*, **5**, 217–237.

—— & THOMSON, M. R. A. 1987. Cretaceous palynomorphs from the James Ross Island area — a pilot study. *British Antarctic Survey Bulletin*, **77**, 13–59.

DOBZHANSKY, T. 1950. Evolution in the tropics. *American Scientist*, **38**, 209–221.

DRUGG, W. S. 1967. Palynology of the upper Moreno Formation (Late Cretaceous-Paleocene), Escarpado Canyon, California. *Palaeontographica*, **120B**, 1–71.

ELLIOT, D. H. 1988. Tectonic setting and evolution of the James Ross Island basin, northern Antarctic Peninsula. *In*: FELDMANN, R. M. & WOODBURNE, M. O. (eds) *Geology and Paleontology of Seymour Island, Antarctic Peninsula*. Geological Society of America Memoir, **169**, 541–555.

FARABEE, M. J. 1986. Numerical analysis of a suspected Upper Cretaceous pollen lineage: *Libopollia* to *Anacolosidites* (in part). Abstract, *Palynology*, **10**, 246–247.

FELDMANN, R. M. & TSHUDY, D. M. 1989. Evolutionary patterns in macrurous decapod crustaceans from Cretaceous to early Cenozoic rocks of James Ross Island region, Antarctica. *In*: CRAME, J. A. (ed.) *Origins and Evolution of the Antarctic Biota*, Geological Society, London, Special Publication, **47**, 183–195.

FOSTER, C. B. 1982. Illustrations of Early Tertiary

(Eocene) plant microfossils from the Yaamba Basin, Queensland. *Geological Survey of Queensland Publication*, **381**, 1–32.

FRANCIS, J. E. 1986. Growth rings in Cretaceous and Tertiary wood from Antarctica and their palaeoclimatic implications. *Palaeontology*, **29**, 665–684.

FREDERIKSEN, N. O. 1979. Paleogene sporomorph biostratigraphy, northeastern Virginia. *Palynology*, **3**, 129–167.

GRUNOW, A. M., KENT, D. V. & DALZIEL, I. W. D. 1987. Mesozoic evolution of West Antarctica and the Weddell Sea Basin: new paleomagnetic constraints. *Earth and Planetary Science Letters*, **86**, 16–26.

HARRIS, W. K. 1965. Basal Tertiary microfloras from the Princetown area, Victoria, Australia. *Palaeontographica*, **115 B**, 75–106.

HARWOOD, D. M. 1988. Upper Cretaceous and lower Paleocene diatom and silicoflagellate biostratigraphy of Seymour Island, eastern Antarctic Peninsula. *In*: FELDMANN, R. M. & WOODBURNE, M. O. (eds) *Geology and Paleontology of Seymour Island, Antarctic Peninsula*. Geological Society of America Memoir, **169**, 55–129.

HICKEY, L. J., WEST, R. M., DAWSON, M. R. & CHOI, D. K. 1983. Arctic terrestrial biota: paleomagnetic evidence of age disparity with mid-northern latitudes during the Late Cretaceous and Early Tertiary. *Science*, **221**, 1153–1156.

HILL, R. S. & MACPHAIL, M. K. 1983. Reconstruction of the Oligocene vegetation at Pioneer, northeast Tasmania. *Alcheringa*, **7**, 281–299.

HOOKER, J. D. 1847. *The botany of the Antarctic voyage of H. M. discovery ships* Erebus *and* Terror *in the years 1839–1843. 1. Flora Antarctica*. Reeve Bros., London. Reprinted 1963, J. Cramer, Weinheim.

HUBER, B. T. 1988. Upper Campanian–Paleocene foraminifera from the James Ross Island region, Antarctic Peninsula. *In*: FELDMANN, R. M. & WOODBURNE, M. O. (eds) *Geology and Paleontology of Seymour Island, Antarctic Peninsula*. Geological Society of America Memoir, **169**, 163–252.

LENTIN, J. K. & MANUM, S. B. 1986. A new peridinioid dinoflagellate from Campanian sediments recovered from DSDP Leg 22, Site 217, Indian Ocean. *Palynology*, **10**, 111–116.

MACELLARI, C. E. 1988. Stratigraphy, sedimentology and paleoecology of Upper Cretaceous/Paleocene shelf-deltaic sediments of Seymour Island (Antarctic Peninsula). *In*: FELDMANN, R. M. & WOODBURNE, M. O. (eds) *Geology and Paleontology of Seymour Island, Antarctic Peninsula*. Geological Society of America Memoir, **169**, 25–53.

MILDENHALL, D. C. 1978. *Cranwellia costata* n.sp. and *Podosporites erugatus* n.sp. from Middle Pliocene (? Early Pleistocene) sediments, South Island, New Zealand. *Journal of the Royal Society of New Zealand*, **8**, 253–274.

—— 1980. New Zealand Late Cretaceous and Cenozoic plant biogeography: a contribution.

Palaeogeography, Palaeoclimatology, Palaeoecology, **31**, 197–233.

MULLER, J. 1981. Fossil records of extant angiosperms. *The Botanical Review*, **47**, 1–142.

——, DI GIACOMO, E. DE & VAN ERVE, A. W. 1987. A palynological zonation for the Cretaceous, Tertiary, and Quaternary of northern South America. *American Association of Stratigraphic Palynologists Contributions Series*, **19**, 7–76.

NICHOLS, D. J., JACOBSON, S. R. & TSCHUDY, R. H. 1982. Cretaceous palynomorph biozones for the central and northern Rocky Mountain region of the United States. *In*: POWERS, R. B. (ed.) *Geologic Studies of the Cordilleran Thrust Belt, Volume II*. Rocky Mountain Association of Geologists, Denver, 721–733.

PARRISH, J. T., ZIEGLER, A. M. & SCOTESE, C. R. 1982. Rainfall patterns and the distribution of coals and evaporites in the Mesozoic and Cenozoic. *Palaeogeography, Palaeoclimatology, Palaeoecology*, **40**, 67–101.

PIRRIE, D. & RIDING, J. B. 1988. Sedimentology, palynology and structure of Humps Island, northern Antarctic Peninsula. *British Antarctic Survey Bulletin*, **80**, 1–19.

POCKNALL, D. T. & CROSBIE, Y. M. 1988. Pollen morphology of *Beauprea* (Proteaceae): modern and fossil. *Review of Palaeobotany and Palynology*, **53**, 305–327.

RAINE, J. I. 1984. Outline of a palynological zonation of Cretaceous to Paleogene terrestrial sediments in West Coast region, South Island, New Zealand. *New Zealand Geological Survey Report*, **109**, 1–82.

RAVEN, P. H. & AXELROD, D. I. 1974. Angiosperm biogeography and past continental movements. *Annals of the Missouri Botanical Garden*, **61**, 539–673.

SKOTTSBERG, C. 1953. Influence of the Antarctic continent on the vegetation of southern lands. *Proceedings, 7th Pacific Science Congress (1949)*, **5**, 92–9.

SMITH, A. G., HURLEY, A. M. & BRIDEN, J. C. 1981. *Phanerozoic paleocontinental world maps*. Cambridge University Press, Cambridge.

SRIVASTAVA, S. K. 1978. Cretaceous spore-pollen floras: a global evaluation. *Biological Memoirs*, **3**, 2–130.

STOTT, L. D. & KENNETT, J. P. (& ODP Leg 113 Scientific Party) 1988. Cretaceous/Tertiary boundary in the Antarctic: climatic cooling precedes biotic crisis. *Geological Society of America, Abstracts with Programs*, **20**(7), 1988 Annual Meeting, A251.

STOVER, L. E. & PARTRIDGE, A. D. 1973. Tertiary and Late Cretaceous spores and pollen from the Gippsland Basin, southeastern Australia. *Royal Society of Victoria, Proceedings* **85**, 237–286.

—— & —— 1982. Eocene spore-pollen from the Werillup Formation, Western Australia. *Palynology*, **6**, 69–96.

TRUSWELL, E. M. 1983. Recycled Cretaceous and Tertiary pollen and spores in Antarctic marine sediments: a catalogue. *Palaeontographica*, **186B**,

121–174.

—— & DREWRY, D. J. 1984. Distribution and provenance of recycled palynomorphs in surficial sediments of the Ross Sea, Antarctica. *Marine Geology*, **59**, 187–214.

——, KERSHAW, A. P. & SLUITER, I. R. 1987. The Australian–south-east Asian connection: evidence from the palaeobotanical record. *In*: WHITMORE, T. C. (ed.) *Biogeographical evolution of the Malay Archipelago*. Oxford Monographs on Biogeography, **4**, Oxford Science Publications, 32–49.

WILSON, G. J. 1975. Palynology of deep-sea cores from DSDP Site 275, Southeast Campbell Plateau. *In*: KENNETT, J. P., HOUTZ, R. E., *et al.* (eds) *Initial Reports of the Deep Sea Drilling Project* **29**, U.S. Government Printing Office, Washington, D.C., 1031–1035.

—— 1984. Some new dinoflagellate species from the New Zealand Haumurian and Piripauan Stages (Santonian–Maastrichtian, Late Cretaceous). *New Zealand Journal of Botany*, **22**, 549–556.

WOLFE, J. A. 1975. Some aspects of plant geography of the northern hemisphere during the late Cretaceous and Tertiary. *Annals of the Missouri Botanic Gardens*, **62**, 264–279.

WOODBURNE, M. O. & ZINSMEISTER, W. J. 1984. The first land mammal from Antarctica and its biogeographic implications. *Journal of Paleontology*, **58**, 913–948.

WRENN, J. H., & HART, G. F. 1988. Paleogene dinoflagellate cyst biostratigraphy of Seymour Island, Antarctica. *In*: FELDMANN, R. M. & WOODBURNE, M. O. (eds) *Geology and Paleontology of Seymour Island, Antarctic Peninsula*. Geological Society of America Memoir, **169**, 321–447.

ZINSMEISTER, W. J. 1979. Biogeographic significance of the late Mesozoic and early Tertiary molluscan faunas of Seymour Island (Antarctic Peninsula) to the final breakup of Gondwanaland. *In*: GRAY, J. & BOUCOT, A. J. (eds) *Historical Biogeography, Plate Tectonics and the Changing Environment*. Oregon State University Press, Corvallis, 349–355.

—— 1982. Late Cretaceous–Early Tertiary molluscan biogeography of the southern Circum-Pacific. *Journal of Paleontology*, **56**, 84–102.

—— 1986. Fossil windfall at Antarctica's edge. *Natural History*, **95**, 60–67.

—— & FELDMANN, R. M. 1984. Cenozoic high latitude heterochroneity of Southern Hemisphere marine faunas. *Science*, **224**, 281–283.

Early Cretaceous biota from the northern side of the Australo–Antarctic rift valley

T. H. RICH[1], P. V. RICH[2], B. WAGSTAFF[2], J. MCEWEN-MASON[2], C. B. DOUTHITT[2,3] & R. T. GREGORY[2]

[1]*Museum of Victoria, 285–321 Russell Street, Melbourne, Victoria 3000, Australia*
[2]*Department of Earth Sciences, Monash University, Clayton, Victoria, 3168, Australia*
[3]*Present address: Finnigan MAT, 355 River Oaks Parkway, San Jose, California 95314, USA*

Abstract: A diverse biota including vertebrates, invertebrates, and plants is known from the Early Cretaceous of southeastern Australia. It is preserved in sediments that accumulated in the rift valley formed as Australia began to separate from Antarctica. As there was no significant terrestrial barrier between the two continents at the time, it is likely that the Early Cretaceous biota of the nearest region of East Antarctica may have been quite similar to that of southeastern Australia.

Dominant among the vertebrates are turtles and at least four hypsilophodontid dinosaur species. Three species of theropods, including *Allosaurus* sp., are the only other dinosaurs represented. Other tetrapods include a few scant traces of plesiosaurs, pterosaurs, a lizard, and a labyrinthodont amphibian.

Southeastern Australia was located well inside the Antarctic Circle of the day. Oxygen isotope studies suggest mean annual temperatures for the area between −5 and 8°C. These low temperatures are concordant with the character of both the flora and invertebrate fauna. The presence of juvenile hypsilophodontids implies that these animals were breeding in the area. The large eyes and optic lobes of the brain of these dinosaurs relative to other ornithopods suggests they may have been adapted for active behaviour under low light conditions.

Late last century W. H. Ferguson, while mapping early Cretaceous rock exposures along the Victorian coast of southeastern Australia, found a lungfish tooth and the claw of a small dinosaur. A. Smith Woodward (1906) compared the claw to *Megalosaurus*. Few terrestrial vertebrates were recovered from these late Mesozoic sediments over the next eighty years, with a turtle (Warren 1969), a possible lizard (Molnar 1980), and a few bird feathers (Talent *et al.* 1966) being the sum total found. Osteichthyians (Waldman 1971) were, however, locally common, which indicated that bone concentrations could be preserved if the conditions were right.

In 1978, new discoveries that yielded dinosaurs, including an allosaurid (Molnar *et al.*, 1981, 1985; Welles 1983), rekindled interest in this area. This was especially so because palaeogeographic reconstructions of southern Australia in the Early Cretaceous, utilizing palaeomagnetic data collected on the continent itself, indicated latitudes within the Antarctic Circle (Fig. 11) perhaps as far south as 85° (Barron 1987; Embleton & McElhinny 1982; Idnurm 1985). Earlier reconstructions, which are still frequently cited and did not have access to data from Australia itself, place the southeastern part of the continent immediately north of the Antarctic Circle (Smith *et al.* 1981; Ziegler *et al.* 1983).

Concerted effort between 1979 and 1987 produced terrestrial vertebrate faunas at several different stratigraphic levels within the Early Cretaceous sequence of this area, ranging from the Valanginian to the Albian (Dettmann 1963, 1986; Wagstaff & McEwen-Mason 1989). Dinosaurs and their associated biota were living and being preserved at very high latitudes, where several months of continuous darkness prevailed, and where temperatures apparently were decidedly milder than at similar latitudes today. The survival of dinosaurs under these conditions for more than 30 Ma (from the Early Cretaceous of Victoria to the Late Cretaceous of Alaska) indicates that the stress of a few months of constant darkness may not directly account for extinction of this group in the late Mesozoic. Therefore, some constraint might be placed on theories that utilize bolide impacts, increased vulcanism, or related mechanisms to explain Late Cretaceous extinctions. Additionally, the endemic and relictual nature of biota, coupled with the preponderance of the large-eyed, possibly large-brained hypsilophodontids, may reflect the polar nature of this assemblage.

From Crame, J. A. (ed.), 1989, *Origins and Evolution of the Antarctic Biota,*
Geological Society Special Publication No. 47, pp. 121–130.

Fig. 1. Map of Australia with study area and location of land and seaways (shaded areas) during the late Early Cretaceous (modified from Frakes *et al.* 1987).

Geological setting and age

Two Early Cretaceous basins occur along the southeastern coast of Australia and extend inland. The Otway Basin to the southwest of Melbourne and the Gippsland Basin to the southeast are separated by the Mornington High (Dettmann 1986; Douglas *et al.* 1976; Drinnan & Chambers 1986). These basins formed as part of the process which led to the onset of rifting between Australia and Antarctica early in the Cretaceous (Veevers 1986).

Up to three kilometres of fluviatile sediments, including mudstones, coals and volcanically derived sandstones, were emplaced in these rift basins on a broad floodplain dominated by braided streams that experienced periodic high flows (Wagstaff 1983; Gill 1977; Lees & Cas pers. comm.). Sand deposition was rapid, as indicated by high energy lithofacies and primary structures (Douglas *et al.* 1976). Rapid deposition may have been related to seasonal meltwater input from the volcanic highlands that probably existed along the Lord Howe Rise during its formative stages (Lees & Cas Pers. Comm.), or else were located at sites unknown closer to the rift valley (Gleadow & Duddy 1981). Aggradation in abandoned channels and lakes represented by mudstones and coals was much slower. Lack of desiccation cracks in most channel deposits implies that even abandoned channels never completely dried out. Most tetrapod fossils were recovered from fluviatile deposits 5–10 m wide, 0–30 cm thick containing numerous clay gall clasts. A few were found in larger channel deposits characterized by cross-bedding and lack of clay gall clasts. In rarer locales, vertebrates were preserved in lacustrine, and in one instance, overbank, deposits where a single footprint establishes that at least one species of dinosaur lived in situ.

The Early Cretaceous age of the rocks of the Otway Group in the Otway Basin and the Strzelecki Group in the Gippsland Basin (the two rock units are sometimes combined as the Korumburra Group) was determined palynologically (Fig. 2) (Dettmann 1963, 1986; Wagstaff & McEwen-Mason 1989; Wagstaff 1983). The tetrapod-bearing sites in the Otway Group of the Otway Basin (Fig. 3) range from latest Aptian to early Albian age, whereas those of the Strzelecki Group in the Gippsland Basin range (Fig. 3) from Valanginian to Aptian (Dettmann 1963, 1986). Fission track ages for the Otway Group place it in the latter half of the Early Cretaceous (Gleadow & Duddy 1981). Such dates for the Koonwarra locality in the Gippsland Basin bracket it between 115 ± 6 Ma and 118 ± 5 Ma (Lindsay 1982). Both groups of dates are concordant with palynological determinations.

New oxygen isotope ratio determinations have been made recently on calcite concretions found in sandstones of the Otway and Strzelecki groups (Fig. 4). These concretions have been interpreted as being predominantly pre-compaction (Gill *et al.* 1977), and probably reflect the $^{18}O/^{16}O$ ratios of the local meteoric water at the time of their formation. To date, the concretions have yielded some of the lightest ^{18}O values ever observed in sedimentary rocks, with observed $\delta^{18}O_{SMOW}$ values as light as 3.6‰ (Gregory *et al.*, in press). The majority of the $\delta^{18}O_{SMOW}$ values cluster around 10.5‰ (Gregory *et al.*, in press), indicating that local meteoric waters were very depleted in ^{18}O, with mean $\delta^{18}O$ values between -14 and -20‰. Comparison with modern data on soil carbonates (Cerling 1984) and meteoric waters (Gat 1980) suggests that the Cretaceous polar environment may have been far from equable, with mean annual air temperatures less than 8°C and perhaps as low as -5°C.

Biotic composition (Fig. 5)

Of special interest in these sediments are three, or perhaps four, species of hypsilophodontid dinosaurs (Rich & Rich 1989). Of these, at least two, or possibly three, are new genera. Only one partial skeleton (that of the smallest genus) is known, whereas all other material is disarticulated. Femora dominate, but some dental and cranial material as well as other postcranial elements are known. Accompanying the one skeleton is a partial skull with an endocast of the brain cavity preserved in dorsal view.

Theropods, comprising at least 3 species, are the only other dinosaur group represented.

	OTWAY BASIN								GIPPSLAND BASIN				
	Dinosaur Cove	Eric The Red	Point Franklin	Point Lewis	Marengo	Skenes Creek	Cumberland River	Punch Bowl	Kilcunda	Harmers Haven	Twin Reefs	Shack Bay	Eagles Nest
POLLEN													
Alisporites grandis	O	O	O	O	O	O	O	O	O	O	O	O	O
Alisporites similis	O	O	O	O	O			O	O	O	O	O	O
Araucariacites australis	O	O	O	O	O	O			O	O	O	O	O
Callialasporites trilobatus								O					
Classopollis chateaunovii	O	O	O	O	O	O			O	O	O	O	O
Cycadopites nitidus	O	O		O	O	O			O	O	O		O
Microcachryidites antarcticus	O	O	O	O	O	O			O	O	O	O	O
Podocarpidites cf. *ellipticus*	O	O	O		O				O	O	O	O	O
Podocarpidites multisemus	O	O							O	O	O		
Trichotomosulcites subgranulatus	O	O	O						O	O	O	O	
Vitreisporites pallidus	O									O			
SPORES													
Aequitriradites spinulosus	O			O	O				O	O		O	O
Aequitriradites verrucosus	O	O	O						O	O		O	O
Baculatisporites comaumensis	O	O	O	O	O	O	O	O	O	O	O	O	O
Balmeisporites holodictyus	O						O						
Biretisporites potoniaei	O	O	O	O									
Biretisporites spectabilis	O	O	O										
Ceratosporites equalis	O	O	O	O	O	O	O	O	O	O	O	O	O
Cicatricosisporites australiensis	O	O	O	O	O	O	O	O	O	O	O	O	O
Cicatricosisporites hughesii				?									
Cicatricosisporites ludbrookiae								O					
Contignisporites cooksoniae								O					
Contignisporites fornicatus				O	?								
Coronatispora perforata								O					
Coronatispora telata								O					
Crybelosporites berberoides								O					
Crybelosporites punctatus	O			O	O								
Crybelosporites striatus	O	O		?	O								
Cyathidites asper	O	O	O	O				O		O			
Cyathidites australis	O	O	O	O	O	O	O	O	O	O	O	O	O
Cyathidites concavus	O	O											
Cyathidites minor	O	O	O	O	O			O	O	O	O	O	O
Cyathidites punctatus	O	O					O	O					
Cyathidites rafaelii	O						O	O					
Cyclosporites hughesii		O	O	O	O				O	O	O		
Dictyophyllidites crenatus								O	O		O	O	O
Dictyophyllidites harrisii				O									O
Dictyosporites complex	O							O		O	O		
Dictyosporites filosus		O						O					
Dictyosporites speciosus	O	O	O	O				O	O	O	O		O
Foraminisporis asymmetricus	O	O	O	O	O		O	O					
Foraminisporis dailyi	O	O	O	O					O	O	O		
Foraminisporis wonthaggiensis	O	O	O	O			O	O	O	O			
Foveosporites canalis	O			O					O	O	O	O	
Foveotriletes parviretus	O								O	O			
Gleicheniidites circinidites	O	O	O	O	O	O	?		O	O	O	O	
Ischyosporites punctatus	O	O							O				
Klukisporites scaberis	O		O	O	O	O			O	O	O	O	O
Laevigatosporites ovatus	O			O	O								
Leptolepidites major	O		O			?		O					O
Leptolepidites verrucatus	O	O	O	O	O			O	O	O	O	O	
Lycopodiacidites asperatus	O	O	O	O	O			O	O	O	O	O	
Microfoveolatisporites canaliculatus	O	O											
Neoraistrikia truncata	O	O	O	O	O			O	O	O	O	O	
Obtusisporites canadiensis													O
Osmundacidites mollis	O	O											
Osmundacidites wellmanii	O			?	O			O					
Perotriletes linearis	O	O	O	O				O	O				
Pilosisporites notensis	O	O		O	O				O	O	O		
Pilosisporites parvispinosus				O					O		O	O	
Polycingulatisporites densatus													O
Punctatosporites scabratus	O				O								
Reticulatisporites pudens	O	O	O	O		?		O	O	O	O	O	O
Reticuloidosporites arcus		O											
Retitriletes austroclavatidites	O	O	O	O				O	O	O			O
Retitriletes circolumensus	O		O	O	O			O	O	O	O	O	
Retitriletes eminulus	O	O	O	O		?	O	?	O	O	O	O	O
Retitriletes facetus	O	O	O	O				O	O	O	O	O	
Retitriletes nodosus	O	O	O	O	O			O	O	O	O	O	O
Retitriletes reticulumsporites	O	O	O		O				O		O	O	
Retitriletes watherooensis													O
Sestrosporites pseudoalveolatus	O		O									O	O
Stereisporites antiquasporites	O	O	O	O	O	O	O	O	O	O	O	O	O
Stereisporites pocockii	O												
Stoverisporites lunaris									O				O
Trilites cf. *T. tuberculiformis*	O	O		O									
Triporoletes radiatus	O												
Triporoletes reticulatus	O		O	O				O					
Triporoletes simplex	O												
Velosporites triquetrus	O												?
ALGAE													
Leiosphaeridia spp.	O	O	O	O	O	O			O	O	O	O	O
Michrystridium spp.					O			O					O
Microfasta evansii													?
Schizosporis reticulatus									O	O	O	O	
No. of samples the species list is based on	11	4	4	4	5	1	4	4	38	3	11	6	13

○ Species identification positive ? Species identification tentative

Allosaurus sp. (Molnar *et al.* 1981, 1985; Welles 1983) is the largest of the 3, but the Australian specimen was a small individual, presumably a juvenile, that would have been between 4 and 5 m long.

Associated with these dinosaurs was a varied biota of vertebrates, invertebrates, and plants comprising more than 150 taxa (Fig. 2). Forests of both deciduous and evergreen plants were dominated by ginkgoes (*Ginkgo australis*), podocarps, and araucarian conifers, whereas the understory contained *Taeniopteris daintreei*, pentoxylaleans, ferns, sphenopsids, and both ground and epiphytic bryophytes (Dettmann 1986; Drinnan & Chambers 1986). Sclerophyllous ferns characterized forest fringes, and *Lycopodium* and sphagnalean mosses were found in more open moorlands (Dettman 1986). *Isoetales*, hepatics, and algae dominated aquatic environments, reflecting the moderately nutrient rich, circulating nature of the waters containing these associations (Dettmann 1986).

Invertebrates have also been preserved in these Early Cretaceous sediments of southeastern Australia. More than 80 species are known, primarily from the Koonwarra locality in the Gippsland Basin, most of which are insects. Ostracodes, syncarids, anostracans, cladocerans (all crustaceans), spiders, possibly earthworms, freshwater bryozoans, and unionoid bivalves were also present (Jell & Duncan 1986). Most diverse among the twelve orders of insects were the hemipterans, coleopterans, and dipterans. Numerically dominant were immature individuals of the aquatic Ephemeroptera and Diptera.

Plant (Dettmann 1986; Drinnan & Chambers 1986) and invertebrate assemblages (Jell & Duncan 1986) are consistent with a cool, humid climate, and the presence of distinct tree rings reflects some seasonality. Our work on the Otway and Strzelecki group vertebrates (Rich & Rich 1989) indicates that at present they offer little palaeoclimatic information, despite claims made by previous workers (Douglas & Williams 1982).

Vertebrate faunas include a variety of fish (dipnoans, actinopterygians), turtles, lepidosaurs, pterosaurs, plesiosaurs (presumably fresh water; Fig. 6), birds, and possibly a labyrinthodont amphibian (Fig. 7). Apart from dinosaurs, turtles are the most common specimens in the terrestrial faunas, being represented

Fig. 2. Occurrence of pollen and spores critical to the dating of the dinosaur-bearing Early Cretaceous rocks in the southeast of Australia.

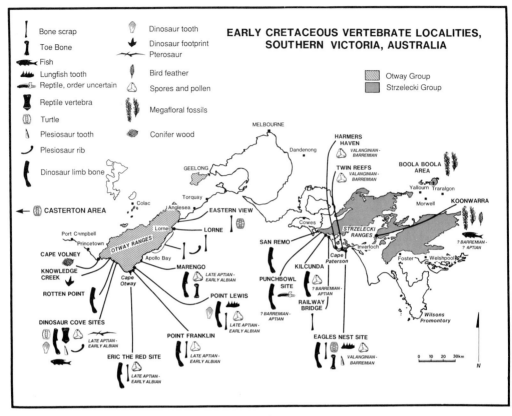

Fig. 3. Localities where fossil biota have been recovered.

by limb bones, vertebrae, and a lower jaw in addition to two partial carapaces (Rich & Rich 1989; Williams & Douglas 1985). These isolated elements represent turtles so primitive that they cannot be assigned to either cryptodires or pleurodires (E. S. Gaffney, pers. comm.). A small partial dermal scute may belong to a crocodilian; otherwise, this group is unknown.

Fish are well represented, especially from the Koonwarra locality. Ceratodontid lungfish, coccolepidid palaeonisciforms, archaeomaenid pholidophoriforms, and two clupeiform groups, the unique Australian koonwarrids as well as an endemic species of leptolepid, were all present (Waldman 1971).

Also known from the Koonwarra locality are five feathers, the only indication that birds were present. Unfortunately, because of the type of preservation, little can be discerned about their family level identity, although they have been restudied recently (Talent *et al.* 1966; Rogers 1987).

Dinosaur remains

The late Early Cretaceous vertebrate faunas of Victoria are unusual in being dominated by hypsilophodontid dinosaurs (Fig. 8). The Australian hypsilophodontids are endemic at the generic level, but clearly belong in the Hypsilophodontidae. This endemicity indicates that some form of isolating barrier, whether climatic, geographic, or topographic, was active at this time.

The numerical dominance of the hypsilophodontids among the dinosaurs could be a result of taphonomic effects related to their small size and the limited nature of the total sample (a few hundred reasonably complete isolated bones, teeth, and one skeleton). The only other dinosaur remains are rare fragments of *Allosaurus* sp. and smaller theropods (Rich & Rich 1989).

Another explanation may be that these two groups of dinosaurs were among the few that were particularly well adapted to a high latitude

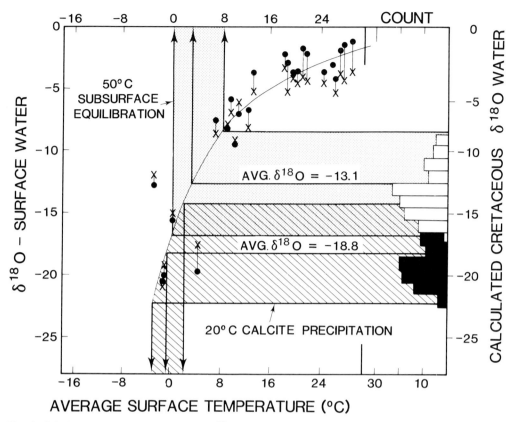

Fig. 4. Calculated Cretaceous surface water $\delta^{18}O$ values for the Otway and Gippsland basins are used to estimate average surface air temperature for the Early Cretaceous. The histograms correspond to calculated $\delta^{18}O$ values for two models: (1) precipitation of concretion calcites at 20°C (black) and (2) subsurface equilibration or formation of calcite at 50°C (unshaded). Arrows point to estimates of the mean annual air temperature derived from the two models represented by the histograms of calculated $\delta^{18}O$ water values. The stippled (50°C) and diagonally ruled (20°C) fields are bounded by lines one standard deviation from the means of the distribution. Also shown are the data (Gat 1980) for the International Atomic Energy Agency continental precipitation stations. The dots correspond to mean monthly $\delta^{18}O$ values whereas the crosses correspond to the means weighted by the amount of rainfall.

existence. On the North Slope of Alaska, at a comparable northern palaeolatitude to the southern one of the Victorian sites, hadrosaurs are dominant and theropods and ceratopsians rare in the Late Cretaceous fauna (Brouwers et al. 1987; Parrish et al. 1987). The Alaskan hadrosaurs are an order of magnitude larger in linear dimensions than the Victorian hypsilophodontids, but both families of dinosaurs were bipedal herbivores and are grouped together as ornithopods. However, this apparent bias in the Alaskan fauna may be simply due to the fact that much less collecting has been carried out there to date than further south in North America where the Late Cretaceous dinosaur fauna is both abundant and diverse.

One of the new Victorian hypsilophodontid genera is represented by an associated partial skeleton and skull. The material is from a juvenile of no more than one-third the adult weight, based on comparison of femoral sizes within the species. An endocast of the brain is preserved. The estimated archosaur encephalization quotient (EQ of of Jerison 1973) ranges between 1.1 and 1.8, depending on the adult size of the animal. Hopson (1979) reported a maximum archosaur EQ of 1.5 for ornithopods. The form of the brain in dorsal view is similar to that of the small theropod *Stenonychosaurus inequalis* (Rich & Rich 1989; Hopson 1979; Russell 1969). The two differ in that *S. inequalis* has a higher archosaur EQ (5.8), and the new Victorian hyp-

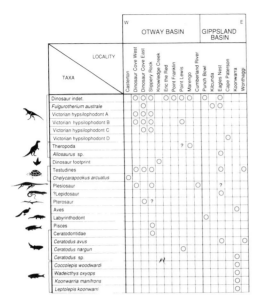

Fig. 5. Known occurrences of fossil vertebrates in the Otway and Strzelecki groups, southeastern Australia, late Early Cretaceous.

Fig. 6. Plesiosaur tooth from the Otway Group, southeastern Australia, late Early Cretaceous.

silophodontid has a distinct pineal body, whereas *S. inequalis* does not (unlike some other small theropods). The prominent optic lobes of the brain and large eyes of the Victorian hypsilophodontid may have been an adaptation for coping with low light conditions associated with polar habitats.

More than half of the hypsilophodontid specimens from Victoria are juveniles. No eggshell has yet been found, but it seems that southeastern Australia in the early Cretaceous was possibly being utilized as a dinosaur nursery. Whether dinosaurs only occupied the area seasonally (like caribou, etc.) and then migrated away from the pole during the winter, or whether they remained permanent residents, is yet unresolved (Rich & Rich 1989; Douglas & Williams 1982; Paul 1988). Because the most recent estimates of the palaeolatitudinal position of southern Victoria vary between 70–85°S, hypotheses regarding distances to be travelled by migratory animals cannot be evaluated with certainty. The minimum round-trip distance northward from 70°S to the Antarctic Circle, without regard to obstacles, is 800 km.

Fig. 7. Possible labyrinthodont lower jaw from the Strzelecki Group, late Early Cretaceous, southeastern Australia.

10 mm

Fig. 8. Skull of a new, tiny hypsilophodontid dinosaur, with brain endocast preserved, from the later Early Cretaceous of southeastern Australia (Photo by S. Morton).

From 85°S, the corresponding distance is 4000 km. Caribou and polar bears annually migrate 2000–2500 km (Paul 1988) but smaller terrestrial vertebrates do not. If the latitude of southern Victoria was near the maximum estimate, it is unlikely that small hypsilophodontids could have annually migrated the distance to areas of winter daylight. What is certain,

though, is that plants, invertebrates, and many vertebrates (ceratodont lungfish, actinopterygians, turtles, etc.) must have overwintered. With the possible exception of the *Allosaurus* sp., all of the animals known in this fauna were small enough to have readily found shelter by burrowing.

Discussion and conclusions

Late Early Cretaceous (Valanginian – early Albian) braided river sediments deposited in a graben created by the breakup of Gondwana, and exposed now along Australia's southeast coast, contain a varied biota, comprising some 150 taxa. Amongst these are at least 7 taxa of dinosaurs, predominantly small hypsilophodontids, mostly endemic and new genera. Three small to medium-sized theropods are represented, one being the youngest occurrence of the genus *Allosaurus*. The survival of a ?labyrinthodont amphibian (Jupp & Warren 1986) and certain elements of the flora (Drinnan & Chambers 1986) into the late Early Cretaceous of southeastern Australia suggests that these relicts may have been able to survive longer in this polar safe area (Vermeij 1987).

The palaeolatitude of southeastern Australia in the late Early Cretaceous was between 70° and 85°S, well within the Antarctic Circle. Oxygen isotope studies suggest that the mean annual temperature was quite cold, between −5°C and 8°C and probably <5°C. As pointed out by Axelrod (1984), however, mean annual temperature data estimated in this fashion do not contain information on the annual temperature range, which may have a greater influence on the biota.

Although not entirely certain, it is likely that the biota of southern Victoria in the late Early Cretaceous inhabited Antarctic polar latitudes that experienced temperate humid conditions subject to seasonal extremes. Together with the Late Cretaceous Alaskan occurrence, the presence of Australian dinosaurs at high palaeolatitudes for several million years during the Early Cretaceous firmly establishes that some ornithopod and theropod dinosaurs managed well under such conditions for at least 30 Ma. Their presence near the poles was an enduring association, not a single random event. It follows that any theory regarding dinosaur extinction must account for this adaptability in some dinosaurs. A prolonged period of darkness alone could have brought about the K–T boundary extinction event only if its duration was greater than 3 to 5 months or if the dinosaurs were only occupying high latitudes during part of the year;

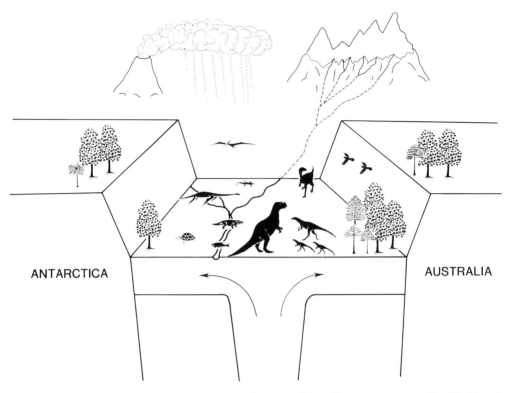

Fig. 9. A cartoon diagram of the setting of the Early Cretaceous biota of southeastern Australia. The biota is preserved in sediments accumulated in the graben formed when Antartica and Australia began to separate from one another with the onset of rifting between them in the Early Cretaceous. Traces of the distal upland flora are recorded in the pollen record as well as that of the more proximal lowland floral community. A volcanic source, the location of which is uncertain, (perhaps the Lord Howe Rise to the east or on the Australia and/or Antarctic plate to the north and south), contributed approximately 50 000 km³ of volcanic detritus to fill the basin. High mountains, again locality unknown, provided extremely cold water, perhaps, melt water, into the graben area. It may be that the volcanic peaks and those from which the meltwater came were one and the same.

the latter may be unlikely for some of the smaller, large-eyed, large-brained Victorian hypsilophodontids.

There were no major terrestrial barriers between East Antarctica and Australia in the Early Cretaceous (Fig. 9). Therefore, it is likely that the Early Cretaceous biota of southeastern Australia and the area of East Antarctica adjacent to it were very similar. Surface outcrops of Early Cretaceous rift sediments from the southern side of the Australo−Antarctic rift valley are unknown. However, a marine core taken close to the coast of East Antarctica at a point opposite southeastern Australia (across the spreading axis) yielded an Aptian palynoflora and lithologies strikingly similar to those from the Otway and Strzelecki groups (Domack *et al.* 1980).

We wish to express our thanks to Atlas Copco, the National Geographic Society, the Australian Research Grants Scheme, David Holdings Ltd., Mobil Oil, Imperial Chemical Industries, J. Herman, W. Loads, J. Chessells, the Friends and Council of the Museum of Victoria, the Sunshine Foundation and the Danks Trust for supporting this research. We also thank L. Kool for preparing much of the material and D. Gelt, S. Morton and F. Coffa for graphics and photography. The isotopic analyses were made possible through the support from the Australian Research Council, N.E.R.D.D.C., and Esso Australia. I. Duddy kindly allowed access to his extensive concretion collection.

References

AXELROD, D. I. 1984. An interpretation of Cretaceous and Tertiary biota in polar regions. *Palaeogeography, Palaeoclimatology, Palaeoecology*, **45**, 105–147.

BARRON, E. J. 1987. Cretaceous plate tectonic reconstructions. *Palaeogeography, Palaeoclimatology, Palaeoecology*, **59**, 3–30.

BROUWERS, E. M., CLEMENS, W. A., SPICER, R. A., AGER, T. A., CARTER, T. A. & SLITER, W. V. 1987. Dinosaurs on the North Slope, Alaska: high latitude, latest Cretaceous environments. *Science*, **237**, 1608–1610.

CERLING, T. E. 1984. The stable isotopic composition of modern soil carbonates and its relationship to climate. *Earth and Planetary Science Letters*, **71**, 229–240.

DETTMANN, M. E. 1963. Upper Mesozoic microfloras from south-eastern Australia. *Proceedings of the Royal Society of Victoria*, **77**, 1–148.

—— 1986. Early Cretaceous palynoflora of subsurface strata correlative with the Koonwarra Fossil Bed, Victoria. *In*: JELL, P. A. & ROBERTS J., (eds) *Plants and invertebrates from the Lower Cretaceous Koonwarra Fossil Bed, South Gippsland, Victoria*. Association of Australasian Palaeontologists Memoir, **3**, 79–110.

DOMACK, E. W., FAIRCHILD, W. W., & ANDERSON, J. B. 1980. Lower Cretaceous sediment from the East Antarctic continental shelf. *Nature*, **287**, 625–626. Erratum. *Nature*, **288**, 626.

DOUGLAS, J. G. & WILLIAMS, G. E. 1982. Southern polar forests: the early Cretaceous floras of Victoria and their palaeoclimatic significance. *Palaeogeography, Palaeoclimatology, Palaeoecology*, **39**, 171–185.

——, ABELE, C., BENEDEK, S., DETTMANN, M. E., KENLEY, P. R. & LAWRENCE, C. R. 1976. Mesozoic. *In*: DOUGLAS, J. G. & FERGUSON, J. A., (eds) *Geology of Victoria*. Special Publication of the Geological Society of Australia, **5**, 143–176.

DRINNAN, A. N. & CHAMBERS, T. C. 1986. Flora of the Lower Koonwarra Fossil Bed (Korumburra Group), South Gippsland, Victoria, *In*: JELL, P. A. & ROBERTS, J., (eds) *Plants and invertebrates from the Lower Cretaceous Koonwarra Fossil Bed, South Gippsland, Victoria*. Association of Australasian Palaeontologists Memoir, **3**, 1–77.

EMBLETON, B. J. J. & MCELHINNY, M. W. 1982. Marine magnetic anomalies, palaeomagnetism and the drift history of Gondwanaland. *Earth and Planetary Science Letters*, **58**, 141–150.

FRAKES, L. A., BURGER, D., APTHORPE, M., *et al.* 1987. Australian Cretaceous shorelines, Stage by Stage, *Palaeogeography, Palaeoclimatology, Palaeoecology*, **59**, 31–48.

GAT, J. R. 1980. *Handbook of Environmental Isotope Geochemistry*, **1**, 21–47. Elsevier, Amsterdam.

GILL, E. D. 1977. Evolution of the Otway Coast, Australia, from the last interglacial to the present. *Proceedings of the Royal Society of Victoria*, **89**, 7–18.

——, SEGNIT, E. R. & MCNEILL, N. H. 1977. Concretions in Otway Group sediments, southeastern Australia. *Proceedings of the Royal Society of Victoria*, **89**, 51–6.

GLEADOW, A. J. W. & DUDDY, I. R. 1981. Early Cretaceous volcanism and the early breakup history of southeastern Australia: evidence from fission-track dating of volcaniclastic sediments. *In*: CRESWELL, M. M. & VELLA, P. (eds), *Gondwana Five*, A. A. Balkema, Rotterdam, 295–300.

GREGORY, R. T., DOUTHITT, C. B., DUDDY, I. R., RICH, P. V. & RICH, T. H. in press. Oxygen isotopic composition of carbonate concretions from the lower Cretaceous of Victoria, Australia: implications for the evolution of meteoric waters on the Australian continent in a paleopolar environment. *Earth and Planetary Science Letters*.

HOPSON, J. A. 1979. Paleoneurology. *In*: GANS, C., NORTHCUTT, R. G. & ULINSKI, P., (eds) *Biology of the Reptilia, Volume 9, Neurology* A. Academic Press, San Francisco, 39–146.

IDNURM, M. 1985. Late Mesozoic and Cenozoic palaeomagnetism of Australia — I. A redetermined apparent polar wander path. *Geophysical Journal of the Royal Astronomical Society*, **83**, 399–418.

JELL, P. A. & Duncan, P. M. 1986. Invertebrates, mainly insects, from the freshwater, lower Cretaceous, Koonwarra Fossil Bed (Korumburra Group), South Gippsland, Victoria. *In*: JELL, P. A. & ROBERTS, J., (eds) *Association of Australasian Palaeontologists Memoir*, **3**, 111–205.

JERISON, H. J. 1973. *Evolution of the brain and intelligence*. Academic Press, New York.

JUPP, R. & WARREN, A. A. 1986. The mandibles of the Triassic temnospondyli amphibians. *Alcheringa*, **10**, 99–124.

LINDSAY, N. M. 1982. The burial history of the Strzelecki Group sandstones, S. E. Australia: a petrographic and fission track study. *MSc Thesis, University of Melbourne*.

MOLNAR, R. E. 1980. Australian late Mesozoic terrestrial tetrapods: some implications. *Mémoires de la Société géologique de France*. Paris, NS, **59**, 131–143.

——, FLANNERY, T. F. & RICH, T. H. V. 1981. An allosaurid theropod from the Early Cretaceous of Victoria, Australia. *Alcheringa*, **5**, 141–6.

——, ——, & —— 1985. Aussie *Allosaurus* after all. *Journal of Paleontology*, **59**, 1511–1513.

PARRISH, M. J., PARRISH, J. T., HUTCHISON, J. H. & SPICER, R. A. 1987. Late Cretaceous vertebrate fossils from the North Slope of Alaska and implications for dinosaur ecology. *Palaios*, **2**, 377–389.

PAUL, G. S. 1988. Physiological, migratorial, climatological, geophysical, survival and evolutionary implications of Cretaceous polar dinosaurs. *Journal of Paleontology*, **62**, 640–652.

RICH, T. H. & RICH, P. V. 1989. Polar dinosaurs and

biotas of the early Cretaceous of southeastern Australia. *National Geographic Research* **5**, 15–52.

Rogers, D. 1987. *Description and interpretation of a Lower Cretaceous fossil feather from Koonwarra, South Gippsland, Victoria.* Third Year Project, Monash University Earth Sciences Department, Clayton, Victoria, Australia.

RUSSELL, D. A. 1969. A new specimen of *Stenonychosaurus* from the Oldman Formation (Cretaceous) of Alberta. *Canadian Journal of Earth Sciences*, **6** 595–612.

SMITH, A. G., HURLEY, A. M. & BRIDEN, J. C. 1981. *Phanerozoic Paleocontinental World Maps.* Cambridge University Press, Cambridge.

TALENT, J. A., DUNCAN, P. M. & HANBY, P. L. 1966. Early Cretaceous feathers from Victoria, *Emu*, **66**, 81–6.

VEEVERS, J. J. 1986. Breakup of Australia and Antarctica estimated as mid-Cretaceous (95± 5 Ma) from magnetic and seismic data at the continental margin. *Earth and Planetary Science Letters*, **71**, 91–9.

VERMEIJ, G. J. 1987. *Evolution and Escalation.* Princeton University Press, Princeton.

WAGSTAFF, B. 1983. *The significance of palynology to studies of the sedimentology and post-depositional history of the Cretaceous Strzelecki Group.* Honours thesis, Department of Earth Sciences, Monash University, Clayton, Victoria, Australia.

— — & McEWEN MASON, J. 1989. Palynological dating of some lower Cretaceous coastal vertebrate fossil localities in Victoria, Australia. *National Geographic Research*. **5**, 54–63.

WALDMAN, M. 1971. Fish from the freshwater Lower Cretaceous of Victoria, Australia with comments on the palaeo-environment. *Special Papers in Palaeontology*, **9**, 1–124.

WARREN, J. W. 1969. A fossil chelonian of probable Lower Cretaceous age from Victoria, Australia. *Memoirs of the National Museum of Victoria*, **29**, 23–28.

WELLES, S. P. 1983. *Allosaurus* (Saurischia, Theropoda) not yet in Australia. *Journal of Paleontology*, **57**, 196.

WILLIAMS, G. E. & DOUGLAS, J. G. 1985. Comments on Cretaceous climatic equability in polar regions. *Palaeogeography, Palaeoclimatology, Palaeoecology*, **49**, 355–359.

WOODWARD, A. S. 1906. On a tooth of *Ceratodus* and a dinosaurian claw from the Lower Jurassic of Victoria, Australia. *Annals and Magazine of Natural History*, (7)**18**, 1–3.

ZIEGLER, A. M., SCOTESE, C. R., & BARRETT, S. F. 1983. Mesozoic and Cenozoic paleogeographic maps. *In*: BROSCHE, P. & SUNDERMANN, J., (eds) *Tidal Friction and the Earth's Rotation*, II. Springer-Verlag, Berlin, 240–252.

Note added a proof: An intercentrum of a labyrinthodont was recently found in the Strzelecki Group. This corroborates the range extension of these amphibians into the Early Cretaceous (see p. 127 and Fig. 7).

Terrestrial tetrapods in Cretaceous Antarctica

R. E. MOLNAR

*Queensland Museum, Queensland Cultural Centre, PO Box 300 South Brisbane,
Queensland, 4101 Australia*

Abstract: The fossil record of continental vertebrates from Antarctica is practically
nonexistent, hence the composition of the supposed Mesozoic terrestrial vertebrate fauna
must be inferred indirectly. This may be done using the reconstructed positions of the
continents during the Mesozoic. These positions suggest the presence in Antarctica of
primitive sauropods, large theropods, hypsilophodontian and iguanodontid ornithopods,
ankylosaurs, pterosaurs, crocodilians, lungfish and possibly birds. In the absence of
evidence to the contrary, monotremes are considered endemic Australian forms during
the Cretaceous. The discovery of hypsilophodontian material in New Zealand corroborates
the inference of Antarctic hypsilophodontians from their occurrence in Australia. The
existence in the Australian Early Cretaceous of relict taxa and taxa that seem unusual
compared to their overseas relatives suggests that the south polar tetrapod fauna was
characterized by singular forms.

The Mesozoic terrestrial tetrapods of Antarctica
are poorly known. There are specimens from
Lower Triassic rocks (Colbert 1974, 1987;
Colbert & Cosgriff 1974, Colbert & Kitching
1975, 1977, 1981; Cosgriff 1983; DeFauw 1989)
and a single specimen of an ankylosaurian dinosaur from the Upper Cretaceous (Gasparini *et
al.* 1987; Olivero *et al.* in press). The Early
Triassic (Scythian) fauna includes one or
more thecodonts, *Prolacerta*, at least seven
therapsids, *Procolophon* and at least two laby-
rinthodonts. This appears to be an unexcep-
tional Scythian tetrapod fauna (from data given
by Thulborn 1986).

It is, however, possible to deduce what forms
may have inhabited Antarctica at certain times
during the Mesozoic. Such deductions are possible because the geographical pattern of the
continents during this time is well understood.
Antarctica lay between, on the one hand,
Australia and New Zealand, and on the other,
Africa and South America (Smith *et al.* 1981).
Thus Antarctica was a 'bridge' providing the
only land link between Australasia and the other
continents (Fig. 1).

Continental taxa that dispersed into Australasia from the other Gondwanan continents or
Laurasia must have either dispersed overland
across Antarctica or across the sea. Thus comparison of Australasian Mesozoic terrestrial
tetrapod taxa with those of Africa and South
America provides insight into which taxa inhabited Antarctica. Future discoveries of Antarctic fossils can (in principle) provide a test of
this.

Australasian continental tetrapods either
originated in Australasia or immigrated from
elsewhere. The Permo-Carboniferous glaciations may have overrun most, or all, of
Australia (Johnson 1980). If so, most or all
post-Carboniferous Australian terrestrial tetrapod lineages must have been immigrants.

This endeavour to deduce the Mesozoic terrestrial tetrapods of Antarctica is limited by the
poor record of terrestrial tetrapod fossils for
pre-Cenozoic Australasia. Triassic continental
tetrapods are unknown in New Zealand, and
occur only in the Scythian of Australia.
Although these Australian continental tetrapod
faunas are of unusual aspect (Thulborn 1986),
their constituent taxa are similar to contemporaneous taxa known elsewhere. Because Jurassic
continental tetrapods are poorly known in
Australia (Molnar 1980*a*), and unknown in New
Zealand, only Cretaceous terrestrial tetrapods
will be considered.

Many Cretaceous tetrapods from Gondwanaland (especially Australasia) are represented by
few and incomplete specimens. Although the
specimens often represent distinct species, their
higher level affiliations are sometimes unclear.
For this reason I have opted in some instances
for a category such as 'large theropod' rather
than use a potentially misleading or incorrect
category such as carnosaur. Recently discovered specimens from Argentina will clarify
many of these poorly understood forms, but
until that work is completed taxonomic vague-
ness at higher levels will persist. The approach
adopted here is to compare the faunas at
sequentially lower taxonomic levels.

From Crame, J. A. (ed.), 1989, *Origins and Evolution of the Antarctic Biota*,
Geological Society Special Publication No. 47, pp. 131–140.

Early Cretaceous tetrapods expected for Antarctica

To begin at the level of Order, large and small theropods, sauropods, large and small ornithopods, pterosaurs, crocodilians, and chelonians are found in Lower Cretaceous rocks on both sides of Antarctica (Fig. 1). Thus we would expect that these inhabited Antarctica during the Early Cretaceous. It is conceivable that pterosaurs, chelonians and crocodilians dispersed, by air and sea respectively, without crossing Antarctica. The other forms would have had some Antarctic populations.

Some useful conjectures may be made regarding lower taxonomic levels, although only a single genus seems to be known from both Australia and other continents (Molnar et al. 1981, 1985). The Australian continental crocodilian ('Crocodylus' selaslophensis) is not allocated at the familial level (Molnar 1980b) while the continental chelonian material is either not allocated at the familial level or (in the case of Chelycarapookus arcuatus) is assigned to a unique monotypic family of uncertain higher taxonomic affinities. The taxonomy of sauropods and therapods is in a state of flux, so consideration at any level below Order is not useful.

The Australasian ornithischian material seems to represent three families, the Hypsilophodontidae, Iguanodontidae and Nodosauridae. The identification of hypsilophodontids in Australia is sound (Molnar & Galton 1986). The family Iguanodontidae is under revision (Norman & Weishampel pers. comm.) and while the allocation of the Australian genus is secure under the currently accepted definition of this family (Bartholomai & Molnar 1981), the future allocation will depend on the results of this revision. The Australian nodosaurid is based on incomplete material and hence this is a tentative identification (Molnar & Frey 1987). Both iguanodontids and hypsilophodontids are known from Africa (Taquet 1976; Cooper 1985). Nodosaurids occur in Europe, North America and Asia, but are not recognized in either South America or Africa (Molnar & Frey 1987). Nonetheless, they presumably dispersed through Antarctica via Africa or South America. We may thus presume that hypsilophodontids, iguanodontids and nodosaurids occurred in Antarctica (and probably in South America as well, where Early Cretaceous ornithischians are represented only by footprints, Leonardi 1984).

The only Cretaceous terrestrial tetrapod occurring both in Australasia and elsewhere (North America) is the carnosaur Allosaurus (Molnar et al. 1981, 1985). This genus has been reported in the Jurassic of Africa (Janensch 1925), and a member of the same family in the Jurassic of Argentina (Bonaparte 1986a). This suggests that allosaurids occurred in Antarctica.

What appears to be part of the jaw of a temnospondyl amphibian has been found in Lower Cretaceous rocks of the southern coast of Victoria (Jupp & Warren 1986). Before rifting north, this region was adjacent to what is now Antarctica. Thus such creatures would be expected to have lived in Antarctica as well.

The report of a sphenodontian from the Lower Cretaceous of South Africa (Rich et al. 1983), together with their widespread distribution earlier in the Mesozoic, suggests that they dispersed across Antarctica to reach what is now New Zealand. If so, why did they not reach Australia? This apparent absence has yet to be explained.

The occurrence of enantiornithine birds in the Early Cretaceous of Australia (Molnar 1986) and Late Cretaceous of Argentina suggests that these may also have inhabited the Antarctic, especially as there is doubt about their flying ability (Walker 1981), and hence their potential for dispersal by air. Their early appearance in Australia suggests that they *may* have originated in Australia.

Early Cretaceous mammals are known from both South America and Australia, but these pertain to different orders. A eupantothere has been found in South America (Bonaparte 1986b) and a monotreme in Australia. In the absence of evidence to the contrary, monotremes are here considered to be endemic Australian forms.

The foregoing considerations suggest that the Early Cretaceous fauna of Antarctica would include: sauropods, Allosaurus (or at least allosaurids), hypsilophodontians, iguanodontids, nodosaurids, a probable temnospondyl, and (less confidently) chelonians, crocodilians and pterosaurs. Enantiornithine birds were probably present in Early or Late Cretaceous.

Late Cretaceous tetrapods expected for Antarctica

Late Cretaceous terrestrial tetrapods, with the exception of the sauropods and trackways of the Winton Fm. (Coombs & Molnar 1981; Thulborn & Wade 1984) and an unstudied chelonian shell, are unknown in Australia (Fig. 1). Because sauropods and turtles are also known from the Australian Lower Cretaceous, and because the Winton Fm. is of Cenomanian

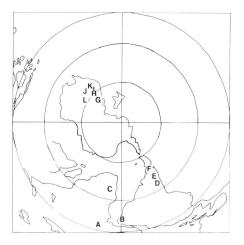 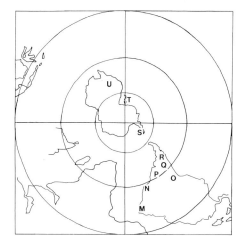

Fig. 1. Positions of the southern continents during the Cenomanian (left) and Campanian (right), in south polar projection. The continents are represented by modern shorelines; terrestrial tetrapod localities by letters. Those for the Early Cretaceous are indicated on the Cenomanian map, and for the Late Cretaceous on the Campanian map. A, Gadoufaoua, Niger (small theropods, sauropods, iguanodonts, hypsilophodonts, crocodiles, chelonians); B, Ceara & Bahia, Brazil (pterosaurs, crocodiles, chelonians); C, Namaqaland, South Africa (hypsilophodont); D, San Luis, Argentina (pterosaurs); E, Neuquen, Argentina (sauropods); F, Chubut, Argentina (large theropods, sauropods); G, Victoria, Australia (small & large theropods, hypsilophodonts, pterosaur, chelonians); H, New South Wales, Australia (large theropod, sauropod, iguanodont, hypsilophodont, crocodile, chelonians); I, southeast Queensland, Australia (ankylosaur); J, west Queensland, Australia (pterosaur); K, central Queensland, Australia (sauropod, iguanodont); L, South Australia, Australia (small theropod); M, Paraiba, Brazil (pterosaurs); N, Parana & Sao Paulo, Brazil (sauropods, crocodiles, chelonians); O, Salta, Brazil (small & large theropods, sauropod, crocodile); P, Uruguay & Entre Rios, Argentina (sauropods, crocodiles); Q, Neuquen & Rio Negro, Argentina (large theropod, sauropods, crocodiles, chelonians); R, Chubut & Santa Cruz, Argentina (sauropods, crocodiles, chelonians); S, James Ross Island (ankylosaur); T, North Island, New Zealand (hypsilophodont, pterosaur); U, central Queensland, Australia (sauropod, chelonian). (Cenomanian map from Smith *et al.* 1981, Campanian from Wiffen & Molnar, in press).

(earliest Late Cretaceous) age, this does not add significantly to what has already been deduced for the Early Cretaceous.

Before separation from Antarctica, New Zealand was a peninsula and tetrapods dispersing into Australia need not have traversed New Zealand. If the terrestrial tetrapods of Campanian–Maastrichtian New Zealand did not arrive by dispersal across the sea, but instead were 'carried along' on New Zealand as it drifted north, then they should provide a sample of the Late Cretaceous continental fauna of Antarctica independent of the deductions based on comparison with Australian forms.

Late Cretaceous terrestrial tetrapods from North Island, New Zealand comprise three specimens, two of which are too incomplete for use here (Molnar 1981*a*; Scarlett & Molnar 1984). The third is an incomplete ilium that pertains to a dryosaur-like ornithopod (Wiffen & Molnar, in press). This occurrence corroborates the inference of Antarctic hypsilopho-

dontians (dryosaurids or hypsilophodontids) derived previously.

Pterosaurs have also been discovered in New Zealand (Wiffen & Molnar 1988 and undescribed material). These indicate that, although pterosaurs need not have dispersed across Antarctica to reach Australia, they did occupy at least the periphery of Antarctica. Given the flight capabilities of pterosaurs this is not surprising.

Other forms may be expected to have inhabited Antarctica from consideration of the modern terrestrial tetrapod fauna of New Zealand. These include ratites, sphenodontians and leiopelmatid frogs (Fleming 1975; Molnar 1981*a*). Evidence of a large terrestrial bird, a phorusrhacoid not a ratite, has been found in Lower Cenozoic rocks of the Antarctic (Case *et al.* 1987). This shows that large ground-dwelling birds could survive in Antarctica at that time.

Although this paper is concerned with terrestrial tetrapods, the logic should apply as well to

continental (freshwater) fishes that cannot disperse through the sea. The occurrence of ceratodont lungfish in Australia in both Lower
Cretaceous and Miocene (and later) beds, and
in the Upper Cretaceous of Argentina (Pascual
& Bondesio 1976), leads us to expect Antarctic
ceratodontid lungfish. In fact, a fragment
of ceratodont lungfish has been discovered in
Triassic rocks in Antarctica (Dziewa 1980).

So the Late Cretaceous fauna of Antarctica
may be expected to include: hypsilophodontian ornithopods, pterosaurs, ratites, sphenodontians, leiopelmatid frogs, and ceratodont
lungfish.

Discussion

At least some elements of the Cretaceous
Antarctic terrestrial tetrapod fauna may be pre-
dicted from comparison of the Australasian and
other Gondwanan terrestrial tetrapod faunas.
However, the Australasian tetrapod record
shows unusual features that deserve further
comment. For example, the prominent role
of marsupials in Australia and (until recently)
ratites in New Zealand is unmatched elsewhere.
These features stem from the isolation of these
lands through most of the Cenozoic.

The Australian Mesozoic fauna is distinctive
in two respects: it includes taxa that appear to
be relicts, and others that seem to have diverged
to a significantly greater extent than related
forms elsewhere. The unusual character of the
Australian terrestrial tetrapod fauna seemingly was already established in the Triassic
(Thulborn 1986), although at this time it was
the composition of the fauna that was odd and
not the relationship of any of its constituent
taxa to those of other continents.

Relict taxa are represented in the fossil record
(cf. Halstead 1987), although to my knowledge
no systematic reviews of their occurrences in
the record have been made. An example among
Mesozoic terrestrial tetrapods is the Chinese
Jurassic temnospondyl (Dong 1985). Relicts are
not prominent in the well known Northern
Hemisphere Cretaceous tetrapod faunas (western North America, eastern Asia), although
they occur among mammals (Fox 1969, 1976,
1984; Lillegraven & McKenna 1986) and dinosaurs (a Late Cretaceous fabrosaurid reported
by Weishampel & Weishampel 1983, and probably the segnosaurs, discussed by Paul 1984).

In the Australian Mesozoic, relicts make up a
considerable proportion of the known terrestrial
vertebrate fauna. Fifteen genera are known
from the Jurassic–Cretaceous of Australia of
which three, *Allosaurus*, *Austrosaurus*, and

Siderops, seem to be relicts. In addition there is
the apparent temnospondyl from the Victorian
Lower Cretaceous (Jupp & Warren 1986).

Siderops kehli (Warren & Hutchinson 1983)
and the Victorian temnospondyl persisted well
after the majority of temnospondyls became
extinct at the end of the Triassic. The Australian
Allosaurus is Valanginian to Aptian in age
(Dettmann 1986). The well known *Allosaurus
fragilis* is from the Morrison Formation, usually
cited as of Kimmeridgian or Tithonian age.
Assuming the latter, the Victorian specimen
would be at least 5 Ma younger.

The sauropod *Austrosaurus* shows several
characters that seem to be plesiomorphic for
sauropods (Coombs & Molnar 1981): small
dorsal pleurocoels, absence of both caudal
pleurocoels and supporting buttresses of caudal
transverse processes, anteriorly placed caudal
neural arches, and elevated femoral head. All
but the first of these are also found among
the contemporaneous titanosaurids; in fact
Austrosaurus seems to share almost all features
with the titanosaurids, except for their diagnostic procoelous caudals. However, Jurassic sauropods are markedly less plesiomorphic than the
Cretaceous titanosaurids (von Huene 1929).
Thus it is in order to speak of *Austrosaurus* as a
relict of Middle Jurassic sauropods of cetio-
saurian aspect. So, to almost as great an extent,
are the titanosaurids. Why sauropods of primitive aspect substantially outlasted more derived
taxa is not known.

No relict forms are obvious among the 22
South American, or 23 African Early Cretaceous terrestrial tetrapods. This may well
reflect some barrier between Australia, on the
one hand, and South America and Africa on
the other, (such as mountains, epeiric seas or
glaciated highlands, cf. Molnar 1981*b*). However Vermeij (1987) has suggested that certain
taxa may survive preferentially in polar regions
after becoming extinct in more equatorial areas.
Thus it is interesting that the Australian Lower
Cretaceous sites are generally less distant from
the South Pole than are the South American
sites (Smith *et al.* 1981). Only the central
Queensland sites in Australia did not lie south
of the southernmost South American sites (in
Chubut, Argentina). The New Zealand ornithopod ilium shows more resemblance to ilia of
Upper Jurassic and Early Cretaceous dryosaurid
ornithopods, than to ilia of contemporaneous
Late Cretaceous taxa (Wiffen & Molnar in
press). Thus some Australasian terrestrial tetrapods are consistent with Vermeij's proposal.

The extent of unusual adaptations is difficult
to estimate. Australian Cretaceous terrestrial

tetrapods seem to have diverged more substantially from the common ancestors shared
with non-Australian forms than have most
of the non-Australian forms. Many of these
features have already been pointed out by
Bonaparte (1986b) in his survey of the endemic
Gondwanan continental vertebrates. The nodosaurid *Minmi paravertebra* shows peculiar
ossified aponeuroses and tendons (Molnar &
Frey 1987) not found among any other known
ankylosaurians. The slenderness of the tibia of
the theropod *Kakuru kujani* is matched only by
those of *Avimimus portentosus* and *Borogovia
gracilicrus*. The raised hollow rostral bulla, the
maxillary tooth form and the tooth rows lacking
any indication of replacement teeth of *Muttaburrasaurus langdoni* are unmatched in the
other iguanodontids. This has led to the suggestion that *Muttaburrasaurus* is not an iguanodontid, which only emphasizes the degree of
difference of this form from contemporaneous
non-Australian ornithopods. *Chelycarapookus
arcuatus* is unusual in certain structures, especially the increased length of the posterior
two pairs of ribs and the correspondingly increased size of the posterior neurals (Warren
1969). These features are not matched in other
chelonians. However, the internal morphology
of the carapace of chelonians is poorly known
(Gaffney 1981), so it is not clear that this constitutes as great a divergence from the condition
of the common ancestor (as seems to be the
case for the dinosaurs discussed above).

Four of the 13 known Early Cretaceous
Australian continental tetrapods appear to show
substantially greater morphological differences
than contemporaneous overseas taxa. Early
Cretaceous African tetrapods seem to show
nothing out of the ordinary in this regard. However, the remarkable theropod *Spinosaurus
aegyptiacus* from Egypt deserves mention.
Spinosaurus is early Cenomanian in age and
hence may well have persisted from the Early
Cretaceous; indeed there is a report of an Early
Cretaceous *Spinosaurus* (Bouaziz *et al.* 1988).
One of the 23 Early Cretaceous African tetrapods appears to have morphological structures
(elongate dorsal neural spines) comparable, in
their degree of divergence from the presumed
ancestral condition, to some of the Australian
forms. The South American Early Cretaceous
theropod *Carnotaurus sastrei* is comparably out-
landish in its development of a deep, short
snout and supraorbital horn cores (Bonaparte
1985). This gives ratios of four out of 13
Australian, one out of 23 African and one out
of eight South American terrestrial tetrapod
species (excluding flying forms) with unusually

divergent morphological structures. Again there
seems to be an unusual fauna in the Mesozoic
south polar regions.

The small sample sizes of the Early Cretaceous faunas of these continents implies that
this may be nothing more than sampling error.
However, an unusual high-latitude mammalian
fauna has been reported in North America
(Hickey *et al.* 1983). The Australian evidence
suggests that the Cretaceous terrestrial tetrapod
fauna of the Antarctic regions may have been
unique. The role of future discoveries in
Antarctica will be critical in corroborating the
Australian evidence, and revealing the extent
to which these faunal features extended through
the south polar regions.

Attendance at the meeting 'Origins and Evolution of
the Antarctic Biota' was made possible by a timely
and much-appreciated grant from the Trans-Antarctic
Association. Assistance in various aspects was also
provided by W. Clemens, T. H. Rich, M. Thompson,
J. Wiffen and two anonymous referees.

Appendix: list of continental tetrapods of Cretaceous Gondwanaland (exclusive of India) used in text comparisons

LOWER CRETACEOUS
Brazil:
Pterosauria
  Anhangueridae
    Anhanguera blittersdorffi
  Criorhynchidae
    Tropeognathus mesembrinus
    Tropeognathus robustus
  Ornithocheiridae
    Araripedactylus dehmi
    Araripesaurus castilhoi
    Araripesaurus santanae
    Brasileodactylus araripensis
    Cearadactylus atrox
    Santanadactylus araripensis
    Santanadactylus brasiliensis
    Santanadactylus pricei
    Santanadactylus spixi
Crocodylia
  Pholidosauridae
    Sarcosuchus hartii
  Uruguaysuchidae
    Araripesuchus gomesi
Chelonia
  Araripemyidae
    Araripemys barretoi
Uruguay:
Crocodylia
  Pholidosauridae
    Meridiosaurus vallisparadisi

Argentina:
Theropoda
  Abelisauridae
    Carnotaurus sastrei
Sauropoda
  Brachiosauridae
    Chubutisaurus insignis
  family *incertae sedis*
    Amargasaurus groeberi
Pterosauria
  Pterodaustriidae
    Pterodaustro guinazui
  family *incertae sedis*
    Puntanipterus globosus
Mammalia
  Vincelestidae
    Vincelestes neuquenianus
Algeria:
Theropoda
  family *incertae sedis*
    Carcharodontosaurus saharicus
  Ornithomimidae
    Elaphrosaurus iguidiensis
Sauropoda
  Brachiosauridae
    Rebbachisaurus tamesnensis
Crocodylia
  Dyrosauridae
    Dyrosaurus sp.
  Pholidosauridae
    Sarcosuchus imperator
Squamata
  Simoliophidae
    Lapparentophis defrennei
Tunisia:
Theropoda
  family *incertae sedis*
    Carcharodontosaurus saharicus
    Spinosaurus sp.
Morocca:
Theropoda
  family *incertae sedis*
    Carcharodontosaurus saharicus
Sauropoda
  Brachiosauridae
    Rebbachisaurus garasbae
Crocodylia
  Libycosuchidae
    Libycosuchus sp.
  Thoracosauridae
    Thoracosaurus cherifiensis
Mali:
Sauropoda
  Brachiosauridae
    Rebbachisaurus tamesnensis
Crocodylia
  Dyrosauridae
    Dyrosaurus sp.
Niger:
Theropoda
  family *incertae sedis*
    Bahariasaurus ingens
    Carcharodontosaurus saharicus
    Inosaurus tedreftensis

Ornithomimidae
  Elaphrosaurus gautieri
  Elaphrosaurus iguidiensis
Sauropoda
  Brachiosauridae
    Rebbachisaurus tamesnensis
  Titanosauridae
    Aegyptosaurus baharijensis
Ornithopoda
  Iguanodontidae
    Ouranosaurus nigeriensis
  Dryosauridae
    Valdosaurus nigeriensis
Crocodylia
  Pholidosauridae
    Sarcosuchus imperator
  Uruguaysuchidae
    Araripesuchus wegeneri
Chelonia
  Araripemyidae
    Taquetochelys decorata
  Pelomedusidae
    Platycheloides cf. *P. nyasae*
    Teneremys lapparenti
Zaire:
Pterosauria
  Ornithocheiridae
    un-named species
South Africa:
Sauropoda
  Camarasauridae?
    Algoasaurus bauri
Ornithopoda
  Hypsilophodontidae?
    Kangnasaurus coetzeei
Stegosauria
  Stegosauridae
    Paranthodon africanus
Sphenodontia
  un-named species
Australia:
Theropoda
  Allosauridae
    Allosaurus sp.
  family *incertae sedis*
    Kakuru kujani
    Rapator ornitholestoides
Sauropoda
  family *incertae sedis*
    Austrosaurus mckillopi
Ornithopoda
  Iguanodontidae
    Muttaburrasaurus langdoni
  Hypsilophodontidae
    Fulgurotherium australe
    2 new spp. in press by Rich & Rich
Ankylosauria
  Nodosauridae
    Minmi paravertebra
Pterosauria
  Ornithocheiridae
    un-named species
Crocodylia
  family *incertae sedis*
    '*Crocodylus*' *selaslophensis*

Aves
  Enantiornithidae
    Nanantius eos
Chelonia
  Chelycarapookidae
    Chelycarapookus arcuatus
Mammalia
  Steropodontidae
    Steropodon galmani
UPPER CRETACEOUS
Brazil:
Sauropoda
  Titanosauridae
    Antarctosaurus brasiliensis
    Titanosaurus sp.
Pterosauria
  Nyctosauridae?
    Nyctosaurus? lamegoi
Crocodylia
  Baurusuchidae
    Baurusuchus pachecoi
  Dyrosauridae
    Hyposaurus derbianus
  Trematochampsidae
    Itasuchus jesuinoi
  Peirosauridae
    Peirosaurus torminni
  family *incertae sedis*
    Sphagesaurus huenei
Squamata
  Iguanidae
    Pristiguana brasiliensis
Chelonia
  Pelomedusidae
    Apodichelys lucianoi
    Podocnemis barrisi
    Podocnemis brasiliensis
    Podocnemis elegans
    Roxochelys wanderleyi
Uruguay:
Sauropoda
  Titanosauridae
    Antarctosaurus wichmannianus
    Argyrosaurus superbus
    Laplatasaurus araukanicus
    Titanosaurus australis
Crocodylia
  Uruguaysuchidae
    Uruguaysuchus aznaresi
    Uruguaysuchus terrai
Bolivia:
Crocodylia
  Baurusuchidae
    Cynodontosuchus rothi
Chelonia
  Pelomedusidae
    Roxochelys? vilavilensis
Mammalia
  Didelphidae
    Roberthoffstetteria nationalgeographica
  Hyopsodontidae
    Molinodus suarezi
    Tiuclaenus minutus

  Pantolambdidae
    Alcidedorbignya inopinata
  Eutheria? *incertae sedis*
    Andinodus boliviensis
Peru:
Chelonia
  Pelomedusidae
    Roxochelys? vilavilensis
Mammalia
  Didelphidae
    Alphadon austrinum
    Peradectes austrinum
  Eutheria? *incertae sedis*
    Perutherium altiplanense
Argentina:
Theropoda
  Abelisauridae
    Abelisaurus comahuensis
    Xenotarsosaurus bonapartei
  Noasauridae
    Noasaurus leali
  family *incertae sedis*
    Loncosaurus argentinus
    Unquillosaurus ceibalensis
Sauropoda
  Titanosauridae
    Antarctosaurus giganteus
    Antarctosaurus wichmannianus
    Argyrosaurus superbus
    Clasmodosaurus spatula
    Laplatasaurus araukanicus
    Saltasaurus loricatus
    Titanosaurus australis
    Titanosaurus nanus
Ornithopoda
  Hadrosauridae
    Kritosaurus australis
    Secernosaurus koeneri
Ceratopsia
  Protoceratopsidae?
    Notoceratops bonarelli
Crocodylia
  Baurusuchidae
    Cynodontosuchus rothi
  Dolichochampsidae
    Dolichochampsa minima
  Notosuchidae
    Notosuchus terrestris
  family *incertae sedis*
    Microsuchus schilleri
Aves
  Enantiornithidae
    Avisaurus sp.
    Enantiornis leali
Squamata
  Dinilysidae
    Dinilysia patagonica
  Madtsoidae
    Alamitophis argentinus
    Patagoniophis parvus
    Rionegrophis madtsoioides
  family *incertae sedis*
    Dicarlesia incognita

Chelonia
  Meiolaniidae
    Niolamia argentina
  Pelomedusidae
    Naiadochelys maior
    Naiadochelys patagonica
Mammalia
  Dryolestidae
    Groebertherium novasi
    Groebertherium stipanicici
    Mesungulatum houssayi
  Ferugliotheriidae
    Ferugliotherium windhauseni
  Gondwanatheriidae
    Gondwanatherium patagonicus
  Triconodontidae
    Austrotriconodon ferox
Anura
  Pipidae
    Saltenia ibanezi
Chile:
Sauropoda
  Titanosauridae
    Antarctosaurus sp.
Aves
  Baptornithidae
    Neogaeornis wetzeli
Egypt:
Theropoda
  Spinosauridae
    Spinosaurus aegyptiacus
  family *incertae sedis*
    Bahariasaurus ingens
    Carcharodontosaurus saharicus
    un-named small theropods
Sauropoda
  Titanosauridae
    Aegyptosaurus baharijensis
Crocodylia
  Aegyptosuchidae
    Aegyptosuchus peyeri
    Stromerosuchus aegyptiacus
  Dyrosauridae
    Dyrosaurus sp.
  Libycosuchidae
    Libycosuchus brevirostris
  Stomatosuchidae
    Stomatosuchus inermis
  Trematochampsidae
    Trematochampsa sp.
Squamata
  Simoliophidae
    Simoliophis sp.
Chelonia
  family *incertae sedis*
    Apertotemporalis baharijensis
Sudan:
Crocodylia
  Dyrosauridae
    Dyrosaurus sp.
Morocco:
Squamata
  family *incertae sedis*
    Pachyvaranus crassispondylus

Senegal:
Pterosauria
  Azhdarchidae?
    un-named species
Niger:
Theropoda
  family *incertae sedis*
    Bahariasaurus? sp.
Crocodylia
  Libycosuchidae
    Libycosuchus sp.
  Trematochampsidae
    Trematochampsa taqueti
Squamata
  Madtsoidae
    Madtsoia sp.
Nigeria:
Crocodylia
  Dyrosauridae
    Dyrosaurus sp.
    Sokotosuchus ianwilsoni
Chelonia
  Chelydridae
    Nigeremys gigantea
  Pelomedusidae
    Sokotochelys lawanbungudui
    Sokotochelys umarumohammedi
Madagascar:
Theropoda
  Abelisauridae?
    Majungasaurus crenatissimus
Sauropoda
  Titanosauridae
    Laplatasaurus madagascariensis
Pachycephalosauria
  Pachycephalosauridae
    Majungatholus atopus
Crocodylia
  Trematochampsidae
    Trematochampsa oblita
Squamata
  Madtsoiidae
    Madtsoia madagascariensis
South Africa:
Anura
  Pipidae
    Eoxenopoides reuningi
Australia:
Sauropoda
  family *incertae sedis*
    Austrosaurus sp.
Chelonia
  un-named species
New Zealand:
Theropoda
  un-named large theropod?
Ornithopoda
  Dryosauridae?
    un-named small ornithopod
Pterosauria
  un-named species

To avoid an excessively long list of references the sources for this table have not been cited. They have been deposited in the Geological Society Library and with the British Library at Boston Spa, W. Yorkshire, UK as Supplementary Publication No. SUP 18057 (also available from author on request).

References

BARTHOLOMAI, A. & MOLNAR, R. E. 1981. *Muttaburrasaurus*, a new iguanodontid (Ornithischia: Ornithopoda) dinosaur from the Lower Cretaceous of Queensland. *Memoirs of the Queensland Museum*, **20**, 319–349.

BONAPARTE, J. F. 1985. A horned Cretaceous carnosaur from Patagonia. *National Geographic Research*, **1**, 149–351.

—— 1986*a*. Les dinosaures (Carnosaures, Allosauridés, Sauropodes, Cétiosauridés) du Jurassique moyen de Cerro Cóndor (Chubut, Argentine). Premiere partie. *Annales de Paléontologie (Vertébrés-Invertébrés)*, **72**, 247–289.

—— 1986*b*. History of the terrestrial Cretaceous vertebrates of Gondwana. *IV Congreso Argentino de Paleontologia y Bioestrategrafia.*, **2**, 63–95.

BOUAZIZ, S., BUFFETAUT, E., GHANMI, M., JAEGER, J.-J., MARTIN, M., MAZIN, J.-M. & TONG, H. 1986. Nouvelles découvertes de vertébrés fossiles dans l'Albien du Sud tunisien. *Bulletin de la Societe géologigue de France*, **8**, 4, 335–359.

CASE, J. A., WOODBURNE, M. O. & CHANEY, D. S. 1987. A gigantic phororhacoid(?) bird from Antarctica. *Journal of Paleontology*, **61**, 1280–4.

COLBERT, E. H. 1974. *Lystrosaurus* from Antarctica. *American Museum Novitates*, **2535**, 1–44.

—— 1987. The Triassic reptile *Prolacerta* in Antarctica. *American Museum Novitates*, **2882**, 1–14.

—— & COSGRIFF, J. W. 1974. Labyrinthodont amphibians from Antarctica. *American Museum Novitates*, **2552**, 1–30

—— & KITCHING, J. W. 1975. The Triassic reptile *Procolophon* in Antarctica. *American Museum Novitates*, **2566**, 1–23.

—— & —— 1977. Triassic cynodont reptiles from Antarctica. *American Museum Novitates*, **2611**, 1–30.

—— & —— 1981. Triassic scaloposaurian reptiles from Antarctica. *American Museum Novitates*, **2709**, 1–22.

COOMBS, W. P., Jr. & MOLNAR, R. E. 1981. Sauropoda (Reptilia, Saurischia) from the Cretaceous of Queensland. *Memoirs of the Queensland Museum*, **20**, 351–373.

COOPER, M. R. 1985. A revision of the ornithischian dinosaur *Kangnasaurus coetzeei* Haughton, with a classification of the Ornithischia. *Annals of the South African Museum*, **95**, 281–317.

COSGRIFF, J. W. 1983. Large thecodont reptiles from the Fremouw Formation. *Antarctic Journal of the United States*, **18**, (5), 52–55.

DEFAUW, S. L. 1989. Evolution of the Dicynodontia (Reptilia, Therapsida) with special reference to Austral taxa. *In:* CRAME, J. A. (ed.) *Origins and Evolution of the Antarctic Biota.* Geological

Society, London, Special Publication, **47**, 63–84.

DETTMANN, M. E. 1986. Early Cretaceous palynoflora of subsurface strata correlative with the Koonwarra Fossil Bed, Victoria. *Association of Australasian Palaeontologists, Memoirs*, **3**, 79–110.

DONG Z. 1985. A middle Jurassic labyrinthodont (*Sinobrachyops placenticephalus* gen. et sp. nov.) from Dashanpu, Zigong, Sichuan Province. *Vertebrata Palasiatica*, **23**, 301–306.

DZIEWA, T. J. 1980. Note on a dipnoan fish from the Triassic of Antarctica. *Journal of Paleontology*, **54**, 488–490.

FLEMING, C. A. 1975. The geological history of New Zealand and its biota. *In:* KUSCHEL, G. (ed.) *Biogeography and ecology in New Zealand.* W. Junk, The Hague, 1–86.

FOX, R. C. 1969. Studies of Late Cretaceous vertebrates. III. A triconodont mammal from Alberta. *Canadian Journal of Zoology*, 47: 1253–6.

—— 1976. Additions to the mammalian local fauna from the Upper Milk River Formation (Upper Cretaceous), Alberta, *Canadian Journal of Earth Sciences*, **13**, 1105–1118.

—— 1984. A primitive, "obtuse-angled" symmetrodont (Mammalia) from the Upper Cretaceous of Alberta, Canada. *Canadian Journal of Earth Sciences*, **21**, 1204–1207.

GAFFNEY, E. S. 1981. A review of the fossil turtles of Australia. *American Museum Novitates*, **2720**, 1–38.

GASPARINI, Z., OLIVERO, E., SCASSO, R. & RINALDI, C. 1987. Un ankylosaurio (Reptilia, Ornithischia) Campaniano en el continente Antarctico. *Anais do X Congreso brasilerio de Palaeontologia* Rio de Janiero, 131–141.

HALSTEAD, L. B. 1987. Agnathan extinctions in the Devonian. *Mémoires de la Societe géologique de France*, **150**, 7–11.

HICKEY, L. J. WEST, R. M. DAWSON, M. R. & CHOI, D. K. 1983. Arctic terrestrial biota: paleomagnetic evidence of age disparity with mid-northern latitudes during the Late Cretaceous and Early Tertiary. *Science*, **221**, 1153–1156.

HUENE, F. VON 1929. Die Besonderheit der Titanosaurier. *Centralblatt Mineralogie, Geologie, Palaeontologie*, **B, 10**, 493–499.

JANENSCH, W. 1925. Die Coelurosaurier und Theropoden der Tendaguru-Schichten Deutsch-Ostafrikas. *Palaeontographica*, Suppl. 7, 1 Reihe, Teil 1, Lieferung **1**, 1–99.

JOHNSON, G. A. L. 1980. Carboniferous geography and terrestrial migration routes. *In:* PANCHEN, A. L., (ed.) *The terrestrial environment and the* origin of land vertebrates. Academic Press, London, 39–54.

JUPP, R. & WARREN, A. A. 1986. The mandibles of Triassic temnospondyl amphibians. Alcheringa, 10, 99–124.

LEONARDI, G. 1984. Le impronte fossili di dinosauri. In: LIGABUE, G., (ed.) Sulla Orme dei Dinosauri. Erizzo. Venice, 163–186.

LILLEGRAVEN, J. A. & McKENNA, M. C. 1986. Fossil mammals from the "Mesaverde" Formation (Late Cretaceous, Judithian) of the Bighorn and Wind River Basins, Wyoming, with definitions of Late Cretaceous North American land-mammal "ages". American Museum Novitates, 2840: 1–68.

MOLNAR, R. E. 1980a. Australian late Mesozoic terrestrial tetrapods: some implications. Mémoires de la Societe géologigue de France, 139, 131–143.

—— 1980b. Procoelous crocodile from Lower Cretaceous of Lightning Ridge, N. S. W. Memoirs of the Queensland Museum, 20, 65–75.

—— 1981a. A dinosaur from New Zealand. In: VELLA, P., & CRESWELL, M., (eds) Gondwana five, A. A. Balkema, Rotterdam, 91–96.

—— 1981b. Reflections on the Mesozoic of Australia. Mesozoic Vertebrate Life, 1, 47–60.

—— 1986. An enantiornithine bird from the Lower Cretaceous of Queensland, Australia. Nature, 322, 736–8.

—— & FREY, E. 1987. The paravertebral elements of the Australian ankylosaur Minmi (Reptilia: Ornithischia, Cretaceous). Neues Jahrbuch der Geologie und Paläontologie, Abhandlungen, 175, 19–37.

—— & GALTON, P. M. 1986. Hypsilophodontid dinosaurs from Lightning Ridge, New South Wales, Australia. Geobios, 19, 231–239.

—— & THULBORN, R. A. 1980. First pterosaur from Australia. Nature, 288, 361–363.

—— FLANNERY, T. F. & RICH, T. H. 1981. An allosaurid theropod dinosaur from the Early Cretaceous of Victoria, Australia. Alcheringa, 5, 141–146.

—— —— & —— 1985. Aussie Allosaurus after all. Journal of Paleontology, 59, 1511–1513.

OLIVERO, E. B. GASPARINI, Z., RINALDI, C. A. & SCASSO, R. In press. First record of dinosaurs in Antarctica (Upper Cretaceous James Ross Island): palaeogeographical implications. In: THOMSON, M. R. A., CRAME, J. A & THOMSON, J. W. (eds) Geological evolution of Antarctica. Cambridge University Press, Cambridge.

PASCUAL, R. & BONDESIO, Y. P. 1976. Notas sobre vertebrado de la frontera Cretaceo-Terciaria. Actas del Sexto Congreso Geologico Argentino, 1, 565–577.

PAUL, G. 1989. The segnosaurian dinosaurs: relicts of the prosauropod-ornithischian transition? Journal of Vertebrate Paleontology, 4, 507–515.

RICH, T. H., MOLNAR, R. E. & RICH, P. V. 1983. Fossil vertebrates from the late Jurassic or early Cretaceous Kirkwood Formation, Algoa Basin, southern Africa. Transactions of the Geological Society of South Africa, 86, 281–91.

—— RICH, P. V., WAGSTAFF, B., McEWEN-MASON, J., DOUTHITT, C. B. & GREGORY, R. T. 1989. Mid-Cretaceous biota from the northern side of the Antarctic Peninsula region. In: CRAME, J. A. (ed.) Origins and Evolution of the Antarctic Biota. Geological Society London Special Publication, 47, 121–130.

SCARLETT, R. J. & MOLNAR, R. E. 1984. Terrestrial bird or dinosaur phalanx from the New Zealand Cretaceous. New Zealand Journal of Zoology, 11, 271–275.

SMITH, A. G., HURLEY, A. M. & BRIDEN, J. C. 1981. Phanerozoic Paleocontinental World Maps. Cambridge University Press, Cambridge.

STEVENS, G. R. 1989. The nature and timing of biotic links between New Zealand and Antarctica in Mesozoic and early Cenozoic times. In: CRAME, J. A. (ed.) Origins and Evolution of the Antarctic Biota. Geological Society, London, Special Publication, 47, 141–166.

TAQUET, P. 1976. Géologie et paléontologie du gisement de Gadoufaoua (Aptien du Niger). Cahiers de Paléontologie, 1976, 1–91.

THULBORN, R. A. 1986. Early Triassic tetrapod faunas of southeastern Gondwana. Alcheringa, 10, 297–313.

—— & WADE, M. 1984. Dinosaur trackways in the Winton Formation (mid-Cretaceous) of Queensland. Memoirs of the Queensland Museums, 21, 413–517.

VERMEIJ, G. J. 1987. Evolution and Escalation. Princeton University Press, Princeton.

WALKER, C. A. 1981. New subclass of birds from the Cretaceous of South America. Nature, 292, 51–53.

WARREN, A. A. & HUTCHINSON, M. N. 1983. The last labyrinthodont? A new brachyopoid (Amphibia, Temnospondyli) from the early Jurassic Evergreen Formation of Queensland, Australia. Philosophical Transactions of the Royal Society of London, B, 303, 1–62.

WARREN, J. W. 1969. A fossil chelonian of probable Lower Cretaceous age from Victoria, Australia. Memoirs of the National Museum of Australia, 29, 23–28.

WEISHAMPEL, D. B. & WEISHAMPEL, J. B. 1983. Annotated localities of ornithischian dinosaurs: implications to Mesozoic paleobiogeography. The Mosasaur, 1, 43–87.

WIFFEN, J. & MOLNAR, R. E. 1988. First pterosaur from New Zealand. Alcheringa, 12, 53–9.

WIFFEN, J. & MOLNAR, R. E. In press. An Upper Cretaceous ornithopod from New Zealand. Geobios.

The nature and timing of biotic links between New Zealand and Antarctica in Mesozoic and early Cenozoic times

G. R. STEVENS

New Zealand Geological Survey, Department of Scientific and Industrial Research, Lower Hutt, New Zealand

Abstract: Until the Early Cretaceous, when the first phases of rifting that preceded the eventual breakup of the Australasian sector of Gondwana occurred, New Zealand and Antarctica had a close geographic relationship on the eastern margin of Gondwana.

After having been influenced by proximity to the Permian ice sheet, during which New Zealand and Antarctica shared elements of the *Glossopteris* flora, the eastern margin of Gondwana entered into a phase of predominantly cool-temperate climates that lasted throughout the Triassic.

In the Jurassic, rotation of Gondwana had moved New Zealand and Antarctica into mid-latitudes and they experienced warm-temperate climates. These climatic conditions facilitated the dispersal of Tethyan marine faunas into the SW Pacific sector of eastern Gondwana. New Zealand and Antarctica shared many elements of the Tethyan marine faunas.

Starting in Middle Jurassic times, earth movements of the Rangitata Orogeny began to create considerable areas of land in the New Zealand region. Eventually, by Late Jurassic and Early Cretaceous times, a large landmass had been developed, extending northwards towards New Caledonia, eastwards to the Chatham Islands, westwards to the Lord Howe Rise and southwards to the edge of the Campbell Plateau. Creation of this landmass, and its associated land and marine links, together with the equable warm-temperate conditions then apparent over large areas of Gondwana, provided favourable conditions for the dispersal of Gondwana elements into New Zealand. Ancestral stocks of at least some of New Zealand's archaic endemic terrestrial biota may have dispersed from eastern Gondwana into New Zealand at this time.

In the Early Cretaceous, land connections to eastern Gondwana became weakened by the onset of rifting along the future sites of the Tasman Sea and Southern Ocean. The eastern edge of Gondwana was rotated into high latitudes, and cool-temperate climates returned to New Zealand and Antarctica. Cool-temperate Austral marine faunas developed and were shared between New Zealand, Antarctica and southern South America. Land links between New Zealand and West Antarctica provided access for early angiosperms. However, all land links between New Zealand and Australia/Antarctica were broken after 85 Ma. From this time onwards the ancestral Tasman Sea and Southern Ocean became effective barriers to overland dispersal between eastern Gondwana and New Zealand. Subsequently, all terrestrial colonists had to arrive by either flying, swimming or floating. Many birds did so, but no terrestrial snakes and no mammals, except bats.

Although land connections had been broken in the Late Cretaceous, shallow-water marine links continued to exist between New Zealand and West Antarctica until the late Paleocene (the Weddellian Province). However, as New Zealand moved progressively northwards and the Southern Ocean widened, such southern links gave way in latest Paleocene and early Eocene times to northern (Australian and Malayo-Pacific) links.

Until it became increasingly ice-bound from the late Paleogene onwards, Antarctica played an important role in SW Pacific palaeobiogeography and, in particular, had a major influence on New Zealand biotas. It is therefore appropriate in the context of this symposium to examine the nature and timing of the biotic links that existed between New Zealand and Antarctica in the past.

Although the topic of southern palaeobiogeography and New Zealand/Antarctic relationships has been reviewed by Fleming (1963, 1975, 1979 *a*, *b*) and Stevens (1980 *a*, *b*, 1985), nonetheless a re-examination is timely because of the progress that has recently been achieved in a number of relevant areas of study, viz.: improved geological and palaeontological knowledge of Antarctica (e.g. Thomson 1983 *a*, *b*; Mutterlose 1986; Crame 1986, 1987); improved reconstructions of Gondwana (e.g. Lawver & Scotese 1987; de Wit *et al*. 1988); new palaeomagnetic results and syntheses (e.g. Grunow *et al*. 1987a; Grunow *et al*. 1987b; and development of 'molecular clocks' as a method

From Crame, J. A. (ed.), 1989, *Origins and Evolution of the Antarctic Biota*, Geological Society Special Publication No. 47, pp. 141–166.

of determining times of separation of animals from ancestral stocks (e.g. Sibley & Ahlquist 1987).

Much of the information presented in this paper is interpretative, particularly relating to dispersal routes and land links. The interpretations are largely palaeobiogeographic, being based on fossil occurrences and distribution patterns observed from the fossil record and on the affinities of the modern biota (as summarized in Fleming 1975; Suggate *et al.* 1978; Stevens & Speden 1978; Stevens 1980*a*).

The New Zealand marine record is reasonably representative, and as it has been documented by Stevens (1980*a*), it is not discussed in any detail in this paper. However, it should be noted that some major gaps exist in the marine record, with, for example, most of the time missing between middle Tithonian and late Aptian (Stevens & Speden 1978; Edwards *et al.* 1988; Stevens in press).

Apart from a sparse floral record, often substantially biased towards coastal and estuarine environments (e.g. Retallack 1987), a record of terrestrial life in New Zealand is non-existent and fossils of many of the archaic groups (e.g. *Sphenodon*, *Leiopelma*, Dinornithiformes etc.) that have probably lived in New Zealand since early Mesozoic times are known only from the latest Cenozoic (e.g. Fleming 1975, p. 67; Worthy 1987 *a*, *b*).

In the absence of fossil evidence to the contrary, it has been assumed that most of the taxa discussed in this paper were immigrants into the New Zealand region, probably from elsewhere in Gondwana (cf. Nelson 1975). However, as is now recognized in the Proteaceae, for example (Pocknall & Crosbie 1988), future discoveries may show that at least some of these taxa may have originated in Australasia. Therefore the possibility of in situ evolution should not be entirely discounted in the ensuing discussions.

Judging from its geological record, New Zealand had a long history of southern or more generally 'eastern Gondwana' biotic links during the Palaeozoic and Mesozoic. As summarized by Fleming (1975) and Stevens (1980*a*), these links were maintained throughout periods of major geographic and climatic change, with New Zealand's position varying from low to high latitudes.

Within the context of such southern and 'eastern Gondwana' biotic links, New Zealand and Antarctica have had close marine connections, first as part of Gondwana, sharing cool-temperate Palaeoaustral faunas and, at times, warm-temperate Tethyan faunas; then, with the fragmentation of Gondwana and the opening up of ocean between New Zealand and Antarctica, Neoaustral marine faunas appeared, reliant upon the newly-developed circum-Antarctic current systems for their dispersal (Fleming 1975, 1979*b*).

Before the fragmentation of Gondwana, New Zealand and Antarctica also undoubtedly had close terrestrial links. However, because of the paucity of the terrestrial fossil record, such links are difficult if not impossible to document and can only be postulated by extrapolating from the affinities of the marine faunas and relationships of the present New Zealand flora and fauna.

The rafting-in and docking of exotic blocks (terranes) has introduced complexities into the geology and palaeobiogeography of the New Zealand Mesozoic (e.g. Howell *et al.* 1984; Hallam 1986; Cooper 1987; Archbold 1987). Notwithstanding these complexities, the following account is based primarily on the faunas of the Murihiku Supergroup, formed in an island arc situation marginal to eastern Gondwana. The faunas of the Torlesse Supergroup, although sometimes sharing elements with the Murihiku, have a significant number of out-of-place or 'exotic' elements. Although it is generally recognized that the Torlesse contains a number of displaced terranes, their exact number and disposition is open to debate (e.g. Retallack 1984, 1987; Spörli 1987; Norris & Craw 1987). The Torlesse terranes form a mosaic or collage along the eastern margin of New Zealand (e.g. Bishop *et al.* 1983; Coombs 1985). Most if not all of these displaced terranes are allochthonous relative to the Murihiku sequences and have been rafted into the New Zealand region from elsewhere. As far as is currently known, the accretion of displaced terranes ended in the latest Jurassic or earliest Cretaceous, and from this time onwards the observed faunal affinities relate entirely to the autochthonous New Zealand situation.

Permian–Triassic

In the Permian the slow drifting of Gondwana across the South Pole had progressively moved the centre of Carboniferous–Permian glaciation from southern Africa and southern South America to Antarctica and Australia (Crowell & Frakes 1975; Frakes 1979, 1986). At this time New Zealand was situated at 60°–70° S and was part of an arc and forearc system that had developed immediately off-shore from the eastern Gondwana coastline (i.e. the cratonic areas of modern Australia, Tasmania and Marie Byrd Land, Antarctica) (Bradshaw *et al.*, 1981;

Korsch & Wellman 1988). As little, if any, land existed in the New Zealand area at this time, it probably escaped being glaciated. Nonetheless, during the successive glacial periods of the Permian it experienced cold climates, and at these times its seas were populated by cold water organisms (Waterhouse 1978; Waterhouse & Bonham-Carter 1975; Archbold 1987).

Although the land present in the New Zealand area throughout the Permian was largely ephemeral in nature, being mainly in the form of volcanic islands and archipelagoes, it nonetheless had reasonably close links to the adjacent eastern Gondwana cratonic areas (Australia, Tasmania and Antarctica), as indicated by the occurrence of *Glossopteris* at a number of localities (Mildenhall 1970, 1976; H. J. Campbell pers. comm.) and similar palynofloras (Crosbie 1985). However, these New Zealand occurrences are in a marine setting and the *Glossopteris* specimens occur as detrital elements in marine sediments.

In the Triassic the same global movements continued that had earlier rotated the continents of eastern Gondwana towards the South Pole. However, in the Triassic the movements gradually moved eastern Gondwana into mid-latitudes and, as the amount of land diminished in the polar regions, so climate improved. Nonetheless, New Zealand apparently remained in high latitudes (70°–80°S) for much of Triassic time. As it is likely that the world was ice-free during the Triassic (Hallam 1985), New Zealand climates were probably not cold but cool-temperate (Stevens 1980a, 1985).

The comparative coolness of New Zealand Triassic climates, compared to those of the remainder of eastern Gondwana, is reflected in the development of distinctive 'Maorian' marine faunas. These faunas had links with New Caledonia, Papua New Guinea and, presumably via Antarctica, to southern South America. The existence of such links has been interpreted as indicating that these countries were at similar southern latitudes and were joined by shallow-water marine routes (Fleming 1979b; Stevens 1980a; Grant-Mackie 1985).

As in the Permian, western New Zealand continued throughout the Triassic to be the site of an active island arc system, flanking the coastline of eastern Gondwana. Nonetheless, judging from general facies evidence, and more specifically, from sequences of non-marine beds, there was probably more land present in the New Zealand region in the Triassic than in the Permian (Stevens & Suggate 1978; Retallack 1987).

The new land areas that developed in the New Zealand region during the Triassic may have facilitated the establishment of more permanent land links to eastern Gondwana (in contrast to the probably more ephemeral links of the Permian). Evidence of the strength and continuity of such Triassic land links is provided by the occurrence of the Gondwana plant *Dicroidium* and other taxa in New Zealand (Retallack 1987). Also, assuming origins elsewhere than New Zealand (virtually impossible to prove or disprove), it is likely that influxes of other ancestral plant stocks came to New Zealand at this time, via the same land routes. These may have included the predecessors, if not the ancestors, of groups such as psilotales, ferns, lycopods, araucarians, podocarps and ginkgos (Fleming 1975, p. 10., 1977a, 1978).

Although in the Triassic New Zealand shared plant stocks such as *Dicroidium* with Australia and Antarctica, its high latitude position throughout much of Triassic time may have been a deterrent to many of the reptilian groups, for example, that populated other regions of Gondwana at this time (Colbert 1980, 1981; Rage 1988). Ancestral stocks of the New Zealand tuatara (*Sphenodon*) may have dispersed to New Zealand in the Jurassic, when climate was more equable and land links better than those of the Triassic.

Jurassic

Rotation of eastern Gondwana in a direction away from the South Pole continued in the Jurassic. The effect of this rotation changed New Zealand's geographical position from 70°–80°S in the Triassic to 60°–70°S in the Middle Jurassic (Stevens 1980a, 1985). As was evident in the Triassic, the shift of land away from the polar regions substantially improved world climates and Jurassic climates were appreciably more equable than those of the present day (Hallam 1975, 1985; Frakes 1986). The advent of equable climatic conditions over large areas of the world except for the northern Boreal regions (e.g. Brandt 1986; Doyle 1987a), and the opening-up of shallow water marine routes during phases of marine transgression, facilitated the spread of warm-temperate 'Tethyan' marine faunas (e.g. McKenzie 1987; Audley-Charles 1988). In eastern Gondwana these 'Tethyan' marine faunas became widely distributed around the Jurassic shorelines of Australia, New Caledonia, New Zealand and Antarctica (Stevens 1980a, 1985). Particularly strong links were established between New Zealand and West Antarctica in Late Jurassic time (Willey 1973; Mutterlose 1986; Crame 1986, 1987; Doyle 1987a).

Evolution of Greater New Zealand

Throughout Jurassic times the earth movements of the Rangitata Orogeny, the initial stirrings of which had been evident in the Middle and Late Triassic, became even more marked (Fleming 1967a, 1970; Stevens 1978; Suggate 1978; Bradshaw et al. 1981). These movements progressively folded and elevated above sea level much of the material that had been deposited in the basin structures of the Permian–Triassic arcs. As Jurassic time continued, more and more land appeared in the New Zealand region. Eventually, in Middle and Late Jurassic time, and extending into the Early Cretaceous, an extensive new landmass appeared and, although it is impossible to define its boundaries with precision, it is likely that it extended well beyond the shorelines of modern New Zealand: northwards to New Caledonia; westwards to the Lord Howe Rise; eastwards to the Chatham Islands; and southwards to the southern edge of the Campbell Plateau. As rifting had not yet begun on what was to become the sites of the Tasman Sea and Southern Ocean, such a landmass was probably also in close physical contact with Australia and Antarctica, which then had substantial areas of land (e.g. Storey et al. 1987; St John 1984). Long fingers of land and interlinked chains of islands probably also extended northwards from the new landmass towards Papua New Guinea and Indonesia, although exactly how far is not known.

The formation of such large areas of land in the Middle/Late Jurassic and Early Cretaceous probably constituted the greatest development of new land that ever occurred in the New Zealand region. To provide some indication of its size, this ancestral landmass ('Greater New Zealand') was about half the size of the modern Australian continent. By comparison, modern New Zealand occupies only 10% of the total area of 'Greater New Zealand', the remainder having been submerged after a Cretaceous and early Cenozoic history of topographic lowering by a combination of thermal subsidence, peneplanation and marine erosion (e.g. Stevens 1985).

Dispersal of 'archaic' elements in New Zealand biota

Development of 'Greater New Zealand' and its associated links to adjacent areas of eastern Gondwana, together with widespread equable climatic conditions, provided optimal dispersal opportunities for both marine and terrestrial organisms. Thus in Middle and Late Jurassic times the New Zealand area had substantial influxes of marine 'Tethyan' faunas, with wide-ranging affinities (Fig. 1). Also, it is likely that this was probably the most favourable time in both the Mesozoic and Cenozoic for New Zealand to receive the ancestors of many of the 'archaic' elements present in the modern New Zealand terrestrial biota, assuming that they originated in areas elsewhere than New Zealand. Among the plants, these Jurassic immigrants may have included ancestral stocks of araucarians, podocarps and early conifers. However, as already noted, influxes of at least some of these stocks may have also occurred in the Triassic.

Among the animals, it is probable that ancestral stocks of at least some of the following groups radiated into New Zealand from eastern Gondwana in the Jurassic: Amphibia (the New Zealand frog, Leiopelma), Reptilia (the tuatara Sphenodon, geckos and skinks); Onychophora (Peripatus); Insecta (e.g. the giraffe weevil Lasiorhynchus barbicornis; the wetas Deinacrida and Hemideina; various spider taxa); Annelida (e.g. the giant New Zealand earthworm Spenceriella gigantea); terrestrial gastropods (e.g. Paryphanta, Powelliphanta, Rhytida, Wainuia, Placostylus); other groups including slugs, freshwater crayfish, freshwater mussels and freshwater fish (see Keast 1973; Fleming 1975; Bull & Whitaker 1975; King 1987; Worthy 1987 a, b) (Fig. 2).

Although contrary opinions are held (e.g. Eskov 1987), it is proposed that ancestors of these modern organisms were originally widely distributed across eastern Gondwana and that establishment of land links between 'Greater New Zealand' and the cratonic areas of Australia and Antarctica facilitated their dispersion into New Zealand (and probably also New Caledonia). As emphasized by many authors (e.g. Fleming 1975; Briggs 1987), there is no fossil record to support such a proposal; however, sphenodontians are known from the Late Triassic, Jurassic and Early Cretaceous of both Gondwana and Laurasia (Fraser 1986; Benton 1986) and fossils related to Leiopelma are known from the late Jurassic of Patagonia (Estes & Reig 1973).

It is generally accepted that the Rangitata Orogeny reached its climax in the Late Jurassic and Early Cretaceous (Suggate 1978). From this it follows that 'Greater New Zealand' reached its maximum geographical extension

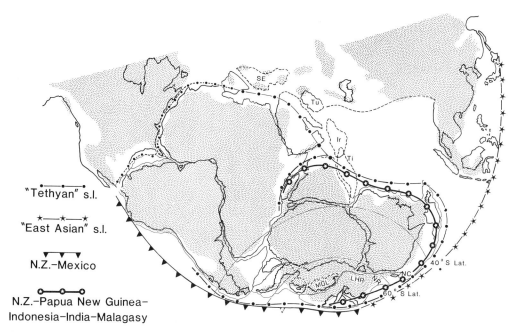

"Tethyan" s.l.

"East Asian" s.l.

N.Z.–Mexico

N.Z.–Papua New Guinea–
Indonesia–India–Malagasy

Fig. 1. In the Middle and Late Jurassic the positioning of New Zealand in mid-latitudes and the opening-up of shallow-water marine dispersal routes by the developing Rangitata Orogeny stimulated major influxes of 'Tethyan' taxa into the New Zealand region. This diagram shows a summary of the geographic affinities of the immigrant taxa that arrived in New Zealand during Bajocian–Tithonian times (see Stevens in press, for further details). The continental reconstruction is modified from Howarth (1981) and Lawver & Scotese (1987). The approximate distribution of land is indicated by the stipple pattern. SE, Southern Europe; Tu, Turkey; Ir, Iranian block; Ti, Tibetan block; NC, New Caledonia; NR, Norfolk Rise; LHR, Lord Howe Rise; MBL, Marie Byrd Land (Antarctica).

between late Tithonian and early Aptian. Also, at this time continuing rotation of Gondwana swung New Zealand even further northwards from its Middle Jurassic position. In Late Jurassic, and possibly also in earliest Cretaceous times, New Zealand was lying at about 55°S and was influenced by sub-tropical/warm-temperate conditions (Stevens 1980a, 1985). This period, extending from the Oxfordian/Kimmeridgian into the Tithonian and perhaps the Neocomian, is considered to have provided the last opportunity for sub-tropical/warm-temperate terrestrial organisms to disperse from eastern Gondwana to New Zealand and New Caledonia using continuous land routes. Towards the end of this time, rifting in areas that were later to open out to become the sites of the Tasman Sea and the Southern Ocean had probably begun to disrupt land connections, making land dispersion difficult.

A number of authors have suggested that the ancestral ratites used these land routes to enter New Zealand (Fleming 1982; Stevens 1985) (Fig. 3). This is based on the thesis that the ancestral ratites were already flightless at this time and needed continuous land routes for dispersion. However, opinion varies on when ratites lost the power of flight; if, for example, the ancestors of the moa and kiwi were able to fly to New Zealand then it is entirely possible that they dispersed across the Gondwana lands at a later time than the latest Jurassic and Early Cretaceous, when land links were at their optimum (Appendix 1).

Early Cretaceous

Onset of rifting

The phase in New Zealand's geological history when extensive land links existed between 'Greater New Zealand' and eastern Gondwana was brought to an end by the onset of extensive rifting along the western margin of New Zealand and along the western flank of the Lord Howe Rise, signalling the beginning of movements that eventually, in the Late Cretaceous, were to lead to development of sea floor in the Tasman Sea and Southern Ocean. These rifting movements, beginning in the late Neocomian and becoming particularly marked in Aptian–Albian times (Laird 1981; Whitworth *et al.* 1985; Spörli 1987; Tulloch & Kimbrough in press), probably disrupted many of the pre-existing land routes, although it is not possible to determine with any accuracy the nature and extent of such disruption. However, judging

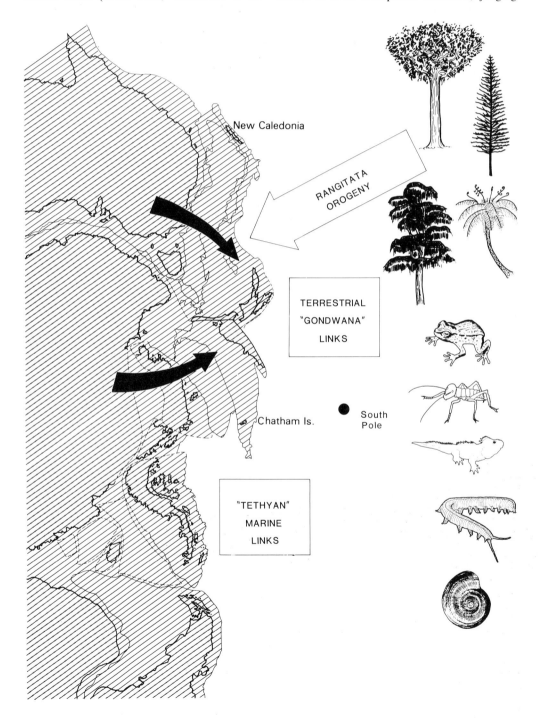

from the extent of Early Cretaceous rifting present both on- and off-shore along the western margin of New Zealand (King & Robinson 1988), and along the southern margin of Australia (Hegarty *et al.* 1988), there was probably substantial disruption of land links between New Zealand and Australia. Some semblance of land links to the south, to West Antarctica and South America, probably persisted until Late Cretaceous times, to provide routes for the radiation of early angiosperms, Proteaceae and *Nothofagus* (Mildenhall 1980; Askin 1989; Dettmann 1989).

During the Late Jurassic and Early Cretaceous, continental movements began in western Gondwana that were to have major effects in eastern Gondwana. The separation of Antarctica and India from Africa to form the Indian Ocean, and separation of Africa from South America to form the South Atlantic Ocean reversed the movement of eastern Gondwana that in the Triassic and Jurassic had been in a direction away from the South Pole (Stevens 1980*b*). As a result, New Zealand's geographical position changed from mid-latitude in the Jurassic to high latitude in the Early Cretaceous. It is postulated (Stevens 1971, 1980*a*, 1985) that the climatic change accompanying this rotation was probably equivalent in terms of modern climate to a change from sub-tropical/warm-temperate in the Jurassic to cool-/cold-temperate in the Early Cretaceous. Although there are indications of ice at the Cretaceous North Pole (Kemper 1987), there are no indications that Cretaceous climates in New Zealand were ever colder than cold-temperate, despite New Zealand being within

$5°-10°$ of the South Pole in the middle Cretaceous (Oliver *et al.* 1979).

Ice-rafted pebbles have been postulated from the Early Cretaceous of both South Australia and New Zealand (Waterhouse & Flood 1981; Frakes & Francis 1988), but the presence of a varied biota, including forest vegetation, (e.g. Stevens & Speden 1978; Raine *et al.* 1981; Rich & Rich 1988) indicates a cool/cold temperate climate, with marked seasonality.

The change in geographical position of the eastern Gondwana margin and the accompanying climatic cooling is reflected in the shallow-water marine biota by the re-appearance of southern 'Austral' groups; the first since Permian–Triassic times (Stevens & Fleming 1978; Stevens 1980*a*).

The extensive landmass of 'Greater New Zealand' resulting from the earth movements associated with the Rangitata Orogeny was immediately exposed to the effects of both terrestrial and marine erosion, and of thermal subsidence. By late Aptian times large areas of the land had been worn down to such low levels that marine transgression was taking place, particularly along the eastern margin of New Zealand (Stevens & Suggate 1978; Stevens & Speden 1978). This erosion and diminution of the 'Greater New Zealand' landmass, when combined with the development of active rifting zones, as noted above, contributed to the steady deterioration of the land links that had been a feature of Jurassic times. However, judging from the entry of ancestral angiosperms into New Zealand in late Albian times (Fleming 1975, p. 22) (Fig. 4), and of ancestral Proteaceae and *Nothofagus* in the Late Cretaceous

Fig. 2. Diagrammatic reconstruction of the eastern margin of Gondwana in the Middle Jurassic, *c.* 180–165 Ma, showing terrestrial 'Gondwanian' links. Base map for this and the succeeding diagrams, Figs. 3–5, modified from Grunow *et al.* (1987*a*); Grunow *et al.* (1987*b*), Lawver & Scotese (1987), de Wit *et al.* (1988).

The diagonal line pattern indicates regions in which extensive areas of land were present at the time, although its exact distribution is not known. At least some of the terranes of eastern New Zealand had not yet docked with the remainder of the country.

At this time, precursor movements of the Rangitata Orogeny were creating new land in the New Zealand region, although its extent is unknown (Speden 1971; Fleming 1967, 1970). Development of this new land provided opportunities for the dispersal of terrestrial biota from adjacent Gondwana lands, as indicated by the solid arrows. Biota that entered New Zealand at this time (and perhaps also earlier in the Triassic, when land links were also present), included at least some of the ancestral stocks of the 'archaic' elements present in the modern New Zealand flora and fauna. Representatives of such immigrant stocks are shown along the right-hand edge of the diagram: (from top to bottom) the Kauri, *Agathis australis*; araucarian stocks; podocarp stocks; tree-fern stocks; the New Zealand frog, *Leiopelma*; the weta, *Deinacrida heteracantha* (Rhaphidophoridae); the tuatara *Sphenodon punctatus*; *Peripatus*: the giant New Zealand land snail, *Paryphanta*.

The text placed in the boxes towards the right-hand side of each diagram (e.g. 'Rangitata Orogeny') highlights the factors that influenced biotic developments in the New Zealand region at the times covered by the reconstructions.

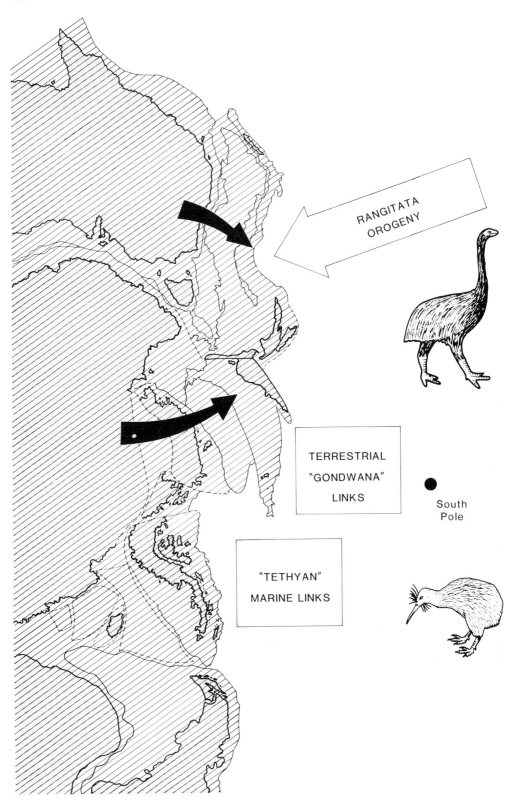

RANGITATA OROGENY

TERRESTRIAL "GONDWANA" LINKS

South Pole

"TETHYAN" MARINE LINKS

(Mildenhall 1980; Pocknall & Crosbie 1988; Dettmann & Jarzen 1988; Dettmann 1989) (Fig. 5), some reasonably continuous land links persisted, particularly towards the south (as rifting and penetration by marine embayments was active in the region between New Zealand and Australia; Veevers 1984, p. 80). The 'Austral' marine affinities evident at this time indicate that the climate along such land links was probably cool- or cold-temperate and the close affinities of New Zealand and South American *Nothofagus*, for example (Poole 1987), suggest a strong South American linkage, presumably via western Antarctica (Case 1988; Dettmann 1989). However, ancestral monotremes that were present in eastern Australia in Aptian–Albian times (Archer *et al.* 1985) may have been prevented from entering New Zealand because of climatic limitations or lack of suitable land connections. Alternatively, they may have been originally part of the New Zealand Mesozoic terrestrial fauna, but did not survive later times of severe habitat restriction (Appendix 2).

A phase of widespread rifting and subsidence had commenced along the south coast of Australia at about 125 Ma (Hegarty *et al.* 1988). This rifting phase gave way at about 90 Ma to an episode of slow sea-floor spreading between Australia and Antarctica (Cande & Mutter 1982; Veevers 1986, 1987; Hegarty *et al.* 1988). However, the onset of rapid sea-floor spreading and substantial separation of the two continents did not begin until 55 Ma (Weissel & Hayes 1972). The impact of this tectonic activity on biotic interchanges between Australia and Antarctica (and by extension, South America) during the period between 90 and 55 Ma is difficult to evaluate because of the extremely sparse terrestrial fossil record. However, occurrences of fossil Proteaceae, *Nothofagus* and other plants (Dettmann & Thomson 1987;

Pocknall & Crosbie 1988; Dettmann & Jarzen 1988; Dettmann 1989) indicate that terrestrial routes persisted across the area of rifting, probably in the region between Tasmania and North Victoria Land (Veevers 1987), and that these routes continued into at least the latest Cretaceous. The occurrence of fossil marsupials in the late Eocene of Seymour Island is also indicative of the persistence of southern terrestrial dispersal routes, presumably linking Australia, Antarctica and South America (Woodburne 1982; Woodburne & Zinsmeister 1982, 1983, 1984; Case *et al.* 1988; Case 1988).

Late Cretaceous

Opening of the Atlantic and Indian Oceans continued at a steady rate throughout Late Cretaceous time. These opening movements affected many sectors of southern Gondwana and had the effect of swinging New Zealand northwards away from the South Pole, so that in the Late Cretaceous it was between 65°–55°S (compared with 85°S in the middle Cretaceous) (Stevens 1980a, b, 1985).

The same movements had also rotated West Antarctica, so that it now straddled the South Pole. Nonetheless, conditions were evidently still not optimal for the accumulation of ice, as traces of Late Cretaceous glaciation are unknown (although the presence of ice-rafted detritus has been postulated for the Early Cretaceous of South Australia by Frakes & Francis 1988). On the contrary, there is abundant evidence from plant beds and fossil wood preserved in West Antarctica that large areas of the continent were probably clothed in forests (presumably of *Nothofagus* and other cool-temperate taxa; Dettmann 1989). Therefore, as in Early and middle Cretaceous times, the climate of New Zealand and other 'southern' lands was cool- or cold-temperate and (judging from tree

Fig. 3. Diagrammatic reconstruction of eastern Gondwana in the Late Jurassic, *c.* 150–145 Ma. The diagonal line pattern indicates regions in which extensive areas of land were present, although it should be noted that in Western Antarctica and southernmost South America, for example, there were also large areas of sea (Riccardi 1983; Thomson 1983b). In New Zealand the main movements of the Rangitata Orogeny were commencing at this time and substantial areas of land were present in the New Zealand region, although the exact distribution of such land is not known. In eastern New Zealand some terrane blocks were still being rafted in from the Pacific margin.

Although land links to adjacent areas of Gondwana had been present in earlier times (Late Triassic and Early-Middle Jurassic), such links were substantially improved in Late Jurassic and Early Cretaceous times (as indicated by the solid arrows), allowing the possibility for ancestral ratite stocks to disperse by walking between southern lands, although if they were dispersed by flying, a later date is possible (in the early Cenozoic, as indicated by biochemical data). The moa, *Dinornis*, and the kiwi, *Apteryx*, the two ratite taxa that colonised New Zealand, are illustrated to the right of the diagram.

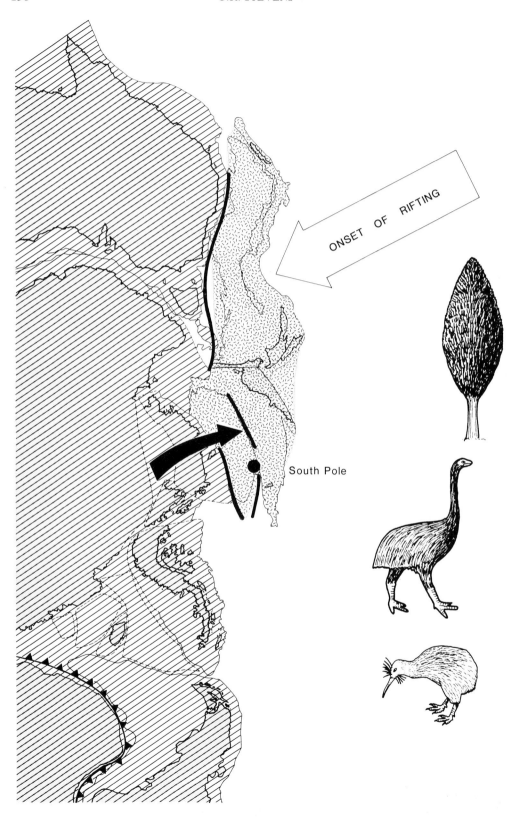

ONSET OF RIFTING

South Pole

rings preserved in fossil wood) had well-defined seasons (Jefferson 1982, 1983; Francis 1986; Parrish & Spicer 1988a, b; Case 1988).

Rifting movements that continued around the primaeval New Zealand landmass throughout Early and middle Cretaceous times culminated about 85 Ma in the establishment of open ocean in the Tasman Sea and in the Southern Ocean south of New Zealand (Fig. 6). The development of open oceanic conditions brought to an end any likelihood of land routes into New Zealand and New Caledonia from the remainder of Gondwana. Shallow-water marine links to the north and west were probably also severed. Nevertheless shallow-water marine routes continued to link New Zealand and New Caledonia (Stevens 1980a, b; Grant-Mackie 1985; Spörli 1987) and southern South America via West Antarctica, thus enabling New Zealand, New Caledonia, West Antarctica, and southern South America to share a 'southern' marine fauna (Fig. 7). Some elements of this southern marine fauna also extended to southern India, Malagasy, and southern Africa (e.g. Doyle 1987b, 1988; Stinnesbeck 1986).

The marine links between New Zealand, West Antarctica, and southern South America, although very strong, were not accompanied by the terrestrial links that had been evident in Early and middle Cretaceous times (i.e. those routes used by the ancestral angiosperms to enter New Zealand and New Caledonia). This is known because, in the Late Cretaceous, ancestral marsupials probably spread from South America to Australia (presumably by using Antarctica as a stepping stone), but apparently were not able to reach New Zealand (Woodburne 1982; Woodburne & Zinsmeister 1982, 1983, 1984; Benton 1985; Zinsmeister 1986; Case et al. 1988).

As the Tasman Sea and the Southern Ocean opened, and as sea-floor spreading moved the New Zealand landmass away from the remainder of eastern Gondwana, its topographic elevation was lowered still further by thermal subsidence (McKenzie 1978; Norton et al. 1985; Kamp 1986 a, b; Spörli 1987; Korsch & Wellman 1988). This subsidence was in addition to the rapid erosion, both terrestrial and marine, taking place simultaneously, so that by Late Cretaceous time many areas had been reduced to low relief. Widespread peneplanation, followed by marine incursion, occurred over large areas of the New Zealand landmass.

Biochemical data indicate that the ancestral stocks of the New Zealand 'wrens' (Xenicidae: rifleman, bush wren, rock wren, Stephens Island wren; Bull & Whitaker 1975; Fleming 1982) radiated to New Zealand between 85 and 90 Ma (Sibley & Ahlquist 1983b). As noted in Appendix 1, such biochemical 'dating' should be regarded with caution. Nonetheless, if this timing is accepted as having some validity, it means that the radiation of the ancestral stocks of the New Zealand 'wrens' may have occurred just before the commencement of sea-floor spreading in the Tasman Sea and the Southern Ocean, and that they were able to overfly the rift zones and marine embayments then actively developing around the New Zealand landmass.

Paleocene

The episode of sea-floor spreading that had commenced in the Tasman Sea in Late Cretaceous times, c. 85 Ma, ceased in the late Paleocene (c. 60 Ma) when the Tasman had probably reached its present width (1850 km), or slightly more (as its width was reduced by the late Cenozoic spreading of the Lord Howe Ridge; Stock & Molnar 1982). Sea-floor spreading continued south of New Zealand, between Antarctica and the southern edge of the Campbell Plateau. From the time when the ancestral

Fig. 4. Diagrammatic reconstruction of the eastern margin of Gondwana in the Early Cretaceous, c. 120–110 Ma. The main uplift phases of the Rangitata Orogeny were under way at this time. The Early Cretaceous, together with the Late Jurassic constitutes the greatest extension of land in New Zealand's geological history: — the development of the New Zealand microcontinent or 'Greater New Zealand' (indicated by the speckled pattern).

In the Aptian–Albian a rifting phase (indicated by thick lines) commenced along the west coast of New Zealand, along the western side of the Lord Howe Rise, in the Great South Basin and in the Bounty Trough. Major rifting occurred between the African and South American plates (indicated by the toothed pattern). Ancestral angiosperms appeared at about this time and radiated throughout Gondwana. Because of the disruption of land routes by the onset of rifting along the western edge of 'Greater New Zealand', heralding formation of the Tasman Sea, the ancestral angiosperms may have used a land route into New Zealand via Antarctica (as indicated by the solid arrow). Ancestors of the moas and kiwis of New Zealand may have also used this route (as also in the Late Jurassic, Fig. 3), if they were flightless at this time.

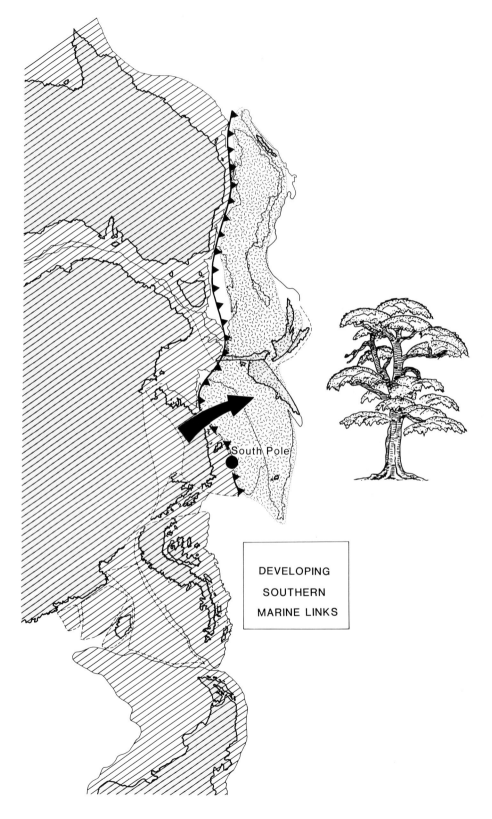

DEVELOPING

SOUTHERN

MARINE LINKS

South Pole

Tasman Sea and Southern Ocean became effective barriers to overland dispersal, all terrestrial colonists had to arrive by either flying or swimming/floating. Many birds did so, but no terrestrial snakes and no mammals except bats made the journey (Fig. 8).

Despite its increasing isolation, the changing archipelago of New Zealand continued to receive bird colonists from nearby lands, especially from Australia, where a rich and diverse avifauna was developing throughout the Cenozoic (Rich 1976).

After the ratites, which may have come by land (see above), the next bird colonists known to have arrived were the endemic families of New Zealand passerines: the New Zealand 'wrens', wattlebirds and 'thrushes' (Xenicidae, Callaeatidae and Turnagridae) of uncertain relationships, but some showing 'Gondwana' affinities with the ovenbirds and antbirds of South America, the lyrebirds of Australia and the pittas of the Old World tropics (Bull & Whitaker 1975; Fleming 1977b, 1982). The ancestral stocks of these endemic bird families may have arrived in the Paleocene, before the seas then opening around New Zealand became too wide and stormy. However, as mentioned earlier in this paper, biochemical data point to the possibility that the ancestral stocks of at least one endemic bird family, the New Zealand 'wrens', may have radiated to New Zealand in Late Cretaceous times (85–90 Ma, according to Sibley & Ahlquist 1983b).

At the same time as the ancestral stocks of the New Zealand endemic passerines were traversing the opening seas around New Zealand, the ancestors of the New Zealand short-tailed bat Mystacina may have made a similar journey. Although the two species of Mystacina are competent fliers, they have markedly terrestrial habits, are rodent-like and have a number of unique morphological adaptations (Daniel & Baker 1986). Possession of these distinctive features strongly suggests lengthy geographical isolation and, as a consequence, the relationship of Mystacina to other bats has hitherto been rather obscure. However, biochemical studies of Mystacina have demonstrated that it has close phylogenetic affinities with Central and South American phyllostomoid bats and have suggested (assuming that bats originated in the Paleocene) that the lineages separated about 35 Ma (Daniel & Baker 1986; Pierson et al. 1982, 1986). It is therefore likely that the ancestral stocks of Mystacina followed a route from South America via Antarctica, and, although as noted earlier, such biochemical 'dates' should be used with caution, radiation may have occurred sometime in the early Paleogene.

Judging from deep sea drilling results (e.g. Barker et al. 1987), large parts of Antarctica, particularly western Antarctica, were forested during the early Paleogene. These forested areas may have offered a variety of trans-Antarctic routes for ancestral Mystacina, which on encountering the steadily widening oceanic gaps then present around New Zealand, flew across them (perhaps with wind assistance), to colonise the 'Greater New Zealand' landmass. Such a passage may have been via the Campbell Plateau, areas of which were emergent at this time (Fig. 8). An alternative route may have involved radiation across Antarctica, thence into Australia and then via a route into New Zealand by means of island-hopping southwards along the Lord Howe Rise. The possible use of this alternative route is suggested by the presence on Mystacina of a tick normally parasitic on Australian bats (Daniel 1979).

The volant immigrants that came to New Zealand in early Paleogene times by flying across the newly created surrounding oceans

Fig. 5. Diagrammatic reconstruction of eastern Gondwana in the mid-Cretaceous, c. 100–95 Ma. By this time the altitude and extent of the 'Greater New Zealand' landmass uplifted by the Rangitata Orogeny (Figs 3 & 4) had been markedly reduced by thermal subsidence and by erosion, both marine and terrestrial; marine transgression was commencing along the east coast of New Zealand.

Active rifting, accompanied by development of finger-like marine embayments along the rift zones was occurring along the western margin of 'Greater New Zealand', marking the sites of the future Tasman Sea and Southern Ocean (indicated by the toothed pattern). The approximate distribution of land in the New Zealand region is indicated by the speckled pattern. The ancestral stocks of Nothofagus and Proteaceae were radiating to southern lands at about this time, using Antarctica as a stepping stone (Mildenhall 1980; Pocknall & Crosbie 1988; Dettmann & Jarzen 1988; Dettmann 1989). Entry of these stocks to New Zealand was probably via a southern route, as indicated by the solid arrow. Palaeogeographic information for the New Zealand region for this figure and for Figs 6 & 8 has been derived from Stevens & Suggate (1978), King & Robinson (1988), and Wood et al. (in press).

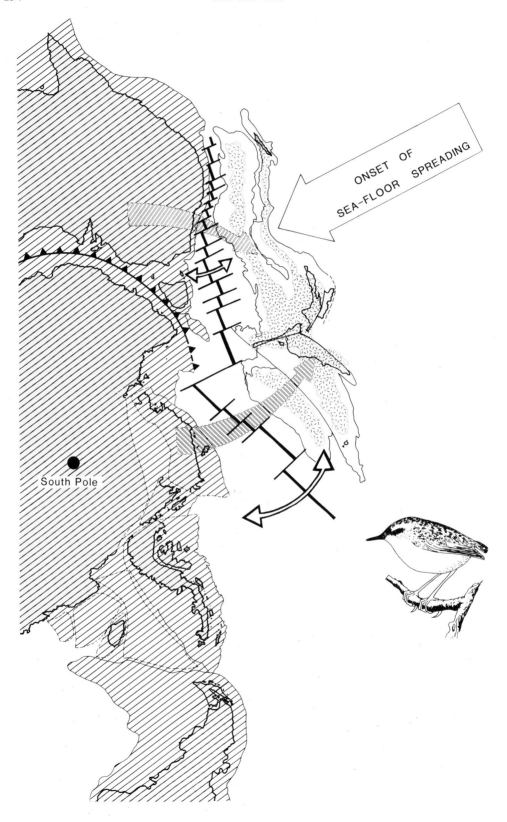

ONSET OF SEA-FLOOR SPREADING

South Pole

could have also included the flying ancestors of the ratites, if the views of Houde (1986) and the biochemical data of Sibley & Ahlquist (1987) are accepted.

The opening-up of ocean between New Zealand and the eastern margin of Gondwana had the effect of moving New Zealand northwards away from the South Pole, so that in the Paleocene it occupied a latitudinal position of 50°−45°S, compared with 65°−55°S in the Late Cretaceous (Stevens 1985).

The northwards movement of New Zealand brought it closer to regions influenced by warm-temperate oceanic currents. However, cool-temperate oceanic conditions continued at least into the early Paleocene, and many molluscan taxa were shared at this time between southern South America, the Antarctic Peninsula and New Zealand (the Weddellian Province of Zinsmeister 1979, 1982). The New Zealand 'Wangaloan' molluscan faunas (of the New Zealand Teurian stage) of cool-water aspect do not appear to continue in New Zealand after the late Teurian, whereas many 'Wangaloan' taxa remained in the Antarctic Peninsula region until the late Eocene (Zinsmeister 1976 a, b, 1979, 1982, 1987; Zinsmeister & Feldmann 1984).

The apparent absence of cool-water molluscan taxa in New Zealand after the late Teurian (i.e. late Paleocene; Edwards 1987) may merely be a reflection of the incomplete knowledge of late Teurian−Porangan molluscan faunas (P. A. Maxwell, pers. comm.). Alternatively, the absence may be a result of the extinction of cool-water taxa as New Zealand moved northwards and encountered warmer water masses. The progressive influx of warm water molluscan taxa, with Malayo-Pacific affinities, has been documented by Fleming (1967 b, 1975) Beu & Maxwell (1968), and Hornibrook (1978). The occurrence of the warm-water foraminiferan *Asterocyclina* in the latest Paleocene and earliest Eocene of the

Chathams Islands has been recorded by Cole (1967) and Hornibrook (in press).

Eocene

As sea-floor spreading drew to a close in the Tasman Sea in the late Paleocene, c. 60 Ma, it commenced in the segments of the Southern Ocean south of New Zealand and Australia. Although there had been slow spreading between Australia and Antarctica since the middle Cretaceous, c. 95 Ma (Cande & Mutter 1982; Veevers 1986, 1987), rapid spreading commenced in the early Eocene, c. 55 Ma. This increased rate of sea-floor spreading had an immediate and significant effect on the palaeo-latitudes of New Zealand, Australia and Antarctica. Antarctica moved southwards and Australia and New Zealand moved northwards. As a consequence, New Zealand's geographic position in the late Eocene was 45°−40°S latitude, compared with 50°−45°S latitude in the Paleocene (Stevens 1985) (Fig. 9). The gradual build-up of ice on Antarctica, commencing in East Antarctica in the earliest Oligocene, or earlier, and in West Antarctica in the late Miocene, has been recently documented by Legs 113 and 119 of the Ocean Drilling Programme (Barker *et al.* 1987; Barron *et al.* 1988). The build-up of ice and the opening-up of seaways around the continent put an end to Antarctica's role as a stepping stone for southern migrants. The opening of the Southern Ocean progressively weakened the shallow-water marine links between New Zealand, West Antarctica and South America. The 'Weddellian' provincial affinities that had been apparent in the marine molluscan faunas of these countries in the Late Cretaceous and Paleocene contracted in the latest Paleocene and early Eocene to include only West Antarctica and South America (Zinsmeister 1976a, b, 1979, 1982). Although, as already noted, knowledge is incomplete for New

Fig. 6. Diagrammatic reconstruction of eastern Gondwana in the Late Cretaceous, c. 75 Ma. Base map modified from Kamp (1986a). Rifting and slow spreading had been occurring between Australia and Antarctica since c. 95 Ma (indicated by the toothed pattern). Active sea-floor spreading had been occurring since c. 85 Ma in the Tasman Sea and in the sector of the Southern Ocean south of the southern edge of the Campbell Plateau (as indicated by the generalized sea-floor spreading pattern).

Although land connections between the New Zealand region and the remainder of Gondwana had been broken, it was still possible for volant animals to overfly the developing oceans before they became too wide and stormy (as indicated by the two broad-striped arrows). Biochemical data suggest that ancestral stocks of the New Zealand 'wrens' may have dispersed to New Zealand at this time. The approximate distribution of land in the New Zealand region is indicated by the speckled pattern.

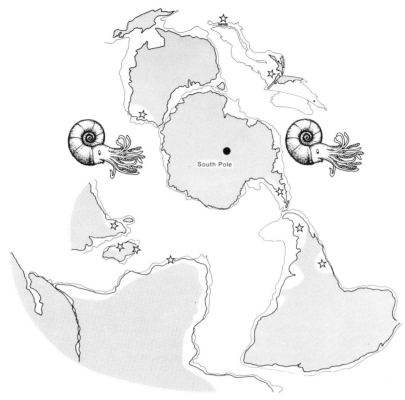

Fig. 7. The positioning of New Zealand and other southern Gondwana lands in high latitudes in the Cretaceous facilitated the development of distinctive 'southern' or 'Austral' groups in shallow-water marine faunas. Illustrated in this diagram is an example of one such 'Austral' group, the ammonite family Kossmaticeratidae, that achieved a southern distribution in the Late Cretaceous (stars). The distributional data are from Basse (1953) Thomson (1981) and Macellari (1986, 1987).

Zealand molluscan faunas for the latest Paleocene−early middle Eocene interval, there appears to be a major presence of warm-water taxa and an absence of cool-water taxa (which is probably a reflection of New Zealand's northwards movement into regions of warmer water masses).

The better known Bortonian (late middle Eocene) faunas of New Zealand, although predominantly of sub-tropical aspect, also include some 'Austral' taxa (e.g. *Speightia*, *Monalaria*). However, these are thought to be relict stragglers which became extinct at the end of the Bortonian. It may be concluded therefore, that, whereas continuous land links between New Zealand and Antarctica had probably virtually disappeared sometime during Campanian−Maastrichtian times (c. 85 Ma), shallow water marine links persisted for perhaps another 25−30 Ma, into the late Paleocene, before being finally severed by the continuance of sea-floor spreading in the developing Southern Ocean.

Although 'southern' Mollusca continued to be shared by New Zealand and South America at times throughout much of the remainder of the Cenozoic (e.g. Beu & Dell in press), they were types with free-swimming larval or adult stages (the 'Neoaustral' elements of Fleming 1975) capable of dispersion across deep oceanic areas; their distribution was undoubtedly assisted by the Circum-Antarctic oceanic current.

Conclusions

New Zealand and Antarctica had a close geographical relationship on the eastern margin of Gondwana until Early Cretaceous times, when the first rifting that eventually culminated in break-up in the Late Cretaceous, took place.

In the Triassic, New Zealand and Antarctica were sited in high southern latitudes, experienced cool-temperate climates, and shared elements of the southern terrestrial *Dicroidium* flora. Although there is no fossil evidence,

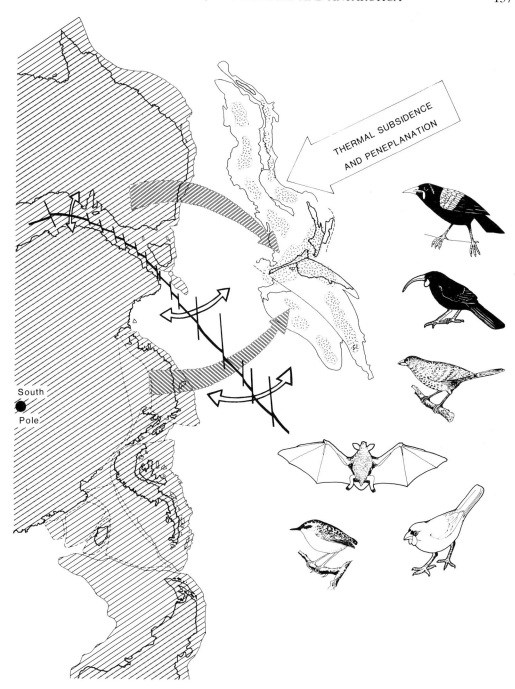

Fig. 8. Diagrammatic reconstruction of eastern Gondwana in the late Paleocene, *c.* 57 Ma. Base map modified from Kamp (1986*a*). Active sea-floor spreading had ceased in the Tasman Sea at *c.* 60 Ma. Sea-floor spreading was continuing in the Southern Ocean south of the Campbell Plateau and commenced between Australia and Antarctica at *c.* 55 Ma (as indicated by the generalised sea-floor spreading pattern).

Although land connections between New Zealand and the remainder of Gondwana had been broken in the Late Cretaceous, *c.* 85 Ma, it is likely that flying animals could still cross the developing oceans (as indicated by the two broad-striped arrows). Such animals may have included the ancestors of the New Zealand short-tailed bat (*Mystacina*) and the stocks that gave rise to the endemic families of New Zealand passerine birds. The drawings along the right-hand edge of the diagram are of representatives of families that probably arrived at about Paleocene time. From top to bottom: saddleback (*Philesturnus carunculatus*), huia (*Heteralocha acutirostris*), New Zealand thrush (*Turnagra capensis*), New Zealand short-tailed bat (*Mystacina tuberculata*), kokako or blue-wattled crow (*Callaeas cinerea*), rock wren *(Xenicus gilviventris).*

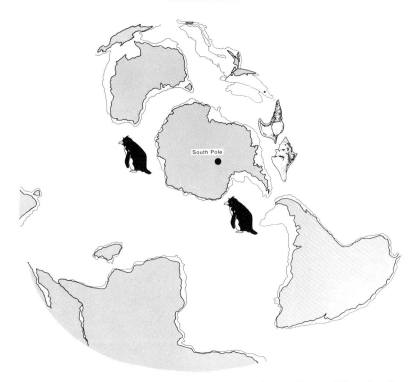

Fig. 9. By late Eocene times, *c.* 42 Ma, widening oceanic gaps had made the possibility of shallow-water marine links between the southern lands very remote. Southern seas were beginning to be stirred by marine current systems that were the forerunners of the modern Circum-Antarctic Current, although full development of an integrated system was delayed until the South Tasman Rise had cleared the east Antarctic shelf at *c.* 34 Ma (Kennett *et al.* 1975) and the Drake Passage had opened up between South America and the Antarctic Peninsula at *c.* 29 Ma (Barker & Burrell 1977). Nonetheless, sufficient migration opportunities were available for ancestral penguins to achieve a broad southern distribution, and many other marine groups with good dispersal capabilities, either as adults or juveniles, achieved similar distributions by utilizing the developing Circum-Antarctic marine currents. The illustrations depict the southern gastropods *Struthioptera* (above) and *Speightia* (below). Base map from Smith *et al.,* 1981.

ancestral stocks of at least some of New Zealand's archaic endemic terrestrial biota may have radiated from eastern Gondwana into New Zealand during Triassic–Jurassic times. These ancestral stocks may have included the forerunners of modern plants such as araucarians, podocarps, tree ferns, kauri (*Agathis*) and animals such as the New Zealand frog (*Leiopelma*), tuatara (*Sphenodon*), *Peripatus*, various insects, annelids, land snails etc.

In the Jurassic, rotation of Gondwana had moved New Zealand and Antarctica into mid-latitudes and they experienced warm-temperate climates. As a result, Tethyan marine faunas dispersed into the south-west Pacific sector of eastern Gondwana. At this time, New Zealand and Antarctica had very close Tethyan marine relationships.

Starting in Middle Jurassic times, earth movements of the Rangitata Orogeny began to create

considerable areas of land in the New Zealand region. Eventually, by Late Jurassic and Early Cretaceous times, a large landmass had been developed, extending northwards towards New Caledonia, eastwards to the Chatham Islands, westwards to the Lord Howe Rise, and southwards to the edge of the Campbell Plateau. Creation of this landmass, and its associated land and marine links, together with the equable warm-temperate conditions then apparent over large areas of Gondwana, provided favourable conditions for the dispersal of Gondwanan elements into New Zealand.

In the Early Cretaceous, land connections to eastern Gondwana became weakened by the onset of rifting along the future sites of the Tasman Sea and Southern Ocean. Creation of new sea floor in the South Atlantic and Indian Oceans rotated the eastern edge of Gondwana into high latitudes, and cool-temperate climates returned to New Zealand and Antarctica. Cool-temperate Austral marine faunas developed. The timing and extent of the onset of rifting and accompanying marine embayment probably varied around the New Zealand landmass, and although some disruption of land disperal routes occurred, particularly those between New Zealand and Australia, some land links still existed, notably between New Zealand and West Antarctica, to provide access for the early angiosperms in the Aptian−Albian, and *Nothofagus* and Proteaceae in the Santonian-Campanian. However, the initiation, at *c.* 85 Ma, of sea-floor spreading on the sites of the Tasman Sea and the Southern Ocean effectively put an end to the possibility of land links between New Zealand and the adjacent Gondwana lands (Australia and Antarctica). After 85 Ma, potential terrestrial colonists of New Zealand had to traverse the steadily widening Tasman Sea and Southern Ocean. Many birds did so, but no terrestrial snakes and no mammals except bats.

Although land connections had been broken, New Zealand and Antarctica continued to have close shallow-water marine links until the late Paleocene (Weddellian Province). However, such links were progressively supplanted in latest Paleocene and early Eocene times by strengthening northern links (Australian and Malayo-Pacific), as New Zealand moved progressively northwards and came under the influence of warm-temperate water masses.

I wish to thank the Convenor of the Symposium, J. A. Crame, for the invitation to present this paper at the Symposium and the Trans-Antarctic Association and NZ Geological Survey for financial support. The paper has benefitted considerably from reviews by NZ Geological Survey colleagues (H. J. Campbell, R. A. Cooper, N. de B. Hornibrook, P. A. Maxwell, D. T. Pocknall, G. H. Scott), J. A. Case (University of California, Riverside) and C. Macellari (University of South Carolina). The manuscript has been typed by M. Cheesman, NZ Geological Survey. The diagrams have been prepared by my wife, D. Stevens (cartography) and R. C. Brazier (illustrations of flora and fauna). All the abovenamed people are warmly thanked for their much appreciated efforts.

Appendix 1

Origins of the Ratites

The origin of the ancestral stocks of the flightless birds or ratites, including the extant kiwis (Apterygiformes) and the extinct moas (Dinornithiformes), the two ratite groups endemic to New Zealand, has been the subject of considerable debate. The rarity of fossil ratites, combined with the fact that they have not

been found in strata older than the Miocene (Fleming 1975, p. 67; Howgate 1986), has meant that emphasis has been placed on interpretations based on embryology, comparative morphology and biochemistry (e.g. McGowan 1982, 1984; Rich & Balouet 1984).

There are two alternative views of the origin of the ratites: (i) they were derived originally from the flying birds and later became flightless after dispersal across the old landmasses, (ii) they have always been flightless, or at the most, poor fliers.

The original flight capabilities of the ratites, and the timing of their loss of flight is of major biogeographical significance, as it provides the key to their dispersal patterns in the geological past.

It has been suggested by a number of authors (e.g. Cracraft 1974 *a*, *b*, 1976, 1986; Carroll 1988: 347−349) that an ancestral flightless ratite stock became widely distributed across Gondwana at a stage when it was a continuous landmass and that this stock gave rise to the various lineages of modern ratite birds. However, studies in embryology and comparative morphology have suggested that although the ratite condition was closer to that of reptiles than to carinates, ratites do share some features with carinates that appear to be correlated with flying. This suggests that ratites evolved from a volant ancestor, but one in which the flying mechanism was not as advanced as in modern flying birds (McGowan 1982, 1984, 1986). Studies of the *Lithornis* group of fossil birds in the Paleocene of Europe and North America also support a flighted ancestry for the ratites (Houde 1986).

Biochemical (i.e. 'molecular clock') analyses of ratites, although supporting the concept that the ancestors of these birds made their way across Gondwana before it broke up, also indicate that the ancestral stocks of the kiwi (and presumably also those of the moa) split off from those of the emu and cassowary at a time (40−45 Ma) much later than that indicated by geological evidence. The commonly accepted view (e.g. Rich & Balouet 1984) is that the ancestors of the moa and kiwi walked into New Zealand about 140−120 Ma, in the latest Jurassic and earliest Cretaceous at a time when land connections and climatic conditions in the Southwest Pacific were at their most favourable for dispersal of land creatures (Fig. 3). The biochemical data indicate, however, that 40−45 Ma may be a more likely date for the arrival of ratites in New Zealand. If this date is confirmed by future work the implication is that ancestral ratites reached New Zealand from eastern Gondwana after the opening of the Tasman Sea and the Southern Ocean by directly overflying the newly-created oceanic areas, probably as in the manner proposed for other endemic New Zealand bird groups (see Paleocene section above). An alternative suggested by Diamond (1983) and Sibley & Ahlquist (1981, 1983*a* 1987), is that the ancestral ratites dispersed by island-hopping, via intervening island arcs or emergent land, presumably along the now submerged Norfolk and Lord Howe ridges. The availability of such a route to facilitate island-hopping is not known.

As a concluding comment on the use of biochemical data, the mathematical basis for the so-

160 G.R. STEVENS

called 'molecular clock' has been questioned (Lewin 1983; Schwartz 1984; Howgate 1986; Houde, 1986). Therefore in the light of geological as well as mathematical difficulties, an age of 40–45 Ma for the dispersal of the ratites should not be accepted unreservedly.

Appendix 2

Radiation of the Monotremes

The absence of monotremes from New Zealand (Fleming 1975, p. 15; Daniel & Baker 1986, p. 2) appears to be anomalous, when contrasted with the fact that as noted earlier in this paper, both New Zealand and Australia shared in the radiation of various terrestrial organisms across the Gondwana continents in Mesozoic time. Furthermore, this anomaly has been reinforced by the recent discovery of a fossil monotreme, *Steropodon galmani*, from the Aptian–Albian of Australia (Archer *et al.* 1985). This new discovery indicates that monotremes had inhabited Australia from at least Early Cretaceous times, when land connections were probably present between Australia and New Zealand (see Early Cretaceous section above).

Because the monotremes have left a fragmentary fossil record, at least for the critical early stages of their evolutionary development, discussion of monotreme origins and dispersal routes involves substantial speculation (Augee 1984; Rich 1982).

A major factor in any discussion of monotreme origins is the divergence of zoological opinion on whether monotremes are mammal-like reptiles or reptile-like mammals. If viewed as reptilian they may be considered to be the only living survivors of the world-wide radiation of mammal-like reptiles that occurred in Triassic times (Lillegraven 1979; Kemp 1982, 1983). If, on the other hand, monotremes are viewed as mammals they may have evolved much later, perhaps in early Cretaceous times (Kemp 1982, 1983; Archer *et al.* 1985). Because palaeogeographical patterns and opportunities for land dispersal varied markedly between Triassic and Cretaceous times, the timing of radiation of the monotremes is of crucial importance. In the Triassic, the availability of land routes enabled a number of land animals to radiate widely across both Laurasia and Gondwana (e.g. Colbert 1980, 1981; Rage 1988). By Cretaceous times, however, many of the land routes had been disrupted by areas of newly formed ocean.

Recent publications favour a mammalian classification for the monotremes (Archer 1984; Archer *et al.* 1985; Griffiths 1983; Strahan 1983). If this view gains wide currency the monotremes, because they are so different from other living mammal groups, can be considered to have diverged at a very early stage from the ancestral mammalian stock. In this context, it has been suggested that they were part of the radiation of the Multituberculata, a group of mammals that became widely distributed in the Northern Hemisphere in Jurassic and Cretaceous times (Clemens 1979; Murray 1984; Savage & Long 1986). In this scenario, radiation of ancestral monotremes into the Southern Hemisphere probably took place in Early Cretaceous, using the land connections that had been developed between many of the Gondwana continents at that time. However, the warm-temperate trans-Tasman route into New Zealand had been disrupted by this time, and the only available route (used by the early angiosperms) was a southern one, probably with cool-temperate or cold-temperate characteristics. Climatic conditions along such a route, and the presence of links with South America and West Antarctica, rather than with Africa and Eurasia, may have been factors militating against its use by ancestral monotremes. As a consequence, ancestral monotremes may have been discouraged from entering New Zealand.

However, judging from data presented by Dettmann (1981), White (1986, pp. 174–5), Frakes & Francis (1988) and Rich & Rich (1988) it is probable that *Steropodon galmani*, the fossil monotreme from the Aptian–Albian of New South Wales, Australia (Archer *et al.* 1985) was adapted to life in cool- or cold-temperate habitats. Therefore the cool- or cold-temperate nature of the terrestrial southern dispersal routes into New Zealand in the Early Cretaceous may not have been in itself a major deterrent to the movement of ancestral monotremes.

An alternative and perhaps more likely hypothesis is that monotremes were originally present in New Zealand, but were eliminated during times of severe habitat restriction: notably during the Oligocene marine transgression, when more than two-thirds of the area of modern New Zealand was submerged and the remaining land eroded to a peneplain (Stevens & Suggate 1978). Drastic environmental changes during the Pleistocene (and at the Cretaceous/Cenozoic boundary?) may also have been contributory factors.

References

ARCHBOLD, N. W. 1987. Southwest Pacific Permian and Triassic marine faunas: their distribution and implications for terrane identification. *In*: LEITCH, E. C. & SCHEIBNER, E. (eds) *Terrane Accretion and Orogenic Belts*. American Geophysical Union Geodynamics Series, **19**, 119–127.

ARCHER, M. 1984. Origins and early radiations of mammals. *In*: ARCHER, M. & CLAYTON, G. (eds) *Vertebrate Zoogeography and Evolution in Australasia*. Hesperian Press, Perth, 477–515.

——, FLANNERY, T. F., RITCHIE, A. & MOLNAR, R. E. 1985. First Mesozoic Mammal from Australia – an early Cretaceous Monotreme. *Nature*, **318**, 363–366.

ASKIN, R. A. 1989. Endemism and Heterochroneity

in the Seymour Island Campanian to Paleocene palynofloras. *In*: CRAME, J. A. (ed.) *Origins and Evolution of the Antarctic Biota*. Geological Society, London, Special Publication, **47**, 107–120.

AUDLEY-CHARLES, M. G. 1988: Evolution of the southern margin of Tethys (North Australian region) from early Permian to late Cretaceous. *In*: AUDLEY-CHARLES, M. G. & HALLAM, A. (eds) *Gondwana and Tethys*. Geological Society, London, Special Publication, **37**, 79–100.

AUGEE, M. 1984. Quills and Bills: The curious problem of Monotreme zoogeography. *In*: ARCHER, M. & CLAYTON, G. (eds) *Vertebrate Zoogeography and Evolution in Australasia*. Hesperian Press, Perth. 567–570.

BARKER, P. A. & BURRELL, J. 1977. The opening of the Drake Passage. *Marine Geology*, **25**, 15–34.

— — *et al.* 1987. Leg 113 explores climatic changes. *Geotimes*, **32 (7)**, 12–15.

BARRON, J. *et al.* 1988. Early glaciation of Antarctica. *Nature*, **333**, 303–304.

BASSE, E. 1953. L'extension des *Kossmaticeras* dans les Mers Antarctico-Indo-Pacifiques au Neocretace. *Proceedings 7th Pacific Science Congress* **2**, 130–135.

BENTON, M. J. 1985. First marsupial fossil from Asia. *Nature*, **318**, 313.

— — 1986. The demise of a living fossil? *Nature*, **323**, 762.

BEU, A. G. & DELL, R. K. in press. Oligocene Mollusca in CIROS-1 Cores, McMurdo Sound, Antarctica *In*: BARRETT, P. J. (ed.) *Antarctic Cenozoic History: CIROS-1 Drillhole*. New Zealand Department of Scientific & Industrial Research Information Series.

— — & MAXWELL, P. A. 1968. Molluscan evidence for Tertiary sea temperatures in New Zealand: a reconsideration. *Tuatara*, **16**, 68–74.

BISHOP, D. G., BRADSHAW, J. D., & LANDIS, C. A. 1983. Provisional terrane map of South Island, New Zealand. *In*: HOWELL, D. G. (ed.) *Proceedings of the Circum-Pacific Terrane Conference*. Standford University Publications Geological Sciences, **18**, 24–31.

BRADSHAW, J. D., ADAMS, C. J. & ANDREWS, P. B. 1981. Carboniferous to Cretaceous of the Pacific margin of Gondwana: The Rangitata Phase of New Zealand *In*: CRESSWELL, M. M. & VELLA, P. (eds) *Gondwana Five*. A. A. Balkema, Rotterdam, 217–221.

BRANDT, K. 1986. Glacioeustatic cycles in the Early Jurassic? *Neues Jahrbuch für Geologie und Paläontologie Monatshefte* **1986 (5)**, 257–274.

BRIGGS, J. C. 1987. *Biogeography and Plate Tectonics* Elsevier, Amsterdam.

BULL, P. C. & WHITAKER, A. H. 1975. The Amphibians, Reptiles, Birds & Mammals. *In*: KUSCHEL, G. (ed). *Biogeography & Ecology in New Zealand*. W. Junk, The Hague, 231–236.

CANDE, S. C. & MUTTER, J. C. 1982. A revised identification of the oldest sea floor spreading anomalies between Australia and Antarctica. *Earth & Planetary Science Letters*, **58**, 151–160.

CARROLL, R. L. 1988. *Vertebrate Paleontology and Evolution*. W. H. Freeman, New York.

CASE, J. A. 1988. Paleogene floras from Seymour Island, Antarctic Peninsula. *In*: FELDMAN, R. M. & WOODBURNE, M. O. (eds) *Geology and Paleontology of Seymour, Island, Antarctic Peninsula*. Geological Society of America Memoir, **169**, 523–530.

— —, WOODBURNE, M. O. & CHANEY, D. S. 1988. A new genus of polydolopid marsupial from Antarctica. *In*: FELDMANN, R. M. & WOODBURNE M. O. (eds) *Geology and Paleontology of Seymour Island, Antarctic Peninsula*. Geological Society of America Memoir, **169**, 505–521.

CLEMENS, W. A. 1979. Notes on the Monotremata. *In*: LILLEGRAVEN, J. A., KIELAN-JAWOROWSKA, Z. & CLEMENS, W. A. (eds) *Mesozoic Mammals: The first two-thirds of Mammalian History*. University of California Press, Berkeley and Los Angeles, 309–311.

COLBERT, E. H. 1980. *Evolution of the Vertebrates*. 3rd Edition. John Wiley, New York.

— — 1981. The distribution of tetrapods and the break-up of Gondwana. *In*: CRESSWELL, M. M. & VELLA, P. (eds) *Gondwana Five*. A. A. Balkema, Rotterdam, 277–282.

COLE, W. STORRS. 1967. Additional data on New Zealand *Asterocyclina* (Forminifera). *Bulletins of American Paleontology*, **52**, 233.

COOMBS, D. S. 1985. New Zealand Terranes. *Geological Society of Australia Abstracts*, **14**, 45–48.

COOPER, R. A. 1987. Are terranes real or imaginary? *Geological Society of New Zealand Newsletter*, **78**, 44–54.

CRACRAFT, J. 1974*a*. Phylogeny and evolution of the ratite birds. *Ibis*, **116**, 494–521.

— — 1974*b*. Continental drift, palaeoclimatology and the evolution and biogeography of birds. *Journal of Zoology*, **169**, 455–545.

— — 1976. The species of Moa (Dinornithidae). *Smithsonian Contributions in Paleobiology*, **27**, 189–205.

— — 1986. The origin and early diversification of birds. *Paleobiology* **12**, 383–399.

CRAME, J. A. 1986. Late Mesozoic bipolar bivalve faunas. *Geological Magazine*, **123**, 611–618.

— — 1987. Late Mesozoic bivalve biogeography of Antarctica. *In*: McKENZIE, G. D. (ed.) *Gondwana Six: Sedimentology & Paleontology* American Geophysical Union Geophysical Monograph, **41**, 93–102.

CROSBIE, Y. M. 1985. Permian palynomorphs from the Kuriwao Group, Southland, New Zealand. *New Zealand Geological Survey Record*, **8**, 109–119.

CROWELL, J. C. & FRAKES, L. A. 1975. The Late Palaeozoic Glaciation. *In*: CAMPBELL, K. S. W. (ed.) *Gondwana Geology*. Australian National University Press, Canberra, 313–331.

DANIEL, M. J. 1979. The New Zealand short-tailed bat, *Mystacina tuberculata*; a review of present knowledge. *New Zealand Journal of Zoology*, **6**, 357–70.

— — & BAKER, A. 1986. *Collins Guide to the Mammals*

of New Zealand. Collins Publishers, Auckland & London.

DETTMANN, M. E. 1981. The Cretaceous Flora. *In:* KEAST, A. (ed.) *Ecological Biogeography of Australia.* W. Junk, The Hague, 357–375.

—— 1989. Antarctica: Cretaceous Cradle of Austral temperate rainforests? *In:* CRAME, J. A. (ed.). *Origins and Evolution of the Antarctic Biota.* Geological Society, London Special Publication, **47**, 89–106.

—— & JARZEN, ‚D. M. 1988. Angiosperm pollen from uppermost Cretaceous strata of southeastern Australia and the Antarctic Peninsula. *Memoir of the Association of Australasian Palaeontologists,* **5**, 217–237.

—— & THOMSON, M. R. A. 1987. Cretaceous palynomorphs from the James Ross Island area, Antarctica — A pilot study. *British Antarctic Survey Bulletin,* **77**, 13–59.

DE WIT, M. *et al.* 1988. *Geological Map of Sectors of Gondwana reconstructed to their disposition 150 Ma* (2 map sheets). American Association of Petroleum Geologists, Tulsa, Oklahoma.

DIAMOND, J. A. 1983. Taxonomy by nucleotides. *Nature,* **305**, 17–18.

DOYLE, P. 1987a. Lower Jurassic–Lower Cretaceous Belemnite Biogeography and the development of the Mesozoic Boreal Realm. *Palaeogeography, Palaeoclimatology, Palaeoecology,* **61**, 237–54.

—— 1987b. The Cretaceous Dimitobelidae (Belemnitida) of the Antarctic Peninsula region. *Palaeontology,* **30**, 147–77.

—— 1988. The Belemnite family Dimitobelidae in the Cretaceous of Gondwana. *In:* WIEDMANN, J. & KULLMANN, J. (eds) *Cephalopods — Present and Past.* Schweizerbart'sche Verlagsbuchhandlung, Stuttgart, 539–552.

EDWARDS, A. R. 1987. Geological Time Scale (Cenozoic). *New Zealand Geological Survey Record,* **20**, 8.

——, HORNIBROOK, N. deB., RAINE, J. I. SCOTT, G. H., STEVENS, G. R. STRONG, C. P. & WILSON, G. J. 1988. A New Zealand Cretaceous-Cenozoic Geological Time Scale. *New Zealand Geological Survey Record,* **35**, 135–149.

ESKOV, K. 1987. A new archaeid spider (Chelicerata: Araneae) from the Jurassic of Kazakhstan, with notes on the so-called "Gondwanian" ranges of recent taxa. *Neues Jahrbuch für Geologie und Palaeontologie Abhandlungen,* **175**, 81–106.

ESTES, R. & REIG, O. A. 1973. The early fossil record of frogs. *In:* VAIL, J. E. (ed.) *Evolutionary Biology of the Anurans.* University of Missouri Press, Columbia, Missouri, 11–63.

FLEMING, C. A. 1963. Paleontology and Southern Biogeography. *In:* GRESSITT, J. L. (ed.) *Pacific Basin Biogeography.* Bishop Museum Press, Honolulu, 369–385.

—— 1967a. Biogeographic change related to Mesozoic orogenic history in the southwest Pacific: *Tectonophysics,* **4**, 419–427.

—— 1967b. Cenozoic history of Indo-Pacific and other warm-water elements in the marine Mollusca of New Zealand. *Venus,* **25**, 105–117.

—— 1970. The Mesozoic of New Zealand: Chapters in the history of the Circum-Pacific Mobile Belt. *Quarterly Journal of the Geological Society of London,* **125**, 125–170.

—— 1975. The Geological History of New Zealand and its Biota. *In:* KUSCHEL, G. (ed.) *Biogeography and Ecology in New Zealand.* Monographiae Biologicae Vol. 27. W. Junk, The Hague, 1–86.

—— 1977a. The History of Life in New Zealand Forests. *New Zealand Journal of Forestry,* **22**: 249–262.

—— 1977b. Review: Feduccia, A. 1976. 'A Model for the Evolution of Perching Birds'. *Systematic Zoology,* **26**, 19–31. Also: *Notornis* **24**, 297.

—— 1978. The History of Life in New Zealand Forests. *Forest and Bird,* **210**, 2–10.

—— 1979a. *The Geological History of New Zealand and its Life.* Auckland University Press and Oxford University Press, Auckland, New Zealand.

—— 1979b. Evolution of the South Pacific Marine Biota: The Expanding Fossil Record. Proceedings of the International Symposium on Marine Biogeography and Evolution in the Southern Hemisphere. *NZ Department of Scientific and Industrial Research Information Series,* **137**, 5–26.

—— 1982. *George Edward Lodge: The unpublished New Zealand Bird Paintings.* Nova Pacifica, Wellington.

FRAKES, L. A. 1979. *Climates Through Geologic Time.* Elsevier, New York.

—— 1986. Mesozoic–Cenozoic climatic history and causes of glaciation. *In:* HSU, K. J. (ed.) *Mesozoic and Cenozoic Oceans.* American Geophysical Union, Geodynamics Series, **15**, 33–48.

—— & FRANCIS, J. E. 1988. A guide to Phanerozoic cold polar climates from high-latitude ice-rafting in the Cretaceous. *Nature,* **333**, 547–9.

FRANCIS, J. E. 1986. Growth rings in Cretaceous & Tertiary wood from Antarctica and their palaeoclimatic implications. *Palaeontology,* **29**, 665–84.

FRASER, N. C. 1986. New Triassic sphenodontids from south-west England and a review of their classification. *Palaeontology,* **29**, 165–86.

GRANT-MACKIE, J. A. 1985. New Zealand–New Caledonian Permian–Jurassic faunas, biogeography and terranes. *New Zealand Geological Survey Record,* **9**, 50–2.

GRIFFITHS, M. 1983. Lactation in Monotremata and speculations concerning the nature of lactation in Cretaceous Multituberculata. *Acta Palaeontologica Polonica,* **28**: 93–102.

GRUNOW, A. M., DALZIEL, I. W. D., & KENT, D. V., 1987a. Ellsworth-Whitmore Mountains crustal block, Western Antarctica: new paleomagnetic results and their tectonic significance: *In:* McKENZIE, G. D. (ed.) *Gondwana Six: Structure, Tectonics and Geophysics.* American Geophysical Union Geophysical Monograph, **40**, 161–171.

——, KENT, D. V., & DALZIEL, I. W. D. 1987b. Mesozoic evolution of West Antarctica and the Weddell Sea Basin: new paleomagnetic constraints. *Earth and Planetary Science Letters*, **86**, 16–26.

HALLAM, A. 1975. *Jurassic Environments*. Cambridge University Press, Cambridge.

— — 1985. A review of Mesozoic climates. *Journal of the Geological Society, London*, **142**, 433–445.

— — 1986. Evidence of displaced terranes from Permian to Jurassic faunas around the Pacific margins. *Journal of the Geological Society, London* **143**, 209–216.

HEGARTY, K. A., WEISSEL, J. K. & MUTTER, J. C. 1988. Subsidence history of Australia's southern margin: constraints on basin models. *American Association of Petroleum Geologists Bulletin*, **72**, 615–33.

HORNIBROOK, N. de B. 1978. Tertiary Climate. *In*: SUGGATE, R. P., STEVENS, G. R. & TE PUNGA. M. T. (eds) *The Geology of New Zealand*. Government Printer, Wellington, 436–443.

— — in press. Micropaleontology. *In*: ANDREWS, P. B. *et al.* (eds) *Cretaceous-Cenozoic Geology of the Chatham Islands, New Zealand*. New Zealand Geological Survey Bulletin.

HOUDE, P. 1986. Ostrich ancestors found in the Northern Hemisphere suggest new hypothesis of ratite origins. *Nature*, **324**, 563–565.

HOWARTH, M. K. 1981. Paleogeography of the Mesozoic. *In*: COCKS, L. R. M. (ed.) *The Evolving Earth*. British Museum (Natural History) & Cambridge University Press, 197–220.

HOWELL, D. G., JONES, D. L., COX, A., & NUR, A. (eds) 1984. *Proceedings of the Circum-Pacific Terrane Conference*. Stanford University, Publications Geological Sciences, **18**.

HOWGATE, M. E. 1986. The German "Ostrich" and the molecular clock. *Nature*, **324**, 516.

JEFFERSON, T. H. 1982. The early Cretaceous fossil forests of Alexander Island, Antarctica. *Palaeontology* **25**, 681–708.

— — 1983. Palaeoclimatic significance of some Mesozoic Antarctic fossil floras. *In*: OLIVER, R. L., JAMES, P. R. & JAGO, J. B. (eds) *Antarctic Earth Science*. Australian Academy of Science, Canberra, 593–598.

KAMP, P. J. J. 1986*a*. Late Cretaceous-Cenozoic tectonic development of the southwest Pacific region. *Tectonophysics*, **121**, 225–251.

— — 1986*b*. The mid-Cenozoic Challenger Rift System of western New Zealand & its implications for the age of Alpine Fault inception. *Bulletin of the Geological Society of America*, **97**, 255–81.

KEAST, J. A. 1973. Contemporary biotas and the separation sequence of the southern continents. *In*: TARLING, D. H. & RUNCORN, S. K. (eds) *Implications of Continental Drift to the Earth Sciences*. **1**, Academic Press, London, 309–343.

KEMP, T. S. 1982. *Mammal-like Reptiles and the Origin of Mammals*. Academic Press, London.

— — 1983. The relationships of mammals. *Zoological Journal of the Linnean Society of London* **77**, 353–384.

KEMPER, E. (ed) 1987. Das Klima der Kreide-Zeit. *Geologisches Jahrbuch* Reihe A, Heft **96**.

KENNETT, J. P. *et al.* 1975. Cenozoic paleoceanography in the Southwest Pacific Ocean, Antarctic glaciation and the Development of the Circum-Antarctic Current. *Initial Reports Deep Sea Drilling Project*, **29**, 1155–1169.

KING, M. 1987. Origin of the Gekkonidae: chromosomal & albumin evolution suggests Gondwanaland. *Search*, **18**, 252–254.

KING, P. R. & ROBINSON, P. H. 1988. An overview of Taranaki region geology, New Zealand. *Energy Exploration & Exploitation*, **6**, 213–232.

KORSCH, R. J. & WELLMAN, H. W. 1988. The geological evolution of New Zealand and the New Zealand region. *In*: NAIRN, A. E. M., STEHLI, F. G. & UYEDA, S. (eds) *The Ocean Basins & Margins. Vol. 7B. The Pacific Ocean*. Plenum Publishing Corporation, New York, 411–82.

LAIRD, M. G. 1981. The late Mesozoic fragmentation of the New Zealand fragment of Gondwana. *In*: CRESSWELL, M. M. & VELLA, P. (eds) *Gondwana Five*. A. A. Balkema, Rotterdam, 311–8.

LAWVER, L. A. & SCOTESE, C. R. 1987. A revised reconstruction of Gondwanaland *In*: McKENZIE, G. D. (ed.) *Gondwana Six: Structure, Tectonics and Geophysics*. American Geophysical Union Geophysical Monograph, **40**, 17–23.

LEWIN, R. 1983. Is the Orangutan a living fossil? *Science*, **222**, 1222–1223.

LILLEGRAVEN, J. A. 1979. Reproduction in Mesozoic mammals. *In*: LILLEGRAVEN, J. A., KIELAN-JAWOROWSKA, Z., & CLEMENS, W. A. (eds) *Mesozoic Mammals: The first two thirds of Mammalian History*. University of California Press, Berkeley and Los Angeles, 259–276.

McGOWAN, C. 1982. The wing musculature of the Brown Kiwi *Apteryx australis mantelli* and its bearing on ratite affinities. *Journal of Zoology*, **197**, 173–219.

— — 1984. Evolutionary relationships of ratites and carinates: evidence from ontogeny of the tarsus. *Nature*, **307**, 733–735.

— — 1986. The wing musculature of the Weka (*Gallirallus australis*), a flightless rail endemic to New Zealand. *Journal of the Zoological Society of London*, **210**, 305–346.

McKENZIE, D. 1978. Some remarks on the development of sedimentary basins. *Earth and Planetary Science Letters*, **40**, 25–32.

McKENZIE, K. G. (ed.) 1987. *Shallow Tethys 2*, A. A. Balkema, Rotterdam.

MACELLARI, C. E. 1986. Late Campanian-Maastrichtian Ammonite fauna from Seymour Island (Antarctic Peninsula). *Paleontological Society Memoir*, **18**.

— — 1987. Progressive endemism in the late Cretaceous Ammonite family Kossmaticeratidae and the break-up of Gondwanaland. *In*: McKENZIE, G. D. (ed.) *Gondwana Six: Stratigraphy, Sedimentology & Paleontology*. American Geophysical Union, Geophysical Monograph, **41**, 85–92.

MILDENHALL, D. C. 1970. Discovery of a New Zealand member of the Permian *Glossopteris* flora. *Australian Journal of Science*, **32**, 474–5.

— — 1976. *Glossopteris ampla* Dana from New

Zealand Permian sediments. *New Zealand Journal of Geology & Geophysics*, **19**, 130–132.

— — 1980. New Zealand Late Cretaceous & Cenozoic plant biogeography: A contribution. *Palaeogeography, Palaeoclimatology, Palaeoecology*, **31**, 197–233.

MURRAY, P. 1984. Furry egg-layers: the Monotreme radiation. *In*: ARCHER, M. & CLAYTON, G. (eds) *Vertebrate Zoogeography and Evolution in Australasia*. Hesperian Press, Perth, 571–626.

MUTTERLOSE, J. 1986. Upper Jurassic belemnites from the Orville Coast, Western Antarctica and their paleobiogeographical significance. *British Antarctic Survey Bulletin*, **70**, 1–22.

NELSON, G. 1975. Biogeography, the vicariance paradigm, and continental drift. *Systematic Zoology*, **24**, 490–504.

NORRIS, R. J. & CRAW, D. 1987. Aspiring Terrane: an oceanic assemblage from New Zealand and its implications for terrane accretion in the Southwest Pacific. *In*: LEITCH, E. C. & SCHEIBNER, E. (eds) *Terrane Accretion & Orogenic Belts*. American Geophysical Union Geodynamics Series, **19**, 169–177.

NORTON, I. O., LOUTIT, T. S. & TAPSCOTT, C. R. 1985. Late Cretaceous and Cenozoic plate tectonics of Antarctica and New Zealand. *Abstracts Sixth Gondwana Symposium, Ohio State University*.

OLIVER, P. J., MUMME, T. C., GRINDLEY, G. W. & VELLA, P. 1979. Palaeomagnetism of the Upper Cretaceous Mt Somers Volcanics, Canterbury, New Zealand. *New Zealand Journal of Geology & Geophysics*, **22**, 199–212.

PARRISH, J. T. & SPICER, R. A. 1988a. Middle Cretaceous wood from the Nanushuk group, central north slope, Alaska. *Palaeontology*, **31**, 19–34.

— — 1988b. Late Cretaceous terrestrial vegetation: a near-polar temperature curve. *Geology*, **16**, 22–25.

PIERSON, E. D., SARICH, V. M., LOWENSTEIN, J. A. & DANIEL, M. J. 1982. *Mystacina* is a Phyllostomatoid Bat. *Bat Research News*, **23 (4)**, 78.

— —., — —., — —., — — & RAINEY, W. E. 1986. A molecular link between the bats of New Zealand and South America. *Nature*, **323**, 60–63.

POCKNALL., D. T. & CROSBIE, Y. M. 1988. Pollen morphology of *Beauprea* (Proteaceae): modern & fossil. *Review of Palaeobotany & Palynology*, **53**, 305–327.

POOLE, A. L. 1987. Southern Beeches. *New Zealand Department of Scientific & Industrial Research, Information Series*, **162**.

RAGE, J. C. 1988. Gondwana, Tethys and terrestrial vertebrates during the Mesozoic and Cainozoic. *In*: AUDLEY–CHARLES, M. G. & HALLAM A. (eds) *Gondwana and Tethys*. Geological Society, London, Special Publication, **37**, 255–73.

RAINE, J. I., SPEDEN, I. G., STRONG, C. P. 1981. New Zealand. *In*: REYMENT, R. A. & BENGTSON, P. (eds) *Aspects of Mid-Cretaceous Regional Geology*. Academic Press, London. 221–267.

RETALLACK, G. J. 1984. Origin of the Torlesse terrane & coeval rocks, South Island, New Zealand: Discussion and reply. *Bulletin of the Geological Society of America*, **95**, 980–982.

— — 1987. Triassic vegetation & geography of the New Zealand portion of the Gondwana supercontinent. *In*: McKENZIE, G. D. (ed.) *Gondwana Six: Stratigraphy, Sedimentology & Paleontology*. American Geophysical Union, Geophysical Monograph, **41**, 29–39.

RICCARDI, A. C. 1983. The Jurassic of Argentina and Chile. *In*: MOULLADE, M. & NAIRN, A. E. M. (eds) *The Phanerozoic Geology of the World II. Mesozoic B*. Elsevier, Amsterdam, 201–263.

RICH, P. V. 1976. The History of Birds on the island continent Australia. *Proceedings 16th International Ornithological Congress*, 53–65.

— — & BALOUET, J. 1984. The waifs and strays of the bird world or the ratite problem revisited, one more time. *In*: ARCHER, M. & CLAYTON, G. (eds) *Vertebrate Zoogeography and Evolution in Australasia*, Hesperian Press, Perth, 447–455.

RICH, T. H. 1982. Montremes, placentals and marsupials: their record in Australia and its biases. *In*: RICH, P. V. & THOMPSON, E. M. (eds) *The Fossil Vertebrate Record of Australasia*. Monash University Press, Melbourne, 385–478.

RICH, T. H. & RICH, P. V. 1988. Polar dinosaurs from South Australia. *Episodes*, **11**, 223.

SAVAGE, R. J. G. & LONG, M. R. 1986. *Mammal Evolution: An Illustrated Guide*. British Museum (Natural History), London.

SCHWARTZ, J. H. 1984. The evolutionary relationships of man and orang-utans. *Nature*, **308**, 501–505.

SIBLEY, C. G. & AHLQUIST, J. E. 1981. The phylogeny and relationships of the ratite birds as indicated by DNA-DNA hybridization. *In*: SCUDDER, G. G. E. & REVEA, J. L. (eds) *Evolution today*. Proceedings of the 2nd International Congress on Systematics and Evolutionary Biology, 301–335.

— — & — — 1983a. The phylogeny and relationships of the ratite birds as indicated by DNA-DNA hybridization. *Abstracts, 15th Congress, Pacific Science Association*, 213.

— — & — — 1983b. The phylogeny of the passerine birds of Australia and New Zealand, based on DNA-DNA hybridization data. *Abstracts, 15th Congress, Pacific Science Association*. 213.

— — & — — 1987. Avian phylogeny reconstructed from comparisons of the genetic material DNA. *In*: PATTERSON, C. (ed.) *Molecules & Morphology in Evolution: Conflict or Compromise?* Cambridge University Press, Cambridge, 95–121.

SMITH, A. G., HURLEY, A. M. & BRIDEN, J. C. 1981. *Phanerozoic Paleocontinental World Maps*. Cambridge University Press, Cambridge.

SPEDEN, I. G. 1971. Geology of Papatowai Subdivision, Southeast Otago. *New Zealand Geological Survey Bulletin*, **81**.

SPORLI, K. B. 1987. Development of the New Zealand Microcontinent. *American Geophysical Union Geodynamics Series*, **182**, 115–132.

STEVENS, G. R. 1971. Relationship of isotopic temperatures and faunal realms to Jurassic and Cretaceous palaeogeography, particularly of the Southwest Pacific. *Journal of the Royal Society of*

New Zealand, **1**, 145—158.

— — 1978. Palaeontology (Jurassic). *In*: SUGGATE, R. P., STEVENS, G. R. & TE PUNGA, M. T. (eds) *The Geology of New Zealand*, Government Printer, Wellington, 215—228.

— — 1980*a*. Southwest Pacific faunal palaeobiogeography in Mesozoic and Cenozoic times: A Review. *Palaeogeography, Palaeoclimatology, Palaeoecology*, **31**, 153—196.

— — 1980*b*. *New Zealand Adrift*. A. H. & A. W. Reed Ltd, Wellington.

— — 1985. *Lands in Collision*. N.Z. Dept. Scientific & Industrial Research. Information Series 161.

— — in press. The Influences of Palaeogeography, Tectonism & Eustasy on Faunal Development in the Jurassic of New Zealand. *Proceedings 2nd Pergola Symposium, 'Fossili, Evoluzione, Ambiente', Pergola, Italy*.

— — & FLEMING, C. A. 1978. The Fossil Record and Palaeogeography: Mesozoic. *In*: SUGGATE, R. P., STEVENS, G. R. & TE PUNGA, M. T. (eds) *The Geology of New Zealand*. Government Printer, Wellington, 710—717.

— — & SPEDEN, I. G. 1978. New Zealand. *In*: MOULLADE, M. & NAIRN, A. E. M. (eds), *The Pharnerozoic Geology of the World II. The Mesozoic A*. Elsevier, Amsterdam, 251—328.

— — & SUGGATE, R. P. 1978. Atlas of Paleogeographic maps. *In*: SUGGATE, R. P., STEVENS, G. R. & TE PUNGA, M. T. (eds) *The Geology of New Zealand*. Government Printer, Wellington, 727—745.

STINNESBECK, W. 1986. Zu den Faunistichen und Palökologischen Verhältnissen in der Quiriquina Formation (Maastrichtium) Zentral-Chiles. *Palaeontographica Abt A*, **194**, 99—237.

ST JOHN, W. 1984. Antarctica — Geology and Hydrocarbon Potential. *In*: HALBOUTY, M. T. (ed.) *Future Petroleum Provinces of the World.*. American Association of Petroleum Geologists Memoir, **40**, 55—100.

STOCK, J. & MOLNAR, P. 1982. Uncertainties in the relative positions of the Australia, Antarctica, Lord Howe & Pacific plates since the Late Cretaceous. *Journal of Geophysical Research*, **87**, 4697—4714.

STOREY, B. C., THOMSON, M. R. A. & MENEILLY, A. W. 1987. The Gondwanian Orogeny within the Antarctic Peninsula. A Discussion. *In*: McKENZIE, G. D. (ed.) *Gondwana Six: Structure, Tectonics & Geophysics*. American Geophysical Union Geophysical Monograph, **40**, 191—198.

STRAHAN, R. (ed.) 1983. *The Australian Museum Complete Book of Australian Mammals*. Angus & Robertson, Melbourne.

SUGGATE, R. P. 1978. The Rangitata Orogeny. *In*: SUGGATE, R. P., STEVENS, G. R. & TE PUNGA, M. T. (eds). *The Geology of New Zealand*. Government Printer, Wellington, 317—333.

SUGGATE, R. P., STEVENS, G. R. & TE PUNGA, M. T. (eds) 1978. *The Geology of New Zealand*. Government Printer, Wellington. 2 vols.

THOMSON, M. R. A. 1981. Mesozoic ammonite faunas of Antarctica and the break-up of Gondwana. *In*: CRESSWELL, M. & VELLA, P. (eds) *Gondwana*

Five. A. A. Balkema, Rotterdam, 269—274.

— — 1983*a*. Late Jurassic Ammonites from the Orville Coast, Antarctica. *In*: OLIVER, R. L., JAMES, P. R. & JAGO, J. B. (eds) *Antarctic Earth Science*. Australian Academy of Sciences, Canberra, 315—319.

— — 1983*b*. Antarctica. *In*: MOULLADE, M. & NAIRN, A. E. M. (eds) *The Phanerozoic Geology of the World II. The Mesozoic B*. Elsevier, Amsterdam, 391—422.

TULLOCH, A. J. & KIMBROUGH, D. L. in press. The Paparoa Metamorphic Core Complex, South Island, New Zealand: Cretaceous extension associated with fragmentation of the Pacific margin of Gondwana. *Geology*.

VEEVERS, J. J. (ed.) 1984. *Phanerozoic Earth History of Australia*. Clarendon Press, Oxford.

— — 1986. Break-up of Australia and Antarctica estimated at mid-Cretaceous (95 ± 5 Ma) from magnetic and seismic data at the continental margin. *Earth and Planetary Science Letters*, **77**, 91—99.

— — 1987. Earth history of the Southeast Indian Ocean and the conjugate margins of Australia & Antarctica. *Journal & Proceedings of the Royal Society of New South Wales*, **120**, 57—70.

WATERHOUSE, J. B. 1978. The fossil record and palaeogeography: Late Palaeozoic. *In*: SUGGATE, R. P., STEVENS, G. R. & TE PUNGA, M. T. (eds) *The Geology of New Zealand*. N.Z. Government Printer, Wellington, 706—710.

— — & BONHAM-CARTER, C. F. 1975. Global distribution and character of Permian biomes based on brachiopod assemblages. *Canadian Journal of Earth Sciences*, **12**, 1085—1146.

— — & FLOOD, P. G. 1981. Poorly sorted conglomerates, breccias and diamictites in Late Paleozoic, Mesozoic and Tertiary sediments of New Zealand. *In*: HAMBREY, M. J. & HARLAND, W. B. (eds) *Earth's Pre-Pleistocene Glacial Record*. Cambridge University Press, 438—446.

WEISSEL, J. K. & HAYES. D. E. 1972. Magnetic anomalies in the southeast Indian Ocean. *In*: HAYES, D. E. (ed.) *Antarctic Oceanology, II. The Australian-New Zealand Sector*. American Geophysical Union, Antarctic Research Series, **19**, 165—196.

WHITE, M. E. 1986. *The Greening of Gondwana*. Reed Books, Sydney.

WHITWORTH, R. *et al*. 1985. Rig Seismic Research Cruise 1: Lord Howe Rise, Southwest Pacific Ocean. *Australian Bureau of Mineral Resources, Geology and Geophysics Report*, **266**.

WILLEY, L. E. 1973. Belemnites from southeastern Alexander Island: II, The occurrence of the family Belemnopseidae in the Upper Jurassic and Lower Cretaceous. *British Antarctic Survey Bulletin*, **36**, 33—59.

WOOD, R. A., ANDREWS, P. B. & HERZER, R. H. in press. Cretaceous & Cenozoic geology of the Chatham Rise. *New Zealand Geological Survey Basin Studies*, **2**.

WOODBURNE, M. O. 1982. Newly discovered land mammal from Antarctica. *Antarctic Journal* of the United States, **17(5)**, 64–65.

— — & ZINSMEISTER, W. J. 1982. Fossil land mammal from Antarctica. *Science*, **218**, 284–286.

— — 1983. A new marsupial from Seymour Island, Antarctic Peninsula. *In*: OLIVER, R. L., JAMES, P. R. & JAGO, J. B. (eds). *Antarctic Earth Science*. Australian Academy of Sciences & Cambridge University Press, 320–322.

— — 1984. The first land mammal from Antarctica and its biogeographic implications. *Journal of Paleontology*, **58**, 913–948.

WORTHY, T. H. 1987a. Osteology of *Leiopelma* (Amphibia: Leiopelmatidae) and descriptions of three new subfossil *Leiopelma* species. *Journal of the Royal Society of New Zealand*, **17**, 201–251.

— — 1987b. Palaeoecological information concerning members of the frog genus *Leiopelma*: Leiopelmatidae in New Zealand. *Journal of the Royal Society of New Zealand*, **17**, 409–420.

ZINSMEISTER, W. J. 1976a. A new genus and species of the gastropod family Struthiolariidae, *Antarctodarwinella ellioti*, from Seymour Island, Antarctica. *Ohio Journal of Science*, **76**, 111–114.

— — 1976b. The Struthiolaridae (Gastropoda) fauna from Seymour Island, Antarctic Peninsula. *Actas del Sexto Congreso Geologico Argentino*, **1**, 609–618.

— — 1979. Biogeographic significance of the late Mesozoic and early Tertiary molluscan faunas of Seymour Island (Antarctic Peninsula) to the final breakup of Gondwanaland. *In*: GRAY, J. & BOUCOT, A. J. (eds) *Historical Biogeography, Plate Tectonics and the Changing Environment*. Oregon State University Press, 349–355.

— — 1982. Late Cretaceous–early Tertiary molluscan biogeography of the Southern circum-Pacific. *Journal of Paleontology*, **56**, 84–102.

— — 1986. Fossil windfall at Antarctica's edge. *Natural History*, **1986 (5)**, 60–66.

— — 1987. Cretaceous paleogeography of Antarctica. *Palaeogeography, Palaeoclimatology, Palaeoecology*, **59**, 197–206.

— — & FELDMANN, R. M. 1984. Cenozoic high latitude heterochroneity of Southern Hemisphere marine faunas. *Science*, **224**, 281–283.

Antarctic belemnite biogeography and the break-up of Gondwana

PETER DOYLE & PHILIP HOWLETT

British Antarctic Survey, Natural Environment Research Council,
High Cross, Madingley Road, Cambridge CB3 OET, UK

Abstract: In the late Jurassic, the belemnite genera *Hibolithes* and *Belemnopsis* were abundant and widespread in Tethys, characterizing a Tethyan Realm that extended south from southern Europe and Asia to Antarctica and the rest of Gondwana. Although a distinct Southern Hemisphere 'Austral' belemnite realm counterbalancing the northern Boreal Realm was absent, it is clear that a significant degree of endemicity existed at the species level, with distinct species groups in Indonesia, Madagascar, Australasia, South America and so on.

Trans-Gondwanan faunal links first developed in the late Jurassic shelf seaway between Antarctica, Madagascar and India. *Belemnopsis* became extinct in southern Europe, and was left as a relict, endemic to this trans-Gondwanan seaway. In the Aptian the Belemnopscidae were eventually replaced around Gondwana by the endemic family Dimitobelidae, which developed as the Antarctic−Australasian core of Gondwana began to drift south. In Tethys few genera remained, the Tethyan Realm finally breaking down in the late Cretaceous. Initially the Dimitobelidae were widespread in Gondwanan seas, the trans-Gondwanan links developed in the Jurassic being maintained. However, as Gondwana fragmented further in the late Cretaceous, these links broke down and the Dimitobelidae survived only in the Antarctic−Australasian region which was retreating southwards.

The aim of this paper is to examine the extent of Tethyan and (later in the Cretaceous) Austral belemnites, comparing them to their Boreal counterparts examined in Doyle (1987c). We mean to determine the role of palaeogeography, and more specifically the break-up of Gondwana, in the development of the Mesozoic marine faunal realms.

The sedimentary basins of the Antarctic Peninsula region provide ample material for comparison with other Gondwanan continents, and the belemnites occurring there are broadly representative of the main Tethyan stock which was found throughout the margins of Gondwana in the late Mesozoic. In the discussion below, the distribution of contemporaneous Boreal belemnites is not dealt with in detail. Important reviews of this topic include Saks (1972); Stevens (1973a, b); Christensen (1976); Mutterlose *et al.* (1983) and Doyle (1987c).

Belemnite biogeography

Pre-late Jurassic

There are relatively few records of belemnites from pre-Upper Jurassic sediments in the Southern Hemisphere. Thus, although belemnoids (aulacocerids) are known in Gondwana from Palaeozoic times on, the first Antarctic belemnite record is that of a single phragmocone of

Sinemurian age from central Alexander Island (Thomson & Tranter 1986). However, although there are other early Jurassic belemnites known from the Southern Hemisphere (Stevens 1965, 1973a), information is too scant to allow detailed inter-areal comparison.

There are a few more records of belemnites from Middle Jurassic sediments in Gondwana, although no belemnites of this age have yet been described from Antarctica. However, Quilty (1970) stated that belemnites were found with Middle Jurassic ammonites in the Behrendt Mountains, Ellsworth Land. Elsewhere in Gondwana, belemnites are more common. *Belemnopsis* is the dominant genus, and species similar to, or conspecific with, the European species *bessina*, *depressa* (= *calloviensis*) and *canaliculata* are dominant in these regions. Thus examples are recorded from western Australia (Phillips 1870; Whitehouse 1924; McWhae *et al.* 1948; Stevens 1965), Madagascar (Besairie 1936) and Kachchh (Spath 1927). *B. mackayi* from the Bajocian of New Zealand is regarded by Stevens (1965) as similar to contemporaneous European species. Other relatively narrow grooved forms close to *B. mackayi* are known from New Zealand (Challinor 1977) and New Caledonia (Avias 1953). Stevens (1965, p. 158) has also recorded the presence of *Belemnopsis* close to *B. mackayi* in collections from the Middle Jurassic strata of Neuquen Province, Argentina. The apparent closeness of these

From Crame, J. A. (ed.), 1989, *Origins and Evolution of the Antarctic Biota,*
Geological Society Special Publication No. 47, pp. 167−182.

167

Gondwanan *Belemnopsis* to contemporaneous *Belemnopsis* in southern Europe (e.g. Pugaczewska 1961; Riegraf 1980) suggests close links were maintained throughout Tethys at this time. *Belemnopsis* was unable to penetrate farther north than Britain in the Bajocian and Bathonian, and was absent altogether after the Callovian when the influx of Arctic cylindroteuthids occurred (Doyle 1987c).

Hibolithes is less well known in the Gondwanan Middle Jurassic than *Belemnopsis*, but it has been recorded from Kachchh (Spath 1927) and New Zealand (Stevens 1965). Both forms are essentially similar to southern European species such as *H. semihastatus*, which is common in the Middle Jurassic of southern Germany (Riegraf 1980). This similarity of species in Gondwana and southern Europe helps strengthen the hypothesis of close links throughout Tethys at this time.

Late Jurassic (Figs 1 & 3; Table 1)

Throughout the late Jurassic the belemnites of the Tethyan Realm showed a broad subdivision into two large regions: (a) the Mediterranean or European region, which was dominated by *Hibolithes*, and (b) the Indo-Pacific region, occupying the remainder of the realm, which was dominated by *Belemnopsis* with *Hibolithes* as only a minor element of the fauna (Stevens 1965, 1973a).

During the Oxfordian, a distinct centre of endemism characterised by the *B. orientalis–gerardi* group apparently existed in East Africa, India and Pakistan, close to the divide between the European and Indo–Pacific regions. This group, which had developed from the *B. bessina–depressa* group in the late Bathonian–Callovian of southern Europe, spread into neighbouring Madagascar (Stevens 1965). Species of the *B. orientalis–gerardi* group are characterised by a conical or slightly hastate outline and conical profile, together with a relatively narrow groove extending almost to the apex. In contrast the *B. bessina–depressa* group is hastate in outline with a narrow ventral groove

which extends into the posterior part of the rostrum but not too near the apex. A further endemic species, *B. tanganensis*, characterised by its broad, shallow ventral groove, also originated in East Africa in the late Oxfordian, but it is not known elsewhere.

This development of the East Africa–India–Pakistan endemic centre signified the end of the faunal links throughout Tethys which were apparent in the mid-Jurassic.

The events in the remainder of the Indo-Pacific region during the Oxfordian are unknown or uncertain, though marine rocks of this age occur in Indonesia and possibly in New Zealand and Antarctica. However, during the Oxfordian–Kimmeridgian, representatives of the widespread family Dicoelitidae existed in the Southern Hemisphere. This family ranges in age from the Callovian to Kimmeridgian, with the earliest occurrences in the Middle Jurassic of Europe (Stevens 1964). In the Indo–Pacific region, *Dicoelites* and *Conodicoelites* are recorded from the ?Oxfordian–Kimmeridgian of most areas including East Africa (Stefanini 1925; Spath 1935), the Himalayas (Arkell 1956; Stevens 1964), Indonesia (Stevens 1964, 1965; Challinor & Skwarko 1982), New Zealand (Stevens 1964, 1965), Antarctica (Stevens 1967; Mutterlose 1986) and possibly South America (Stevens 1965). However, they are never common.

In the Kimmeridgian, a further centre of endemism developed around Indonesia, New Guinea, Australia and New Zealand. It is here that the *B. aucklandica–uhligi* group (= the 'uhligi-complex' of Stevens 1965) appears to have originated. The earliest forms of the group are thought to be derived from the *B. orientalis–gerardi* group, and include *B. moluccana* and *B. sularum* (Stevens 1965). The *B. aucklandica–uhligi* group (e.g. Fig. 1a) differs from the *B. orientalis–gerardi* group in consisting of short and robust forms with a typically circular cross-section and a broad, deep ventral groove. Members of the *B. orientalis–gerardi* group are typically elongate and slender, with depressed cross-sections and a narrow ventral

Fig. 1. Representative late Jurassic and earliest Cretaceous belemnites from Antarctica. All specimens are housed in the British Antarctic Survey, Cambridge and were collected from Alexander Island (see Crame & Howlett 1988 for localities). 4 cm scale bar. (**a**) *Belemnopsis* cf. *aucklandica* (Hochstetter) (KG.2905.19), ventral outline and left profile. Lower Tithonian, Himalia Ridge. (**b**) *Belemnopsis gladiatoris* Willey (KG.2942.33), ventral outline and left profile. Valanginian, Himalia Ridge. (**c**) *Belemnopsis* cf. *madagascariensis* (Besairie) (KG.3455.87), ventral outline and right profile. Valanginian, Leda Ridge. (**d**) *Hibolithes argentinus* Fergulio (KG.2909.18), ventral outline and right profile. Upper Tithonian, Himalia Ridge. (**e**) *Hibolithes antarctica* Willey (KG.3234.5), ventral outline and left profile. Berriasian, Europa Cliffs.

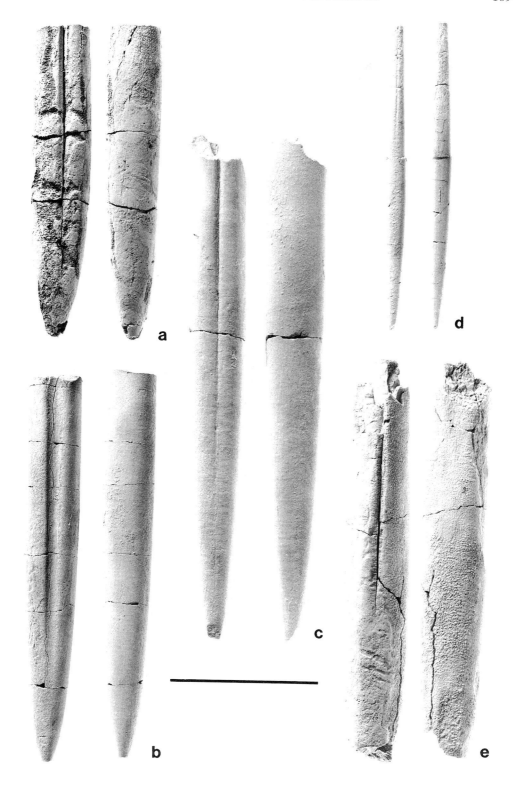

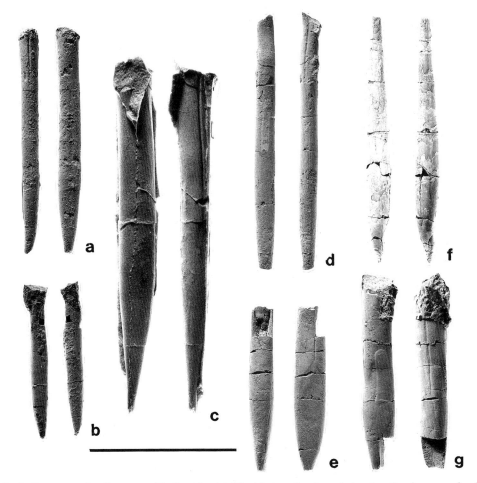

Fig. 2. Representative Cretaceous (Aptian-Maastrichtian) belemnites from Antarctica. Specimens prefixed D are housed in the British Antarctic Survey, Cambridge; that prefixed USNM is in the US National Museum, Washington. See Doyle (1985, 1987a) and Doyle & Zinsmeister (1988) for localities. 4 cm scale bar. (a) *Parahibolites blanfordi* (Spengler) (D.8423.7), ventral outline and left profile. Upper Albian, Lost Valley, James Ross Island. (b) *Neohibolites* sp. (D.8420.40), ventral outline and left profile. Albian, Gin Cove area, James Ross Island. (c) *Peratobelus* cf. *oxys* (Tenison-Woods) (KG.10.61), ventral outline and right profile of silicone rubber cast. Aptian, Succession Cliffs, Alexander Island. (d) *Dimitobelus (Dimitobelus)* cf. *stimulis* var. *extremis* Whitehouse (D.8403.35b), ventral outline and left profile. Albian, Kotick Point, James Ross Island. (e) *Tetrabelus seclusus* (Blanford) (D.8420.37), ventral outline and right profile. Albian, Gin Cove area, James Ross Island. (f) *Dimitobelus (Dimitocamax) seymouriensis* Doyle (USNM 404813), ventral outline and left profile. Campanian−Maastrichtian, Seymour Island (g) *Dimitobelus (Dimitobelus) praelindsayi* Doyle (D.8420.11a), ventral outline and left profile. Albian, Gin Cove area, James Ross Island.

groove. Representatives of the important and widespread *B. aucklandica−uhligi* group and its forerunners (*B. alfurica* and *B. keari*) are recorded from the Himalayas, Iran, South America and Antarctica (Stevens 1965, 1973a; Crame & Howlett 1988).

In Madagascar, East Africa, Ethiopia, Somalia and Kachchh, the distinctive endemic *B. orientalis−gerardi* group and *B. tanganensis* survived throughout the Kimmeridgian.

At about the same time, *Hibolithes* with affinity to some European species, e.g. *H. savornini* and *H. astartinus* (Stevens 1965), began to spread into the Madagascar region, gradually replacing *Belemnopsis* (Stevens 1965, 1973a). As these groups were replaced, the *B. casterasi-madagascariensis* group developed, possibly as an offshoot of the *B. orientalis−gerardi* group. This new group, of which *B. casterasi* is the earliest known representative, is distinctive on

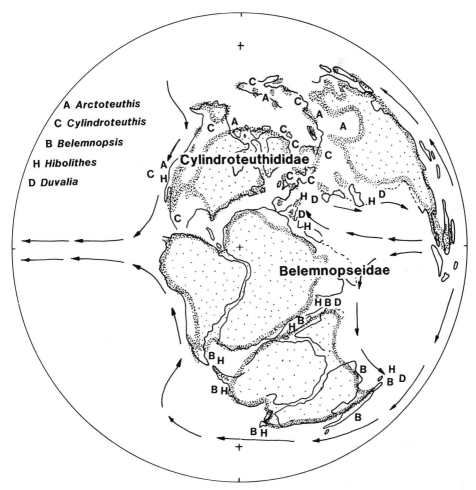

Fig. 3. World distribution of late Jurassic belemnites. Base map and hypothetical ocean currents (arrows) modified after Frakes (1979, fig. 6.3). The Cylindroteuthididae (*Cylindroteuthis* and *Arctoteuthis*) dominated the Arctic basin and north European shelf, whereas the Belemnopseidae (and Duvaliidae) occupied Tethys and the Gondwanan margins.

account of its elongate, cylindrical or conical, frequently robust rostrum (Fig. 1c). The cross-sections are typically circular throughout, and the ventral groove is narrow, but deeply incised, extending almost to the apex. A rostrum, apparently belonging to the *B. casterasi–madagascariensis* group, has been described as *B.* cf. *stolleyi* by Jeletzky (1983, pl. 2, fig. 4A–E) from the Tithonian of the Falklands Plateau.

In the Tithonian, the southern European region contained a diverse *Hibolithes–Duvalia* assemblage (Stevens 1965, 1973a). This fauna extended into Madagascar and East Africa, where *Hibolithes* was the dominant genus,

together with *Duvalia* and rare *Belemnopsis* (Stevens 1965, 1973a), and eventually spread throughout Tethys into the Indo–Pacific region, slowly replacing the *B. aucklandica–uhligi* group. In Madagascar, the *B. casterasi–madagascariensis* group continued to spread farther south along a trans-Gondwanan seaway, with related forms occurring in South America (*B. patagoniensis*). Throughout the Tithonian, the Antarctic Peninsula region was dominated by the *B. aucklandica–uhligi* group with some *Hibolithes* and *Produvalia* (Fig. 1d), thereby showing similarities with New Zealand and South America.

Table 1. *Distribution of late Jurassic species and species-groups in Gondwana and adjoining areas*

	Patagonia/ F. Plateau	Alexander I.	New Zealand	Australia	New Guinea	Madagascar	Himalayas/ Tibet/Kachchh	Indonesia	E. Africa	S. Europe
B. orientalis–gerardi										
orientalis	–	–	–	–	–	–	×	–	×	–
gerardi	–	–	–	–	×	–	×	–	–	–
B. casterasi–madagascariensis										
casterasi	×	–	–	–	–	×	–	–	–	–
madagascariensis	–	–	–	–	–	×	–	–	–	–
patagoniensis	×	–	–	–	–	–	–	–	–	–
B. aucklandica–uhligi										
aucklandica	?	×	×	×	–	–	×	×	–	–
uhligi	–	–	–	–	–	–	×	×	–	–
moluccana	–	–	–	–	–	–	–	×	–	–
sularum	–	–	–	–	–	–	–	×	–	–
(alfurica)	–	×	×	×	–	×	×	–	×	–
B. tanganensis	–	–	–	–	–	–	–	–	–	–
Hibolithes										
argentinus	×	×	–	–	–	–	–	–	–	–
savornini	–	–	–	–	–	×	–	×	–	–
compressus	–	–	–	–	–	×	–	×	–	–
Conodicoelites	–	–	×	–	–	–	×	×	–	–
Dicoelites	×	–	×	–	–	×	×	×	×	×
Duvalia	–	–	–	–	–	–	–	×	×	×

–, absence
×, presence
F. Plateau, Falkland Plateau
compiled from numerous sources.

Earliest Cretaceous (Fig. 1; Table 2)

Lower Cretaceous rocks are less widespread in the Southern Hemisphere than those of the Upper Jurassic, but it appears that belemnites were still common throughout Tethys. The Southern European region continued to be dominated by the *Hibolithes*−duvaliid assemblage, which first appeared in the Tithonian (Stevens 1965, 1973*b*). Throughout the Indo−Pacific region a similar, abundant *Hibolithes*−*Duvalia* assemblage was also present, although *Duvalia* was common only in Madagascar and the Mediterranean region. In general, the species of *Hibolithes* and *Duvalia* present have strong affinities with European forms, e.g. *H. pistilliformis* in Europe (Combémorel 1979), Pakistan (Spath 1939) and Madagascar (Besairie 1930); *D. lata* and *D. dilatata* in Europe (Combémorel 1973), Indonesia (Stolley 1935), Pakistan (Noetling 1897) and Madagascar (Besairie 1930, 1936). *Belemnopsis* is unknown in the European and most of the Indo−Pacific regions in the early Cretaceous, having been totally replaced by the *Hibolithes*−*Duvalia* assemblage in the latest Tithonian. A 'relict' population of *Belemnopsis* existed in the trans-Gondwanan seaway.

In the late Tithonian−Berriasian, the Antarctic Peninsula region had a *Hibolithes* fauna which consisted almost entirely of endemic forms (e.g. *H. antarctica*, Fig. 1e), although they do possess some similarity to Indo−Pacific species. In the Valanginian and ?Hauterivian, *Belemnopsis* was dominant. Some of the species present appear to be derived from the *B. aucklandica*−*uhligi* group, with similar forms being found in South America, South Africa and Madagascar (*B. gladiatoris*; Willey 1973; Howlett 1986), Pakistan and Tibet (*B. gladiatoris* and *B. extenuatus*; Willey 1973; Howlett 1986; Fig. 1b). This distribution outlines the 'relict' *Belemnopsis* fauna around the newly formed southern Atlantic Ocean and Mozambique Channel (i.e. the late Jurassic trans-Gondwanan seaway). This relict population coincides with the distribution of the *B. casterasi*−*madagascariensis* group (Fig. 1c) which first appeared in the Kimmeridgian in Madagascar and spread to South America and Alexander Island in the Tithonian and Valanginian, respectively (Stevens 1965; Crame & Howlett 1988).

Later in the early Cretaceous (possibly Hauterivian, although see discussion in Crame & Howlett 1988, p. 26) an endemic Antarctic *Belemnopsis* group restricted to Alexander Island developed (P. J. H., new data). This new group, which is elongate and cylindrical with a typically circular cross-section and a narrow prominent ventral alveolar groove extending almost to the mid-point of the rostrum, may represent a derivation from the *B. casterasi*−*madagascariensis* group and forms the last known occurrence of the genus *Belemnopsis*.

Belemnites of Barremian age are unknown in the Indo−Pacific region. In the European region, *Mesohibolites* became the dominant form which, however, did not spread into the remainder of Tethys (Stevens 1965, 1973*b*).

Early Cretaceous (Aptian−Albian) to late Cretaceous (Figs 2 & 4; Table 3)

By the Aptian, the Belemnopseidae had largely died out, with only *Neohibolites* and its allies remaining. *Belemnopsis*, abundant in the earlier Cretaceous of the Madagascar−Antarctic Peninsula region, made its last appearance in the probable Hauterivian of Alexander Island, and the last recorded occurrence of *Hibolithes* in the Southern Hemisphere is also from sediments of Hauterivian age (Stevens 1965). However, *Hibolithes* survived into the Aptian in northwest Europe (Mutterlose *et al.* 1983) with the single, short-lived species *Hibolithes minutus*.

Neohibolites first appeared in the Aptian, where it is particularly abundant in northwest Europe (e.g. Stolley 1911; Swinnerton 1955). Forms conspecific with these European examples are recorded from Aptian and Albian sediments of Antarctica (Willey 1973; Doyle 1987*a*; Fig. 2*b*), South America (Liddle 1946) and Mozambique (Doyle 1987*b*), while other species are known from southern India (Blanford 1861), possibly Australia (= *Belemnites liversidgei*; Etheridge *in* Jack & Etheridge 1892), New Guinea (= *Parahibolites blanfordi*; Glaessner 1945) and Madagascar (Lemoine 1906). *Parahibolites* shares a similar distribution, but did not manage to penetrate as widely. Nevertheless, species are recorded from southern India (Blanford 1861), Antarctica (Doyle 1985; Fig. 2a), Patagonia (Stolley *in* Richter 1925) and possibly Madagascar (Lemoine 1906). *Mesohibolites*, however, is uncommon and restricted largely to southern Europe (Stevens 1973*b*). *Neohibolites*−*Parahibolites* was able to migrate widely in the Southern Hemisphere, and often co-existed with the initial members of the Dimitobelidae (Stevens 1973*b*; Doyle 1985, 1987*b*; Fig. 4). The last *Neohibolites*−*Parahibolites* appeared in the Cenomanian, in both Southern and Northern hemispheres.

The Dimitobelidae composes a striking part

Table 2. *Distribution of early Cretaceous (Berriasian–Barremian) belemnite species in Gondwana and adjoining areas.*

	Patagonia	Alexander I.	South Africa	Madagascar	Himalayas/ Tibet/Kachchh	Indonesia	Southern Europe
B. aucklandica–uhligi							
alexandri	–	×	–	–	–	–	–
gladiatoris	×	×	×	×	×	–	–
extenuatus	–	×	–	–	×	–	–
africana	–	–	×	–	–	–	–
B. casterasi–madagascariensis							
madagascariensis	×	×	–	×	–	–	–
patagoniensis	×	–	–	–	–	–	–
Hibolithes							
antarctica (IP)	–	×	–	–	–	–	–
subfusiformis (Eu)	–	–	–	×	×	×	×
Duvalia							
lata	×	–	–	×	×	×	×
dilatata	–	–	–	×	×	×	×

–, absence
×, presence
IP, Indo-Pacific
Eu, European
compiled from numerous sources.

Table 3. *Distribution of Cretaceous (Aptian–Albian and Cenomanian–Maastrichtian) belemnite species in Gondwana and adjoining areas*

	Patagonia	Alexander I.	James Ross I. Group	New Zealand	W. Australia	E. Australia	New Guinea	S. India	Mozambique
APTIAN–ALBIAN									
Peratobelus									
oxys	–	×	–	–	–	×	–	–	–
australis	–	–	–	–	–	×	–	–	–
bauhinianus	–	–	–	–	–	×	–	–	–
foersteri	–	–	–	–	–	–	–	–	×
Tetrabelus									
seclusus	–	–	×	–	–	–	–	×	–
willeyi	–	×	×	–	–	–	–	–	–
whitehousei	–	–	×	–	–	–	–	–	–
Dimitobelus									
diptychus	?	×	×	–	–	×	–	–	–
stimulus	?	×	×	–	–	×	–	–	–
dayi	–	–	×	–	–	×	–	–	–
praelindsayi	–	×	–	–	–	×	–	–	–
kleini	–	–	–	–	–	×	–	–	–
macgregori	–	–	–	–	–	–	×	–	–
superstes	–	–	–	×	–	–	–	–	–
CENOMANIAN–MAASTRICHTIAN									
Dimitobelus									
superstes	–	–	×	×	–	–	–	–	–
ongyleyi	–	–	×	×	×	–	–	–	–
lindsayi	–	–	–	×	–	–	–	–	–
Dimitocanax									
hectori	–	–	–	×	–	–	–	–	–
seymouriensis	–	–	×	–	–	–	–	–	–

–, absence
×, presence
compiled from numerous sources. New records from James Ross Island not mentioned in text added in proof.

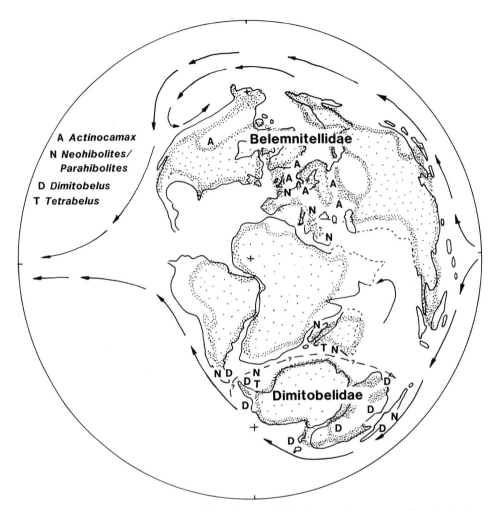

Fig. 4. World distribution of Cretaceous (Albian–Cenomanian) belemnites. Base map and hypothetical ocean currents (solid arrows) modified after Frakes (1979, fig. 6.8). A peri-Antarctic water mass (dashed-arrow) probably developed at this time with the generation of ocean crust between Antarctica and Africa. The Belemnitellidae (e.g. *Actinocamax*) replaced the Cylindroteuthididae, whereas *Neohibolites* and *Parahibolites* are the last representatives of the Belemnopseidae. An endemic Gondwanan family, the Dimitobelidae (*Dimitobelus* and *Tetrabelus*) developed in the Aptian.

of the Southern Hemisphere fauna, as it is entirely endemic to the Gondwanan margins (Fig. 4). The dimitobelids are distinctive with their double-grooved morphology (Fig. 2c–g), and they made their first appearance in the Aptian. *Peratobelus* (Fig. 2d) appears to have been the first and most primitive member of the family and its distribution is largely circum-Gondwanan, as species have been recorded from Antarctica, Australia and Mozambique (Doyle 1987*b*, 1988). *Tetrabelus* also appeared at this time, and was restricted to Alexander Island (Doyle 1987*a*).

In the Albian, *Tetrabelus* (Fig. 2c) had a trans-Gondwanan distribution mirroring that of *Belemnopsis* of the earlier Cretaceous, with similar species occurring in Antarctica (James Ross Island) and southern India (Doyle 1985). However, *Dimitobelus*, and more specifically the nominal subgenus (Figs 2e & f), was the dominant belemnite throughout the Gondwanan margins. The species which occur in the Antarctic Peninsula and Australia are largely the same, indicating the maintenance of close links, although there are some species endemic to both areas. The real enigma is the fauna of

New Zealand, which is largely endemic and morphologically separate from the circum-Gondwanan *Dimitobelus diptychus*-like forms. In the Cenomanian, the situation was similar to that of the Albian, although trans-Gondwanan *Tetrabelus* was no longer apparent.

Records of dimitobelid belemnites are sparse from younger sediments, particularly from the Turonian—Santonian interval. *Dimitobelus* s. s. continued from the Albian into the Coniacian—Santonian in New Zealand with *D. (Dimitobelus) superstes*, and a second species of this subgenus, *D. (Dimitobelus) ongleyi*, first appeared in the Santonian sediments of the same region. This species was previously unknown outside New Zealand, but is here recorded from the Santonian Toolonga Calcilutite of the Carnarvon Basin, Western Australia (specimen BMNH C.89544, to be described elsewhere). *Dimitobelus* s. s. last appeared in the Campanian with *D. (Dimitobelus) lindsayi* in New Zealand, and an undescribed species from James Ross Island, Antarctica. Also in the Campanian, the first representatives of the subgenus *Dimitocamax* (Doyle & Zinsmeister 1988) appeared, probably with an undescribed species from the Santa Marta Formation of James Ross Island. Endemic species of *D. (Dimitocamax)* occur in younger sediments in Seymour Island *(D. (Dimitocamax) seymouriensis*, Campanian—Maastrichtian; Fig. 2g) and New Zealand *(D. (Dimitocamax) hectori*, Maastrichtian) (Doyle & Zinsmeister 1988).

Synopsis

Belemnite records are sparse in the Lower Jurassic of the Southern Hemisphere, but there are more records in the Middle Jurassic. *Belemnopsis* with close morphological affinities to European Tethyan species are known from sediments of mid-Jurassic age from several southern continents. Far greater records are available for the late Jurassic, and a picture emerges of a Boreal fauna in the Arctic basin—North Europe composed of Cylindroteuthididae, in contrast to a *Belemnopsis—Hibolithes—Duvalia* fauna extending south from European Tethys to the circum-Gondwanan margins (Fig. 3). Local endemism within *Belemnopsis* is a feature of the Gondwanan seas, and several distinct species groups appear. This scenario remains valid for the early Cretaceous, except that *Belemnopsis* was then present only in the trans-Gondwanan seaway.

In the Aptian, the southern fauna polarised into an endemic family, the Dimitobelidae, which inhabited the margins of the core Gond-

wana region. Tethyan elements were reduced, and in the north the Cylindroteuthididae and Oxyteuthididae were replaced by the Belemnitellidae (Fig. 4). The closely analogous Belemnitellidae and Dimitobelidae remained until the end of the Maastrichtian.

Discussion

The pattern of Jurassic Boreal, and an areally more extensive Tethyan, realms has been recognized to a varying degree in many marine invertebrates (see papers in Middlemiss *et al.* 1971; Casey & Rawson 1973; Hallam 1973). An Austral fauna has been recorded by some authors in the Jurassic (Enay 1972; Crame 1986), but most recognise that a distinct southern fauna developed only in the late Cretaceous (Stevens 1965, 1973*b*; Kauffman 1973; Scheibnerova 1973; Rawson 1981; Doyle 1988).

This disjunct distribution (Fig. 3) has led some authors to conclude that latitudinal climatic zonation was not a major force in the delimitation of Mesozoic marine realms (see Hallam 1969, 1971, 1975, 1984; Fürsich & Sykes 1977; Doyle 1987*c*). Recent work by these authors suggests that the picture is a complex one, and that environmental factors controlling distributions are multi-facetted. These factors have been exhaustively discussed elsewhere and are not dealt with here (see references in Doyle 1987*c*). However, in comparing the distributions discussed above with palaeogeographic reconstructions for the Mesozoic (Howarth 1981), it is perhaps possible to distinguish a broad pattern that suggests they were intimately related. A similar pattern has also been observed in the ammonites (Hallam 1971; Pòzaryska & Brochwicz-Lewiniski 1975). The Boreal Cylindroteuthididae were largely confined to the almost land-locked Arctic basin and north European epicontinental sea (Doyle 1987*c*). This contrasts strongly with the distribution of the Tethyan Belemnopseidae (Fig. 3).

Fürsich & Sykes (1977) have described a model of an environmentally unstable European epicontinental sea with reduced diversity, acting as a buffer between the two major faunal realms. Migrations and mingling of Tethyan and Arctic Boreal faunas in this area appear to have occurred only at times of high sea-level stand (Mutterlose *et al.* 1983, 1987; Doyle 1987*c*; Mutterlose 1988). Thus, in the early Cretaceous (Valanginian—Hauterivian), Arctic *Boreioteuthis* was able to extend southwards into Europe while at the same time Tethyan *Hibolithes* managed to reach East Greenland and Svalbard (Doyle 1987*c*). Duvaliids are also present at this time with *Boreioteuthis* in East Greenland (J. A. Jeletzky pers. comm. 1987). In Tethys the belemnopseids were widespread, with many records of *Hibolithes* and *Belemnopsis* from European Tethys to Gondwana (Fig. 3). The species which occur in these widely spaced areas are held by some workers to be identical, and it is clear that these genera were able to migrate freely in the expanse of Tethys and the Tethyan borders of the palaeo-Pacific. Many authors (e.g. Enay 1972; Gordon 1974; Frakes 1979; Westermann 1981; Lloyd 1982, etc.) have modelled prevailing ocean currents from the equator towards Gondwana (Fig. 3). As present-day oceanic squid are dependent on ocean currents (see discussion below), the latter may also have played a large part in the dispersal of the belemnopseids.

Why then, did an Austral realm develop? Trans-Gondwanan migration of marine organisms was initiated in the late Jurassic (Enay 1972; Thomson 1981; Mutterlose 1986) and continued into the Cretaceous (Doyle 1985, 1988). By the mid-Cretaceous, these marine connections across Gondwana were enhanced by rifting and production of oceanic crust between Antarctica and Africa (Martin & Hartnady 1986) (Fig. 4). Although probable intermittent land-links were maintained between South America and Antarctica through the Antarctic Peninsula (Woodburne & Zinsmeister 1984), Antarctica−Australasia was migrating southwards. Stevens (1973b) has suggested that this move southwards created an Austral fauna through entering a cooler climatic zone. Seen from a palaeogeographic viewpoint, the now almost continuous seaways surrounding Antarctica and Australasia provided an area of shelf seas separated from the main oceanic areas of the east Pacific border (Fig. 4). The Dimitobelidae were initially found throughout the Gondwanan seas (Doyle 1987b, 1988) in the Aptian, but retreated to the Antarctica−Australasian region later in the Cretaceous, as this block moved south. Tethyan belemnopseids *Neohibolites* and *Parahibolites*, widespread in the Aptian−Albian, finally died out in the Cenomanian.

The Dimitobelidae most probably favoured a shelf life or at least were restricted to the waters surrounding Antarctica−Australasia due to the development of a peri-Antarctic watermass (see Lloyd 1982). *Tetrabelus* is first recorded from Antarctica in the Aptian (Doyle 1987a), and seemingly spread eastwards in the Albian to India (Doyle 1985). In the Antarctic Peninsula region, dimitobelids are known from shallow water, relatively coarse grained sediments

in the Alexander Island fore-arc basin exposed at Succession Cliffs (71°10'S, 68°15'W) (Butterworth & Macdonald in press). Similarly, dimitobelids are common in the proximal volcaniclastic sediments of the Gustav Group, and in the shallow marine sediments of the Marambio Group of the James Ross Island back-arc basin (Ineson *et al.* 1986; D. Pirrie pers. comm. 1988). Dimitobelids are also present in coarse, shallow water sediments in New Zealand (Stevens 1965). The Belemnitellidae, successors to the Cylindroteuthididae in the Northern Hemisphere and morphologically analogous to the Dimitobelidae (Doyle 1988), are known only from shallow-water shelf sediments (Christensen 1976; Surlyk & Birkelund 1977; W. K. Christensen pers. comm. 1988). These relatively shallow habitats are contrasted with the occurrence of the Belemnopseidae in a deep water setting, in the Antarctic Peninsula region at least. In the Fossil Bluff Group (Butterworth *et al.* 1988) of Alexander Island, belemnopseids (*Belemnopsis* and *Hibolithes*) are commonest in a series of mudstones, siltstones, sandstones and conglomerates exposed north of Fossil Bluff (71°20'S, 68°17'W) and interpreted as deep marine and slope assemblages (Butterworth & Macdonald in press). This example perhaps illustrates that the belemnopseids may have inhabited deep marine as well as shelf environments (Stevens 1965, p. 181).

Mode of life of belemnites: taxonomic uniformitarianism

The principle of interpretation of the ecological tolerances of fossils from their living relatives is well established (taxonomic uniformitarianism; Dodd & Stanton 1981). Although there are no direct correlatives of belemnites living today, recent reconstructions based on preserved soft parts (Riegraf & Hauff 1983; Reitner & Urlichs 1983) indicate a belemnite body-form not unlike that of present-day squid (Teuthida; Bandel & Boletzky 1988, text-figs 4−6; Bandel & Spaeth 1988, p. 248). The function of the calcite rostrum, formerly supposed to be a counterbalance for the phragmocone, is now interpreted by some authors as a firm base for fin musculature (Bandel 1985; Bandel & Spaeth 1988). These authors suggest that the belemnite animal would have had a posture not unlike the modern squid *Loligo*.

Today, teuthids are present in all oceans of the world, and can be divided into two distinct categories based on habitat preference (Voss 1973). Neritic squid (Myopsida) are shallow-water dwellers, inhabiting the inner shelf, while oceanic squid (Oegopsida) inhabit the open ocean, although spending at least part of their life cycle inshore (Clarke 1966; Voss 1973; Arnold 1979; Amaratunga 1983). Both groups exhibit seasonal migrations, and it is accepted that migrations of oceanic squid are intimately associated with water bodies and oceanic currents (Clarke 1966; Jefferts 1988).

Belemnites have been considered as primarily neritic animals restricted to a shelf environment (Stevens 1965, p. 181). Stevens cited their absence from oceanic sediments, coupled with a supposed carnivorous diet restricting them to the shelf, as evidence for this hypothesis. As oceanic squid are also carnivorous, there is little to uphold the latter statement, and given the occurrence of belemnites in mixed facies in the Antarctic Peninsula region, it is more likely that belemnites inhabited a wide range of niches similar to present-day teuthids (cf. Stevens & Clayton 1971). It is feasible to suggest that the belemnopseids may have been analogous to the oegopsids, having a pelagic lifestyle but coming inshore periodically to spawn or feed (Clarke 1966; Voss 1973; Amaratunga 1983). Endemic species groups of *Belemnopsis* may have an analogy in the restricted distributions of some oegopsid squid such as *Illex* and *Todarodes* (Arnold 1979, fig. 3), which are found inshore in specific areas at certain times in their life cycle. *Hibolithes* remained more migratory at the same time. Cylindroteuthids may have been, perhaps, more like the myopsids, having a shelf-dwelling habitat (cf. Stevens & Clayton 1971, p. 877). The Belemnitellidae and Dimitobelidae, closely analogous in body form, may have been neritic descendants from a possible oceanic ancestor such as *Hibolithes* (Doyle 1988).

Summary and conclusions

The Jurassic Boreal−Tethyan belemnite distributions are intimately related to palaeogeography. The Boreal cylindroteuthids may have been primarily shelf dwellers, possibly analogous in lifestyle to present day myopsid teuthids. Conversely, the Tethyan belemnopseids were widespread in the Jurassic and early Cretaceous, similar species often occurring from European Tethys through to Gondwana (Fig. 3). These migratory cephalopods seem to have a present-day analogue in the oegopsid teuthids which are oceanic and often intimately related to the movement of water masses. Endemic species groups developed in broad areas of shelf, similar to the periodic restriction of some present-day oceanic squid to certain shelf areas (Clarke 1966).

In the Aptian−Albian, Antarctica−Australasia was drifting south from the rest of Gondwana by the production of new ocean floor between Africa and Antarctica. This provided an expanse of shelf to which the initially circum-Gondwanan Dimitobelidae were able to retreat. With increasing isolation the Dimitobelidae were restricted to Antarctica−Australasia. In the north, the Belemnitellidae were closely analogous in body form and distribution in shallow sediments. Both are presumed to have developed from a similar migratory ancestor, possibly *Hibolithes* (Doyle 1988), remaining restricted to a shelf habitat until the end of the Cretaceous.

We thank J. A. Crame for discussion, and E. Hatfield for data on present-day squid. J. A. Jeletzky kindly supplied some late Cretaceous Australian belemnites. We also thank our anonymous referees for their suggestions. Part of this work was carried out while one of us (PD) was in receipt of an NERC Fellowship at the British Museum (Natural History).

References

AMARATUNGA, T. 1983. The role of cephalopods in the marine ecosystem. In: CADDY, J. F. (ed.), *Advances in assessment of world cephalopod resources*. FAO Fisheries Technical Paper, **231**, 379−413.

ARKELL, W. J. 1956. *Jurassic Geology of the World*. Oliver and Boyd, Edinburgh.

ARNOLD, G. P. 1979. *Squid. A review of their biology and fisheries*. Ministry of Agriculture, Fisheries and Food. Directorate of Fisheries Research, Lowestoft. Laboratory Leaflet, 48.

AVIAS, J. 1953. Contribution a l'étude stratigraphique et paléontologique de la Nouvelle-Calédonie centrale. *Science de la Terre*, **1**, 1−276.

BANDEL, K. 1985. Composition and ontogeny of *Dictyoconites* (Aulacocerida, Cephalopoda). *Paläontologische Zeitschrift*, **59**, 223−244.

−− & BOLETZKY, S. VON 1988. Features of development and functional morphology required in the reconstruction of early Coleoid Cephalopods. *In*: WIEDMANN, J. & KULLMANN, J. (eds), *Cephalopods − Present and Past*. Schweizerbart'sche Verlagsbuchhandlung, Stuttgart, 229−246.

−− & SPAETH, C. 1988. Structural differences in the ontogeny of some belemnite rostra. *In*: WIEDMANN, J. & KULLMANN, J. (eds), *Cephalopods − Present and Past*. Schweizerbart'sche Verlagsbuchhandlung, Stuttgart, 247–271.

BESAIRIE, H. 1930. Recherches géologiques à Madagascar. *Bulletin de la Société d'histoire naturelle de Toulouse*, **60**, 345–642.

— — 1936. Recherches géologiques à Madagascar. Première suite. La géologie du Nord-Ouest. *Mémoires de la l'Académie Malgache*, **21**, 145–148.

BLANFORD, H. F. 1861–65. The fossil Cephalopoda of the Cretaceous rocks of southern India: Belemnitidae-Nautilidae. *Memoires of the Geological Survey of India, Palaeontologia indica*, **1**, 1–400.

BUTTERWORTH, P. J. & MACDONALD, D. I. M. in press. Basin shallowing from the Mesozoic Fossil Bluff Formation of Alexander Island and its regional significance. *In*: THOMSON, M. R. A., CRAME, J. A. & THOMSON, J. W. (eds), *Geological evolution of Antarctica*. Cambridge University Press, Cambridge.

— —, CRAME, J. A., HOWLETT, P. J. & MACDONALD, D. I. M. 1988. Lithostratigraphy of Upper Jurassic-Lower Cretaceous strata of eastern Alexander Island, Antarctica. *Cretaceous Research*, **9**, 249–264.

CASEY, R. & RAWSON, P. F. (eds) 1973. *The Boreal Lower Cretaceous*. Geological Journal Special Issue, **5**.

CHALLINOR, A. B. 1977. A new Lower or Middle Jurassic belemnite from Southwest Auckland, New Zealand. *New Zealand Journal of Geology and Geophysics*, **20**, 249–262.

— — & SKWARKO, S. K. 1982. Jurassic belemnites from Sula Islands, Moluccas, Indonesia. *Geological Research and Development Centre, Palaeontology*, **3**.

CHRISTENSEN, W. K. 1976. Palaeobiogeography of late Cretaceous belemnites of Europe. *Paläontologische Zeitschrift*, **50**, 113–129.

CLARKE, M. R. 1966. A review of the systematics and ecology of oceanic squids. *Advances in Marine Biology*, **4**, 91–300.

COMBEMOREL, R. 1973. Les Duvaliidae Pavlow (Belemnitida) du Crétacé inférieur français. *Documents des Laboratoires de Géologie de la Faculté des Sciences de Lyon*, **57**, 131–185.

— — 1979. Les bélemnites. *In*: BUSNARDO, R., THIEULOY, J.-P. & MOULLADE, M. (eds), *Les stratotype français, Vol. 6. Hypostratotype mesogéen de l'étage Valanginien (sud-est de la France)*. Editions du C.N.R.S., 69–76.

CRAME, J. A. 1986. Late Mesozoic bipolar bivalve faunas. *Geological Magazine*, **123**, 611–618.

— — & HOWLETT, P. J. 1988. Late Jurassic and early Cretaceous biostratigraphy of the Fossil Bluff Formation, Alexander Island. *British Antarctic Survey Bulletin*, **78**, 1–35.

DODD, J. R. & STANTON, R. J. Jr 1981. *Paleoecology, Concepts and Applications*. John Wiley & Sons, New York.

DOYLE, P. 1985. 'Indian' belemnites from the Albian (Lower Cretaceous) of James Ross Island, Antarctica. *British Antarctic Survey Bulletin*, **69**,

23–34.

— — 1987a. The Cretaceous Dimitobelidae (Belemnitidae) of the Antarctic Peninsula. *Palaeontology*, **30**, 147–177.

— — 1987b. Early Cretaceous belemnites from southern Mozambique. *Palaeontology*, **30**, 311–317.

— — 1987c. Lower Jurassic–Lower Cretaceous belemnite biogeography and the development of the Mesozoic Boreal Realm. *Palaeogeography, Palaeoclimatology, Palaeoecology*, **61**, 237–254.

— — 1988. The belemnite family Dimitobelidae in the Cretaceous of Gondwana. *In*: WIEDMANN, J. & KULLMANN, J. (eds), *Cephalopods — Present and Past*. Schweizerbart'sche Verlagsbuchhandlung, Stuttgart, 539–552.

— — & ZINSMEISTER, W. J. 1988. A new dimitobelid belemnite from the Upper Cretaceous of Seymour Island, Antarctic Peninsula. *Geological Society of America Memoir*, **169**, 285–290.

ENAY, R. 1972. Paléobiogéographie des ammonites du Jurassique terminal (Tithonique/Volgien/Portlandien s.l.) et mobilité continentale. *Géobios*, **5**, 355–407.

FRAKES, L. A. 1979. *Climates throughout Geologic Time*. Elsevier, Amsterdam.

FÜRSICH, F. T. & SYKES, R. M. 1977. Palaeobiogeography of the European Boreal Realm during Oxfordian (Upper Jurassic) times: a quantitative approach. *Neues Jahrbuch für Geologie und Paläontologie Abh.*, **155**, 137–161.

GLAESSNER, M. F. 1945. Mesozoic fossils from the Central Highlands of New Guinea. *Proceedings of the Royal Society of Victoria*, **56**, 151–168.

GORDON, W. A. 1974. Physical controls on marine biotic distribution in the Jurassic period. *In*: Ross, C. A. (ed.), *Paleogeographic provinces and provinciality*. Society of Economic Paleontologists and Mineralogists Special Publication, **21**, 136–147.

HALLAM, A. 1969. Faunal realms and facies in the Jurassic. *Palaeontology*, **12**, 1–18.

— — 1971. Provinciality in Jurassic faunas in relation to facies and palaeogeography. *In*: MIDDLEMISS, F. A., RAWSON, P. F. & NEWALL, G. (eds), *Faunal Provinces in space and time*. Geological Journal Special Issue, **4**, 129–152.

— — (ed.) 1973. *Atlas of Palaeobiogeography*. Elsevier, Amsterdam.

— — 1975. *Jurassic Environments*. Cambridge University Press, Cambridge.

— — 1984. Distribution of fossil marine invertebrates in relation to climate. *In*: BRENCHLEY, P. J. (ed.), *Fossils and Climate*. Geological Journal Special Issue, **11**, 107–125.

HOWARTH, M. K. 1981. Palaeogeography of the Mesozoic. *In*: COCKS, L. R. M. (ed.), *The Evolving Earth*. British Museum (Natural History), London, 197–220.

HOWLETT, P. J. 1986. *Olcostephanus* (Ammonitina) from the Fossil Bluff Formation, Alexander Island, and its stratigraphical significance. *British Antarctic Survey Bulletin*, **70**, 71–7.

INESON, J. R., CRAME, J. A. & THOMSON, M. R. A.

1986. Lithostratigraphy of the Cretaceous strata of west James Ross Island, Antarctica. *Cretaceous Research*, **7**, 141–159.

JACK, R. L. & ETHERIDGE, R. Jr 1892. *The geology and palaeontology of Queensland and New Guinea, with sixty-eight plates and a geological map of Queensland.* J. C. BEAL, Brisbane.

JEFFERTS, K. 1988. Zoogeography of northeastern Pacific cephalopods. In: WIEDMANN, J. & KULLMANN, J. (eds), *Cephalopods — Present and Past.* Schweizerbart'sche Verlagsbuchhandlung, Stuttgart, 317–339.

JELETSKY, J. A. 1983. Macroinvertebrate paleontology, biochronology and paleoenvironments of Lower Cretaceous and Upper Jurassic rocks, Deep Sea Drilling Hole 511, Eastern Falkland Plateau. *In:* LUDWIG, W. J., KRASHENINNKOV, V. A. *et al.* (eds), *Initial Reports of the Deep Sea Drilling Project*, **71**, 951–975.

KAUFFMAN, E. G. 1973. Cretaceous Bivalvia. *In:* HALLAM, A. (ed.), *Atlas of Palaeobiogeography.* Elsevier, Amsterdam, 353–383.

LEMOINE, P. 1906. *Etudes géologiques dans le nord de Madagascar.* Hermann, Paris.

LIDDLE, R. A. 1946. *The Geology of Venezuela and Trinidad.* Palaeontological Research Institute, Ithaca.

LLOYD, C. R. 1982. The mid-Cretaceous Earth. Paleogeography; ocean circulation and temperature; Atmospheric circulation. *Journal of Geology*, **90**, 399–413.

MARTIN, A. K. & HARTNADY, C. J. H. 1986. Plate tectonic development of the southwest Indian Ocean: a revised reconstruction of east Antarctica and Africa. *Journal of Geophysical Research*, **91**, 4767–4786.

MCWHAE, J. R. H., PLAYFORD, P. E., LINDER, A. W., GLENISTER, B. F. & BALME, B. E. 1948. The stratigraphy of Western Australia. *Journal of the Geological Society of Australia*, **4**, 1–161.

MIDDLEMISS, F. A., RAWSON, P. F. & NEWALL, G. (eds) 1971. *Faunal provinces in space and time.* Geological Journal Special Issue, 4.

MUTTERLOSE, J. 1986. Upper Jurassic belemnites from the Orville Coast, Western Antarctica, and their palaeobiogeographic significance. *British Antarctic Survey Bulletin*, **70**, 1–22.

— — 1988. Migration and evolution patterns in Upper Jurassic and Lower Cretaceous belemnites. *In:* WIEDMANN, J. & KULLMANN, J. (eds), *Cephalopods — Present and Past.* Schweizerbart'sche Verlagsbuchhandlung, Stuttgart, 525–537.

— —, SCHMID, F. & SPAETH, C. 1983. Zur Paläobiogeographie von Belemnite der Unter-Kreide in NW-Europa. *Zitteliana*, **10**, 293–307.

— —, PINCKNEY, G. & RAWSON, P. F. 1987. The belemnite *Acroteuthis* in the *Hibolithes* beds (Hauterivian-Barremian) of north-west Europe. *Palaeontology*, **30**, 635–645.

NOETLING, F. 1987. The fauna of Baluchistan and N.W. Frontier of India. Vol 1: The Jurassic Fauna. Part 2: The Fauna of the (Neocomian) Belemnite Beds. *Memoirs of the Geological*

Survey of India, Palaeontologia indica, **1**, 16.

PHILLIPS, J. 1870. *In:* MOORE, C. Australian Mesozoic geology and palaeontology. *Quarterly Journal of the Geological Society of London*, **26**, 258–259

POZARYSKA, K. & BROCHWICZ-LEWINSKI, W. 1975. The nature and origin of Mesozoic and early Cenozoic marine faunal provinces: some reflections. *Mitteilungen aus dem Geologischen Institut der Universität Hamburg*, **44**, 207–216.

PUGACZEWSKA, H. 1961. Belemnoids from the Jurassic of Poland. *Acta Palaeontologica Polonica*, **6**, 105–236.

QUILTY, P. G. 1970. Jurassic ammonites from Ellsworth Land, Antarctica. *Journal of Paleontology*, **44**, 110–116.

RAWSON, P. F. 1981. Early Cretaceous ammonite biostratigraphy and biogeography. *In:* HOUSE, M. R. & SENIOR, J. R. (eds), *The Ammonoidea.* Systematics Association Special Volume, Academic Press, London, **18**, 499–529.

REITNER, J. & URLICHS, M. 1983. Echte Weichteilbelemniten aus dem unter Toarcian (Posidonienschiefer) Südwestdeutschlands. *Neues Jahrbuch für Geologie und Paläontologie Abh.*, **163**, 460–465.

RICHTER, M. 1925. Beiträge zur Kenntnis der Kreide In Feuerland. *Neues Jahrbuch für Mineralogie, Geologie und Paläontologie* Beilbd, **52**, 524–568

RIEGRAF, W. 1980. Revision der Belemniten des Schwäbischen Jura. Teil. 7. *Palaeontographica* Abt. A, **16g**, 128–206.

— — & HAUFF, R. 1983. Belemnitenfunde mit Weichkörper, Fangarmen und Gladius aus dem Untertoarcian (Posidonienschiefer) und Unteraalenium (Opalmuston) Südwestdeutschlands. *Neues Jahrbuch für Geologie und Paläontologie* Abh., **163**, 466–483.

SAKS, V. N. (ed.) 1972. *The Jurassic-Cretaceous boundary and the Berriasian stage in the Boreal Realm.* Nauka, Novosibirsk. (In Russian: translation 1975, by Israel Program for Scientific Translation, Jerusalem).

SCHEIBNEROVA, V. 1973. A comparison of the austral and boreal Lower Cretaceous foraminiferal and ostracodal assemblages. *In:* CASEY, R. & RAWSON, P. F. (eds), *The Boreal Lower Cretaceous.* Geological Journal Special Issue, **5**, 407–414.

SPATH, L. F. 1927. Revision of the Jurassic cephalopod fauna of Kachh (Cutch). *Memoirs of the Geological Survey of India, Palaeontologica indica* N. S. **9**, 279–550.

— — 1935. Jurassic and Cretaceous Cephalopoda. *In: The Mesozoic Palaeontology of British Somaliland*, Chapter 10, 205–228.

— — 1939. The Cephalopoda of the Neocomian Belemnite Beds of the Salt Range. *Memoirs of the Geological Survey of India, Palaeontologica indica* N. S., **25**, 1–154.

STEFANINI, G. 1925. Molluschi del Giuralias della Somalia. *Palaeontographia italica. Memorie dei Paleontologia*, **32**, supplement 1, 1–53.

STEVENS, G. R. 1964. The belemnite genera *Dicoelites* Boehm and *Prodicoelites* Stolley. *Palaeon-*

tology, **7**, 606–20.

— — 1965. The Jurassic and Cretaceous belemnites of New Zealand and a review of the Jurassic and Cretaceous belemnites of the Indo-Pacific region. *New Zealand Geological Survey, Palaeontological Bulletin*, **36**.

— — 1967. Upper Jurassic fossils from Ellsworth Land, West Antarctica, and notes on Upper Jurassic biogeography of the South Pacific region. *New Zealand Journal of Geology and Geophysics*, 10, 345–283.

— — 1973a. Jurassic belemnites. In: HALLAM, A. (ed.) *Atlas of Palaeobiogeography*. Elsevier, Amsterdam, 259–274.

— — 1973b. Cretaceous belemnites. In: HALLAM, A. (ed) *Atlas of Palaeobiogeography*. Elsevier, Amsterdam, 385–401.

— — & CLAYTON, R. N. 1971. Oxygen isotope studies on Jurassic and Cretaceous belemnites from New Zealand and their biogeographic significance. *New Zealand Journal of Geology and Geophysics*, 14, 829–897.

STOLLEY, E. 1911. Studien an den Belemniten der unteren Kreide Norddeutschlands. *Neues Jahrbuch für Geologie und Paläontologie Beilbd*, Abt. B, **73**, 42–69.

SURLYK, F. & BIRKELUND, T. 1977. An integrated stratigraphical study of fossil assemblages from the Maastrichtian White Chalk of Northwestern Europe. *In:* KAUFFMAN, E. & HAZEL, J. E. (eds), *Concepts and Methods of biostratigraphy*. Dowden Hutchinson & Ross, Stroudsberg, 187–212.

SWINNERTON, H. H. 1955. A monograph of British Lower Cretaceous belemnites. *Palaeontographi-*

cal Society (Monograph) Part 5, 63–86.

— —, 1935. Zur Kenntnis des Jura und de Unterkreide von Misol. Paläontologischen Teil. *Neues Jahrbuch Für Mineralogie, Geologie und Paläontologie, Beilage Band*, **73B**, 41–49.

THOMSON, M. R. A. 1981. Mesozoic ammonite faunas of Antarctica and the break-up of Gondwana. *In:* CRESSWELL, M. M. & VELLA, P. (eds), *Gondwana Five*. A. A. Balkema, Rotterdam, 269–275.

— — & TRANTER, T. H. 1986. Early Jurassic fossils from central Alexander Island and their geological setting. *British Antarctic Survey Bulletin*, 70, 23–39.

VOSS, G. L. 1973. Cephalopod resources of the world. *FAO Fisheries Circular*, **149**.

WESTERMANN, G. E. G. 1981. Ammonite biochronology and biogeography of the Circum-Pacific Middle Jurassic. *In:* HOUSE, M. R. & SENIOR, J. R. (eds), *The Ammonoidea*. Systematics Association Special Volume, Academic Press, London, **18**, 459–498.

WHITEHOUSE, F. W. 1924. Dimitobelidae — a new family of Cretaceous belemnites. *Geological Magazine*, **61**, 410–416.

WILLEY, L. E. 1973. Belemnites from southeastern Alexander Island: II. The occurrence of the family Belemnopseidae in the Upper Jurassic and Lower Cretaceous. *British Antarctic Survey Bulletin*, **36**, 33–59.

WOODBURNE, M. O. & ZINSMEISTER. W. J. 1984. The first land mammal from Antarctica and its biogeographic implications. *Journal of Paleontology*, **58**, 913–948.

Evolutionary patterns in macrurous decapod crustaceans from Cretaceous to early Cenozoic rocks of the James Ross Island region, Antarctica

RODNEY M. FELDMANN & DALE M. TSHUDY

Department of Geology, Kent State University Kent, Ohio 44242, USA

Abstract: The fossil record of macrurous decapod crustaceans is unusually rich in the Late Cretaceous to Paleocene rocks of the James Ross Island region, Antarctica. Four species of lobsters, contained in four genera, are known from the region. *Hoploparia stokesi* (Weller) is most abundant and has been collected on James Ross, Vega, Humps, Cockburn, Seymour, and Snow Hill islands from early Campanian through Paleocene rocks. Whereas morphometric analysis of the species shows no evolutionary trends in shape, carapace ornamentation and claw ornamentation become more variable higher in the section. *Metanephrops jenkinsi* Feldmann, has also been identified. *Metanephrops* has previously been known from only one fossil occurrence, in the Pliocene of New Zealand. Its occurrence on Seymour Island may lie within the region of origin of this taxon which currently inhabits deep water, lower latitude habitats in the Atlantic and Indo-Pacific oceans. *Metanephrops jenkinsi* is known in the early Campanian Santa Marta Formation, in the Maastrichtian portion of the Lopez de Bertodano Formation, and in the Paleocene Sobral Formation. Other taxa are less abundant. In early to middle Campanian rocks on James Ross Island, *Meyeria crofti* Ball represents a relatively late occurrence of a conservative group. *Linuparus macellarii* Tshudy & Feldmann represents the highest latitude occurrence of a species within this genus, which is known from numerous Cretaceous and Paleogene sites in the Northern Hemisphere but from only two other fossil occurrences in the Southern Hemisphere. This species ranges from the early to middle Campanian into the Paleocene. There is no apparent effect of the K/T boundary 'event' on the decapod fauna. New taxa appear below the K/T boundary and none of the Antarctic species disappear from the record at that time.

The fossil record of the James Ross Island region, Antarctic Peninsula, was recognized as being robust and diverse from its discovery in the early twentieth century. Results of the Swedish South Polar Expedition of 1901–1903 documented not only the rich molluscan fauna but also the presence of corals, brachiopods, and decapod crustaceans (Weller 1903; Andersson 1906). Subsequent collections and continued study of the decapods have resulted in the description of six brachyuran and two anomuran species from the Eocene La Meseta Formation (Feldmann & Zinsmeister 1984; Feldmann & Wilson 1988) as well as one anomuran and four macruran species from the Lopez de Bertodano Formation (Ball 1960; Del Valle & Rinaldi 1975; Tshudy & Feldmann 1988).

Analysis of the evolutionary significance of decapods from the La Meseta Formation has led to the conclusion (Zinsmeister & Feldmann 1984) that several species of marine invertebrates evolved in high latitude, shallow water habitats and subsequently radiated into lower latitude, deeper water environments, where descendants are found living today. Originally,

this conclusion was based upon the observation that two species of decapods, *Lyreidus antarcticus* Feldmann and Zinsmeister, *Chasmocarcinus seymourensis* Feldmann and Zinsmeister, three species of echinoderms, and four species of molluscs are first represented in the fossil record in the La Meseta Formation and were progenitors of modern forms, four species of which are known only from deep water settings (Zinsmeister & Feldmann 1984). These observations were later reinforced by examination of a larger collection of decapod crustaceans (Feldmann & Wilson 1988) from which six additional species of decapods were described. Five of the taxa, *Munidopsis scabrosa* Feldmann and Wilson, *Homolodromia chaneyi* Feldmann and Wilson, *Calappa zinsmeisteri* Feldmann and Wilson, ?*Callinectes* sp. and ?*Micromithrax miniusculus* Feldmann and Wilson, were shown to have their oldest occurrences in the La Meseta. Three, *Munidopsis scabrosa*, *Homolodromia chaneyi*, and *Lyreidus antarcticus*, have modern, deep water counter-parts. This pattern of high latitude, shallow water origin of deep water organisms may represent an important rider to the hypothesis of Jablonski *et al.* (1983)

From Crame, J. A. (ed.), 1989, *Origins and Evolution of the Antarctic Biota,*
Geological Society Special Publication No. 47, pp. 183–195.

183

that shallow water faunas tend to radiate into deeper water habitats while their original components become displaced by newly evolving shallow water assemblages.

During the austral summer of 1986–1987, Feldmann and others made extensive collections of decapod crustaceans from a variety of sites in the Lopez de Bertodano and Sobral formations in the vicinity of James Ross and Seymour islands (Figs 1 and 2). In the austral summer of 1987–1988, Eduardo B. Olivero made an important collection of decapods from the Santa Marta Formation on James Ross Island (Fig. 1). Results of this research included description of a new species of the genus *Metanephrops* Jenkins (Feldmann 1989), collection of several more specimens of *Linuparus*

macellarii Tshudy and Feldmann, and several hundred specimens of the most common lobster known from the region, *Hoploparia stokesi* (Weller). It is these specimens that form the basis of the present study.

The purposes of this work are to describe morphological changes that can be observed in *Hoploparia stokesi* over its range of existence from early Campanian through Paleocene time; to consider the possibility that *Metanephrops* evolved from *Hoploparia stokesi* in the high southern latitudes in the region surrounding James Ross Island; and to describe the pattern of radiation of *Metanephrops* from that area in the Cretaceous to its present range in deeper water, lower latitude habitats. Finally, *Hoploparia stokesi*, *Linuparus macellarii*, and *Metane-*

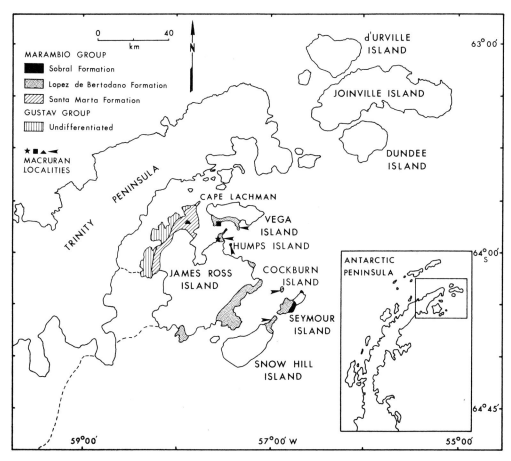

Fig. 1. Index map showing the sites from which macrurous decapod crustaceans have been collected in the region of James Ross Island, Antarctica. Specific localities on Seymour Island are shown in Fig. 2. Sites located by the arrow are localities from which *Hoploparia stokesi* was the sole macruran species collected. This species occurs in combination with other macruran taxa, as follows: with *Meyeria crofti*, star; with *Linuparus macellarii*, square; with *Metanephrops* n. sp., triangle. Maps compiled from Ball 1960, Bibby 1966, Crame 1981, Fleming & Thomson 1979 and Olivero *et al.* 1986.

phrops jenkinsi Feldmann were collected from sites spanning the Cretaceous–Tertiary boundary and, therefore, can be used to consider the effect of the K/T extinction event.

Bedrock of the James Ross and Seymour islands region (Figs 1 & 2) ranges in age from Early Cretaceous through Eocene. The stratigraphy of Seymour Island has been developed in greater detail than for the other islands. Whereas Cretaceous and Paleocene rocks of Seymour Island have been divided into numerous subunits (Macellari 1986, 1988), the stratigraphy of the remainder of the James Ross Basin remains somewhat more generalized (Ineson *et al.* 1986; Olivero *et al.* 1986; Olivero 1988). Early Campanian to Paleocene rocks of the Marambio Group have been subdivided into the Santa Marta, Lopez de Bertodano, and Sobral formations, in ascending order (Fig. 1). The group rests conformably upon the Early to Late Cretaceous Gustav Group which crops out in western James Ross Island (Olivero 1988). Rocks of the Santa Marta Formation tend to be sandstones, while those of the Lopez de Bertodano and Sobral formations are predominantly silty shale (Ineson *et al.* 1986;

Olivero *et al.* 1986). Macrurans have been collected from all formations within the Marambio Group, but none are known from the Gustav Group.

No macrurans have been found in the overlying Cross Valley or La Meseta formations, however. The diverse fauna typically associated with the decapods includes cidarid echinoids and several species of ahermatypic corals, but it is dominated by pelecypods, gastropods and cephalopods. Macellari (1986) concluded that the depositional environments of the Lopez de Bertodano and Sobral formations lay below normal wave base and represent an inner shelf habitat that received almost continuous sedimentation throughout the late Cretaceous and Paleocene. A similar, perhaps even shallower, habitat could be postulated for the coarser grained Santa Marta Formation. Although a minor discontinuity marks the boundary between the Lopez de Bertodano Formation and the overlying Sobral Formation, sedimentation appears to be nearly continuous throughout this timespan. The boundary between the Cretaceous and Tertiary lies not at the formational contact, but near the base of the uppermost unit

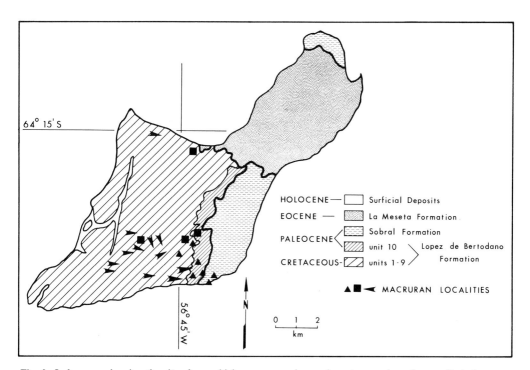

Fig. 2. Index map showing the sites from which macrurous decapod crustaceans have been collected on Seymour Island, Antarctica. Symbols as on Fig. 1. Map base simplified from Sadler 1988. Area of exposure of Cross Valley Formation and several areas of Holocene deposits are not indicated.

(Unit 10) of the Lopez de Bertodano Formation (Fig. 2). The precise position of the boundary is subject to question, dependent upon the group of organisms used as the 'index' group; however, all evidence suggests that the uppermost Lopez de Bertodano Formation was deposited in Paleocene time. The interpretation of the Cretaceous–Tertiary boundary is currently under study.

Hoploparia stokesi (Weller 1903)

The fossil record of the astacid lobster *Hoploparia stokesi* (Weller) (Figs 3 and 4) provides, by virtue of its persistence in rocks throughout the James Ross Island region, an opportunity to observe the nature and rate of phyletic change in this shallow marine species. Since the discovery of *H. stokesi* from lower to middle Campanian rocks by Weller (1903), the species has been identified on James Ross, Vega, Humps, Cockburn, Snow Hill and Seymour islands, extending its known range into the early Paleocene (Tshudy & Feldmann 1988)

Studied material of this taxon included plaster casts of holotype and additional specimens from the British Museum (Natural History), including In. 51772, 51777A, B, 51778–81, and 60125; five specimens from Ohio State University, including OSU 37–3, K-150, K-179, and two unnumbered specimens; and over 400 specimens from the Feldmann collection, including USNM 410841 through 410887 and 410889–410928. The holotype, FMNH 9705, in the Stokes Collection at the Field Museum of Natural History, Chicago, was not examined.

All available specimens were studied, and the better-preserved specimens were measured for the purpose of detecting any morphometric change through time (Fig. 5). Measurements and ratios of these measurements were plotted against one another to differentiate ontogenetic and stratigraphic variation. Variations in the length–height ratio of the carapace (Fig. 5.1), the position of the post-cervical groove at the dorsal midline (Fig. 5.2), and the position of the mandibular external articulation relative to both the orbit and the post-cervical groove at the dorsal midline (Figs 5.3 and 5.4) are not time dependent. It does appear, however, that the position of the mandibular external articulation (Fig. 5.3) does change through ontogeny. This observation, coupled with a much weaker indication 'that the intersection of the post-cervical groove with the dorsum moved forward through ontogeny (Fig. 5.2), suggests that growth of the branchial region is anisometric.

The carapace of geologically older specimens is often more inflated posterior to the post-cervical groove over the branchial and cardiac regions than in younger specimens (Figs 3.1 and 3.2). The post-cervical groove may also be broader in older specimens. These phenomena are most evident in the lower Lopez de Bertodano Formation and less apparent in the Santa Marta Formation. In contrast, on geologically younger specimens a less-marked post-cervical groove divides the carapace into two regions of similar elevation (Fig. 3.3). An unlithified, buried carapace tends to be highly vulnerable to changes in topography, particularly that of an exuvia or a newly-moulted specimen; however, this trend in carapace topography is too consistent stratigraphically to be considered a taphonomic effect.

The adaptive significance of the inflation is not clear. The possibility that the inflation was produced by parasites, such as turbellarian worms or bopyrid isopods, was rejected due to the similarity in inflation between different specimens. Bilateral symmetry is observable in the carapace of at least one specimen in which both sides were preserved. Another specimen has one side more inflated than the other. Parasites in the branchial region, or compaction, may deform the carapace irregularly. However, repeated observation of this inflation in specimens within the lower portion of the Lopez de Bertodano Formation suggests that the inflation was a real morphological character.

In crayfishes, a heightening of the branchial chamber, which is delineated dorsally by the branchiocardiac groove, aids respiration in oxygen-deficient environments (Hobbs 1974). In *Hoploparia stokesi*, the height of the branchial chamber does not appear to vary stratigraphically, but the more vaulted, and therefore more voluminous, branchial chamber in the specimens from the lowermost Lopez de Bertodano Formation might have been similarly related to oxygen availability. Observations on lithology and associated fauna in this part of the section (Ball 1960; Bibby 1966; Macellari 1986), however, do not provide conclusive evidence that these specimens inhabited an environment either richer or poorer in oxygen than that of the younger specimen.

In addition to an inflated branchial region, individuals of this same group possess post-antennal and hepatic spines (Figs 3.1 & 3.2). These spines are typically reduced or absent on carapaces in the younger group (Fig. 3.3). The postantennal spine, although itself subtle, is usually the more prominent of the two.

The adaptive significance of these spines, if any, is also unclear. Like the carapace inflation,

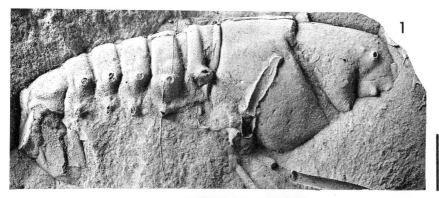

Fig. 3. *Hoploparia stokesi* (Weller). (**1**) Right lateral view of latex cast of USNM 430023 showing generally inflated branchial region, well developed postantennal and hepatic spines and broad postcervical groove. (**2**) left lateral view of latex cast of USNM 430024 showing inflated branchial region, well developed postantennal and hepatic spines and broad postcervical groove. (**3**) right lateral view of latex cast of USNM 430025 showing normal branchial chamber, and reduced hepatic and postantennal spines. Scale bars equal 1 cm.

the spines were ultimately selected against. Examination of Recent specimens shows that spines are reflected on the interiors of carapaces as pits which can receive bundles of muscles. In light of the location of these spines, muscle attachment is a possibility. In the lower Lopez de Bertodano Formation, increase in spinosity may have been a response to a subtle environ-

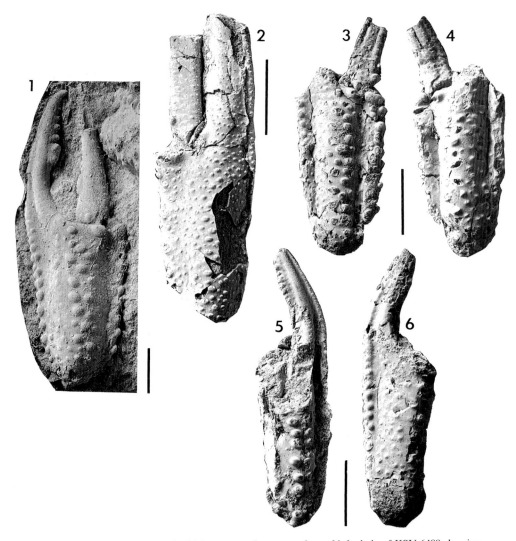

Fig. 4. *Hoploparia stokesi* (Weller). (**1**) latex cast of upper surface of left chela of KSU 6409 showing moderately coarse nodose ornamentation. (**2**) upper surface of right chela of OSU 37−3 showing granular ornamentation. (**3**) & (**4**) upper and lower surfaces of left chela of USNM 430026 showing nodose ornamentation. (**5**) & (**6**) upper and lower surfaces of right chela of same specimen, USNM 430026, showing nodose ornamentation and generally equivalent size and morphology as that of the left chela. Scale bars equal 1 cm.

mental change. Although too small to have aided significantly in protection, the spines might have served to anchor the lobster in its burrow. Fields of spines and terraced ornamentation on arthropod exoskeletons have been interpreted (Stitt 1976; Feldmann *et al.* 1986) as having this function.

Chela ornamentation varies greatly within the species and, to a lesser degree, within collections from the same locality. Chelae on specimens from the Santa Marta Formation tend to be coarsely granulose. Those on all specimens from the lower portion of the Lopez de Bertodano Formation are more finely granulose (Fig. 3.2). In contrast, there is considerable variation in chela ornamentation among the younger specimens, ranging from granulose (Fig. 4.2) to very coarsely nodose (Figs 4.1 and 4.3−4.6) in specimens from the same locality.

On the coarser chelae, both the left and right chelae have broad medial ridges on their upper and lower surfaces (Figs 4.3−4.6). The ridges

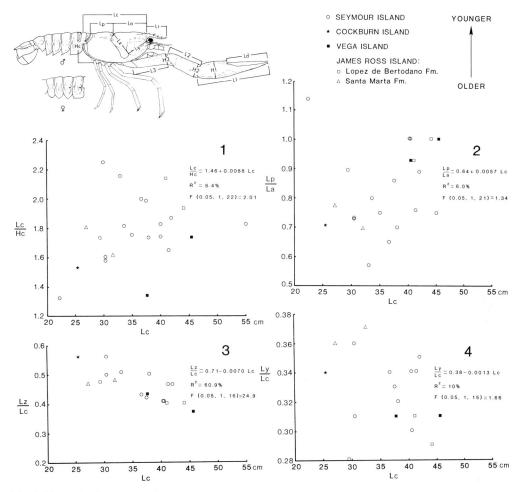

Fig. 5. Plots of ratios computed on specimens of *Hoploparia stokesi* (Weller). Orientation of measurements illustrated on sketch of the species (from Ball 1960). Symbols represent measurements on specimens from sites as indicated on the diagram. Full discussion of parts 1–4 contained in text.

are convex in transverse section, being most accentuated on the ventral surface. The medial ridge on the upper surface continues over the length of the fixed finger. On the lower surface, the medial ridge is less distinct, and bordered externally by a broader furrow.

Ornamentation on the upper surface is coarser than on the lower surface, and coarsest on the medial ridge and inner and outer margins. Nodes on the inner margin of the lower surface become larger distally, and the same is true, but typically less noticeably, on the medial ridge. Ornamentation is uniformly coarse along the outer margin. Stout processes on the upper and lower surfaces are located at the base of articulation of the dactylus. All nodes on the palm

trend distally. On the upper surface ornamentation slightly coarsens distally on the medial ridge, is essentially uniform on the inner margin, and on the outer margin is coarsest over the proximal second quarter of the palm.

It is most probable that intraspecific variation accounts for the range in ornamentation at single collecting localities. Sexual dimorphism seems a less probable explanation, but cannot be tested due to the lack of corresponding abdominal pleurae, which are known to be sexually diagnostic in this species (Ball 1960; Tshudy & Feldmann 1988).

The significance of spines on the claws is, perhaps, straightforward. Distally directed nodes would have served to better anchor the lobster in its burrow. The claws may also have functioned as defensive devices.

The record of *Hoploparia stokesi* extends over a known range of as much as 20 million years within the same, small, depositional basin. During this span, changes can be observed in degree of inflation of the cephalothorax, spinosity in the cephalic region, and ornamentation of the claws. The change in claw ornamentation appears to have been an adaptation to burrow dwelling, although changes in carapace topography do not necessarily corroborate this interpretation. The adaptive significance of the reduction in carapace inflation and spinosity through time is not clear, although they may be related to variation in environmental conditions related to variations in water depth. More work must be done on the interpretation of the physical stratigraphy of the Marambio Group before definitive relationships of functional morphology will be possible.

Metanephrops jenkinsi Feldmann 1989

Twenty-eight specimens of *Metanephrops jenkinsi* Feldmann (Fig. 6) have been collected from eight localities within the upper portion of the Lopez de Bertodano Formation and the Sobral Formation on Seymour Island (Fig. 2). An additional three specimens were collected from the Santa Marta Formation. Based primarily on the distribution of ammonites, the upper part of this sequence has been determined to range from Maastrichtian into the Paleocene (Macellari 1986), whereas the lower portion is early Campanian (Olivero 1988). These occurrences span the entire stratigraphic range of *Hoploparia stokesi* and, without exception, the collecting sites from which *Metanephrops jenkinsi* was collected contained representatives of both species.

There are some underlying morphological similarities between *Metanephrops jenkinsi* and *Hoploparia stokesi*. This suggests that *Hoploparia stokesi* may have been the root stock from which *Metanephrops* evolved. If this is so, the speciation event was not documented by intermediate forms.

Metanephrops is readily recognized on the basis of its having longitudinal carinae well developed in the branchial region and in possessing a series of spinose ridges on the cephalic region (Fig. 6.1). Although there are various spines and subtle ridges present in the cephalic region of *Hoploparia stokesi*, nothing even vaguely resembling the longitudinal carinae on the branchial region is known

from that species or from any other species of *Hoploparia*.

The abdominal regions of the two species tend to be generally similar, although a unique pattern of spines is developed on the lateral margins and on the axial region of the sixth abdominal somite in *Metanephrops jenkinsi*. This spine arrangement is diagnostic of the genus *Metanephrops*. Spines are present on that segment of *Hoploparia stokesi* but the pattern is different.

Claws in the two species tend to be more or less similar, considering the entire range of variation within *Hoploparia stokesi*. As discussed above, representatives of *Hoploparia stokesi* with generally smooth claws are known from part of its stratigraphic range. Claws of *Metanephrops* are characterized by keeled outer margins and a few rows of spines on an otherwise smooth field (Figs 6.3 & 6.4). The claws of *Metanephrops* also tend to be proportionately larger than those of *Hoploparia stokesi*.

These differences permit the two species to be distinguished from one another readily. However, there is an underlying similarity in the carapace groove patterns (cf. Figs 3 & 6.1) suggesting that *Hoploparia stokesi* may have been the ancestral species from which *Metanephrops* evolved.

The fundamental importance of the groove pattern in interpreting the evolution of decapod crustaceans has been well documented (Glaessner 1960, 1969; Secretan 1964) and linked to the architecture of the internal anatomy of the organisms (Secretan 1964). Furthermore, *Hoploparia* has been considered to be the ancestor to modern nephropine lobsters, although the pathway suggested by Glaessner (1960, 1969) suggests a linear ascendancy through the poorly known genus *Paraclythia* Fritsch. It is probable that various species of *Hoploparia* gave rise to representatives of several modern taxa, including *Homarus* Weber (Feldmann 1974), *Nephrops* Leach (Glaessner 1960, 1969), and *Metanephrops*. The modern distribution of these three genera, coupled with their morphological distinctiveness, suggests that the origins of the three taxa are independent of one another. The genus *Hoploparia* had a nearly worldwide distribution in the Cretaceous and, in all probability, served as the basic stock from which the other taxa were derived.

Jenkins (1972) summarized the geographical distribution of the 14 extant species of *Metanephrops*. He arrayed species into four groups in which all but three species were confined to the western Pacific region. One species

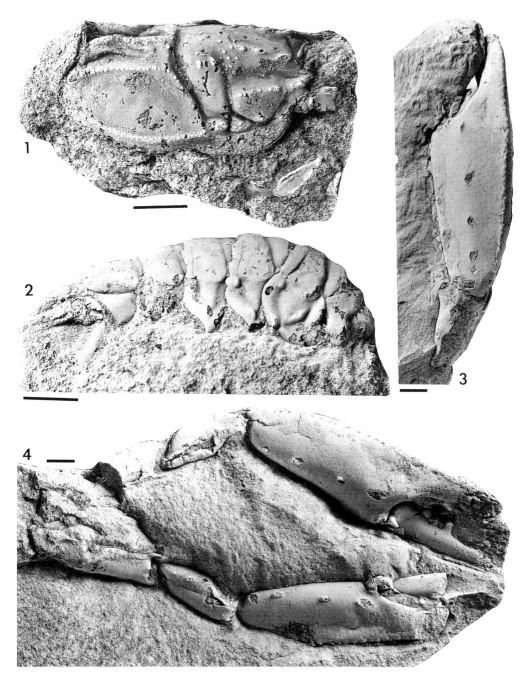

Fig. 6. *Metanephrops jenkinsi* Feldmann. (**1**) right lateral view of cephalothorax of USNM 424598 showing ridges on the branchial region and spinose ridges on the cephalic region. (**2**) right lateral view of abdomen of USNM 424605. (**3**) & (**4**) upper surface of right chela and lower view of right and left chelae of USNM 424602 showing spinose and nodose ridges on a generally smooth field. Scale bars equal 1 cm.

extends from that region into the eastern and western Indian Ocean and the remaining two species, comprising the *M. rubellus* group, are restricted to the western Atlantic Ocean. The organisms were typically collected from fine siliciclastic substrata in depths ranging from 50 to 885 m (Jenkins 1972). Thus, *Metanephrops* presently has a broad range of distribution in outer shelf and slope habitats.

At the same time, Jenkins described the only known fossil species, *M. motunauensis*, from the Pliocene of South Island, New Zealand. Because that represented the sole fossil form, and because of the relative timing of the occurrence, Jenkins suggested that the genus must have had a longer history but he was unable to speculate on the details other than to suggest that the genus evolved in the Indo-West Pacific region and dispersed into the Atlantic through the Tethys. The discovery of *Metanephrops* in Cretaceous and Paleogene rocks of Antarctica, therefore, provides the first material basis for refining this history.

Metanephrops jenkinsi, from Seymour Island, most closely resembles *M. rubellus* (Moreira) (Feldmann 1989) (Fig. 7) known from habitats at shelf depth, between 50 and 150 m, in temperate latitudes along the coast of South America from Brazil to Argentina. Only one other species, *M. japonicus* (Tapparone Canefri), has been collected in water of comparably shallow depths and this species also ranges into mid-temperate latitudes. All other species are known only from outer shelf and slope depths.

This combination of bathymetric and geographic distribution very closely parallels that of modern species of the crab genus *Lyreidus* (Griffin 1970; Feldmann 1986). It has been suggested (Feldmann & Zinsmeister 1984; Feldmann 1986) that *Lyreidus* had its origin in the region of Seymour Island, in the Eocene, and that it subsequently radiated to deeper water, lower latitude habitats, probably following a low temperature thermal gradient. Thus, it appears probable that the radiation pattern for the two genera was similar.

Linuparus macellarii Tshudy & Feldmann 1988

Linuparus macellarii was originally described on the basis of a single specimen from the Lopez de Bertodano Formation (Tshudy & Feldmann 1988). The collection has now expanded to eight specimens and extended in stratigraphic range throughout the Lopez de Bertodano Formation; that is, from early Campanian into the Paleocene. The material basis for the taxon consists of a single abdomen, several sterna, and several partial carapaces. Because the material is limited, it is not possible to describe temporal changes in morphology.

Meyeria crofti Ball 1960

A single, fragmentary, specimen formed the basis of *Meyeria crofti* Ball 1960. The specimen was collected from the Naze, James Ross Island, from early to middle Campanian rocks. Subsequent collecting has produced no additional material and, therefore, there is no possibility of describing patterns of evolution or dispersal within this taxon.

The Cretaceous–Tertiary boundary

Rocks of the Marambio Group on Seymour Island range in age from late Campanian into the Paleocene. Because the fossil record is robust and because the stratigraphic record seems to be unusually complete (Zinsmeister *et al.* 1987), it is possible to determine the position of the Cretaceous–Tertiary boundary based upon a variety of palaeontological criteria. Ammonoid cephalopods (Macellari 1988), foraminiferans (Huber 1988), dinoflagellates and pollens (Askin 1988) have been studied in detail and the position of the K/T boundary has been identified on the basis of each of these groups. The results suggest (Zinsmeister *et al.* 1987) that the change from Cretaceous to Tertiary biota was gradual, rather than abrupt, and that the precise position of the boundary varied through a few metres, depending upon the reference taxon used. This set of observations probably reflects the continuous, relatively rapid, sedimentation rate during this interval of time which resulted in attenuation, rather than truncation, of stratigraphic ranges. As a result, the uppermost ranges of taxa in the section appear to be stepped rather than synchronous. However, recognizing the vagaries of fossil preservation and discovery, these observations must be interpreted cautiously (Signor & Lipps 1982). Regardless of the reference taxon used, certain conclusions can be drawn. The position of the Cretaceous–Tertiary boundary lies within the uppermost part of the Lopez de Bertodano Formation, within Unit 10 (Macellari 1986). The Sobral Formation, therefore, lies entirely within the Paleogene.

Representatives of all macruran species, with the exception of *Meyeria crofti*, have been collected both below and above the transitional zone that defines the Cretaceous–Tertiary boundary. *Hoploparia stokesi* and *Metanephrops jenkinsi* range into the Sobral Formation and *Linuparus macellarii* has been collected within a few metres of the contact between the Lopez de Bertodano and Sobral

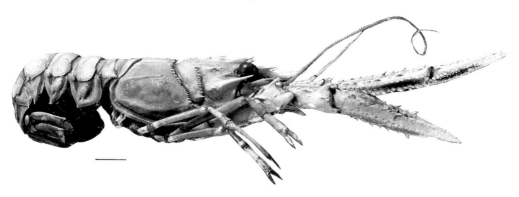

Fig. 7. *Metanephrops rubellus* (Moreira), USNM 170688, collected from the Atlantic Ocean off Santa Catarina State, Brazil. Scale bar equals 1 cm.

formations. None of the species becomes extinct within the boundary interval.

Insufficient material representing *Linuparus macellarii* is available for evaluation of morphological variation that might have resulted from the K/T boundary 'event' of Alvarez *et al.* (1980). However, the fossil record of *Hoploparia stokesi* and *Metanephrops jenkinsi* is adequate to make a quantitative assessment of the former species (Fig. 5) and a detailed qualitative evaluation of the latter. In neither species is there any indication of morphological change or change in general abundance of individuals at this time. If there was an 'effect' in this area at the end of the Cretaceous, it was subtle and not catastrophic.

Conclusions

The rocks of the Marambio Group, exposed throughout the James Ross Island region, contain abundant remains of *Hoploparia stokesi* and *Metanephrops jenkinsi* and sparse remains of *Linuparus macellarii* and *Meyeria crofti*. The span of ages of these rocks, from early Campanian to Paleocene, may exceed 20 million years. There is no evidence for either a catastrophic or an abrupt effect on the decapod crustacean fauna across the Cretaceous–Tertiary boundary. During the entire 20 million year interval of time, the morphology of *H.*

stokesi does not appear to have undergone directional change. Variations in ornamentation of the cephalothorax and claws are best interpreted as ecophenotypic. Although *Metanephrops jenkinsi* is morphologically distinct from all species of *Hoploparia*, there are sufficient basic structural similarities to warrant suggesting that the former species may have evolved from *H. stokesi* in the region of the Antarctic Peninsula and that *Metanephrops jenkinsi* represents the oldest species within the genus. The resulting pattern of radiation of this taxon from a shallow water, inshore origin in high latitudes to the deeper water, lower latitude habitats of living congenors mimics the pattern of origin and dispersal observed in several Cenozoic taxa from Antartica.

In addition to specimens collected by Feldmann during the field seasons of 1984–1985 and 1986–1987, several specimens used in this study were collected on Seymour Island by W. Zinsmeister, M. Woodburne, D. Elliot, D. Reisky, M. Kooser, and H. Zimmerman; on Humps Island by F. Barbus; and on Cockburn Island by J. Stilwell in 1986–1987. E. Olivero collected the decapods from the Santa Marta Formation, on James Ross Island, during the austral summer of 1987–1988. The original draft of this paper was read by A. Williams. Field work by Feldmann was supported by an NSF grant to Zinsmeister and laboratory work was supported by NSF grant DPP8715945 to Feldmann. Contribution 380, Department of Geology, Kent State University, Kent, Ohio 44242, USA.

References

Andersson, J. G. 1906. On the geology of Graham Land. *Geological Institute of the University of Upsala*, *Bulletin*, **7**, 19–71, 6 pls.

Alvarez, L., Alvarez, W., Asaro, F. & Michel, H. 1980. Extraterrestrial cause for the Cretaceous–Tertiary extinctions. *Science*, **208**, 1095–1108.

ASKIN, R. A. 1988. Campanian to Eocene paly-
nological succession of Seymour and adjacent
islands, northeastern Antarctic Peninsula. *In*:
FELDMANN, R. M. & WOODBURNE, M. O. (eds),
*Geology and Paleontology of Seymour Island,
Antarctica, Geological Society of America
Memoir*, **169**, 131–154.

BALL, H. W. 1960. Upper Cretaceous Decapoda and
Serpulidae from James Ross Island, Graham
Land. *Falkland Islands Dependencies Survey
Scientific Reports*, **24**, 1–30.

BIBBY, J. S. 1966. The stratigraphy of part of northeast
Graham Land and the James Ross Island Group.
British Antarctic Survey Scientific Report, **53**,
1–37.

CRAME, J. A. 1981. Upper Cretaceous inoceramids
(Bivalvia) from the James Ross Island Group
and their stratigraphical significance. *British
Antarctic Survey Bulletin*, **53**, 29–56.

DEL VALLE, R. & RINALDI, C. A. 1975. Sobre la
presencia de *Hoploparia stokesi* (Weller) en las
'Snow Hill Island Series' de la Isla Vicecomodoro
Marambio, Antartida. *Instituto Antartico Argentino, Contribucion*, **190**, 1–19.

FELDMANN, R. M. 1974. *Hoploparia riddlensis*, a new
species of lobster (Decapoda: Nephropidae) from
the Days Creek Formation (Hauterivian, Lower
Cretaceous) of Oregon. *Journal of Paleontology*,
48, 586–593.

—— 1986. Paleobiogeography of two decapod crustacean taxa in the Southern Hemisphere. *In*:
GORE, R. H. & HECK, K. L. (eds), *Crustacean
Biogeography*, A. A. Balkema, Rotterdam,
5–20.

—— 1989. *Metanephrops jenkinsi* n.sp. (Decapoda:
Nephropidae) from the Cretaceous and Paleo-
cene of Seymour Island, Antarctica. *Journal of
Paleontology*, **63**, 64–69.

——, BOSWELL, R. M. & KAMMER, T. W. 1986.
Tropidocaris salsiusculus, a new rhinocaridid
(Crustacea: Phyllocarida) from the Upper
Devonian Hampshire Formation of West
Virginia. *Journal of Paleontology*, **60**, 379–383.

—— & WILSON, M. T. 1988. Eocene decapod crustaceans from Antarctica. *In*: FELDMANN, R. M.
& WOODBURNE, M. O. (eds), *Geology and Paleontology of Seymour Island, Antarctica, Geological Society of America Memoir*, **169**, 465–488.

—— & ZINSMEISTER, W. J. 1984. New fossil crabs
(Decapoda: Brachyura) from the La Meseta
Formation (Eocene) of Antarctica: paleogeographic and biogeographic implications. *Journal
of Paleontology*, **58**, 1046–1061.

FLEMING, E. A. & THOMSON, J. W. 1979. Geologic
map of Northern Graham Land and South Shetland Islands. *British Antarctic Survey, 1:500 000
Geologic Map*, Series BAS 500, G, Sheet 2,
Edition 1.

GLAESSNER, M. F. 1960. The fossil decapod Crustacea
of New Zealand and the evolution of the Order
Decapoda. *New Zealand Geological Survey
Paleontological Bulletin*, **31**, 1–78.

—— 1969. Decapoda, Part R4(2), Arthropoda. *In*:
MOORE, R. C. (ed.), *Treatise on Invertebrate

Paleontology, Geological Society of America and
University of Kansas Press, R400–R533.

GRIFFIN, D. J. G. 1970. A revision of the Recent
Indo-Pacific species of the genus *Lyreidus* de
Haan (Crustacea, Decapoda, Raninidae). *Transactions of the Royal Society of New Zealand*,
12(10), 89–112.

HOBBS, H. H. JR, 1974. Adaptations and convergence in North American crayfishes. *Proceedings
of the Second International Crayfish Symposium,
Louisiana State University*, 541–551.

HUBER, B. T. 1988. Upper Campanian–Paleocene
Foraminifera from the James Ross Island region,
Antarctic Peninsula. *In*: FELDMANN, R. M. &
WOODBURNE, M. O. (eds), *Geology and Paleon-
tology of Seymour Island, Antarctica, Geological
Society of America Memoir* **169**, 163–252.

INESON, J. R., CRAME, J. A. & THOMSON, M. R. A.
1986. Lithostratigraphy of the Cretaceous strata
of west James Ross Island, Antarctica. *Cretaceous Research*, **7**, 141–59.

JABLONSKI, D., SEPKOSKI, J. J., JR., BOTTJER, D. J. &
SHEEHAN, P. M. 1983. Onshore–offshore patterns in the evolution of Phanerozoic shelf communities. *Science*, **222**, 1123–1125.

JENKINS, R. J. F. 1972. *Metanephrops*, a new genus of
late Pliocene to Recent lobsters (Decapoda,
Nephropidae). *Crustaceana*, **22**(2), 161–177.

MACELLARI, C. E. 1986. Late Campanian–
Maastrichtian ammonite fauna from Seymour
Island (Antarctic Peninsula) *Paleontological
Society Memoir*, **18**, 1–55.

—— 1988. Stratigraphy, sedimentology and paleoecology of Upper Cretaceous/Paleocene shelf-
deltaic sediments of Seymour Island (Antarctic
Peninsula). *In*: FELDMANN, R. M. & WOODBURNE,
M. O. (eds), *Geology and Paleontology of
Seymour Island, Antarctica, Geological Society
of America Memoir*, **169**, 25–54.

OLIVERO, E. B. 1988. Early Campanian heteromorph
ammonites from James Ross Island, Antarctica.
National Geographic Research, **4**, 259–271.

——, SCASSO, R. A. & RINALDI, C. A. 1986. Revision
of the Marambio Group, James Ross Island,
Antarctica. *Instituto Antarctice Argentino Contribucion*, **331**, 1–29.

SADLER, P. M. 1988. Geometry and stratification
of Paleogene and latest Cretaceous units on
Seymour Island, northern Antarctic Peninsula.
In: FELDMANN, R. M. & WOODBURNE, M. O.
(eds), *Geology and Paleontology of Seymour
Island, Antarctica, Geological Society of America
Memoir*, **169**, 303–320.

SECRETAN, S. 1964. Les crustacés decapodes du Jur-
assique Superieur et du Crétacé de Madagascar.
*Memoires du Museum National d'Histoire
Naturelle, Nouvelle Série*, **14**, 1–226, Pl. 1–20.

SIGNOR, P. W., III & LIPPS, J. H. 1982. Sampling bias,
gradual extinction patterns and catastrophes in
the fossil record. *Geological Society of America
Special Paper*, **190**, 291–296.

STITT, J. H. 1976. Functional morphology and life
habits of the Late Cambrian trilobite *Stenopilus
pronus* Raymond. *Journal of Paleontology*, **50**,

561–577.

TSHUDY, D. M. & FELDMANN, R. M. 1988. Macrurous decapod crustaceans, and their epibionts, from the Lopez de Bertodano Formation (Late Cretaceous), Seymour Island, Antarctica. *In*: FELDMANN, R. M. & WOODBURNE, M. O. (eds), *Geology and Paleontology of Seymour Island, Antarctica, Geological Society of America Memoir*, **169**, 291–302.

WELLER, S. 1903. The Stokes collection of Antarctic fossils. *Journal of Geology*, **11**, 413–419.

ZINSMEISTER, W. J. & FELDMANN, R. M. 1984. Cenozoic high latitude heterochroneity of Southern Hemisphere marine faunas. *Science*, **224**, 281–283.

——, —— WOODBURNE, M. O., KOOSER, M. A., ASKIN, R. A. & ELLIOT, D. E. 1987. Faunal transitions across the K/T boundary in Antarctica. *Geological Society of America, Abstracts with Programs*, **20**, 101.

New plesiosaurs from the Upper Cretaceous of Antarctica

SANKAR CHATTERJEE & BRYAN J. SMALL

The Museum of Texas Tech University, Lubbock, Texas 79409, USA

Abstract: New plesiosaur remains from the Upper Cretaceous Lopez de Bertodano Formation (late Campanian−Maastrichtian) of Seymour Island, Antarctic Peninsula include Cryptoclididae and Elasmosauridae. The occurrence of Cryptoclididae is reported from the Antarctic region for the first time. The taxon represents a new genus and species, based on a skull and associated cervical vertebrae. The long, slender and delicate teeth may have formed a 'trapping' device that enabled cryptoclidids to feed on small fish and crustaceans that abound in the same deposits. The cryptoclidids had a restricted distribution, being known so far from the Middle and Late Jurassic of England, and the Late Cretaceous of Chile, Argentina, and Antarctica.

Other specimens, represented by several postcranial skeletons, are taxonomically indeterminate, but they share some features with other contemporary elasmosaurid genera such as *Hydrotherosaurus*, *Morenosaurus*, *Thalassomedon*, and *Mauisaurus*. Unlike the cryptoclidids, the elasmosaurids had a cosmopolitan distribution during the Jurassic and Cretaceous periods.

Trophic diversity within guilds of marine predators is examined in the Lopez de Bertodano palaeocommunities. Three predator guilds are recognized on the basis of tooth morphology and prey preference. The mosasaurs composed the 'Cut guild', and were the principal predators. The elasmosaurids constituted the 'Pierce guild', and the cryptoclidids formed the 'Trap guild'. These marine reptiles exploited the various pelagic resources such as sharks, bony fish, soft cephalopods and crustaceans, and survived until the end of the Cretaceous. The plesiosaurs were excellent swimmers, and used their hyperphalangic paddles for subaqueous flight in the manner of modern sea lions.

Plesiosaurs were among the most diverse and widespread of the Mesozoic marine reptiles which dominated the seas throughout the Jurassic and Cretaceous periods. They were highly adapted for a marine predaceous existence with their paddle-like, hyperphalangic limbs, which were used for aquatic locomotion. They were the largest of the marine reptiles with some of the Late Cretaceous genera attaining a length of 15 m, and necks over 7 m long.

Despite extensive knowledge of the cranial anatomy of plesiosaurs, their systematic position has long been subject to dispute because of the structure of the temporal region. There is a single, large upper temporal opening, bounded above by the parietal and below by the squamosal and postorbital. The cheek region below the fenestra is open and was thought to be derived by the emargination of a solid cheek like that of *Araeoscelis* (Williston 1925). On the basis of a single upper temporal opening, the plesiosaurs have long been classified among euryapsids (Romer 1956, 1966; Colbert 1965, 1969). However, it is currently believed that plesiosaurs are more appropriately allied with the diapsid reptiles that paralleled the squamates in the breakdown of the lower temporal arcade (Kuhn-Schnyder 1967; Carroll 1981,

1988). Clearly this feature evolved independently in many diverse groups, including protorosaurs (Chatterjee 1986), nothosaurs (Carroll 1981), pleurosaurs (Carroll & Gaskill 1985), sphenodontids (Robinson 1973), millerettids (Gow 1972) and eosuchians. (Evans 1980).

Traditionally the order Plesiosauria is divided into two superfamilies, the Plesiosauroidea and the Pliosauroidea, on the basis of the relative length of the skull and neck. Plesiosauroids had long, slender necks, small heads, and a short mandibular symphysis. On the other hand, pliosauroids had large heads, a long mandibular symphysis, and short necks.

Recently, abundant plesiosauroid remains have been recovered from the Late Cretaceous Lopez de Bertodano Formation of Seymour Island, Antarctic Peninsula. They include members of two different families: Elasmosauridae and Cryptoclididae. Elasmosaurids from Seymour Island have been reported previously (del Valle *et al.* 1977; Chatterjee & Zinsmeister 1982; Chatterjee *et al.* 1984; Gasparini *et al.* 1984). Additional elasmosaur material is discussed here. The cryptoclidid material, reported here for the first time from the Antarctic region, represents a new taxon. So far no pliosauroid remains have been found in

From Crame, J. A. (ed.), 1989, *Origins and Evolution of the Antarctic Biota*,
Geological Society Special Publication No. 47, pp. 197−215.

197

Seymour Island. The plesiosaurs described in this paper are important in that they provide new taxa from Antarctica, as well as demonstrating the wider distribution of cryptoclidids (previously known from only England and South America).

Geological setting

Seymour Island contains a remarkably continuous record of Late Cretaceous through Paleocene nearshore marine and coastal/deltaic sediments of a transgressive–regressive cycle. The plesiosaur fossils were recovered from the Lopez de Bertodano Formation, which is exposed in the southern two-thirds of the island (Fig. 1). The lithology consists of *c.* 1200 m of loosely consolidated sandy siltstones, intercalated with calcareous concretions. Macellari (1986) subdivided this formation into 10 informal lithological units. The lowermost six units (1–6) are grouped into '*Rotularia* units', characterized by the abundance of the annelid worm tube *Rotularia*, but the near absence of macroinvertebrates. The upper 4 units (7–10) are included in the 'molluscan units', characterized by abundant macro-invertebrates such as ammonites, echinoids, bivalves, gastropods, and arthropods. The age of the Lopez de Bertodano

Formation ranges from late Campanian to Maastrichtian on the basis of palynomorphs, invertebrates, and glauconite datings (Huber 1985). The inferred K−T boundary has been recognized primarily by the last appearance of ammonites. It probably occurs between units 9 and 10 of the Lopez de Bertodano Formation, about 45 m below the base of the overlying Sobral Formation (Macellari 1986).

The marine vertebrate material collected from the Lopez de Bertodano Formation is highly varied and includes the remains of teleosts and sharks (Grande & Chatterjee 1987), plesiosaurs and mosasaurs (Chatterjee & Zinsmeister 1982; Chatterjee *et al.* 1984), and a newly discovered bird skeleton. The sharks, mosasaurs, and elasmosaurids are distributed throughout the formation, but the cryptoclidids, teleosts and bird specimen are so far restricted to the 'molluscan units'. Some of the plesiosaurs have been recovered from the uppermost Cretaceous section, close to the K−T boundary.

Material and methods

Although isolated plesiosaur bones were found scattered throughout the Lopez de Bertodano Formation, only eight partially associated skeletons were collected during the three field

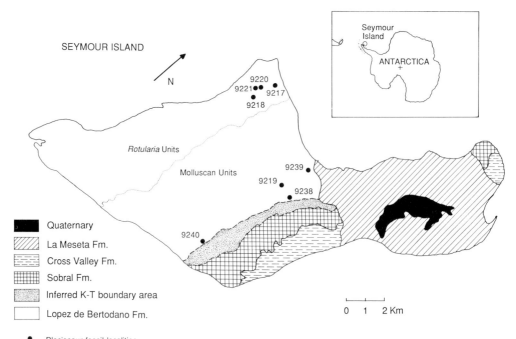

Fig. 1. Geological map of Seymour Island with plesiosaur fossil localities (after Huber 1985 and Macellari 1986).

seasons (1981–82, 1983–84, and 1984–85 austral summers). The specimens from the 'Rotularia units' are often found exposed on the surface, and were virtually free of matrix. On the other hand, specimens collected from the 'molluscan units' were encased in hard calcareous nodules which made the preparation time-consuming and difficult.

In the field we encountered two major problems in retrieving the bones from the ground. While removing the overburden, the bones were often found encased in a hard, permafrost layer, which lay about a foot below the surface. In these situations, the recovery of bones from the permafrost layer was found almost impossible with conventional field tools. We also tried to make plaster jackets on several occasions but were confronted with a major problem: the plaster would not set at such a cold temperature. To overcome this we used a makeshift oven, with the help of camp stoves and aluminum foil to block the cold air. This allowed the jackets to set firmly enough to be moved without damaging the bones.

A combination of mechanical and chemical procedures was employed in the preparation of the plesiosaur bones in the laboratory. A Foredom flexible shaft power tool was used, whenever possible, to grind down the matrix. During acid preparation, the bones were coated with polyvinyl butyral (Butvar B-76) and thoroughly dried before placing them in the acid bath. A weak (10–15%) acetic acid worked very well in the preparation. Once in the bath, frequent checks were made of the progress. After the preparation, the bones were washed and soaked in water for an extended period of time. Insoluble residues attached to the bones were removed mechanically by grinding.

A list of associated plesiosaur specimens from Antarctica in the collection of The Museum of Texas Tech University is given below:

A. *Cryptoclidid specimen*
(1) TTU P 9219: young adult, skull and associated cervicals;

B. *Elasmosaurid specimens*
(2) TTU P 9217: young adult, consisting of 40+ vertebrae, fragments of pubis, ischium, complete ilium, humerus, femur, and pelvic paddles;
(3) TTU P 9218: juvenile, includes cervicals and caudals, fragments of pectoral and pelvic girdles, limb bone fragments, and isolated paddle elements;
(4) TTU P 9220: adult, includes part of cervical, dorsal, sacral, and caudal vertebrae, coracoid, pelvis and associated gastroliths;
(5) TTU P 9221: juvenile, includes some cervi-

cal, pectoral, dorsal, sacral and caudal vertebrae, along with fragments of pectoral and pelvic girdles, and isolated paddle elements;
(6) TTU P 9238: adult, includes part of cervicals, rib fragments, isolated paddles, and gastroliths;
(7) TTU P 9239: adult, includes isolated vertebrae, limb bones, paddle elements, and ribs;
(8) TTU P 9240: adult, consists of dorsal, sacral and caudal vertebrae, limbs, and paddle fragments.

Systematic descriptions
Order Plesiosauria
Superfamily Plesiosauroidea
Family Cryptoclididae Williston, 1925
Genus *Turneria* gen. nov.

Diagnosis. Skull, short, broad and shallow with elongated orbit; frontal bifurcates strongly around external naris; pineal opening exposed as a narrow slit; the parietals extend forward between frontals, excluding posterior-half of the latter from midline; foramen incisivum present on premaxillae; pterygoids meet with a long median suture leaving a small interpterygoid vacuity; paroccipital process narrow and slender, dipping at an angle of 60°; articular not fused with surangular; teeth small, highly recurved, slightly compressed, undifferentiated, and show flutings on lingual side; premaxillary teeth 8, maxillary teeth 38+, and dentary teeth 46+; cervical centra strongly binocular-shaped, platycoelous with distinctive lateral ridges.
Type species. *Turneria seymourensis* sp. nov.
Horizon. Upper 'molluscan units' of the Lopez de Bertodano Formation; Maastrichtian.
Derivation of name. The generic name is given in honour of Dr Mort D. Turner for his keen interest in the Seymour Island project; the specific name refers to Seymour Island from where the type material was discovered.
Specific diagnosis. As for genus.
Holotype. TTU P 9219; skull and associated cervicals in the collection of The Museum of Texas Tech University.
Locality. The specimen was collected at approximately lat. 64° 16'25" S, long. 56° 43' 25" W, at the top of a ravine on Seymour Island, Antarctic Peninsula.

Description of Turneria

Skull: The skull consists of both articulated and isolated bones. They are the right premaxilla, right maxilla, frontals, part of the parietals, right squamosal, right vomer, pterygoids, basioccipital, exoccipitals-opisthotics, portions of both dentaries, and surangular-angular-articular complex. The skull is about 40 cm long, 22 cm wide, and 13 cm high (Figs 2, −6).

There are many structural features in the skull of *Turneria* which are common to all other plesiosauroids. The skull is typically short, broad and of relatively small size; the postorbital portion is short and weakly built relative to the preorbital. The external nares are set well back from the anterior margin of the skull, on the dorsal surface not far in front of the orbits; they are small, elongated and closely spaced (an aquatic adaptation). The orbits are large and directed more laterally. The upper temporal fenestra is enormous and the lower temporal arch is breached with the loss of the quadratojugal. The squamosal is a massive bone between the upper temporal fenestra and the lower temporal embayment. There is a large pineal opening in front of the parietal. The palate is extensive, plate-like and was built for strength and rigidity. The internal nares are anterior to the external nares. The braincase is loosely connected with the skull roof and the palate. The post-temporal fenestra is large. The marginal teeth are small, slender, pointed and recurved with thecodont implantation; palatal teeth are absent.

The identity and homology of cranial bones have been subject to great differences of opinion. It appears that many of the elements which are otherwise present in more primitive diapsid reptiles are conspicuously absent in plesiosaurs. There is no evidence of the presence of the nasal, prefrontal, lacrimal, supratemporal, stapes, tabular and probably prearticular bones as discrete elements. The loss or fusion of these bones may be attributed to their extreme aquatic adaptations, which have caused certain modifications in the structure and relations of the cranial elements that are unique among reptiles (Williston 1925).

The *premaxilla* is a large element which correlates with the posterior shift of the external naris. It extends back as a narrow process to meet the frontal. The suture between the premaxilla and maxilla curves dorsally and posteriorly toward the external naris. The external surface is gently convex from side to side and is punctured by numerous foramina running subparallel to the alveolar margin. Similar foramina are found on the maxilla and dentary, aligned in a longitudinal row. These must be for nerves or blood vessels, or both, and indicate a sensitive facial region. Each premaxilla bears eight tooth sockets of uniform diameter. Ventrally, the bone extends as a wide shelf toward the vomer, but is pierced anteriorly by a large opening, the foramen incisivum. The distribution of this foramen in different genera of plesiosaurs is not well known. In modern crocodilians, this foramen is closed by skin in the intact head and neither blood vessels nor nerve branches pass

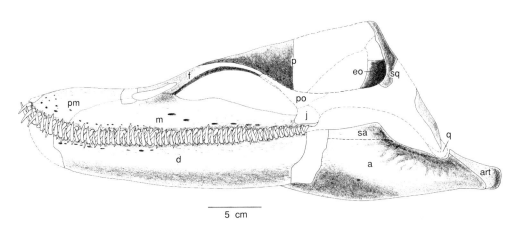

5 cm

Fig. 2. *Turneria seymourensis*, n. gen., n. sp., skull in lateral view, TTU P 9219, holotype. Abbreviations: a, angular; art, articular; bo, basioccipital; bos, basioccipital sinus; bs, basisphenoid; d, dentary; ec, ectopterygoid; eo, exoccipital; f, frontal; fo, fenestra ovalis; hc, horizontal semicircular canal; j, jugal; m, maxilla; mf, metotic foramen; op, opisthotic; p, parietal; pl, palatine; pm, premaxilla; po, postorbital; pt, pterygoid; pvc, posterior vertical semicircular canal; q, quadrate; sa, surangular; so, supraoccipital; v, vomer; vc, vestibular cavity; vf, vascular foramen; XII, foramina for branches of hypoglossal nerve.

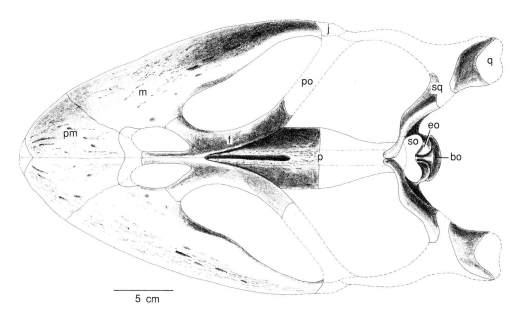

Fig. 3. *Turneria seymourensis*, n. gen., n. sp., skull in dorsal view, TTU P 9219, holotype. Abbreviations as for Fig. 2.

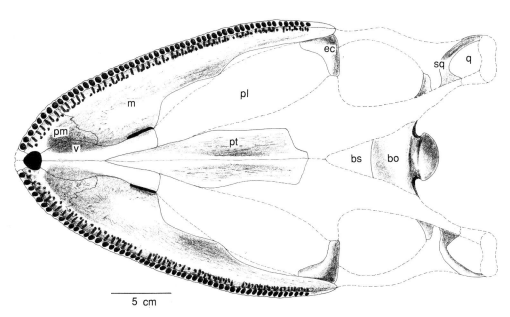

Fig. 4. *Turneria seymourensis*, n. gen., n. sp., skull in ventral view, TTU P 9219, holotype. Abbreviations as for Fig. 2.

through it (Iordansky 1973).

The *maxilla* is triangular in lateral aspect, and extends back as a tapering process to the level of the posterior border of the orbit. Each bone is horizontally flattened to give the skull a low and wide appearance. It forms the antero-ventral margin of the orbit with a prominent ascending process which contacts the frontal

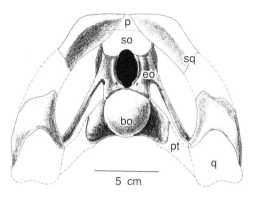

Fig. 5. *Turneria seymourensis*, n. gen., n. sp., skull in occipital view, TTU P 9219, holotype. Abbreviations as for Fig. 2.

behind the external naris. The palatal component of the maxilla overlaps the vomer considerably in front and is notched by the choana. There are more than 38 alveoli which slightly decrease in size posteriorly.

The conjoined *frontals* form a narrow arch between the orbits. Each bone is slender and is bifurcated anteriorly forming the borders of the external naris. The medial process extends forward as a long slender projection which makes contact with the premaxilla. The ventrolateral process is wider and thicker and meets the maxilla. The rest of the frontal is broad and

curved and forms the dorsal and posterior part of the orbital rim. The posterior extension is unusually long and was probably sheathed considerably by the postorbital. Anteriorly the two frontals meet each other along the midline, but are split by the parietals in the posterior-half.

The *parietals* are unusual in *Turneria* in that they extend forward as a narrow pair of conjoined processes between the frontals to separate them from each other. This anterior region accommodates an elongate, slit-like pineal opening, about 65 mm long. A high and narrow sagittal crest about 10 mm thick is present behind the pineal opening, and forms a ridge for muscular attachment. Its posterior extension is unknown as the parietal is broken-off a short distance behind the posterior pineal margin.

The *squamosal* is represented by two fragments. It is a massive, tetraradiate bone, directing dorsally, anteriorly, ventrally, and postero-medially. The dorsal process extends inward behind the temporal fenestra to contact the parietal. Unlike the typical reptilian condition, the parietal in plesiosaurs lacks the lateral flange, so that the two squamosals meet each other along the midline while approaching the parietals. The ventral ramus extends down to the level of the ventral margin of the quadrate and occupies the position of the lost quadratojugal. It forms a powerful bracing device in front of the quadrate making the latter bone entirely monimostylic. Internally there is a large

Fig. 6. *Turneria seymourensis*, n. gen., n. sp., A: Skull in ventral view; B: right exoccipital-opisthotic in medial view; C: skull in dorsal view; D: left mandible in lateral view; TTU P 9219, holotype; all scales 5 cm.

socket for housing the quadrate head. The postero-medial process wraps the quadrate head intimately and receives the paroccipital process on the inner aspect. The anterior process, largely missing, would form a horizontal bar between the upper temporal opening and lower temporal notch. The postorbital, jugal and quadrate bones are not preserved in our specimen.

The *vomer* is a narrow bone that is slightly expanded anteriorly. The two bones meet in the midline anteriorly and form convex sutures with the premaxillae anteriorly. Posteriorly they are separated by the pterygoids. Laterally each bone contacts the maxilla, but the suture is interrupted by a small and narrow choana. The nature of its contact with the palatine is uncertain as the later are missing in the specimen.

The *pterygoids* are medial, forming a large part of the palate. In the midline they meet in symphysis for a long distance except for the very back of the braincase. Contrary to this situation, in many plesiosauroids (e.g. *Kimmerosaurus, Tricleidus*) the pterygoids are delicately built with a large interpterygoid vacuity. In this respect, the pterygoids of *Turneria* are comparable in general to those of pliosaurs. The bone is incomplete laterally and posteriorly.

The *ectopterygoid* has been displaced forward and is partially obscured by matrix, palate and frontals. It is a small, triangular bone with a wide base laterally that articulates with the maxilla. Here the bone is deeply hollowed out. Medially it sends a narrow process to meet the pterygoid.

As in many plesiosaurs, such as *Muraenosaurus, Tricleidus* (Andrews 1910), *Peloneustes* (Andrews 1913) and *Kimmerosaurus* (Brown 1981; Brown *et al.* 1986), individual elements of the braincase are found disarticulated except for the exoccipital-opisthotic bones which are fused together. The braincase is more or less open in front without any evidence of ossification anterior to the sphenoid region. Its articulation with the skull roof and the palate seems to be loose without any strong suture or fusion. The basisphenoid, supraoccipital and prootic are missing in this specimen (Figs 5 & 7).

The supraoccipital would have extended beneath the ventral margin of the parietal, but is not suturally connected. Similarly, the paroccipital process loosely abuts against the squamosal. The ventral articulation of the braincase with the palate is very flexible. Furthermore, the lack of ossification of the laterosphenoid clearly indicates that additional reinforcement of the skull roof with the braincase was not established in plesiosaurs. All these features suggest that a potential metakinetic movement seems to exist between the parietals and the braincase in these groups. Metakinesis is widespread among early diapsids.

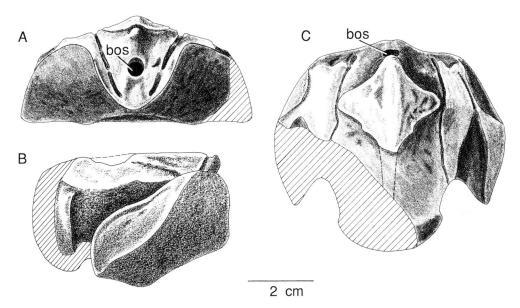

Fig. 7. *Turneria seymourensis*, n.gen., n. sp., basioccipital; (**A**) anterior view; (**B**) lateral view; (**C**) dorsal view, TTU P 9219, holotype. Abbreviations as for Fig. 2.

The *basioccipital* is a massive, compact bone, oriented horizontally. It forms the hemispherical occipital condyle and is partially overlapped by the exoccipitals, which extend all the way to the edges of the condyle. There is no constricting groove on the dorsal surface of the basioccipital. This appears to be an important taxonomic character for the cryptoclidid family (Brown 1981).

Dorsally the basioccipital bears two elongated oval facets for the reception of the exoccipital-opisthotic pedicle. These facets are not concavities as in *Kimmerosaurus*. They are predominantly dorsally oriented with a slight lateral tilt, and are separated in the midline by a narrow ridge of bone. Each facet is traversed by a longitudinal groove.

Anterolaterally to the occipital condyle, the bone constricts to a neck, and then flares again to form a pair of stout tubera. Usually in reptiles these tubera would receive neck muscles; however, in contrast, these structures take the main role in articulation with the pterygoids in plesiosaurs. Their ends are truncated by large oval facets which face outward and forward. The actual basipterygoid processes are considerably reduced, and may provide secondary palatal support.

Anteriorly the basioccipital shows a vertical face for union with the basisphenoid. This face is pierced centrally by the 'hypophyseal−basicrania' fenestra, as in crocodilians (Tarsitano 1985). The fenestra is comprised of two openings, the anterior one leading to the basisphenoid sinus, the posterior one leading to the basioccipital sinus.

The *exoccipital-opisthotic* bones are fused together. However, as in other plesiosaurs, the line of junction of the two elements is represented by distinctive notches on the dorsal and ventral surfaces for union with the supraoccipital and basioccipital respectively. The exoccipital proper forms the sidewall of the foramen magnum and much of the ventral pedicel for the basioccipital. Laterally there is a large metotic foramen, shared between the opisthotic and exoccipital for nerves IX−XI, and the posterior branch of the jugular vein. Behind this, there are two or three (asymmetry between left and right side) small openings for the exits of the hypoglossal (XII) nerve. On the inner surface, above these two foramina, there is a blind pit, probably for a blood vessel (Fig. 8).

Anterior to the metotic foramen lies the stout ventral ramus of the opisthotic, as in many primitive reptiles, which is notched anteriorly by the fenestra ovalis. The stapes is lost as in other plesiosaurs, primarily due to aquatic

adaptations; the middle ear was not sensitive to airborne vibrations. The opisthotic proper largely forms the facet for the supraoccipital and the entire paroccipital process. This process is unusually long and slender for a cryptoclidid. It projects downward considerably at an angle of 60° from the horizontal and terminates in an expanded end for articulation with the squamosal.

Part of the osseous labyrinth of the inner ear is preserved in the medial surface of the opisthotic. There is a large vestibular cavity facing anteriorly for housing the utriculus. The articular surface for the prootic is traversed by the horizontal semicircular canal. At the supraoccipital contact, the opening for the posterior vertical canal can be seen. The development of large semicircular canals indicates refined neuro-muscular coordination associated with subaqueous flight.

The lower jaw is long and slender, and slightly convex outward. The jaw articulation lies well below the alveolar margin. The individual elements of the posterior part of the jaw are difficult to interpret. This arises from the fact that some of the bones of the typical reptilian mandible are either lost (e.g. prearticular) or fused with other elements (e.g. angular-surangular or surangular-articular). Moreover, the splenial is not preserved in our specimen, and coronoid, if present, is represented by a tiny fragment.

Both *dentaries* are represented, except for the symphyseal and posterior regions. The occlusal surface reveals space for more than 47 alveoli of similar size, but most of the mature teeth have been lost leaving the alveoli empty. The sockets are inclined outward, as in the premaxilla and maxilla, and lined with spongy bone of attachment. There is a row of replacement alveoli lying lingual to the true alveoli and running subparallel with them. The Meckelian groove is well developed on the medial surface and would be closed off by the splenial below and the coronoid above. As in *Cryptoclidus*, the groove is mostly dorsally oriented and narrows anteriorly.

The *surangular* is fused to the angular in *Turneria*; the suture between them is unclear, but it would probably coincide with a break running subparallel to the jaw ramus. This suture is clearly visible in *Cryptoclidus* and *Kimmerosaurus* (Brown 1981, figs 4 & 38). The bone is narrow, elongate, triangular in lateral aspect and overlies the angular back to the articular. At its midlength, the bone rises to a tapering process to meet the coronoid. Posteriorly it extends to the articular to coat its

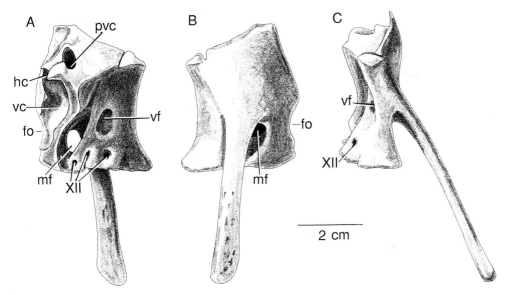

Fig. 8. *Turneria seymourensis*, n. gen., n. sp., right exoccipital-opisthotic; (**A**) medial view; (**B**) lateral view; (**C**) posterior view, TTU 9219, holotype. Abbreviations as for Fig. 2.

lateral surface, and sends a medial flange to form a conjoined glenoid with the articular. Unlike other plesiosaurs, the surangular is not fused to the articular in our specimen.

The *angular* is a very large bone forming the whole lower part of the posterior half of the lower jaw. The prolonged anterior process for the union with the dentary is missing. Posteriorly it wraps the lateral surface of the articular.

The *articular* is found as a separate element. This is somewhat unusual, as in all other plesiosaurs the bone is intimately fused to the surangular. It is a compact bone, differentiated into two parts: the anterior region consisting of glenoid, and the posterior region forming the retroarticular process. The bone is widest in the glenoid region and is sheathed by the surangular and angular. The glenoid is saddle-shaped for the reception of the quadrate. Posteriorly it extends as a strong retroarticular process, but its tip is damaged.

The *teeth* are long and slender cones, recurved with sharply pointed apices. In cross-section, the distal parts of the crowns are oval, being slightly compressed labio-lingually, without any development of true carinae. Longitudinal ridges or striations are conspicuous on the lingual side, indicating some infolding of the enamel. There are about 12 fine ridges on each tooth extending to the tip. The labial side is smooth without ornamentation. The dentition is essentially homodont (Fig. 9).

The teeth do not show any wear facets. This feature, coupled with the pointed apices, suggests that they were not used to masticate prey. Massare (1987) reported this kind of tooth form in certain plesiosauroids where the teeth were presumably used to pierce soft prey items such as fish and cephalopods. Although this interpretation may be true for plesiosaurids, the occlusal pattern of *Turneria* and other cryptoclidids is so unique that it may have some special functional significance.

In cryptoclidids, the sockets are inclined outward at a considerable degree from the vertical position, and so are the crowns. The crowns are slanted outward, so that the opposing teeth intermesh to form an interlocking trap. This kind of tooth occlusion would be ideal for straining a diet of small organisms, such as fish, soft arthropods, and cephalopods, from the water. The animals would gulp a mouthful of water and then spurt it out through the teeth, trapping the pelagic organisms (modern crab-eater seals use similar techniques to procure their food). Also, since the occlusal margin lies outside the alveolar margin, this type of tooth arrangement is not very suitable for seizing, piercing and manipulating prey. The lack of wear facets suggest furthermore that these delicate teeth were not used for processing food.

Implantation is of the thecodont type, where the teeth were probably held in life by soft, non-calcified connective tissue, as in modern

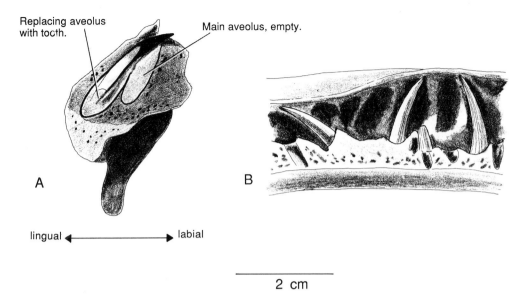

Fig. 9. *Turneria seymourensis*, n. gen., n. sp., dentition. A: transverse section of the left dentary showing a replacing tooth in a secondary alveolus, lingual to the main one; B:occlusal view of the right dentary showing replacing teeth, as well as main and secondary alveoli; TTU P 9219, holotype.

crocodilians. This is evident from the nature of preservation of the teeth. Almost all of the functional teeth were lost postmortemly, leaving the sockets entirely empty. Many of the replacing teeth are preserved intact in their temporary positions. From the number of tooth sockets, the dental formula is: premaxillary teeth 8, maxillary teeth 38+, and dentary teeth 46+.

The mechanism of tooth replacement can be clearly observed in both dentaries. The pattern of tooth replacement is similar to that of nothosaurs (Edinger 1921) and other plesiosaurs (Edmund 1960). Each replacement tooth is developed in a narrow but deep socket, distolingual as well as dorsal to the functional or true alveolus and separated by a bridge of bone. In this position the lingual side of the replacing alveolus remains open, where the young tooth continues to grow. During the earlier stages at least, the growth of the young tooth is such that the tip of the crown remains near the level of the occlusal plane. When replacement is about to occur, the succeeding tooth along with its alveolus moves slowly towards the functioning position (Fig. 9). Eventually, the bridge of bone, separating the two alveoli, breaks down, and the new tooth enters the true alveolus. The old tooth probably becomes loosened by resorption of its base. Brown (1981) reported these resorption pits in several teeth of *Kimmerosaurus*.

Vertebral column. The postcranial skeleton is represented by only a few cervicals. The fused atlas-axis and six cervicals are preserved in *Turneria*. The atlas-axis complex is intact, but retains the sutures, so that the individual elements can be easily demarcated. The atlas intercentrum, the paired neural arches and the odontoid process combine to form a broad cup to receive the occipital condyle. The odontoid separated the atlantal arches dorsally by 5 mm and forms the floor of the neural canal. Behind the atlantal arch, the atlantal centrum is exposed as a small, triangular bone, sitting on top of its intercentrum. The axis is characterized by the large, hatchet-shaped neural spine, sloping forward. The posterior face of the centrum is flat and shows the beginning of the binocular-shaped outline which becomes more prominent in the posterior series (Fig. 10).

The atlantal rib is fused to the intercentrum without any visible suture, and overlaps the axis rib as in *Aristonectes*. The axis rib facet is more elaborate, and is shared among the atlas intercentrum anteriorly, the axis intercentrum ventrally, and the axis centrum posteriorly. As in other posterior cervicals, the ribs are not intimately fused with the axis.

There are six relatively complete cervicals present in the collection, probably representing the mid-cervical region. Fragmentary remains of some anterior and posterior cervicals are also

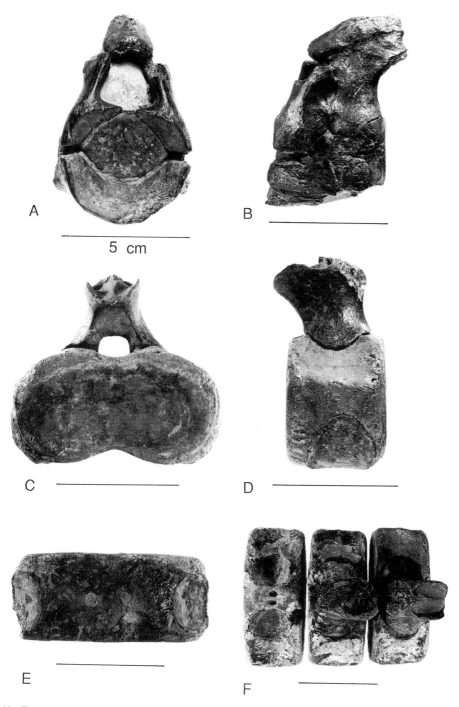

Fig. 10. *Turneria seymourensis*, n. gen., n. sp., A: atlas-axis complex in anterior view; B: atlas-axis in lateral view; C: a middle cervical in anterior view; D: the same in lateral view; E: the same in ventral view; F: three middle cervicals in anterior view; TTU P 9219, holotype; all scales 5 cm.

present in the collection, but they lack any features. The ribs and neural arches are not fused to the centra (a juvenile condition?), but were probably held in life by soft tissues as in many aquatic reptiles. The cervicals are very diagnostic in *Turneria* because of their extreme width and the binocular-shaped outline of the centra. The width of the centra is over twice the length and the ratio increases posteriorly. Also the centra are higher than long. They are platy-coelous and elongated transversely. Such a wide and flat centra indicates a relative lack of flexi-bility of the neck in this region. This is also supported by the tilt of the zygapophyses, which are close to 45° so as to limit lateral bending. The facets for neural arches are deep, oval concavities. The neural canal is notched into the centrum. There is also a ventral notch between the two nutritive foramina. These two notches constrict the centrum in the middle, creating a 'binocular' shape. There is a distinct lateral ridge on the lateral surface. This is indi-cative of long neck and is commonly found in elasmosaurs (Welles 1943). This ridge descends on the lateral surface with each succeeding vertebrae. It is absent in the posterior-most preserved cervical.

The rib facet is single, broad, oval in outline, and is more posteriorly placed on the centra. They face downward and outward (Table 1).

The affinities of Turneria

The relatively small skull and the long, delicate, sharply pointed teeth of *Turneria* suggest that it belongs to the superfamily Plesiosauroidea. The estimated skull length is 40 cm, but in contrast,

in some pliosaur genera, the skull length may exceed 3 m. The closest affinities of *Turneria* lie with the genera *Cryptoclidus*, *Kimmerosaurus* and *Aristonectes* of the family Cryptoclididae, which share the following derived characters (Brown 1981):

(1) premaxillary teeth more than 5;
(2) dentary teeth primitively 24, increasing in number to 58 in advanced forms;
(3) tooth ornament reduced or absent.
(4) opposing teeth form a 'trapping' device;
(5) occipital condyle formed by the basioc-cipital and exoccipitals and not ringed by a constricting groove.

Cryptoclidus is known from the Middle Jurassic (Callovian) and *Kimmerosaurus* from the Late Jurassic (Kimmeridgian) of England (Brown 1981). On the other hand, *Aristonectes* is known from the Upper Cre-taceous (Maastrichtian) of Chile and Argentina (Cabrera 1941; Casamiquela 1969), and *Turneria* from contemporary beds in Antarctica. A comparison of the diagnostic characters among the different genera of cryptoclidids is shown in Table 2. It becomes apparent from the table that *Aristonectes* has the largest number of premaxillary (15) and dentary (58) teeth among cryptoclidids. *Turneria* and *Kimmerosaurus* share the same number of premaxillary (8) teeth.

Turneria exhibits the following apomorphic characters unknown in other cryptoclidids:

(1) frontal bifurcates strongly around ex-ternal naris;
(2) two parietals extend forward between

Table 1. *Main measurements (in mm) of* TURNERIA

Skull

Skull length:	403 (e)				
Skull width (at orbit):	225				
Skull height:	131 (e)				
Mandible length:	445 (e)				
Mandible height:	71				

Vertebral column

	L	H	B	H/L × 100	B/L × 100
Atlas-axis:	54	79	50		
Cervicals:	38	46	85	121	224
Cervicals:	37	47	88	127	238
Cervicals:	37.5	48	88	128	235
Cervicals:	38	49	92	129	242
Cervicals:	39	55	99	141	254
Cervicals:	41	58	102	141	249

e, estimated; L, length; H, height; B, breadth.

Table 2. *Diagnostic characters of different genera of Cryptoclididae*

	Pm. teeth	Mx. teeth	Dentary teeth	Tooth pattern	Tooth ornament	Paroccipital process	Occipital condyle	Cervical Proportion L:W	Lateral ridge on centra	Cervical centra, type	Cervical centra, shape
Cryptoclidus	6	21	24–26	homodont	reduced	short	short, no constricting groove	moderate	absent	amphicoelous	oval
Kimmerosaurus	8	?	36	homodont	absent	short	short, no constricting groove	moderate	absent	amphicoelous	circular
Aristonectes	15	42?	58	homodont	?	?	short, no constricting groove	moderate to short	absent	amphicoelous	oval to binocular
Turneria	8	38+	46+	homodont	present on lingual	long, slender	short, no constricting groove	very short	present	platycoelous	strong binocular

? Unknown characters

frontals to exclude the posterior-half of the latter from the midline;

(3) pterygoids with long median symphysis;
(4) paroccipital process is long and slender; dipping downward at a high angle;
(5) numerous longitudinal ridges on lingual side of teeth;
(6) cervical centra extremely wide, platycoelous, strongly binocular-shaped, with distinctive lateral ridge.

Description of Elasmosauridae

Superfamily Plesiosauroidea
Family Elasmosauridae Cope, 1869

Elasmosaurid remains have been reported previously from Seymour Island, but the material was non-diagnostic (del Valle *et al.* 1977; Chatterjee & Zinsmeister 1982; Chatterjee *et al.* 1984; Gasparini *et al.* 1984). The material described in this paper represents a new collection. Although the elasmosaurids are the most abundant vertebrates in the Late Cretaceous deposits of Seymour Island, none of the 7 skeletons recovered is completely preserved; they lack the critical portions necessary for finer diagnosis.

There is some degree of morphological variation among different individuals. It is difficult to ascertain at this stage whether this variation is due to age or sex, or whether it is taxonomic. A brief, composite description of the plesiosaurid material, which is only identifiable to family level, is presented here.

The following postcranial characters suggest the affinities of the Antarctic material to the Family Elasmosauridae.

(1) large number of cervical vertebrae; more than 30 associated cervicals are preserved in one individual; the estimated number may be more than double. In contemporary pliosaurs, this number is reduced to 13.

(2) cervicals have distinct lateral ridges; they are conspicuously absent in all pliosaurs and most cryptopclidids except for *Turneria*.

(3) cervical centra relatively elongated; in pliosaurs the centra are compressed anteroposteriorly where the width is generally twice the length.

(4) propodials are massive; they are very slender in pliosaurs and to some extent in cryptoclidids too.

(5) ischium is as long as wide; in pliosaurs the ischium is elongated antero-posteriorly.

Genus and species indeterminate

Although Antarctic elasmosaurid material does

not allow for finer diagnosis at a generic and specific level, specimen 9220 shows a general resemblance to *Hydrotherosaurus* from the Late Cretaceous of California (Welles 1943). These similarities can be seen in the coracoid and pelvis. Similarly, specimen 9217 shared some features in the limbs and pelvis with those of *Morenosaurus* from the Late Cretaceous of California (Welles 1943). These identifications are very tentative, because of the incomplete nature of the material (Fig. 11).

Vertebral column. There are 29 anterior cervicals preserved in association in specimen 9218, except for the atlas-axis. Passing backward, the centra steadily increase in size and proportion so that the thirtieth vertebra (length: 45 mm; breadth: 65 mm, height: 41 mm) becomes more than twice the size of the third one (length: 17 mm, breadth: 27 mm, height: 15 mm). The neural arches and ribs are fused in all specimens irrespective of size. There is a distinct lateral ridge in anterior cervicals which fades away from the twentieth cervical backward. In some specimens, such as (9217), there is a prominent separation of the articulating face from the body of some centra, resulting in a 'disc'-like structure. Welles (1943) reported this feature in *Thalassomedon* from the Late Cretaceous of Colorado. A pair of nutritive foramina is ubiquitous in all specimens.

In the dorsal series, the neural arches are missing from most of the specimens and the centra are basically spool-shaped. In one specimen (9221), there is a central swelling with a pronounced pit at the middle on both anterior and posterior faces of the centra. Similar features are known in *Mauisaurus* from the Late Cretaceous of New Zealand (Wiffen & Moisley 1986).

A couple of isolated sacrals are preserved, recognized by their distinctive rib facets. In the caudal series, the chevron facets are conspicuous. The centra decrease gradually in size and proportion in posterior direction.

Shoulder girdle. Three specimens exhibit part of the shoulder girdle. In 9220, the coracoid is beautifully preserved, except for the glenoid region which is compensated by another coracoid specimen, associated with a partial scapula (9221). An interclavicle is present in specimen 9217.

The *interclavicle* is a diamond-shaped bone, weakly arched at the midline. The dorsal surface is concave. Ventrally there is a strong keel that becomes shallower posteriorly.

The *scapula* is represented by two fragments,

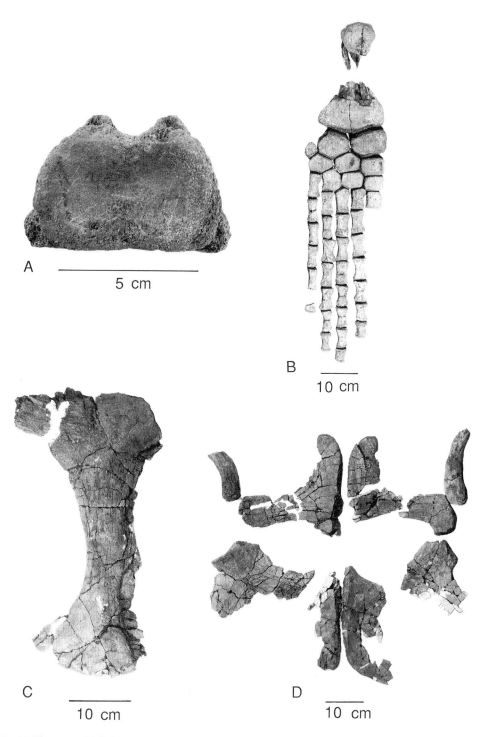

Fig. 11. Elasmosaurids indet. A: TTU P 9217, anterior cervical in anterior view: B: TTU P 9217, left hind limb in medial view; C: TTU P 9220, coracoid; D: TTU P 9220, pelvis in dorsal view.

but the connecting shaft is missing. The anterior part is differentiated into dorsal and ventral surfaces which are set almost at right angles to each other. Posteriorly the bone is thickened and shows two distinct facets. The scapular facet is flat and rough with large pits. The glenoid facet is smooth, concave and triangular.

The *coracoid* is a long, slender and thin plate somewhat resembling that of *Hydrotherosaurus* from the Late Cretaceous of California (Welles 1943). The bone is expanded at both ends with a constricting shaft. The median symphysis is very thick, straight, elongated and deeply pitted. The glenoid surface is smooth, highly concave, unlike many other plesiosaurids, and faces laterally. On the ventral surface a transverse ridge runs from the glenoid to the median symphysis. The shaft is long and narrow. The distal end is expanded with a smooth convex outline.

Fore limb. The only brachial elements preserved in good condition are the right humerus, radius and ulna in specimen 9217. The *humerus* is complete but somewhat eroded distally. The head is a broad, flat unfinished surface, oval in outline and encompassing the entire proximal region. It was probably capped by cartilage in life to form a spherical head. Farther distally, the tuberosity is widely separated from the head in dorsal aspect. It is set at a 50° angle from the shaft, the latter being highly robust. Distally the humerus is highly expanded and marked by numerous pits. There are two prominent facets for the radius and ulna; the former is the more extensive.

The *radius* and *ulna* are as long as wide, and enclose a large epipodial foramen between them. The two bones meet above and below this foramen, without any broad contact surfaces. Each bone would be about one-fourth of the length of the humerus.

Pelvic girdle. The pelvis is partially preserved in three specimens (9220, 9218, 9217). Specimen 9220 shows important features of the three pelvic elements, and is very similar to that of *Hydrotherosaurus* (Welles 1943) in morphology. Specimens 9218 and 9217 are fragmentary; the former may belong to the same taxon as 9220. However, specimen 9217 is distinctive from the other two, as is evident from the preserved part of the ischium. It resembles that of *Morenosaurus* (Welles 1943). The description of the pelvis is based on specimen 9220 (Fig. 11).

The *ilium* is a short, stocky, curved bone, rotated backward as a tapering process. It has a distinct knee midway along the lateral border.

The thickened lower end bears two facets, one large and slightly concave for the union with the ischium, the other small for the acetabulum. As in all plesiosaurids, the ilium does not contact the pubis.

The *ischium* is a relatively slender bone with a massive articular portion, narrow neck and a broad ventral plate. The head bears three articular facets: the anterior one for the pubis, the lateral one for the acetabulum, and the posterior one for the ilium. The median symphyseal facet is very thick and straight, narrowing anteriorly and posteriorly.

The *pubis* is a large plate, probably wider than long. The median symphysis is straight, thick and deepest in the middle; however, it tapers anteriorly. The ischium and pubis do not meet with a midline bar. The postero-lateral extremity is thickened and bears the acetabular and ischial facets at an angle of 140°.

Hind limb. The articulated hind limb is preserved in specimen 9217. The *femur* is represented by proximal and distal ends, but the shaft is missing. Proximally the head is nearly confluent with the trochanter, but is separated by a distinct groove on the anterior side. The head has an oval, rounded surface. The distal end, like that of humerus, is formed by two facets for the reception of the tibia and fibula.

The *tibia* and *fibula* are somewhat similar in structure to the radius and ulna, but the former is stouter and broader. The two bones are separated by the epipodial foramen. Distally the tibia shows two facets for *tibiale* and *intermedium*, while the fibula receives the intermedium and *fibulare*.

The *paddle* has the usual configuration of an advanced elasmosaurid, where the fifth metapodial has shifted proximally into the distal metapodial row. There is one supernumerary ossification contacting the fibula and fibulare. The extent of the hyperphalangy is unknown as most of the digit at the tips are missing. There are eight phalanges preserved in the third digit.

Palaeoecology

In the Late Cretaceous, plesiosaurus and mosasaurs filled the role of large mobile predators in marine communities, in a manner similar to that of seals and whales in the present-day Southern Ocean. These marine reptiles probably fed on various pelagic animals off the coast of Seymour Island, capturing their prey from the water column rather than the sea floor. Neither their ears nor their braincases show specializations comparable to those found in modern marine mammals adapted for deep diving.

Massare (1987) recognized several predator guilds among Mesozoic marine reptiles on the basis of tooth morphology and prey preference. These include 'Crush, Crunch, Smash, Pierce, General, and Cut' guilds. To these various feeding types, another new category, 'Trap guild', may be added, as exemplified by the cryptoclidids. The Late Cretaceous Lopez de Bertodano fauna contains representatives of at least three predatory guilds, exhibited by the mosasaurs, plesiosaurids and the cryptoclidids, respectively.

The mosasaurs belong to the 'Cut guild'. The crowns of the teeth are robust, pointed, compressed sidewise, and characterized by the development of anterior and posterior serrated keels, which functioned in life as sharp cutting edges. These teeth were very effective for seizing or tearing large fleshy animals such as fish or other reptiles. The mosasaurs had powerful jaws and probably ate anything they could catch; they were the top carnivores in the Lopez de Bertodano palaeoguilds.

The elasmosaurs were probably the second-most dominant carnivores and composed the 'Pierce guild'. Their teeth, known from closely related taxa, were more robust than the teeth of cryptoclidids, and were presumably used for piercing fish and soft cephalopods (Massare 1987). Fossilized stomach contents suggest that they were piscivores (Brown 1904).

The cryptoclidids, with their long, slim and delicate teeth, were probably third in order within the marine predators, and composed the 'Trap guild'. Their specialized tooth occlusion was a device for straining and trapping small fish and crustaceans.

Which resources were available to these marine reptiles in the Lopez de Bertodano palaeo-environment? An extensive fish fauna is known from this horizon, including hexanchiform sharks, beryciform teleosts, and various indeterminate taxa (Grande & Chatterjee 1987). The invertebrate fauna is varied and abundant, comprising ammonites, echinoids, bivalves, gastropods, arthropods, serpulid worms, and foraminifera (Macellari 1986). None of the reptiles from Seymour Island, however, show any dental specializations for crushing clams or other thick-shelled invertebrates. On the other hand, the fossil record of decapod crustaceans, such as lobsters, is unusually rich in the Lopez de Bertodano Formation (Feldmann 1984). They might have occurred in huge swarms, similar to those of modern krill and perhaps formed an important food source. *Turneria* could have

strained these lobsters using its long and specialized interlocking teeth. Their gullet width, measured as distance between the jaw articulations, is about 12 cm, indicating that the prey was small in size. The delicate teeth without wear facets suggest that the food was soft and required no additional processing.

The pattern of aquatic locomotion employed by plesiosaurs has been debated for a long time. In crocodiles, active swimming is effected by lateral undulations of the powerful tail. The limbs take no part in propulsion, being closely applied to the flanks. In plesiosaurs, on the other hand, the hyperphalangic paddles were used for propulsion. The tail is short and probably served as a rudder.

Watson (1924) suggested that the plesiosaurs rowed through the water with their paddles, while Robinson (1975) claimed that they 'flew' through the water in the manner of sea turtles and penguins, using their limbs as hydrofoils. Godfrey (1984) pointed out that, in subaqueous fliers, the pectoral girdles are reinforced both dorsally and ventrally, and the propulsion is characterized by dorso-ventral movements of the pectoral limbs. Contrary to this, the dorsal reinforcement of the pectoral girdle is lacking in plesiosaurs, as the scapulae make little or no contact with the ribs. Ventrally, however, the girdles are strongly reinforced in a horizontal plane, thus making the downstroke more powerful than the upstroke. The direction of the limb movement was probably mainly antero-posterior rather than dorso-ventral. Thus the locomotion of plesiosaurs was comparable to that of sea lions, combining elements of rowing and subaqueous flight. In sea lions, however, the pelvic limbs are used exclusively for directional controls, not for swimming. In plesiosaurs, the morphological similarity between pectoral and pelvic limbs indicates that both were probably used for propulsion. The rigidity of the trunk region allowed the effective use of both limbs as paddles (Godfrey 1984; Taylor 1986; Carroll 1988).

The plesiosaurs had hydrodynamically designed, streamlined bodies. Massare (1988) calculated their swimming speeds by estimating the total drag and the amount of energy available through metabolism. She concluded that plesiosaurs were faster sustained swimmers than crocodiles and mosasaurs, and could cruise at a rate of 2.3 m/s^{-1}. With the eyes and nostrils set high in the skull, and a long, flexible neck, the plesiosaurs had a natural advantage over their prey; they could swim almost totally submerged and scan large areas for food before being detected.

Gastroliths have been found associated with two plesiosaurid specimens. (9220, 9238). They probably served in life to increase hydrodynamic stability. Cott (1961) observed that modern Nile crocodiles ingest pebbles to increase their body weight so that they may function with greater hydrodynamic efficiency in their aquatic habitat. These pebbles do not aid in maceration of food. Similarly, the use of gastroliths in plesiosaurs was probably to provide ballast, allowing them to remain almost totally submerged and to inhibit rolling action, especially in turbulent seas (Darby & Ojakangas 1980).

The plesiosaurs were presumably ectothermic like modern reptiles and their distribution was affected by environmental temperatures. Existing crocodilians and turtles are confined to the tropics and some warm edges of the temperate zone. The skeletal remains of plesiosaurs have been recorded from terminal Cretaceous (Maastrichtian) sediments in the Antarctic Peninsula and the Queen Elizabeth Islands of Arctic Canada (Russell 1967). During this period, the last fragmentation of the Pangea occurred and the various continents were separated by shallow seas. Antarctica was still in contact with Australia and South America, and consequently there was no circumantarctic gyre to cool the deep sea (e.g. Stanley 1987). The distribution of plesiosaurs in high latitudes indicate that the ocean temperatures within both the Arctic and Antarctic Circles were equable and probably warm temperate (to support such a rich biota). The equatorial–polar temperature gradients were substantially less in the Late Cretaceous than at the present day.

The disappearance of plesiosaurs coincides with the terminal Cretaceous extinction event which is now believed to have extended over intervals of one to three million years and had a partly stepwise character. The causes of the Late Cretaceous extinctions have long been subject to dispute. Currently they are linked to either multiple cometary impacts (Hut et al. 1987) or large-scale intermittent volcanism such as the Deccan Traps of India (Officer & Drake 1985). Plesiosaurs and mosasaurs survived till the end of the Cretaceous, and their decline was matched by the diversification and proliferation of the teleost fish.

We thank M. D. Turner, W. J. Zinsmeister, M. O. Woodburne, M. W. Nickell, C. E. Macellari, B. Huber, R. Askin, T. Horner, S. Spesshardt, D. Chaney, G. Wilhite and many others who assisted us in the field. We thank M. W. Nickell for drawings, N. L. Olson for photography and L. Lamb for typesetting. We are grateful to A. C. Milner for allowing us to study the English plesiosaurs at the British Museum (Natural History). J. Wiffen and Z. Gasparini provided valuable information on plesiosaurs from the Southern hemisphere. We thank J. A. Crame and anonymous reviewers for their critical review of the manuscript. This research was supported by a series of grants from the National Science Foundation (DPP 81–07152, DPP 82–14686, DPP 84–43847 and DPP 86–13419) and Texas Tech University.

References

ANDREWS, C. W. 1910. A descriptive catalogue of the marine reptiles of the Oxford Clay. Part I, British Museum (Natural History), London.
—— 1913. A descriptive catalogue of the marine reptiles of the Oxford Clay. Part II, 206 pp. British Museum (Natural History), London.
BROWN, B. 1904. Stomach stones and food of plesiosaurs. Science, 20, 184–185.
BROWN, D. S. 1981. The English Upper Jurassic Plesiosauroidea (Reptilia) and a review of the phylogeny and classification of the Plesiosauria. Bulletin of the British Museum of Natural History (Geology), 35(4), 253–347.
——, MILNER, A. C. & TAYLOR, M. A. 1986. New material of the plesiosaur Kimmerosaurus langhami. Brown from the Kimmeridge Clay of Dorset. Bulletin of the British Museum of Natural History (Geology), 40(5), 225–234.
CABRERA, A. 1941. Un plesiosauria nuevo del Cretaceo del Chubut. Revista Museo de la Plata (sec. Paleontologia), 2(8), 113–130.

CARROLL, R. L. 1981. Plesiosaur ancestors from the Upper Permian of Madagascar. Philosophical Transactions of the Royal Society of London. B 293(1066), 315–383.
—— 1988. Vertebrate Paleontology and Evolution. W. H. Freeman, New York.
—— & GASKILL, P. 1985. The nothosaur Pachypleurosaurus and the origin of plesiosaurs. Philosophical Transactions of the Royal Society of London, B 309, 343–393.
CASAMIQUELA, R. 1969. La presencia en Chile de Aristonectes Cabrera (plesiosauria), del maestrichense del Chubut, Argentina. Ecad. y caracter de transgresion "rocanense". Actas de las Cuartas Jornadas Geologicas Argentinas, Mendoza 1, 199–213.
CHATTERJEE, S. 1986. Malerisaurus langstoni, a new diapsid reptile from the Triassic of Texas. Journal of Vertebrate Paleontology, 6(4), 297–312.
—— & ZINSMEISTER, W. J. 1982. Late Cretaceous marine vertebrates from Seymour Island,

Antarctica. *Antarctic Journal of the United States* **17(5)**, 66.

— —, SMALL, B. J., & NICKELL, M. W. 1985. Late Cretaceous marine reptiles from Antarctica. *Antarctic Journal of the United States Annual Review*, **19**(5), 7–8.

COLBERT, E. H. 1965. *The Age of Reptiles*. Weidenfeld & Nicolson, London.

— — 1966. *Evolution of the Vertebrates*. Second ed. John Wiley, New York.

COTT, H. B. 1961. Scientific results of an inquiry into the ecology and economic status of the Nile crocodile (*Crocodilus niloticus*) in Uganda and northern Rhodesia. *Transactions of the Zoological Society of London*, **29**, 211–357.

DARBY, D. G. & OJAKANGAS, R. W. 1980. Gastroliths from an Upper Cretaceous plesiosaur. *Journal of Paleontology*, **54**(3), 548–556.

DEL VALLE, R., MEDINA, F. & DE BRANDONI, Z. 1977. Nova preliminar sobre el hallazgo de reptiles fosiles marinos del suborden Plesiosauria en las islas James Ross y Vega, Antartida. *Instituto Antartico Argentino Contribucion*, **212**, 1–13.

EDINGER, T. 1921. Uber Nothosaurus 11. *Zur Gaumenfrage*, Senckenbergiana, **3**, 193–205.

EDMUND, A. G. 1960. Tooth replacement phenomena in the lower vertebrates. *Contributions Royal Ontario Museum Life Science Division*, **52**, 1–179.

EVANS, S. E. 1980. The skull of a new eosuchian reptile from the Lower Jurassic of South Wales. *Zoological Journal Linnean Society*, **70**, 203–264.

FELDMANN, R. M. 1984. Decapod crustaceans from the Late Cretaceous and the Eocene of Seymour Island, Antarctic Peninsula. *Antarctic Journal of the United States, Annual Review*, **19**(5), 4–5.

GASPARINI, Z., DEL, VALLE, R. & GONI, R. 1984. An elasmosaurus (Reptilia, Plesiosauria) of the Upper Cretaceous in the Antarctic *Instituto Antartico Argentino Contribucion*, **305**, 1–24.

GODFREY, S. J. 1984. Plesiosaur subaqueous locomotion: a reappraisal. *Neues Jahrbuch für Geologie und Paläontologie, Monashefte*, **11**, 661–672.

GOW, C. E. 1972. The osteology and relationships of the Millerettidae (Reptilia: Cotylosauria). *Journal of the Zoological Society of London*, **167**, 210–264.

GRANDE, L. & CHATTERJEE, S. 1987. New Cretaceous fish fossils from Seymour Island, Antarctic Peninsula. *Palaeontology*, **30**(4), 829–837.

HUBER, B. T. 1985. The location of the Cretaceous/Tertiary contact on Seymour Island, Antarctic Peninsula. *Antarctic Journal of the United States*, **20**(5), 46–48.

HUT, P., ALVAREZ, W., ELDER, W. P., HANSEN, T., KAUFFMAN, E. G., KELLER, G., SHOEMAKER, E. &

WEISSMAN, P. R. 1987. Comet showers as a cause of mass extinction. *Nature*, **329**, 118–126.

IORDANSKY, N. N. 1973. The skull of the Crocodilia. *In*: GANS, C. & PARSONS, T. (eds) *Biology of the Reptilia*, **4**, Academic Press, New York, 201–262.

KUHN-SCHNYDER, E. 1967. Das problem der Euryapsida. *Problémes actuels de Paléontologie (Évolution des Vertébrés)* CNRS, Paris, **163**, 335–348.

MACELLARI, C. E. 1986. Late Campanian-Maastrichtian ammonite fauna from Seymour Island (Antarctic Peninsula). *Journal of Paleontology Memoirs*, **18**, 1–55.

MASSARE, J. A. 1987. Tooth morphology and prey preference of Mesozoic marine reptiles. *Journal of vertebrate Paleontology*, **7**(2), 121–137.

— — 1988. Swimming of Mesozoic marine reptiles: implications for method of predation. *Paleobiology*, **14**(2), 187–205.

OFFICER, C. B. & DRAKE, C. L. 1985. Terminal Cretaceous environmental events. *Science*, **227**, 1161–1167.

ROBINSON, J. A. 1975. The locomotion of plesiosaurs. *Neues Jahrbuch für Geologie und Paläontologie, Abhandlungen*, **149**(3), 286–332.

ROBINSON, P. L. 1973. A problematic reptile fauna from the British Upper Trias. *Journal of the Geological Society, London*, **129**, 457–479.

ROMER, A. S. 1956. *Osteology of the Reptiles*. The University of Chicago Press, Chicago.

— — 1966. *Vertebrate Paleontology*. The University of Chicago Press, Chicago.

RUSSELL. D. A. 1967. Cretaceous vertebrates from the Anderson River N.W.T. *Canadian Journal of Earth Sciences*, **4**, 21–38.

STANLEY, S. M. 1987. *Extinction*. Scientific American Books, Inc., New York.

TARSITANO, S. F. 1985. Cranial metamorphosis and the origin of the Eusuchia. *Neues Jahrbuch für Geologie und Paläontologie, Abhandlungen*, **170**(1), 27–44.

TAYLOR, M. A. 1986. Lifestyle of plesiosaurs. *Nature*, **319**, 179.

WATSON, D. M. S. 1924. The elasmosaur shoulder girdle and forelimb. *Proceedings of the Zoological Society of London*, **2**, 885–917.

WELLES, S. P. 1943. Elasmosaurid plesiosaurs with description of new material from California and Colorado. *Memoirs of the University of California*, **13**, 125–254.

WIFFEN, J. & MOISLEY, W. L. 1986. Late Cretaceous reptiles (Families Elasmosauridae and Pliosauridae) from the Mangahouanga stream, North Island. *New Zealand Journal of Geology and Geophysics*, **29**, 205–252.

WILLISTON, S. W. 1925. *The osteology of the Reptiles*. Harvard University Press, Cambridge, Mass.

Antarctica: the effect of high latitude heterochroneity on the origin of the Australian marsupials

JUDD A. CASE

Department of Earth Sciences, University of California Riverside, CA, USA 92521

Abstract: The record of the Antarctic marine fauna during the Paleogene indicates the occurrence of taxa in high latitude regions (>60°) before these same taxa are known in mid- to low latitudes. It has been hypothesized that high latitude regions serve both as 'holding tanks' for taxa and as regions in which novel adaptations leading to new lineages within a taxon can arise. These features of high latitude regions appear to have had a major impact on the origin of the Australian marsupial fauna. The australidelphian marsupial clade (all Australian marsupials plus South American microbiotheriids) and the ameridelphian marsupial clade (all South American marsupials except microbiotheriids) diverged in the Late Cretaceous. The South American marsupial radiation occurred in mid- to low latitude regions during the Late Cretaceous to early Paleocene. The Australian marsupial radiation probably did not occur until the medial or late Eocene. If accurate, this represents a time differential of 20 to 25 Ma between the two marsupial radiations. The heterochroneity of the marsupial radiations results from the fact that during the Late Cretaceous to Eocene, Australia resided in a high latitude region, the cool temperate Weddellian Biogeographical Province. The radiation of Australian marsupials only reaches a level of taxonomic diversity comparable to that in South America after the continent drifted northward into lower latitudes and habitat diversity increased. This is deduced from changes in floral diversity and the projected timing of the diprotodontian divergences based on DNA hybridization data.

The time and place of the origin and initial radiation of the Australian marsupial fauna is unknown as the substantive fossil record of marsupials in Australia only dates back to the medial Miocene (Woodburne *et al.* 1985). During this time (*c.* 15 Ma), deposits from the Etadunna and Namba formations have produced a wide variety of forms, creating a family level diversity that surpasses today's modern fauna on the Australian continent (Marshall *et al.* 1989). However, little is known of the Australian marsupial fossil record prior to the Miocene (Fig. 1).

Past concepts of marsupial relationships have hypothesized a close relationship between the carnivorous marsupials of South America and those in Australia (e.g. Ride 1964; Table 1). Thus, the marsupicarnivores of South America dispersed from South America, through Antarctica, to Australia (see Marshall 1980, for various biogeographic models). South American marsupicarnivores (didelphids and borhyaenids) thus gave rise to the Australian marsupicarnivores (dasyurids and thylacinids), a group from which bandicoots are derived, with a bandicoot-like animal giving rise to the primarily herbivorous clade, the diprotodontians (e.g. kangaroos, possums, koalas, wombats and kin). The timing of the Australian marsupial radiation has been conjectured to be somewhere between the middle Eocene and the Oligocene (approximately 50–35 Ma; Clemens 1977; Keast 1977).

Evaluation of marsupial relationships took on a completely new aspect with the work of Szalay (1982*a, b*) on marsupial tarsal (=ankle) bones, which revealed a fundamental division between the South American marsupials (minus the microbiotheriids), the ameridelphians, and the Australian marsupials (plus the microbiotheriids), the australidelphians (Table 1). The morphology of the australidelphian ankle joint and foot is highly adapted for locomotion in an arboreal habitat. The ankle joint is capable of inverting the hindfoot so that toes are pointed laterally and encircle a tree branch, while the highly divergent hallux (='big toe') is pointed forward along the upper surface of the branch (Szalay 1982*a*). These pedal characteristics produce a strongly grasping arboreal foot capable of supporting and maintaining the balance of the animal, even to the ends of very narrow tree limbs.

Thus, based on the differences of their respective tarsal morphologies, the carnivorous marsupials of South America and Australia have originated independently and have no close phyletic relationships; they represent members of two different marsupial cohorts. In fact, the analysis of marsupial tarsal elements has demonstrated that the marsupicarnivores of South America (Ride 1964; Table 1) are also poly-

From Crame, J. A. (ed.), 1989, *Origins and Evolution of the Antarctic Biota*, Geological Society Special Publication No. 47, pp. 217–226.

217

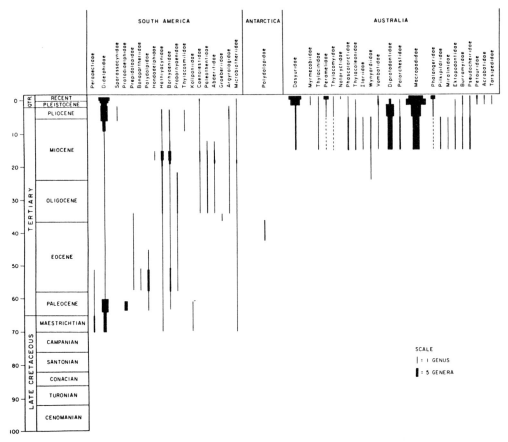

Fig. 1. Generic diversity for the marsupial families recorded from South America, Antarctica and Australia. The substantive fossil record for Australia begins in the medial Miocene from deposits of the Etadunna and Namba formations of South Australia (redrawn from Marshall et al. 1989).

phyletic as borhyaenids are more closely related to Ride's (1964) paucituberculatan caenolestids (=Borhyaeniformes; Szalay 1982a; Table 1) than to didelphids (=Didelphiformes; Szalay 1982a; Table 1).

The dichotomy between South American and Australian marsupials based on tarsal evidence has been corroborated by cytology (only the ameridelphians exhibit the derived character of paired sperm, while all other marsupials have unpaired sperm; Biggers & DeLamater 1965) and immunology on whole blood sera (Kirsch 1977) and on albumin (Lowenstein et al. 1981). Both of the immunological studies indicated that the Australian carnivorous marsupials, the dasyurids and thylacinids, grouped closer to the Australian herbivores (e.g. kangaroos = Macropodidae) than they did to the South American carnivorous marsupials (=Didelphidae, Fig. 2). Subsequent phyletic studies

utilizing two proteins, albumin and transferrin, have demonstrated that the South American microbiotheriid, *Dromiciops australis* (the only extant member of the family), groups closer to other australidelphians (dasyurids and bandicoots) than to any of the other South American marsupials (=ameridelphians; Sarich, pers. comm. in Marshall et al. 1989). The immunological data support the concept of Szalay (1982 a, b) that the Australian marsupials represent a clade distinct from the South American marsupial clade.

A new question arises

The geologically oldest member of the Australidelphia is a microbiotheriid from the Tiupampa Local Fauna, Bolivia of Maastrichtian age (Marshall & de Muizon 1988). This brings a secure minimum age of the apparent fundamental

Table 1. *Classification of the Marsupicarnivora (Ride 1964) versus the Ameridelphia and Australidelphia (Szalay 1982a)*.

RIDE, 1964	SZALAY, 1982a
Order Marsupicarnivora	Cohort Ameridelphia
Superfamily Didelphoidea	Order Didelphida
Family Didelphidae	Suborder Didelphiformes
Superfamily Borhyaenoidea	Family Didelphidae
Family Borhyaenidae	Suborder Borhyaeniformes
Superfamily Dasyuridae	Superfamily Borhyaenoidea
Family Dasyuridae	Family Borhyaenidae
Family Thylacinidae	Superfamily Caenolestoidea
Order Paucituberculata	Family Caenolestidae
Superfamily Caenolestidae	Family Polydolopidae
Family Caenolestidae	Cohort Australidelphia
Family Polydolopidae	Order Dromisciopsia
Order Peramelina	Family Microbiotheriidae
Superfamily Perameloidea	Family Dasyuridae
Family Peramelidae	Family Thylacinidae
Order Diprotodontia	Order Syndactyla
Family Phalangeridae	Suborder Syndactyliformes
Family Wynyardiidae	Superfamily Notoryctoidea
Family Vombatidae	Family Notoryctidae
Family Diprotodontidae	Suborder Perameliformes
Family Macropodidae	Superfamily Perameloidea
Marsupialia *incertae sedis*	Suborder Phalangeriformes
Family Notoryctidae	(=Diprotodontia)

NOTE: Szalay (1982a) has not been duplicated in its entirety, as it only includes families given in Ride (1964) or equivalent taxonomic levels.

dichotomy between the ameridelphian and the australidelphian marsupial cohorts at least back to the Maastrichtian. If the phylogenetic hypotheses presented by Marshall *et al.* (1989) are correct, then a pre-Maastrichtian divergence would be indicated based on the earliest australidelphian (Campanian pediomyids) or even the earlier occurrence of an ameridelphian stagodontid (Cenomanian; Cifelli & Eaton 1987). In South America an extensive radiation of marsupial taxa occurred between the latest Cretaceous (Tiupampa Local Fauna; 70 Ma) and the medial Paleocene (the Itaboraian Local Fauna, Brazil; 60 Ma; Marshall 1987). This resulted in a wide range of family level diversity (Fig. 1). Since a marsupial fossil record older than medial Miocene (c. 15 Ma, Fig. 1) is lacking for Australia, it has been conjectured that an equivalent level of family diversity began somewhere during late Eocene to early Oligocene time (45–35 Ma).

A new question arises from these data. If the split between the two marsupial cohorts had taken place by at least the Maastrichtian, why is there a substantial time difference between the ameridelphian family level radiation in South America (Late Cretaceous to medial Paleocene and the presumed australidelphian family

level radiation in Australia (mid- to late Eocene, a time frame now supported by floral and molecular data; see below)? This is a time differential of approximately 20–25 Ma between the two Southern Hemisphere marsupial radiations.

A new hypothesis on the Australian marsupial radiation

The time difference between the origination and the time of initial radiation of the Australian marsupial fauna and the origin and early radiation of the South American marsupials is proposed here to stem from the fact that Australia was part of a large, high latitude cool temperate landmass (the Weddellian Province) from the latest Cretaceous to the early Eocene (Fig. 3). Under this proposal, the high latitude Weddellian region, with its closed-canopy podocarp and *Nothofagus* forests, served as both a 'holding tank' for the arboreal australidelphian marsupial fauna and a centre for novel adaptations leading to new australidelphian marsupial lineages. Family level radiations within these lineages did not occur until Australia drifted into mid- to low latitudes starting in the mid- to late Eocene, when floristic

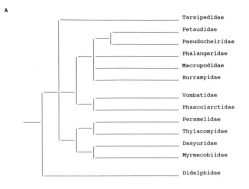

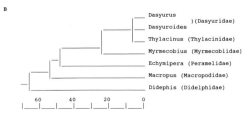

Fig. 2. Topologies of marsupial relationships based on immunological studies of: **(A)** whole blood sera (redrawn from Kirsch 1977); and **(B)** albumin (redrawn from Lowenstein *et al.* 1981). Both topologies indicate that carnivorous Australian marsupials (Dasyuridae and Thylacinidae) are more closely related to the Australian herbivores (e.g. Macropodidae) than to South American carnivorous marsupials (Didelphidae). The scale in B is divergence time in millions of years.

changes stemming from Australia's entry into climatically warmer and drier settings gave rise to an increase in habitat diversity. The Weddellian marsupial fauna that remained on the Antarctic continent after its separation from Australia eventually became extinct with the deteriorating climatic conditions beginning in the Oligocene.

The Weddellian Province

The Weddellian Zoogeographic Province, based on marine molluscan, echinoderm and arthropod faunas, was conceived by Zinsmeister (1979, 1982) as a cool temperate, shallow water region which extended from southern South America, along the Antarctic Peninsula and West Antarctica, to Tasmania and southeastern Australia (New Zealand is also part of this province). Case (1988) expanded the concept of the Weddellian Province to a 'biogeographic

province' with the inclusion of a terrestrial floral component, the Southern Beech (*Nothofagus*) forests, during the Paleocene and Eocene and possibly extending back to the latest Cretaceous. Of particular interest also was the association of marsupials and the *Nothofagus* forests presently in southern South America and southeastern Australia, and during the late Eocene on Seymour Island, Antarctic Peninsula (Woodburne & Zinsmeister 1982, 1984; Case *et al.* 1988).

The *Nothofagus* megafossils (Case 1988) and pollen (Cranwell 1959; Cranwell *et al.* 1960; Askin & Fleming 1982; Askin 1989) recovered from Antarctica are of paramount importance to the status of the continent relative to the disperal of *Nothofagus* in the Southern Hemisphere. *Nothofagus* currently ranges from southern South America, to New Zealand, Australia and associated islands such as New Guinea and New Caledonia. The Weddellian Province brings together into a single biogeographic region those areas which for the most part have two unique aspects of their terrestrial biota, marsupials and *Nothofagus*. This long, continuous coastal region would facilitate the dispersal of both plants and animals throughout the area, neither of which are known for long range dispersal capabilities across seaways; *Nothofagus* is definitely, and marsupials most likely, obligate overland dispersers (Case 1988). The australidelphian marsupials are considered by Szalay (1982a. b) to be especially adapted for arboreal habitats and these marsupials would then have the abilities to disperse through the dense beech forest (Case *et al.* 1988).

High latitude heterochroneity

Hickey *et al.* (1983) and Zinsmeister & Feldmann (1984) have both put forward the hypothesis that taxa may make their initial appearance in high latitude (i.e. >60°) regions before those same taxa are known in mid- to low latitudes. Zinsmeister & Feldmann (1984) analyzed Antarctic marine faunas from Seymour Island and pointed out two important aspects of 'high latitude heterochroneity' (i.e. differential appearance of taxa between high and mid- to low latitudes): (1) that the high latitude regions serve as 'holding tanks' for taxa which remain isolated until suitable conditions exist for their dispersal out of high latitude regions; or (2) as centres of origin for new taxonomic groups which can escape the high latitude region under existing conditions due to new adaptations.

Jablonski *et al.* (1983) indirectly, and Lewin

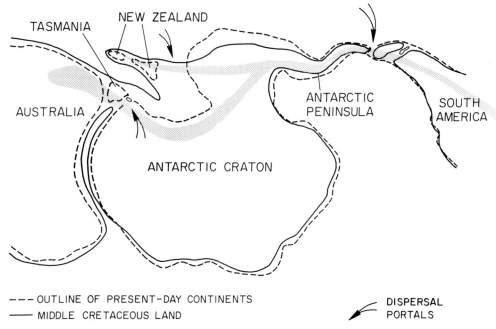

Fig. 3. The Weddellian Province, a terrestrial and marine biogeographic region which extended from southern South America to the Antarctic Peninsula, Marie Byrd Land, New Zealand, Tasmania and southeastern Australia, is presented in this mid-Cretaceous reconstruction (redrawn from Woodburne & Zinsmeister 1984). The stippled areas represent the overland dispersal routes within the province, with arrows indicating dispersal portals which may have acted as biogeographic filters.

(1983) directly, also hypothesized that high latitude regions may be the sites for the origination of novel taxa leading to new taxonomic groups, rather than areas of low latitude, such as the tropics. Compared to the tropics (characteristics in parentheses), high latitude regions exhibit low species diversity (versus high species diversity), large population sizes (versus low population sizes), low rates of speciation (versus high rates) and low rates of extinction (versus high rates). It is the low extinction rates that are believed to be the key factor here (Jablonski *et al.* 1983). Once a novel adaptation arises, even at a lower rate of occurrence, it is more likely to persist and not become extinct because of the lower rate of extinction in the high latitude region. In the tropics, novel adaptations are likely to arise more frequently due to the higher rates of speciation, but would be less likely to persist because the extinction rates would be higher.

Timing of the Australian marsupial radiations

Since the substantive marsupial record in Australia only goes back to the medial Miocene

(Fig. 1) (Woodburne *et al.* 1985; Marshall *et al.* 1989), it is not known precisely when the marsupial familial level radiations took place. It is the contention here that the familial diversity is tied to habitat diversity and that through an examination of floristic (i.e. primarily palynomorph) data during the Cenozoic of Australia, insights can be achieved into when these radiations began and when different major taxonomic groups within the australidelphian clade may have diversified.

In the latest Cretaceous and early Paleocene, when Australia was firmly connected to Antarctica, cool temperate, closed rainforests composed of podocarps, araucarians, Proteaceae and *Nothofagus* (*brassi* group) were widespread, reaching the centre of the continent in the region of Alice Springs (Dettmann 1981; Kemp 1981; Truswell & Harris 1982). During the mid- to late Paleocene and into the early Eocene Australian floras increased in diversity (Fig. 4), as indicated by the recovery of fossil pollen of *Anacolosa* (Olacaceae), *Beauprea* (Proteaceae), Cupanieae and Cunoniaceae (Kemp 1981; Sluiter 1988).

In the mid- to late Eocene, a change in the major closed forest components obviously oc-

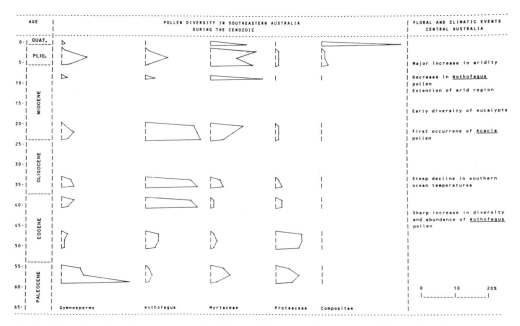

Fig. 4. Pollen diversity of gymnosperms, *Nothofagus*, Myrtaceae, Proteaceae and Compositae of southeastern Australia (from Martin 1978, 1982) and the floral and climatic events of central Australia (from Truswell & Harris 1982) during the Cenozoic. Scale at the lower right pertains to the pollen diversity diagrams and represents percent of total sample for the pollen data at each time interval.

curred as there is an increase in the diversity and abundance of *Nothofagus* species (includes the *fusca* and *menziesii* pollen types). The increased presence of *Nothofagus* results in beeches becoming the dominant tree taxon and podocarps declining in abundance (Fig. 4). The cool temperate, *Nothofagus*-dominated, closed forest of present day southeastern Australia (i.e. in Victoria) has a dense and thick canopy that can extend for more than half, and up to two-thirds, of the tree's height (10->30 m.; Howard 1981). The understory in such a forest is cool and moist and primarily composed of non-angiosperm taxa such as treeferns, ground ferns and bryophytes. This agrees with spore palynomorph data from deposits of the Otway basin (Victoria) and Nerriga (southern New South Wales) during the Paleocene through middle Eocene (Martin 1978, 1981). The same extreme density of structure within extant *Nothofagus* forests has also been documented for the beech forests of southern Argentina (Pearson & Pearson 1982).

Pollen of *Nothofagus menziesii* has its first occurrence in the Maastrichtian of the Antarctic Peninsula and subsequently has a first occurrence in Australia in the Eocene (Dettmann 1989; Askin, 1989). Other Gondwana plant taxa, such as *Anacolosa*, *Beauprea* and members of the Cupanieae and Bombacaceae, exhibit this same dispersal and temporal pattern as the *N. menziesii* pollen type (Askin 1989), indicating a dispersal route from the Antarctic Peninsula to southeastern Australia by way of West Antarctica (from the Maastrichtian to the Eocene).

A possible response to the sharp drop in ocean temperatures during the Eocene–Oligocene transition (Shackelton & Kennett 1975) would be a lowering of precipitation and the beginning of the arid zone in central Australia (Truswell & Harris 1982). This would also signal the beginning of open forest habitats with a more diverse herbaceous understory due to the addition of angiosperm components, which probably achieve a greater regional expansion with the increase of aridity during the Miocene (Truswell 1984). The increased aridity also results in an increase in the diversity of the Myrtaceae, with a special focus on the early radiation of the eucalypts. Eventually, the Myrtaceae replaced the beeches as the dominant tree taxon on the Australian continent (Fig. 4).

In the Miocene, a highly abundant and diverse herbaceous vegetation developed both as an understory in the open forests and in open lands (possibly grasslands) between water courses (Callen & Tedford 1976; Kemp 1978).

The vegetation on the Australian continent underwent numerous changes from the end of the Cretaceous to the present and can be summarized in the following sequence (Fig. 5): the Late Cretaceous to early Eocene, podocarp-dominated closed forest is widespread throughout the Weddellian Province (Dettmann 1981); in Australia this forest is replaced in the mid- to late Eocene by the more diverse *Nothofagus*-dominated closed forest where many of the floral components had dispersed from the Antarctic Peninsula in the Maastrichtian (however, the understory remains unchanged in composition from that of the podocarp understory); during the Oligocene, a drop in ocean temperatures and reduced precipitation lead to a more open forest structure, with an increase in the Myrtaceae and in a herbaceous angiosperm understory; this pattern continues throughout the Miocene as eucalypts radiate and grasslands spread in the arid regions.

Habitat diversity and marsupial radiations

It is possible to speculate that, with the changes of the vegetation described above for Australia in the Eocene, a diversity of new habitats would be generated. These new habitats resulted in new marsupial adaptive zones leading to an array of familial lineages within the australidelphian orders (equalling the diversity level in South America some 20 Ma earlier). Thus within the Maastrictian–early Eocene closed podocarp and beech forests of the Weddellian Province, the basal Weddellian australidelphians probably diverged in three lineages representing the three primary marsupial orders (Dasyuromorphia, Peramelina and Diprotodontia, excluding *Notoryctes*, the marsupial mole; nomenclature according to Marshall *et al.* 1989) present in Australia today as new adaptive complexes arose. From the arboreal insectivorous to omnivorous ancestral stock, dasyuromorphs became terrestrial insectivores and peramelinan terrestrial omnivores in the fern/bryophyte understory, while the diprotodontians initially remained arboreal, but switched to an herbivorous diet. These three lineages would constitute the low diversity, Weddellian marsupial fauna among the widespread yet somewhat uniform forested habitats of the region. Diversity within each of these orders would remain low until Australia separated from Antarctica and drifted into lower latitudes. Presumably these forms tracked the spread of the newly developing components of the more diverse *Nothofagus* forest from their Maastrichtian occurrence in the Antarctic Peninsula to their Eocene occurrence in Australia.

Beginning with the middle Eocene, the change from podocarp- to *Nothofagus*-dominated closed forests would result in an increase of arboreal habitats facilitating a radiation of arboreal herbivore (or omnivore) marsupial forms. However, the understory remained unchanged in composition (i.e. non-angiosperm; Martin 1978, 1981) which thus would not facilitate an extensive radiation of terrestrial forms. The opening of the forests in the Oligocene and on into the Miocene, with a concomitant increase in floral diversity of the herbaceous understory, most likely provided new habitats for terrestrial diprotodontian herbivores. The greater development of the herbaceous understory and grasslands in the medial Miocene would facilitate development of more specialized herbivores such as diprotodontids and macropodine kangaroos (such as the forms in the upper part of the Etadunna Formation and in the Wipajiri Formation of South Australian; JAC, pers. observ.).

The Weddellian marsupial fauna remaining in Antarctica after the separation of Australia was doomed to extinction with the deterioration of climatic conditions which resulted in the formation of the East Antarctic ice cap.

The scenario for the tempo and mode of the marsupial radiations in Australia presented above has recently been given support with the development of a molecular clock based on DNA hybridization studies of Australian marsupials (Springer 1988). The DNA clock uses the divergence between macropodine and potoroine kangaroo clades at approximately 15 Ma (a split seen in taxa from the Etadunna Fm.; Woodburne *et al.* 1989) as the calibration point. The divergence within the arboreal and primarily herbivorous phalangiform clade (which includes kangaroos and various possums) is projected to occur at about 45–50 Ma, which coincides with the development of the diverse *Nothofagus*-dominated closed forests in the middle Eocene (Fig. 5). Divergences within the terrestrial herbivorous vombatiform clade (which includes koalas and wombats) are projected at approximately 30 Ma, which coincides with the development of open forest habitats in the Oligocene. Extensive radiations within the more specialized herbivore families, Diprotodontidae and Macropodidae (Marshall *et al.* 1989), coincide with the development of an extensive herbaceous understory and grasslands during the medial Miocene. This too, coincides with the divergence of the more herbivorous macropodine kangaroos from the more om-

224 J.A. CASE

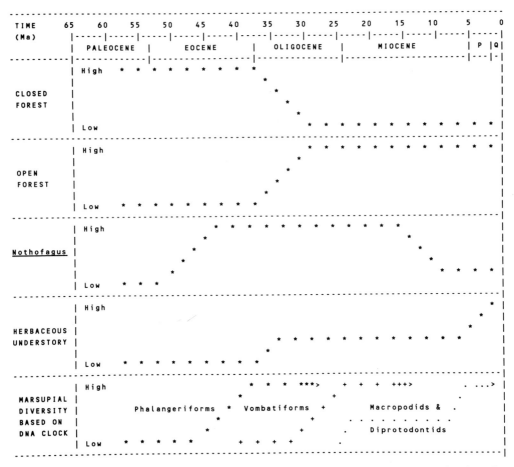

Fig. 5. Vegetation trends for Australia through the Cenozoic. The high and low scales represent abundance for the Closed and Open Forests and also diversity for *Nothofagus* and Herbaceous Understory. Temporal patterns for phalangeriform, vombatiform and macropodid marsupial diversity based on DNA clock data of Springer (1988). Macropodid and diprotodontid temporal patterns also based on generic diversity of Marshall *et al.* (1989).

nivorous potoroine kangaroos at about 15 Ma (Springer 1988).

Recent evidence suggests that the age of the Etadunna Formation may be older than medial Miocene by some 10 Ma (ca. 25 Ma ago or latest Oligocene; Woodburne *et al.* 1989). This would also shift the divergence points of the Australian marsupial clades back some 10 Ma as the divergence between the two kangaroo subfamilies would be older (i.e. at 25 Ma ago). The earlier projected dates for the age of the Etadunna Formation would not affect the validity of the scenario proposed here as the precision in the dating of the various floral events interpreted from fossil pollen data is broad (with the dates being couched in terms of late Eocene

or medial Miocene rather than specific radiometric dates). Thus the dating of the floral events can easily accommodate a backwards time shift in the marsupial cladogenic events and still exhibit a high incidence of correlation.

Summary

Antarctica's position during the Maastrichtian to early Eocene within the high latitude Weddellian Biogeographic Province played an important role in retarding the radiation of the Australian marsupials relative to the extensive marsupial radiations in the mid- to low latitude regions of South America in the same time interval. The high latitude province served as holding tank for the Weddellian marsupials due to low diversity of habitats resulting from the widespread podocarp/*Nothofagus* closed forests with non-angiosperm understories. Extensive familial level cladogenesis probably did not occur among the Australian marsupials until the mid- to late Eocene when the continent had sufficiently separated from Antarctica and generated a more diverse habitat structure. The Weddellian marsupial fauna remaining on the Antarctic continent subsequent to the separation of Australia from Antarctica became extinct sometime after the beginning of the Oligocene due to the deteriorating climatic conditions and the development of the polar ice cap.

I would like to acknowledge the National Science Foundation which provided travel funds for my participation in this Antarctic symposium. I would like to thank L. G. Marshall, R. A. Askin, M. O. Woodburne and an anonymous referee for their constructive comments on the manuscript. Illustrative and photographic assistance by L. Bobbit, U. C. Riverside, was in part supported by a National Science Foundation, Division of Polar Programs grant, DPP 8521368, to M. O. Woodburne.

References

ASKIN, R. A. 1989. Endemism and heterochroneity in the Seymour Island Campanian to Paleocene palynofloras: implications for origins, dispersal and palaeoclimates of southern floras *In*: CRAME, J. A. (ed.) *The origins and evolution of the Antarctic biota*. Geological Society, London. Special Publication, **47**, 107–120.

—— & FLEMING, R. F. 1982. Palynological investigations of Campanian to lower Oligocene sediments on Seymour Island, Antarctic Peninsula. *Antarctic Journal of the United States*, **17**, 70–71.

BIGGERS, J. D., & DE LAMATER, E. D. 1965. Marsupial spermatozoa pairing in the epididymis of American forms. *Nature*, **208**, 1602–1603.

CALLEN, R. A. & TEDFORD, R. H. 1976. New late Cainozoic rock units and depositional environments, Lake Frome area, South Australia. *Transactions of the Royal Society of South Australia*, **100**, 125–167.

CASE, J. A. 1988. Paleogene floras from Seymour Island, Antarctic Peninsula, *In*: FELDMANN, R. M. & WOODBURNE, M. O. (eds), *Geology and Paleontology of Seymour Island, Antarctic Peninsula*. Geological Society of America, Memoir, **169**, 523–530.

——, WOODBURNE, M. O., & CHANEY, D. S., 1988. A new genus of polydolopid marsupial from Antarctica. *In*: FELDMANN, R. M. & WOODBURNE, M. O. (eds), *Geology and Paleontology of Seymour Island, Antarctic Peninsula*. Geological Society of America, Memoir, **169**, 505–521.

CIFELLI, R. L. & EATON, J. G. 1987. Marsupial from the earliest Late Cretaceous of Western U.S. *Nature*, **325**, 520–522.

CLEMENS, W. A. 1977. Phylogeny of the marsupials. *In*: STONEHOUSE, B. & GILMORE, D. (eds), *The Biology of Marsupials*. MacMillan Press Ltd, London, 51–68.

CRANWELL, L. M. 1959. Fossil pollen from Seymour Island, Antarctica. *Nature*, **184**, 1782–1785.

——, HARRINGTON, H. J. & SPEDEN, I. G. 1960. Lower Tertiary microfossils from McMurdo Sound, Antarctica. *Nature*, **186**, 700–702.

DETTMANN, M. E. 1981. The Cretaceous flora. *In*:

KEAST, A. (ed.), *Ecological Biogeography of Australia*. Dr. W. JUNK by Publishers, The Hague, 355–375.

—— 1989. Antarctica: Cretaceous cradle of austral temperate rainforests? *In*: CRAME, J. A. (ed.) *The origins and evolution of the Antarctic biota*. Geological Society, London. Special Publication **47**, 89–106.

HICKEY, L. J., WEST, R. M., DAWSON, M. R. & CHOI, D. K. 1983. Arctic terrestrial biota: paleomagnetic evidence of age disparity with midnorthern latitudes during the Late Cretaceous and early Tertiary. *Science*, **221**, 1153–1156.

HOWARD, T. M. 1981. Southern closed forests. *In*: GROVES, R. H. (ed.), *Australian vegetation*. Cambridge University Press, Cambridge, 102–120.

JABLONSKI, D., SEPKOSKI, J. J., BOTTJER, D. J. & SHEEHAN, P. M. 1983. Onshore–offshore patterns in the evolution of Phanerozoic shelf communities. *Science*, **222**, 1123–1125.

KEAST, A. 1977. Historical biogeography of the marsupials. *In*: STONEHOUSE, B. & GILMORE, D. (eds), *The Biology of Marsupials*. MacMillan Press Ltd, London, 69–95.

KEMP, E. M. 1978. Tertiary climatic evolution and vegetation history in the southeast Indian Ocean region. *Palaeogeography, Palaeoclimatology, Palaeoecology*, **24**, 169–208.

—— 1981. Tertiary palaeogeography and the evolution of Australian climate. *In*: KEAST, A. (ed.), *Ecological Biogeography of Australia*. Dr. W. Junk by Publishers, The Hague, 33–49.

KIRSCH, J. A. W. 1977. The comparative serology of Marsupialia, and a classification of Marsupials. *Australian Journal of Zoology, supplementary series*, **52**, 1–152.

LEWIN, R. 1983. Origin of species in stressed environments. *Science*, **222**, 1112.

LOWENSTEIN, J. M., SARICH, V. M. & RICHARDSON, B. J. 1981. Albumin systematics of the extinct mammoth and Tasmanian wolf. *Nature*, **291**, 409–411.

MARSHALL, L. G. 1980. Marsupial biogeography. *In*: JACOBS, L. L. (ed.), *Aspects of Vertebrate History*.

Museum of Northern Arizona Press, Flagstaff, 345–386.

—— 1987. Systematics of Itaboraian (Middle Paleocene) age 'opossum-like' marsupials from the limestone quarry at São José de Itaborai, Brasil. *In*: ARCHER, M. (ed.), *Possums and Opossums: Studies in Evolution*. Royal Zoological Society of New South Wales, Sydney, 91–160.

——, CASE, J. A. & WOODBURNE, M. O. 1989. Phylogenetic relationships of the families of marsupials. *Current Mammalogy*, **2**, (in press).

—— & de MUIZON, C. 1988. The dawn of the age of mammals in South America, *National Geographic Research*, **4**, 23–55.

MARTIN, H. A. 1978. Evolution of the Australian flora and vegetation through the Tertiary: evidence from pollen. *Alcheringa*, **2**, 181–202.

—— 1981. The Tertiary flora. *In*: KEAST, A. (ed.), *Ecological Biogeography of Australia*. Dr. W. Junk by Publishers, The Hague, 393–406.

PEARSON, O. P., & PEARSON, A. K. 1982. Ecology and biogeography of the southern rainforests of Argentina. *In*: MARES, M. A. & GENOWAYS, H. A. (eds), *Mammalian Biology in South America*. Pymatuning Laboratory of Ecology, Special Publication, **6**, 129–142.

RIDE, W. D. L. 1964. A review of the Australian fossil marsupials, *Journal and Proceedings of the Royal Society of Western Australia*, **47**, 97–131.

SHACKELTON, N. J. & KENNETT, J. P. 1975. Palaeotemperature history of the Cenozoic and the initiation of Antarctic glaciation: oxygen and carbon isotope analyses in DSDP sites 227, 279, 281. *In: Initial Reports of the Deep Sea Drilling Project*. U.S. Government Printing Office, Washington, **29**, 743–755.

SLUITER, I. R. K. 1988. Early Tertiary vegetation and palaeoclimates, Lake Eyre region, northeastern South Australia, (in press).

SPRINGER, M. S. 1988. *The phylogeny of diprotodontian marsupials based on single-copy nuclear DNA-DNA hybridization and craniodental anatomy*. Unpublished Ph.D. thesis, University of California, Riverside.

SZALAY, F. S. 1982a. A new appraisal of marsupial phylogeny and classification. *In*: ARCHER, M. (ed.), *Carnivorous Marsupials*. Royal Zoological Society of New South Wales, 621–640.

—— 1982b. Phylogenetic relationships of the Marsupials. *Geobios, Mémoire Special*, **6**, 177–190.

TRUSWELL, E. M. 1984. The temperate southern beech/podocarp rainforest; Eocene to Miocene evolution in Australia, Antarctica and New Zealand. *Abstract; Sixth International Palynological Conference, Calgary*, 167.

—— & HARRIS, W. K. 1982. The Cainozoic palaeobotanical record in arid Australia: fossil evidence for the origins of an arid-adapted flora. *In*: BARKER, W. R. & GREENSLADE, P. J. M. (eds), *Evolution of the Flora and Fauna of Arid Australia*. Peacock Publications, Adelaide, 67–76.

WOODBURNE, M. O., & ZINSMEISTER, W. J. 1982. Fossil land mammal from Antarctica. *Science*, **218**, 284–286.

—— & —— 1984. The first land mammal from Antarctica and its biogeographic implications. *Journal of Paleontology*, **58**, 913–948.

——, TEDFORD, R. H., ARCHER, M., TURNBULL, W. D., PLANE, M. D. & LUNEDIUS, E. L. 1985. Biochronology of the continental mammal record of Australia and New Guinea. *South Australia Department of Mines and Energy Special Publication*, **5**, 347–363.

——, MC FADDEN, B. J., CASE, J. A., SPRINGER, M. S., PLEDGE, N. S. *et al.* 1989. Land mammal biostratigraphy and magnetostratigraphy of the Etadunna Formation biostratigraphy and magnetostratigraphy of the Etadunna Formation (medial Miocene) of South Australia. *University of California Publications in Geological Sciences*, (in press).

ZINSMEISTER, W. J. 1979. Biogeographic significance of the late Mesozoic and early Tertiary molluscan faunas of Seymour Island (Antarctic Peninsula) to the final breakup of Gondwanaland. *In*: GRAY, J. & BOUCOT, A. (eds), *Historical Biogeography, Plate Tectonics and the Changing Environment, Proceedings, 37th Annual Biological Colloquium and Selected Papers*. Oregon State University Press, Corvallis, 349–355.

—— 1982. Late Cretaceous–Early Tertiary molluscan biogeography of southern circum-Pacific. *Journal of Paleontology*, **56**, 84–102.

—— & FELDMANN, R. M. 1984. Cenozoic high latitude heterochroneity of Southern Hemisphere marine faunas. *Science*, **224**, 281–283.

Late Cretaceous–early Tertiary floras of King George Island, West Antarctica: their stratigraphic distribution and palaeoclimatic significance

K. BIRKENMAJER[1] & E. ZASTAWNIAK[2]

[1]Institute of Geological Sciences, Polish Academy of Sciences, Senacka 3, 31–002 Kraków, Poland

[2]W. Szafer Institute of Botany, Polish Academy of Sciences, Lubicz 46, 31–512 Kraków, Poland

Abstract: Stratigraphic positions of Late Cretaceous and Tertiary floras have been established on King George Island, West Antarctica, based on K-Ar dating of associated volcanics. The oldest dated floras, of Late Cretaceous and Paleocene ages, provide evidence for a warm climate phase preceding the early Eocene glacial event termed the Kraków Glaciation. Late Eocene through early Oligocene floras indicate another warm period termed the Arctowski Interglacial, which post-dates the Kraków Glaciation. The youngest floras, of Oligocene–Miocene boundary age, indicate a temperate (or cool-temperate) climate, corresponding to the Wawel Interglacial which separated late Oligocene glacial and interglacial epochs (Polonez Glaciation, Wesele Interglacial and Legru Glaciation) from early Miocene glaciation (Melville Glaciation).

There is a long record of late Mesozoic through Tertiary terrestrial floras in King George Island, South Shetland Islands (West Antarctica) (Figs 1 and 2). The plant remains occur as petrified wood, imprints of leaves and shoots and, less commonly, as seeds or spore-pollen assemblages, in volcaniclastic and related sediment intercalations within thick volcanic piles of late Mesozoic through late Paleogene-early Neogene ages. Based on radiometric dating of associated volcanics, these floras have been grouped into the following assemblages: pre-Late Cretaceous; Late Cretaceous; ?Late Cretaceous–early Paleogene; middle Eocene–early Oligocene; and Oligocene–Miocene boundary.

The description of fossil plant assemblages presented here is based mainly on collections assembled during the Polish Antarctic Expeditions of 1977–1986, and on data published by Barton (1964), Orlando (1964), del Valle *et al.* (1984), Czajkowski & Rösler (1986), Lyra (1986) and Troncoso (1986).

Stratigraphic distribution of fossil plants

Pre-Late Cretaceous plant assemblages

The oldest Mesozoic plant remains occur on King George Island within stratiform terrestrial volcanic–volcaniclastic complexes of the Barton Horst, distinguished as the Cardozo Cove and the Martel Inlet groups (Birkenmajer

1982). The lavas associated with plant-bearing strata yielded widely scattered K-Ar dates, between 66.7 and 26.4 Ma, showing no positive correlation with relative stratigraphic ages of particular formations (Birkenmajer *et al.* 1983 *a*, 1986*b*). This suggested reheating by Tertiary volcanic–plutonic activity and resultant argon loss. These rocks are pre-Tertiary, probably late Jurassic to early Cretaceous in age.

The plant remains are represented mainly by fragments of petrified wood showing annual growth rings (Martel Inlet Group; Keller Peninsula and Precious Peaks; Fig. 2, Loc. 1 and 2) and by conifer shoots attributed by Barton (1964) to the genus *Araucaria* (corrected to *Pagiophyllum* Heer by Zastawniak 1981, p. 100) (Cardozo Cove Group; Admiralen Peak; Fig. 2, Loc. 3). Detailed descriptions of these fossil plant localities have been presented by Birkenmajer (1982).

Late Cretaceous plant assemblages

The Late Cretaceous plant assemblages occur in the downthrown Warszawa Block to the south of the Barton Horst, within stratiform volcanic-sedimentary complexes distinguished as the Paradise Cove and the Baranowski Glacier groups (Birkenmajer 1980).

Paradise Cove flora. Plant remains occur mainly in the middle part of the Paradise Cove Group within the Creeping Slope Formation (Fig. 2, Loc. 4). This is a terrestrial sedimentary unit

From Crame, J. A. (ed.), 1989, *Origins and Evolution of the Antarctic Biota*, Geological Society Special Publication No. 47, pp. 227–240.

227

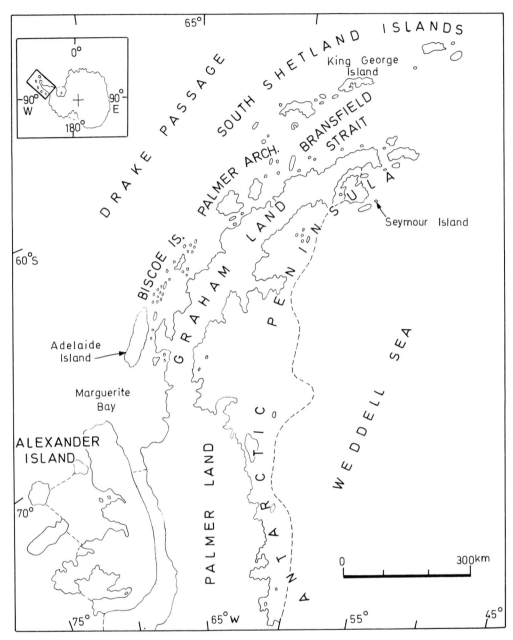

Fig. 1. Position of King George Island and Seymour Island in the Antarctic Peninsula sector, and in Antarctica (inset)

about 60 m thick consisting of red shale and a tuffite-pellet conglomerate horizon; the latter has yielded large fragments of silicified wood showing well preserved annual growth rings, determined as *Nothofagoxylon* by Jagmin (1987). K-Ar dating of underlying basaltic lavas yielded a date of 67.7 ± 3.5 Ma. This apparently Late Cretaceous date is considered to be a result of reheating and argon loss by younger igneous activity, as the basaltic lavas of the succeeding Baranowski Glacier Group yielded a much older Late Cretaceous date of 77 ± 4 Ma (Birkenmajer *et al.* 1983*b*).

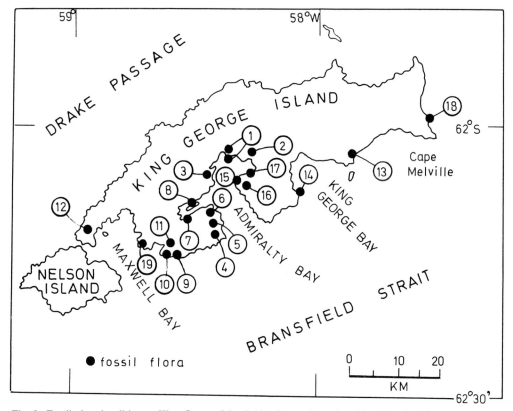

Fig. 2. Fossil plant localities on King George Island. Numbers refer to localities described in the text

Zamek flora. A rich assemblage of fossil leaves has been recovered from a moraine below the Zamek cliff belonging to the Zamek Formation, Baranowski Glacier Group (Fig. 2, Loc. 5). The Late Cretaceous age of this group has been established on K-Ar dating of its basal basaltic lavas (77 ± 4 Ma, Santonian−Campanian), and basaltic andesite lavas of the succeeding Arctowski Cove Formation (Rakusa Point Member) (66.7 ± 1.5 Ma, Maastrichtian). The Zamek Formation forms the uppermost part of the Baranowski Glacier Group; it consists of basaltic andesite lavas alternating with scoria and tuff (40 m thick), and a plant-bearing tuff horizon about 1 m thick. The latter yielded mainly various dicotyledonous leaf impressions belonging principally to the genus *Nothofagus* (Fig. 3(2), (5)) and various laurophyllous plants (Fig. 3(6)). Notophyll or even mesophyll, unlobed, brochidodromous-veined leaves with prominent, regular, orthogonal−reticulate tertiary venation and thick texture (Fig. 3(1)) are particularly characteristic for this taphocoenosis; their systematic position is unknown. Infrequent leaves with either actinodromous or acrodromous venation (Fig. 3(3)) are undeterminable, a few stenophyll leaves of Myrtaceae-type and some small leaves with brochidodromous venation (Fig. 3(4)) have also been found. These plants are associated with several frond impressions and conifer shoot imprints (Araucariaceae or Podocarpaceae).

A very poorly preserved leaf fragment of *Nothofagus*-type has been illustrated from tuffs intercalated in andesite lavas of the Mazurek Point Formation at Three Sisters Point (Fig. 2, Loc. 3) by Paulo & Tokarski (1982, pl. III, fig. 1). This formation, at its type locality, is now regarded as representing Upper Cretaceous (cf. Birkenmajer & Gaździcki 1986).

?Late Cretaceous−early Paleogene plant assemblages

There are three fossil plant assemblages of inadequately known stratigraphic position, either Late Cretaceous or early Paleogene in age: the Dufayel Island and Barton Peninsula floras within the Barton Horst, and the Fildes

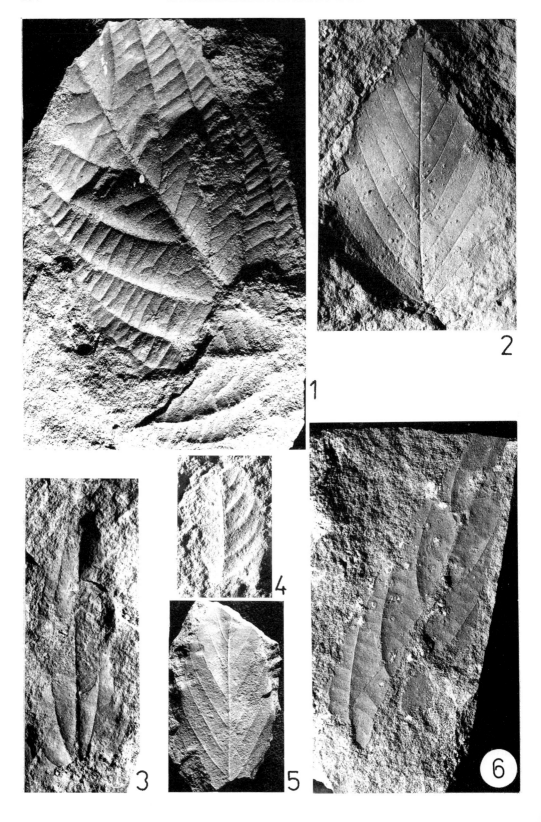

Peninsula flora within the downthrown Fildes Block, to the north of the Barton Horst.

Dufayel Island flora. (Fig. 2, Loc. 8). The fossil plant assemblage consists mainly of angiosperms, particularly dicotyledonous leaf impressions. It is characterized by the presence of leaves of *Nothofagus* sp., Myrtaceae family, and numerous impressions of laurophyllous plant leaves accompanied by lobed leaves of aff. *Cochlospermum* and *Sterculia*-type. No gymnosperms have been found in this taphocoenosis, and only two specimens may belong to pteridophytes (Birkenmajer & Zastawniak 1986). The plant impressions occur in tuffs of the Dalmor Bank Formation beneath basaltic lavas K-Ar-dated at 51.9 ± 1.5 Ma, and above altered and folded andesite lavas of the Cardozo Cove Group, K-Ar-dated at 56.8 ± 1.2 Ma (Birkenmajer *et al.* 1983a, b). The Dufayel Island Group is separated from the underlying Cardozo Cove Group by an important angular unconformity.

The above K-Ar dates may suggest an early Eocene–late Paleocene age of the fossil plant assemblage. However such an age should be treated with caution, as considerable alteration of the dated rocks has occurred, suggesting reheating and argon loss. An even older, Late Cretaceous or early Paleocene age cannot be ruled out (see Fig. 4), as suggested also by similarities between the Dufayel Island and the Zamek floras.

Barton Peninsula flora. (Fig. 2, Loc. 19). Five fossil plant taxa considered to represent an early Eocene–early Oligocene time span, have been described but not illustrated from Barton Peninsula by Del Valle *et al.* (1984). In our small collection of leaves assembled in 1986, only two taxa were determinable: *Nothofagus subferruginea* (Dusén) Tanai, and *N. densinervosa* Dusén. Other leaf impressions with brochidodromous venation represent undeterminable dicotyledonous plants (Tokarski *et al.* 1987). The leaf impressions occur in red tuffs belonging to a tectonic unit crossed by the Noel Hill gabbro–diorite pluton which yielded K-Ar

ages between 60.7 and 46 Ma (cf. Birkenmajer *et al.* 1983a). The oldest date of this pluton indicates that the plant-bearing beds pre-date late Paleocene, and may be early Paleocene or Late Cretaceous in age; a correlation with the Dufayel Island plant-bearing beds (Dufayel Island Group) cannot be ruled out (Fig. 4).

Fildes Peninsula flora. (Fig. 2, Loc. 12). The main fossil plant locality at Fildes Peninsula, known as 'Mount Flora' or 'Leaves Hill', yielded numerous impressions of fossil plants described by Orlando (1964), Czajkowski & Rösler (1986) and Troncoso (1986). Palaeobotanical investigation of collections assembled during the Polish Antarctic Expeditions between 1978 and 1981 permit classification of this taphocoenosis as a palaeoassemblage with *Monimiophyllum antarcticum* and *Sterculia*-type leaves. It is characterized by the presence of thick, non-entire, pinnately veined leaves with slightly irregular semicraspedodromous venation, determined as *Monimiophyllum antarcticum* Zastawniak (in Birkenmajer & Zastawniak, in press) (Fig. 5(6A–C); Fig. 6) and palmately-lobed leaves with actinodromous venation of *Sterculia*-type (Fig. 5(1), (2)). They are associated with impressions of entire-margined leaves with acrodromous venation determined as *Dicotylophyllum duseni* Zastawniak (in Birkenmajer & Zastawniak, in press) (Fig. 5 (5)) and with other dicotyledonous leaves which are entire or non-entire, microphyll or notophyll (mesophyll leaf-class sizes are absent). This palaeoassemblage is poor in *Nothofagus*-type leaves. The gymnosperms are represented, according to Orlando (1964), Czajkowski & Rösler (1986) and Troncoso (1986), by Araucariaceae, Cupressaceae and Podocarpaceae.

Palynological investigation of Fildes Peninsula sediments from 'Leaves Hill' and Suffield Point by Lyra (1986) indicated the presence of Pteridophyta (5 taxa), Gymnospermae (exclusively Podocarpaceae, 5 taxa) and Angiospermae (8 taxa) including the genus *Nothofagidites* Potonié (3 taxa).

The stratigraphic age of the Fildes Peninsula flora based on plant remains was initially ac-

Fig. 3. Plant impressions from the Zamek Formation, Baranowski Glacier Group. Moraine below Zamek, Admiralty Bay, King George Island (Phot. M. Kleiberowa). Natural size (**1**) dicotyledonous leaf, mesophyll, brochidodromous venation, thick texture (No 73 ZPAL); (**2**) *Nothofagus subferruginea* (Dusén) Tanai (No 70 ZPAL); (**3**) dicotyledonous leaf with acrodromous venation (No 97 ZPAL); (**4**) dicotyledonous leaf, nanophyll, brochidodromous venation, thick texture (No 97 ZPAL); (**5**) *Nothofagus subferruginea* (Dusén) Tanai (No 66 ZPAL); (**6**) dicotyledonous leaf of a laurophyllous plant, notophyll, thick texture (No 77 ZPAL)

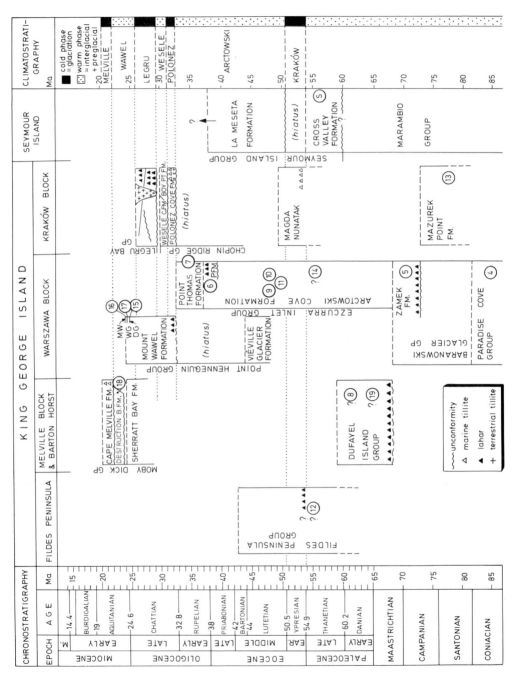

Fig. 4. Stratigraphic position of fossil floras of the Antarctic Peninsula sector: Late Cretaceous and Tertiary. Numbers refer to localities described in the text and shown in Fig. 2. DG. Dragon Glacier plant beds; WG, Wanda Glacier plant beds; MW, Mount Wawel plant beds; S, Paleocene plant site on Seymour Island

Fig. 5. Plant impressions from the Fildes Peninsula Group. 'Mount Flora', Fildes Peninsula, King George Island (Phot. M. Kleiberowa and A. Pachoński). All remains natural size except when stated (1) palmately lobed leaf of *Sterculia*-type (No 174 ZPAL); (2) same (No 127 ZPAL); (3) Pteridophyta indet., detached pinna or pinnula of a frond, × 1.5 (No 173/480); (4) fragment of conifer shoot? (No 173/535); (4a) same, × 3; (5) *Dicotylophyllum duseni* Zastawniak (No 173/454); (6A−C) *Monimiophyllum antarcticum* Zastawniak; (6D) very peculiar dicotyledonous palmately lobed leaf with very shortened middle lobe, entire margined (platanoid type); (6a) same, × 1.5 (6A-D, 6a; No 120 ZPAL); (7) dicotyledonous leaf palmately compound, with serrate margins, × 1.5 (No. 173/551)

Fig. 6. Impressions of *Monimiophyllum antarcticum* Zastawniak from 'Mount Flora', Fildes Peninsula, King George Island (Phot. M. Kleiberowa) (No 190 ZPAL)

cepted as Miocene (Orlando 1964), but later corrected to late Paleocene–middle Eocene (Romero 1978) and ?late Paleocene–early Eocene (Troncoso 1986). The Fildes Peninsula Group lavas range in age between 59 Ma (late Paleocene) in the south and 43 Ma (middle Eocene) in the north (Smellie *et al.* 1984, table XVIII). The plant-bearing sediments of 'Mount Flora' occur approximately in the middle of the volcanic complex, thus probably close to the Paleocene–Eocene boundary (Fig. 4). According to Czajkowski & Rösler (1986), the present assemblage greatly resembles that of the upper Paleocene Cross Valley Formation of Seymour Island (Zinsmeister 1982). According to Troncoso (1986), the Fildes Peninsula flora is most similar to the Río Pichileufu flora from Argentina (Berry 1938), attributed to the early Eocene by Romero (1986).

Middle Eocene–early Oligocene plant assemblages

Potter Peninsula floras. (Fig. 2, Loc. 9– 11). Rich plant assemblages collected mainly from morainic blocks (Fig. 2, Loc. 11), and partly from volcaniclastic sediments in situ (Fig. 2, Loc. 9 & 10) at Potter Peninsula, Maxwell Bay, are yet to be elaborated. Radiometric dating of associated volcanics indicates a middle Eocene age for these rocks which probably correlate with the Arctowski Cove Formation of Admiralty Bay (Birkenmajer 1985, p. 12, fig. 10).

Two late Eocene–early Oligocene plant assemblages have been determined, and their age well established on K-Ar dating of associated lavas: the Petrified Forest Creek and the Cytadela floras.

Petrified Forest Creek flora. (Fig. 2, Loc. 6). Fresh-water clays of the Petrified Forest Member (Arctowski Cove Formation, Ezcurra Inlet Group) yielded a pollen-spore *Nothofagus*-pteridophyte assemblage. Nine types of *Nothofagus* pollen, including the three Recent groups, *fusca*, *brassi* and *menziesii*, have been recognized, the most common being the *fusca* group. Numerous and diversified Pteridophyta are represented by sporomorphs of Cyathaceae, Dennstaedtiaceae, Gleicheniaceae, Hymenophyllaceae, Polypodiaceae, Salviniaceae and Schizeaceae, as well as several taxa *incertae sedis* (Stuchlik 1981).

Petrified wood fragments showing annual growth-rings which occur within the Petrified Forest Member have been attributed to the genus *Araucaria* and forms intermediate between *Fagus* and *Nothofagus* (Cortemiglia *et al.* 1981).

The base of the Arctowski Cove Formation (Rakusa Point Member; basaltic andesite lavas) has been K-Ar-dated at 66.7 ± 1.5 Ma (Maastrichtian; Birkenmajer *et al.* 1983b), and the base of the succeeding Point Thomas Formation (high-Al basaltic lavas) at 37.4 ± 1.1 Ma (Eocene–Oligocene boundary; Birkenmajer *et al.* 1986a). According to Stuchlik (1981), the stratigraphic age of the pollen–spore assemblage should not be older than late Eocene–early Oligocene. An Eocene–Oligocene boundary age has therefore been accepted for the Petrified Forest Member flora (Fig. 4).

Cytadela flora. (Fig. 2, Loc. 7). A tuff intercalation within high-Al basaltic lavas of the Point Thomas Formation (Ezcurra Inlet Group) at Cytadela, Ezcurra Inlet, yielded infrequent leaf imprints of various ferns, including *Blechnum* sp., dicotyledonous plants of *Nothofagus*-type and pinnately-veined leaves of other types, and probably also conifer remains (Podocarpaceae ?).

A radiometric K-Ar age of 37.4 ± 1.1 Ma obtained from the basal basaltic lavas of the Point Thomas Formation (some 150 m below the Cytadela plant beds) indicates the Eocene–Oligocene boundary. As based on other radiometric dates from the same formation (see Birkenmajer 1985; in press *a*; Birkenmajer *et al.* 1986a), the age of the formation corresponds to early Oligocene (Fig. 4).

Oligocene–Miocene boundary plant assemblages

Point Hennequin floras. (Fig. 2, Loc. 15–17). Three sites with fossil plant remains have been studied in the vicinity of Point Hennequin, all belonging to the Mount Wawel Formation of the Point Hennequin Group: the Dragon Glacier moraine site (Fig. 2, Loc. 15); the Wanda Glacier moraine site (Fig. 2, Loc. 17); and the Mount Wawel site (sediment exposure, Fig. 2, Loc. 16) (Birkenmajer 1981; Zastawniak 1981; Zastawniak *et al.* 1985). These sites correspond to three successive plant-bearing sediment intercalations within an upper part of the Mount Wawel Formation which consists mainly of andesite lavas. These lavas yielded a 24.5 ± 0.5 Ma K-Ar date (Birkenmajer *et al.* 1983b), indicating an Oligocene–Miocene boundary age of the fossil flora.

The Dragon Glacier (Fig. 7) and the Mount Wawel floras represent the *Nothofagus-*

Fig. 7. Plant impressions from the Point Hennequin Group, Dragon Glacier moraine, Admiralty Bay, King George Island (phot. M. Kleiberowa). All figures natural size, except when stated
(**1**) dicotyledonous pinnately compound leaf with entire margins (No 173/170); (**2**) Coniferae indet. (Podocarpaceae or Araucariaceae; No 173/115), × 2; (**3**) cladodes (leaf-like branches) of *Phyllocladus* sp. (Phyllocladaceae, Coniferae; No 173/133), × 2; (**4**) dicotyledonous leaf with serrate margin (No 173/44); (**5, 6**) *Nothofagus subferruginea* (Dusén) Tanai (specimens not numbered)

Podocarpaceae assemblage. The remains of podocarps, comparable with the Recent genera *Dacrydium*, *Dacrycarpus*, *Podocarpus* and *Stachycarpus*, and leaf-like branches of *Phyllocladus* sp. (Fig. 7(3)), are associated with numerous leaf impressions belonging to several species of the genus *Nothofagus* and infrequent leaves of other dicotyledonous plants; e.g. *Cupania grosse-serrata* (Engelhardt) Berry, *Cochlospermum* sp., *Dicotylophyllum* sp. div. (Zastawniak 1981; Zastawniak *et al.* 1985).

The youngest plant remains in the Tertiary succession of King George Island are represented by wood (driftwood) fragments found in the lower Miocene marine Destruction Bay Formation which underlies glacio-marine strata of the Cape Melville Formation at Wrona Buttress near Cape Melville (Fig. 2, Loc. 18). The lower Miocene age of the enclosing sediment is well established on both micro- and macrofauna (Birkenmajer 1987; Birkenmajer & Luczkowska 1987) and on K-Ar dating of associated volcanics (Birkenmajer *et al.* 1985; in press).

Palaeoclimatic significance of fossil plant communities

Palaeoclimatic aspects of the Late Cretaceous and Tertiary epochs in the Antarctic Peninsula sector, based on available palaeobotanical, palaeozoological, sedimentological and pedological evidence from terrestrial environments, and on fauna of shallow-marine environments of King George Island and Seymour Island, have been discussed by Birkenmajer (1985). The conclusion was reached that the Late Cretaceous climate of this sector of West Antarctica was generally warm and differentiated into summer and winter seasons; snow or ice caps appeared only on the tops of higher volcanoes, especially close to the Cretaceous−Tertiary boundary. The Paleogene climate of this sector was warm to temperate and also differentiated into summer and winter seasons, favouring local accumulation of coal beds but with areally restricted river delta systems; snow or ice caps occurred mainly on the tops of higher volcanoes.

Further radiometric and palaeontological dating of Tertiary strata of King George Island (e.g. Birkenmajer & Gaździcki 1986; Birkenmajer & Luczkowska 1987; Birkenmajer *et al.* 1985, 1986a, in press) formed the basis for constructing a climatostratigraphic scale of recurrent warm-phase (interglacial-type) and cold-phase (glacial) events for the Eocene through early Miocene time span. Four glaci-

ations, separated by three interglacials, have been distinguished (Birkenmajer in press *a*, *b*; Fig. 4). This had a bearing on the interpretation of the vegetational history of the Antarctic Peninsula sector from Late Cretaceous through early Miocene time. Discontinuities caused by retreat, decimation or disappearance of vegetation cover during glaciations alternated with the return of plants from refugia during amelioration of climate (interglacials). The best known fossil plant assemblages have been selected to serve the purpose of the palaeoclimatic review given below.

Late Cretaceous−Paleocene warm phase

The best known and considerably well dated fossil plant assemblage from Zamek (Fig. 2, Loc. 5) is taken as an example of a Late Cretaceous plant community. Analysis of leaf physiognomy indicates that this community was represented by broad-leaved forests, possibly partly evergreen as indicated by imprints of thick, coriaceous leaves. Plants with *Nothofagus*-type leaves represent a distinct deciduous element. Low values of a leaf-size index, expressed as predominance of microphyll and notophyll leaf-size classes, along with a large proportion of coriaceous leaves, suggest a subhumid mesothermal climate (cf. Wolfe & Upchurch 1987).

The Dufayel Island plant community (Late Cretaceous or Paleocene) resembles Recent broad-leaved forests of temperate South America with requirements for mean annual temperature of 10−12°C and precipitation of 1000−4000 mm (cf. Hueck & Seibert 1972).

The leaf-size index of the Fildes Peninsula plant community is similar to that of Zamek; however, the share of coriaceous leaves is higher in the former. The Fildes Peninsula flora represents broad-leaved evergreen forests with an admixture of conifers which grew in drier climatic conditions than broad-leaved evergreen and deciduous forests of the Zamek palaeoassemblage.

Early Eocene cold phase

K-Ar dating of basaltic lavas (49.4 ± 5 Ma) capping basaltic hyaloclastites which pass downward into glacio-marine sediments at Magda Nunatak, King George Island (Birkenmajer *et al.* 1986a), and age-ranges of calcareous nannoplankton contained in the matrix of hyaloclastite, suggest an early Eocene age of this glacial event, termed the Kraków Glaciation. This cold phase may have been expressed in the Seymour Island sequence as a sedimentary hiatus and erosion surface between the terrestrial plant-bearing non-glacial Paleocene Cross Valley Formation and the marine fossil-bearing non-glacially controlled middle Eocene−?lower Oligocene La Meseta Formation (Birkenmajer in press *a, b*).

Middle Eocene−early Oligocene warm phase

A warm climatic phase about 18 Ma long, lasting from the middle Eocene through early Oligocene, and characterized by several well developed terrestrial plant communities, has been distinguished as the Arctowski Interglacial (Birkenmajer in press *b*). Two fossil floras from King George Island may be cited as representatives of the younger part of this climatic phase: the Petrified Forest Creek flora and the Cytadela flora.

The spore-pollen spectrum from the Petrified Forest Member, dated at the Eocene−Oligocene boundary, suggests the presence of *Nothofagus* forests with well developed undergrowth in which ferns, including probably also tree ferns, played the most important role. The abundance of pteridophyte spores and their high frequencies in the pollen-spore diagram generally indicate moist climatic conditions comparable to the present-day frost-free Auckland Province lowlands (Stuchlik 1981).

The early Oligocene plant community of the Cytadela flora also points to broad-leaved *Nothofagus* forests with ferns in the undergrowth, and possibly also fern communities similar to those of the Petrified Forest Creek flora. The plant communities discussed may be compared with Recent fern bush communities of the southern oceanic islands, with mean annual temperatures between 11.7 and 15°C and mean precipitation between 3225 and 1220 mm for Gough and Auckland islands, respectively (cf. Wace 1960; Walter *et al.* 1975).

Late Oligocene cold and warm phases

Two glaciations, the older Polonez Glaciation (Antarctic-wide), and the younger Legru Glaciation (local, in the South Shetland Islands), separated by the Wesele Interglacial, span the late Oligocene from about 32 Ma to about 25 Ma BP (Birkenmajer in press *a, b*) (Fig. 4). Glacio-marine deposits of the Polonez Glaciation yielded rich shallow-marine fauna and stratigraphically valuable calcareous nannoplankton; the succeeding freshwater deposits of the Wesele Interglacial, and glacial deposits

of the Legru Glaciation are barren of fossils. There is no evidence for any vegetation cover on King George Island during this time.

Oligocene−Miocene boundary warm phase

The plant remains of the Point Hennequin Group (Mount Wawel Formation), dated at the Oligocene−Miocene boundary, represent *Nothofagus*-Podocarpaceae forest communities. Similar plants grow today in evergreen or deciduous forests of southern lands, particularly in southern South America and New Zealand (Zastawniak 1981; Zastawniak *et al.* 1985). Climatic requirements for the Point Hennequin floras could be similar to those of the Recent Patagonian-Magellanian forests: mean annual temperature 5−8°C and precipitation of 600−4300 mm (cf. Hueck & Seibert 1972).

Attribution of the Point Hennequin plant-bearing beds to the Wawel Interglacial (Birkenmajer in press *a, b*), which succeeded a long cold period of two glaciations separated by a probably relatively short interglacial (Fig. 4), explains strong differences in composition of these floras with respect to older ones. At about 25 Ma BP, during the Wawel Interglacial, trees of *Nothofagus*, Podocarpaceae and some other plants (including *Cochlospermum*) recolonized West Antarctica, probably using an island-arc bridge linking Tierra del Fuego with the Antarctic Peninsula at the initial stage of opening of the Drake Passage. It is of interest to note that fossil floras resembling those of Point Hennequin, and equally rich in *Nothofagus* leaves, have been described from Río de las Minas near Puntas Arenas by Dusén (1907). Tanai (1986) considered this assemblage to be of Oligocene−early Miocene age.

Dicotyledonous leaves of the Wawel Interglacial are visibly smaller than before, and there is no trace of laurophyllous plants and such leaves as *Sterculia*-type. These plants did not survive the cold period of the Polonez and Legru glaciations.

The warm phase of the Wawel Interglacial lasted for about 3−3.5 Ma. Its terminal stage is witnessed by non-glacial shallow−marine Destruction Bay Formation deposits with pieces of wood (driftwood). These deposits underlie glacio-marine strata of the Cape Melville Formation dated at 22−20 Ma BP, which comprise evidence for the early Miocene Melville Glaciation in West Antarctica (Birkenmajer 1984, 1987, in press *a, b*). No middle or late Miocene and Pliocene plant-bearing strata have so far been found in the South Shetland Islands.

References

BARTON, C. M. 1964. Significance of the Tertiary fossil floras of King George Island, South Shetland Islands. *In*: ADIE, R. J. (ed.) *Antarctic Geology*, North-Holland Publishing Company, Amsterdam, 603−609.

BERRY, E. W. 1938. Tertiary floras from the Río Pichileufu, Argentina. *Geological Society of America, Special Paper*, **12**, 1−140.

BIRKENMAJER, K. 1980. Tertiary volcanic-sedimentary succession at Admiralty Bay, King George Island (South Shetland Islands, Antarctica). *Studia Geologica Polonica*, **64**, 7−65.

—— 1981. Lithostratigraphy of the Point Hennequin Group (Miocene volcanics and sediments) at King George Island (South Shetland Islands, Antarctica). *Studia Geologica Polonica*, **72**, 59−73.

—— 1982. Mesozoic stratiform volcanic-sedimentary succession and Andean intrusions at Admiralty Bay, King George Island (South Shetland Islands Antarctica). *Studia Geologica Polonica*, **74**, 105−154.

—— 1984. Geology of the Cape Melville area, King George Island (South Shetland Islands, Antarctica): pre-Pliocene glaciomarine deposits and their substratum. *Studia Geologica Polonica*, **79**, 7−36.

—— 1985. Onset of Tertiary continental glaciation in the Antarctic Peninsula sector (West Antarctica). *Acta Geologica Polonica*, **35**, 1−31.

—— 1987. Oligocene−Miocene glacio-marine sequences of King George Island (South Shetland Islands), Antarctica. *Palaeontologia Polonica*, **49**, 9−36.

—— in press a. Tertiary glaciation in the South Shetland Islands, West Antarctica: evolution of data. *In*: THOMSON, M. R. A., CRAME, J. A. & THOMSON, J. W. (eds) *The geological evolution of Antarctica*. Cambridge University Press, Cambridge.

—— in press b. Tertiary glacial and interglacial deposits, South Shetland Islands, Antarctica: geochronology versus biostratigraphy (a progress report). *Bulletin of the Polish Academy of Sciences, Earth Sciences*.

—— & GAŹDZICKI, A. 1986. Oligocene age of the *Pecten* conglomerate on King George Island, West Antarctica. *Bulletin of the Polish Academy of Sciences, Earth Sciences*, **34**, 219−26.

—— & LUCZKOWSKA, E. 1987. Foraminiferal evidence for a Lower Miocene age of glaciomarine and related strata, Moby Dick Group, King George Island (South Shetland Islands, Antarctica). *Studia Geologica Polonica*, **90**, 81−123.

—— & ZASTAWNIAK, E., 1986. Plant remains of the Dufayel Island Group (Early Tertiary ?), King George Island, South Shetland Islands (West Antarctica). *Acta Palaeobotanica*, **26**, 33−54.

—— & ——, in press. Late Cretaceous−Early Neogene vegetation history of the Antarctic Peninsula sector, Gondwana break-up and Ter-

tiary glaciations. *Bulletin of the Polish Academy of Sciences, Earth Sciences*.

—— NARĘBSKI, W., NICOLETTI, M. −PETRUCCIANI, C. 1983a. K-Ar ages of 'Jurassic volcanics' and 'Andean' intrusions of King George Island, South Shetland Islands (West Antarctica). *Bulletin de l'Académie Polonaise des Sciences, Sciences de la Terre*, **30**, 121−31.

——, ——, ——, ——, & —— 1983b. Late Cretaceous through Late Oligocene K-Ar ages of the King George Island Supergroup volcanics, South Shetland Islands (West Antarctica). *Bulletin de l'Académie Polonaise des Sciences, Sciences de la Terre*, **30**, 133−143.

—— GAŹDZICKI, A., KREUZER, H. & MÜLLER, P. 1985. K-Ar dating of the Melville Glaciation (Early Miocene) in West Antarctica. *Bulletin de l'Académie Polonaise des Sciences, Sciences de la Terre*, **33**, 15−23.

——, DELITALA, M. C., NARĘBSKI, W., NICOLETTI, M. & PETRUCCIANI, C. 1986a. Geochronology of Tertiary island-arc volcanics and glacigenic deposits, King George Island, South Shetland Islands (West Antarctica). *Bulletin of the Polish Academy of Sciences, Earth Sciences*, **34**, 257−272.

——, KAISER, G., NARĘBSKI, W., PILOT, J. & RÖSLER, H. J. 1986b. The age of magmatic complexes of the Barton Horst, King George Island (South Shetland Islands, West Antarctica), by K-Ar dating. *Bulletin of the Polish Academy of Sciences, Earth Sciences*, **34**, 139−155.

——, SOLIANI, E., Jr. & KAWASHITA, K., in press. Early Miocene K-Ar age of volcanic basement of the Melville Glaciation deposits, King George Island, West Antarctica. *Bulletin of the Polish Academy of Sciences, Earth Sciences*.

CORTEMIGLIA, G. C., GASTALDO, P. & TERRANOVA, R. 1981. Studio di piante fossili trovate nella King George Island delle Isole Shetland del Sud (Antartide). *Atti Società Italiana Sci. nat. Museo civ. Stor. nat. Milano*, **122**, 37−61.

CZAJKOWSKI, S. & RÖSLER, O. 1986. Plantas fósseis da Península Fildes; Ilha Rei Jorge (Shetlands do Sul): morfografia das impressões foliares. *Anais da Academia Brasileira de Ciências*, **58**, (Supplemento), 99−110.

DEL VALLE, R. A., DÍAZ, M. T. & ROMERO, E. J. 1984. Preliminary report on the sedimentites of Barton Peninsula, 25 de Mayo Island (King George Island), South Shetland Islands, Argentine Antarctica. *Instituto Antártico, Argentino Contribution*, **308**, 1−19.

DUSÉN, P. 1907. Über die Tertiäre Flora der Magellansländer. *Wissenschaftliche Ergebnisse der Schwedischen Expedition nach den Magellansländern 1895−1897*, **1**, (4), 87−107, **1** (5), 241−8.

HUECK, K. & SEIBERT, P. 1972. Vegetationskarte von Südamerika. *In*: WALTER, H. (ed.) *Vegetations-monographien der einzelnen Gross-*

räume. G. Fischer Verlag, Stuttgart.

JAGMIN, N. I. B. 1987. Estudo anatômico dos troncos fósseis de Admiralty Bay, King George Island (Peninsula Antártica). *Acta Biologica Leopoldensia*, **9**, 81–98.

LYRA, C. S. 1986. Palinologia de sedimentos terciários da Península Fildes, Ilha Rei George (Ilhas Shetland do Sul, Antártica) e algumas considerações paleoambientais. *Anais da Academia Brasileira de Ciências*, **58** (Suplemento), 137–47.

ORLANDO, H. A. 1964. The fossil flora of the surroundings of Ardley Peninsula (Ardley Island), 25 de Mayo Island (King George Island), South Shetland Islands. *In*: ADIE, R. J. (ed.) *Antarctic Geology*, North-Holland Publishing Company, Amsterdam, 629–36.

PAULO, A. & TOKARSKI, A. K. 1982. Geology of the Turret Point–Three Sisters Point area, King George Island (South Shetland Islands, Antarctica). *Studia Geologica Polonica*, **74**, 81–103.

ROMERO, E. J. 1978. Paleoecológia y paleofitografia de las tafofloras del Cenafitico de Argentina y áreas vecinas. *Ameghiniana*, **15**, 209–227.

—— 1986. Paleogene phytogeography and climatology of South America. *Annals of the Missouri Botanical Garden*, **73**, 449–461.

SMELLIE, J. L., PANKHURST, R. J., THOMSON, M. R. A. & DAVIES, R. E. S. 1984. The geology of the South Shetland Islands: VI. Stratigraphy, geochemistry and evolution. *British Antarctic Survey Scientific Reports*, **87**, 1–85.

STUCHLIK, 1981. Tertiary pollen spectra from the Ezcurra Inlet Group of Admiralty Bay, King George Island (South Shetland Islands, Antarctica). *Studia Geologica Polonica*, **72**, 109–132.

TANAI, T. 1986. Phytogeographic and phylogenetic history of the genus *Nothofagus* Bl. (Fagaceae) in the Southern Hemisphere. *Journal of the Faculty of Science, Hokkaido University, Series IV: Geology and Mineralogy*, **21**, 505–582.

TOKARSKI, A. K., DANOWSKI, W. & ZASTAWNIAK, E. 1987. On the age of fossil flora from Barton Peninsula, King George Island, West Antarctica. *Polish Polar Research*, **8**, 293–302.

TRONCOSO, A. A. 1986. Nuevas órgano-especies en la tafoflora terciaria inferior de Península Fildes, Isla Rei Jorge, Antárctica. *Instituto Antártico Chileno, Seria Cientifica* **34**, 23–46.

WACE, N. M. 1960. The botany of the southern oceanic islands. *Proceedings of the Royal Society of London*, **B 152**, 475–90.

WALTER, H., HARNICKELL, E. & MUELLER-DOMBOIS, D. 1975. *Klimadiagramm-Karten der einzelnen Kontinente und der ökologische Klimagliederung der Erde.* G. Fischer Verlag, Stuttgart.

WOLFE, J. A. & UPCHURCH, G. R., Jr. 1987. North American nonmarine climates and vegetation during the Late Cretaceous. *Palaeogeography, Palaeoclimatology, Palaeoecology*, **61**, 33–77.

ZASTAWNIAK, E. 1981. Tertiary leaf flora from the Point Hennequin Group of King George Island (South Shetland Islands, Antarctica). Preliminary report. *Studia Geologica Polonica*, **72**, 97–108.

——, WRONA, R., GAŹDZICKI, A. & BIRKENMAJER, K. 1985. Plant remains from the top part of the Point Hennequin Group (Upper Oligocene), King George Island (South Shetland Islands, Antarctica). *Studia Geologica Polonica*, **81**, 143–164.

ZINSMEISTER, W. J. 1982. Review of the Upper Cretaceous–Lower Tertiary sequence on Seymour Island, Antarctica. *Journal of the Geological Society, London*, **139**, 779–786.

Evolution of the Antarctic fish fauna with emphasis on the Recent notothenioids

J. T. EASTMAN[1] & L. GRANDE[2]

[1] *Department of Zoological and Biomedical Sciences and College of Osteopathic Medicine, Ohio University, Athens, Ohio 45701−2979, USA* [2] *Department of Geology, Field Museum of Natural History, Roosevelt Road at Lake Shore Drive, Chicago, Illinois 60605−2496, USA*

Abstract: The composition of the Antarctic fish fauna has undergone remarkable changes through time. Fossil fishes are known from Devonian, Jurassic, Cretaceous and early Tertiary deposits of the region. The Recent fauna does not appear to be derived from any part of any of the known fossil faunas. The Devonian fish fauna includes agnathans, placoderms, acanthodians, chondrichthyans and osteichthyans all belonging to families now extinct. The Jurassic fauna is known by only one species, a neopterygian of the now extinct family Archaeomaenidae. Both the Palaeozoic and early Mesozoic fish faunas known from Antarctica indicate Australian biogeographic affinities based on other (*i.e.*, non-Antarctic) known Palaeozoic and Mesozoic fish faunas. Late Cretaceous and early Tertiary species from the Antarctic region (Seymour Island) belong largely to families of fishes that today live in other regions of the world, but are extinct in the Antarctic region. The Recent (extant) fish fauna is drastically different from any of the fossil faunas and is dominated by perciforms of the suborder Notothenioidei. To date the fossil record has provided no fossils identifiable as notothenioids, or even species closely related to the group. It is unlikely that low water temperatures were directly responsible for the local extinction of the Tertiary Seymour Island ichthyofauna or for the lack of diversity in the Recent fauna. The decline of suitable substrate and trophic factors may have been more important in changing the composition of the fish fauna. There is considerable morphological, physiological and ecological diversification within the Notothenioidei. Although the emergence of this group was probably not a direct response to cooling, the subsequent radiation of notothenioids was associated with a variety of specializations related to low water temperature.

The Recent Antarctic fish fauna is neither speciose nor diverse, unlike the situation in the shelf waters of other southern continents. Most species belong to the highly endemic perciform suborder Notothenioidei that dominates Antarctic waters. For decades ichthyologists surmised that the notothenioid stock had been isolated on the continental shelf of Antarctica for most of the Tertiary; and that the fauna had evolved in situ and in isolation, adapting gradually to cooling conditions in the Southern Ocean (Regan 1914; Norman 1938; DeWitt 1971). The only modification of this scenario has been recognition of a possible role for tectonic movements in notothenioid evolution (Andersen 1984; Miller 1987).

There is unfortunately no Antarctic fossil record known for the Notothenioidei. There are taxonomically diverse fossil fishes from Antarctic deposits formed about the time the notothenioid stock probably emerged (late Cretaceous or early Tertiary). These fishes are not closely related to notothenioids, and are also unlike any other component of the Recent fauna. In addition, the ichthyofauna shows an Australian biogeographic affinity during the Palaeozoic and early Mesozoic that is not evident in the Cenozoic and Recent faunas. This paper is a primarily theoretical consideration of the evolution of notothenioids. While this will be our focus, we also have broader goals, including (1) briefly reviewing the Antarctic fish fauna through the Phanerozoic, (2) considering the importance of temperature and other factors in producing faunal change and (3) discussing the diversification of notothenioids.

Palaeontological history of the Antarctic fish fauna through the Phanerozoic

In their review of the Antarctic fossil fish fauna, Grande & Eastman (1986) indicated that the biogeographical significance of the fauna is variable through geological time. Their conclusions are briefly summarized below and updated with recent additions to the fauna.

From Crame, J. A. (ed.), 1989, *Origins and Evolution of the Antarctic Biota*, Geological Society Special Publication No. 47, pp. 241−252.

241

Palaeozoic

All Palaeozoic fishes from Antarctica are in Devonian rocks. These fishes inhabited fresh-water ecosystems that were free of ice (Tasch 1977). Several diverse groups are represented: the thelodont agnathan *Turinia* sp., placoderms, acanthodians, xenacanthid elasmobranchs, osteolepid crossopterygians and palaeonisciform actinopterygians. Most of these specimens are too fragmentary for use in phylogenetic analysis. The placoderm *Antarctaspis mcmurdoensis* and the xenacanthid *Antarctilamna prisca* reflect an Antarctic–Australian biogeographical relationship during the Devonian, a period predating the breakup of Gondwana.

In evaluating all vertebrate faunas from Palaeozoic Gondwana, Young (1987) refined biogeographic relationships for fishes. He recognizes strong biogeographic affinities between faunas in southeastern Australia and Victoria Land, Antarctica, and between faunas in South Africa–South America and the Ohio Range, Transantarctic Mountains, Antarctica. Young (1989) also indicates that, as the most persistent Palaeozoic continental region, initial colonisation of non-marine aquatic environments took place in Gondwana.

Mesozoic

Lower Jurassic (179–161 Ma) fishes consist of several complete individuals of the osteichthyan pholidophoriform *Oreochima ellioti* from lacustrine interbeds in the Transantarctic Mountains, Victoria Land (Schaeffer 1972). A member of the freshwater family Archaeomaenidae, *O. ellioti* also has Australian affinities as all other species in this family are confined to Australia.

Seymour Island, near the tip of the Antarctic Peninsula (Fig. 1), has been an exceptionally rich locality for late Cretaceous fishes. Late Cretaceous material consists largely of fragmentary elasmobranch teeth and undetermined vertebral centra. Based on teeth, Grande & Eastman (1986) recognized the lamnid shark *Isurus* sp., and Grande & Chatterjee (1987) reported hexanchiform sharks including *Notidanodon* sp. and possibly *Sphenodus* sp.

A specimen from the late Cretaceous of Seymour Island also provided Grande & Chatterjee (1987) with material for a description of the oldest teleost from Antarctica. This species, *Antarctiberyx seymouri*, is a member of the beryciform family Trachichthyidae, an extant group with a wide marine distribution. Although modern representatives of this group are restricted to deep-sea habitats, late Cretaceous beryciform fishes may have been present in inshore waters — the ecological equivalents of modern Perciformes (Moyle & Cech 1988).

The late Cretaceous was an important time in the evolutionary history of fishes with teleosts radiating during this time. Unlike the situation hypothesized for some invertebrate groups from Seymour Island (Zinsmeister & Feldmann 1984), we see no evidence suggesting that Seymour Island served as a centre of origin and dispersal for fishes that later occupied deep and shallow water ecosystems at lower latitudes.

Cenozoic

All known Antarctic Cenozoic fishes are from Seymour Island. Most fossils are confined to the late Eocene or early Oligocene La Meseta Formation, a relatively shallow water beach (Woodburne & Zinsmeister 1984) or deltaic (Pezzetti & Krissek 1986) deposit *c.* 40 Ma. Included in this material are a diverse array of chondrichthyans including sharks, saw sharks, rays and ratfish. The sharks were cosmopolitan in temperate oceans during the mid-Eocene to early Oligocene. A ratfish, mentioned by Grande & Eastman (1986) as a new addition to the Antarctic fauna, is being described by Ward and Grande.

Teleosts are represented in the La Meseta Formation by a siluriform pectoral spine, by a variety of fragmentary and unidentifiable vertebral centra and by unidentifiable jaw bones from large and small individuals. Thus the Cenozoic material contributes little biogeographic information as the represented taxa have wide distributions outside Antarctica.

Woodward (1908) described isolated vertebral centra from Eocene deposits on Seymour Island. He attributed these to the Nototheniidae; however his identification is probably not accurate. Centra are not diagnostic for most nototheniids, and furthermore the centra described by Woodward are indistinguishable from those of a majority of teleosts. Stinton's (1957) report of *Notothenia* from the mid–late Miocene of New Zealand is also a misidentification (Fordyce 1982). Thus while over 70 distinct perciform families are recognizable in Eocene deposits around the world (60 reported by Romer 1966, and several others since then in various publications), fossil notothenioids have not yet been discovered.

Grande & Chatterjee (1987) indicated that artifactual preservation is probably responsible for the greater diversity of chondrichthyans relative to osteichthyans in the late Cretaceous/early Tertiary Seymour Island fauna. Most of

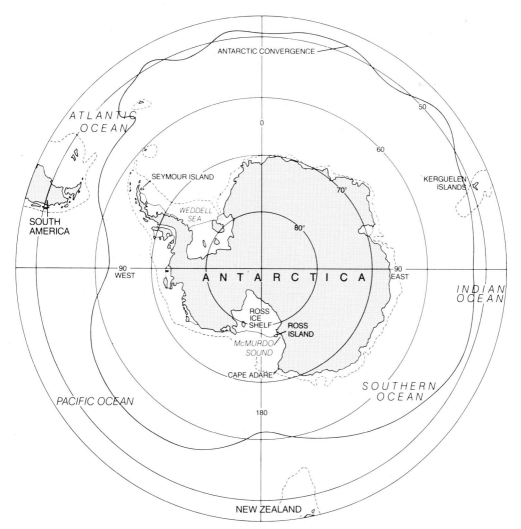

Fig. 1. Antarctica and the Southern Ocean showing Seymour Island near the tip of the Antarctic Peninsula. Limited shallow water habitat is indicated by the dashed line of the 1000 m isobath. Heavy line is the Antarctic Convergence, a natural zoogeographic boundary for some Recent species. Base map from De Witt (1971) and location of the Convergence from Hedgpeth (1969). From Eastman & DeVries (1986a), copyright © 1986 by Scientific American, Inc. All rights reserved.

the fish fossils known from Seymour Island are preserved as isolated teeth, vertebrae, spines and other fragments. Isolated teeth are diagnostic for both fossil and Recent chondrichthyans, but not for most teleosts. Therefore the teleost fauna appears less diverse as isolated teeth (and other isolated bone bits) are assigned to indeterminate species.

Recent

In summarizing information concerning the distribution and endemism of the Recent Antarctic fish fauna, DeWitt (1971) notes that the fauna comprises 120 species including about 80 species of the highly endemic perciform suborder Notothenioidei. Since 1971 many new taxa have been described; there are now 203 species (including about 110 notothenioids) representing 28 families in the Antarctic region (Andriashev 1987). Andriashev mentions that discovery of many new non-notothenioid species has reduced the dominance of notothenioids; however they still constitute 53% of the species in the Antarctic region. He notes that endemism is also high within the Notothenioidei; 97% for species and

244   J.T. EASTMAN & L. GRANDE

85% for genera.

The Recent Antarctic fauna is therefore less diverse than might be expected given the considerable age (discussed below) and large size of the marine ecosystem. This fauna is also markedly different in composition from the fauna that preceded it in geologic time. The Tertiary marine fauna of Australia, for example, is similar to the Recent fish fauna (Long 1982). This is not true in the Antarctic as formerly diverse and abundant groups such as the Chondrichthyes are represented today by only a few species of rajids. Moreover, the Recent fauna is unusual in that it is dominated by a single perciform suborder, the Notothenioidei.

Taxonomy and systematics of notothenioids

The classifications of Regan (1914) and Norman (1937, 1938) persisted largely intact until recently. Today the systematic relationships of notothenioids are areas of active research (Eakin 1981; Andersen 1984; Balushkin 1984; Iwami 1985). Monophyly had been assumed but not proven until Iwami (1985) used cladistic methodology to evaluate phylogenetic relationships among notothenioids (Fig. 2). The notothenioid sister group has not been definitely identified among the Perciformes, although blennioids are generally thought (Gosline 1968; Eakin 1981) to be closely related to them, and a likely candidate. No unambiguous character evidence has been presented to clearly establish blennioids or any other perciform subgroup as the sister group. Eakin (1981) presents a good discussion of the problems encountered in separating convergences from true relationships and in searching for a notothenioid sister group.

Cenozoic geology and palaeoclimatology

Breakup of Gondwana

The breakup of Gondwana and the subsequent development of cold ocean currents around Antarctica certainly influenced the composition of the Antarctic fish fauna. These events also provide a time frame for considering the evolution of notothenioids. Key elements may be briefly summarized as follows (Craddock 1982; Woodburne & Zinsmeister 1984; Zinsmeister 1987). Antarctica, at the centre of the Gondwana landmass, attained a south polar position during the Cretaceous. During the late Cretaceous, South America, Antarctica, New

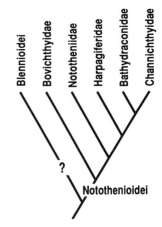

Fig. 2. Cladogram of notothenioids modified from Iwami (1985). Use of blennioids as the sister group is from Gosline (1968) and Eakin (1981). Iwami (1985) provides character information supporting the cladogram.

Zealand and Australia were still continuous, although there were prominent breaks between all crustal blocks. Separation of Australia from East Antarctica may have begun by the late Jurassic, with the development of deep-sea conditions in this narrow trough c. 80 Ma. Final separation took place between late Eocene and early Oligocene (38 Ma). Separation of West Antarctica from South America occurred during the early Tertiary, however deep-sea conditions have prevailed in the Drake Passage only since the Oligocene/Miocene boundary 23.5 Ma. Once Antarctica was fully isolated by seafloor spreading, the unrestricted circum-Antarctic current began to reach full development. By decoupling warm subtropical gyres from the continent, this current served as a barrier to heat flow and thermally isolated Antarctica. Thereafter glaciers developed, the polar ice cap began to form and the Antarctic Convergence formed and expanded northward (Kennett 1978, 1980). The most recent data suggest that the ice cap may have initially appeared 30 Ma, disappeared, and then reformed 13 Ma to persist to the present (Kerr 1984, 1987).

Antarctic Convergence

The Antarctic Convergence, located between 50° and 60°S (Fig. 1), represents an approximate northern boundary for the Antarctic Ocean. Here cold, north flowing Antarctic surface water meets and sinks beneath warmer, less dense Subantarctic water. While there are few physical barriers to migration and dispersal in marine fishes, the Convergence delimits, at least for the Recent fauna, a natural biogeographic province. The Convergence is characterized by a 3° to 4° change in water temperature, as well as by changes in a number of other oceanographic parameters. This region of abrupt thermal change has had a marked effect on the shallow water fauna by preventing southern migration and colonization of Antarctic waters by most pelagic fishes.

Southern Ocean palaeotemperatures

Although there is not complete agreement on all details, Southern Ocean palaeotemperatures during the Cenozoic provide a basis for discussion of water temperatures in relation to the changing Antarctic fish fauna and the emergence of notothenioids. Palaeotemperature variations are inferred from the isotopic ratio of $^{18}O/^{16}O$ in the calcium carbonate of foraminiferal tests obtained from cores of the ocean bottom (Kennett 1977). The amount of ^{18}O in the tests is dependent on the temperature of seawater at the time tests were formed.

Clarke (1983) has a good summary of palaeotemperature information as it relates to the physiological adaptation of Antarctic marine organisms. He indicates that, based on isotopic data, during the late Cretaceous and Paleocene Antarctic waters were considerably warmer than they are today. A general cooling trend began in the late Paleocene/early Eocene and has continued into the Recent. During this period of c. 50 Ma, water temperatures fell from c. 15°C to less than 0°C. This trend has been interspersed with periods of warming, and there have been occasional drops in water temperature that deviated sharply from the general decline. For example, over a period of a few million years at the Eocene/Oligocene boundary, temperatures declined from 10° to 6°C. Bottom temperatures have been less than 5°C for c. 12 Ma. Shelf waters today are below 0°C and in some areas, McMurdo Sound for example, the mean annual water temperature is only −1.86°C.

Factors producing faunal change or limiting diversity

Comparison of the early Tertiary temperate fauna from Seymour Island with the Recent fauna conveys the impression of marked change in species composition. While in part artifactual, as explained previously, it is true that the Seymour Island fossil faunas have no representatives among the Recent fauna. Furthermore, the Recent fauna is not diverse considering the large size of the ecosystem. Unfortunately there is no fossil record from the late Tertiary, leaving a gap of 38 Ma between the Eocene/Oligocene and the present, and no direct knowledge of the transition fauna (from late Tertiary to Recent).

Low water temperature

It is tempting to attribute change in the fish faunal composition to low water temperature, perhaps the most obvious environmental feature of the ecosystem. It is unlikely, however, that this parameter was directly responsible for the disappearance of the Seymour Island fauna or the lack of diversity in the Recent fauna. Low temperature should not have been an insurmountable problem in the evolutionary adaptation of fishes given a decrease of 15°C over a period of c. 50 Ma. When averaged out, the change is 0.03°C per 100 000 years. Even the sharp drop of 5°C over 2 million years at the Eocene/Oligocene boundary amounts to only 0.25°C per 100 000 years. Schopf (1980, p. 246) considers a change of 1°C per 100 000 years as 'well within the adaptive capabilities for every species ever examined'.

On page one of his thought-provoking monograph, Dunbar (1968) cautions against 'preoccupation with temperature' when considering the evolution of polar ecosystems. He notes (1968, p. 56) 'that adaptation to low temperature as such presents few evolutionary difficulties and has been accomplished by thousands of species'.

Clarke (1983) has also argued convincingly that low temperature is not a limiting factor for biosynthetic processes, that evolutionary adaptation to low temperature has occurred repeatedly and that the majority of life history patterns observed in invertebrates and fishes are reflections of ecological constraints rather than low temperature.

We think that both these viewpoints are valid in considering the evolution of the Antarctic fish fauna. Low water temperature cannot singularly account for the disappearance of the Seymour Island fauna or for the paucity of non-notothenioids among the Recent fauna. Factors in the realm of ecological constraints (Clarke 1983) provide suitable alternative hypotheses relating to available substrate and to food supply.

Limited habitat

As a result of isostatic depression of Antarctica by the ice sheet, the average depth of the continental shelf is 500–900 m. This is four times greater than that of the other continents, and twice as great as that of the Arctic (Johnson *et al*. 1982). As a result of glacial erosion the shelf is also narrow. Furthermore, Antarctica lacks the extensive archipelagos characteristic of many continents. Thus with deep water close to the continental margin (Fig. 1), the prime habitat for fish diversity is limited.

Glaciation also eliminated existing riverine and estuarine habitats on the continent. These habitats have not existed since the onset of glaciation *c.* 25 Ma (Johnson *et al*. 1982). In addition, glaciation on the shelves reduced potential marine habitat for shallow water benthic species (Miller 1987).

Trophic considerations

In considering the pattern of diversification in fossil and Recent fishes, Thomson (1977) suggested that there may be a ceiling on diversity imposed largely by trophic resources. Although Antarctic waters are productive during the summer, seasonal oscillation in the food supply is marked in some areas and may have constrained the evolution of certain trophic types among fishes. There are, for example, no phytoplankton feeders or obligatory mass consumers of planktonic crustaceans. Filter feeders might not be able to tolerate the low productivity in the ecosystem during the austral winter. Since seawater has a high kinematic viscosity at low temperatures (Vogel 1981), it is also possible that the evolution of continuously swimming, filter feeding fishes was hampered by the energetic cost of these activities in subzero seawater.

Benthos is a seasonally stable resource that is theoretically available to fishes at all times of the year. On the east side of McMurdo Sound infaunal densities are among the highest in the world (Dayton & Oliver 1977). However in some areas the Antarctic benthos consists largely of sessile particle feeders or scavengers (Hedgpeth 1969; Dell 1972) that are mostly inedible (sponges, barnacles, sedentary polychaetes, sea urchins, sea stars, sea spiders and brittle stars). The taxonomic composition of the invertebrate fauna may also contribute to a reduced number of niches for fishes. Grande & Eastman (1986) suggested that the diversification of rajids in the Recent chondrichthyan fauna may have been restricted by the absence of suitable food. Molluscs, an important food group for many rajids, are poorly represented in the Antarctic fauna (Dell 1969; Dayton & Oliver 1977).

Many aspects of the trophodynamics of the Antarctic marine ecosystem are poorly understood, and the food web itself is more complex than previously realized (Clarke 1985). Another level of complexity is added in attempting to assess the influence of trophic factors on the evolution of the fish fauna. Some research, for example, indicates that the distinction between the midwater and benthic productivity may be artificial. At McMurdo Sound, near the southerly limit of marine life, Berkman *et al*. (1986) found bottom sediments containing viable algal material throughout the period of winter darkness. They suggested that such primary detritus, when resuspended by currents, could serve as a 'food source for planktonic herbivores feeding during the austral winter in nearshore environments'.

Origin of notothenioids

Nonexistent fossil record

Identification of fossil notothenioids would probably be based on only osteological evidence. Unfortunately there is not a unique osteological feature known that characterizes the suborder. A morphological diagnosis of the Notothenioidei includes (Eakin 1981): (1) three flat, platelike pectoral radials, (2) pleural ribs poorly developed and floating or absent, (3) one nostril on each side of the head, (4) nonpungent fin spines, (5) no swim bladder, (6) two or three lateral lines, occasionally one, (7) jugular pelvic fins and (8) usually fewer than 15 principal caudal rays (10–19). Individually none of these characters is unique to notothenioids and therefore as characters they are somewhat ambiguous. The group is diagnosed by a proposed unique combination of characters rather than by one or more synapomorphies. Hence it will be difficult to recognize as a notothenioid any specimen not possessing the entire suite of characters. Also some are features of the soft anatomy that will not be represented in the fossil record.

Traits essential for survival at low temperature

We have advanced the hypothesis that low water temperature was not primarily responsible for either eliminating the Seymour Island fish fauna or for constraining the diversity of the Recent fauna. Similarly the emergence of notothenioids was probably not a direct response to cooling, but the subsequent radiation of this group under cold conditions was necessarily associated with a variety of specializations related to low water temperature.

Antifreeze. The acquisition of antifreeze glycopeptides by notothenioids was absolutely essential for survival in certain habitats in the Southern Ocean. Over the past 20 years DeVries (1982) has elucidated the structure, mode of action and distribution of antifreezes in notothenioids.

Nothing is known about the evolutionary origin of antifreezes. Surface mucus may have acted as an antifreeze by preventing propagation of ice across the thin epithelial surfaces of the gills. This may have provided adequate protection before waters reached temperatures below the body's freezing point. As climatic cooling continued, more effective internally synthesized and systemically distributed antifreezes may have become necessary.

While antifreezes were probably not present (or at least not necessary for survival) in the ancestral notothenioids living in the warmer waters of the late Cretaceous/early Tertiary, the acquisition of antifreezes allowed the radiation of some notothenioids into ice-laden habitats later in the Tertiary. Not all Recent notothenioids require antifreezes, only those species living in portions of the water column colder than $-0.8°C$ and where ice is present, or species liable to encounter ice during latitudinal or vertical migrations. *Notothenia kempi*, for example, inhabits a $+1°C$ layer of water near the Balleny Islands and does not possess antifreezes (DeVries & Lin 1977). The New Zealand black cod (*Notothenia angustata*) does not have antifreeze translation products and does not synthesize antifreezes (DeVries *et al.* 1982). While cold resistance is ancestral for the suborder (Andriashev 1987), the possession of definitive antifreeze glycopeptides may be a derived condition associated with other derived conditions such as aglomerular kidneys (Eastman & DeVries 1986*b*).

We do not agree with the theory of Scott *et al.* (1986) that antifreezes first appeared in notothenioids as the stock responded to the Eocene/Oligocene cooling event at 38 Ma. Waters at this time were simply too warm ($5°-7°C$) to require the presence of definitive antifreezes in a benthic stock not likely to be exposed to sea ice. Water temperatures did not approach $0°C$ until the late Miocene, *c*. 10 Ma. As the notothenioid stock diversified, cryopelagic offshoots like *Pagothenia borchgrevinki* might have required antifreezes early in their evolutionary history as they became associated with ice. On the other hand, some benthic species, confined to ice free habitats or to Subantarctic waters, might never require antifreezes. We suspect that antifreezes evolved rapidly in various notothenioids during the past 10 Ma. They have been isolated in only 15 species to date (Ahlgren & DeVries 1984; Eastman & DeVries 1986*b*).

Enzyme adaptations. Low environmental temperatures have a rate-depressing effect on biochemical processes in poikilotherms. Notothenioids are stenothermal with some species from McMurdo Sound having upper lethal temperatures of $+6°C$ (DeVries 1977). Since upper and lower lethal temperatures cannot be raised or lowered by warm or cold acclimation, notothenioid enzyme systems are obviously specialized for function under constantly cold conditions. While far from complete, research on notothenioids to date suggests that only one set of isozymes is necessary for an unchanging thermal environment like McMurdo Sound (Clarke 1987). Clarke also indicates that notothenioid enzyme systems function at constantly low temperatures through genetic expression of enzyme variants with lower free energies of activation and by fine control of the enzyme microenvironment.

Membrane adaptations. Normal function of plasma and organelle membranes is necessary for survival of notothenioids at low temperatures. That vital functions like ionic transport and synaptic transmission occur is evidence that there has been adaptation in the membrane systems of these fishes. Conservation of the physical state of membrane lipids and proteins in poikilotherms is known as homeoviscous adaptation (Hochachka & Somero 1984). The fluidity of membranes is influenced by the ratio of saturated/unsaturated fatty acids in membrane phospholipids (Prosser 1986). Increased unsaturation of fatty acids ensures membrane fluidity at low temperatures. Work on notothenioids indicates that both sensory and motor nerves are resistant to blockade at low temperatures, and show compensatory increases in excitability and conduction velocity compared with values extrapolated from temperate fishes (Macdonald 1981; Montgomery & Macdonald 1984).

Conclusion

Biochemical specializations have predominated over gross morphological specializations in allowing notothenioids to successfully exploit this environment. Recent work on the eyes of notothenioids is a case in point (Eastman 1988*b*).

McMurdo Sound notothenioids are subject to a unique photic regime. Four month periods of continual darkness in the winter and continual light in the summer are separated by two month transition periods. However, there is no obvious correlation between the unusual light conditions in McMurdo Sound and ocular morphology among ecologically diverse notothenioids. The key evolutionary adaptations for visual function in this habitat are biochemical. Antifreeze glycopeptides prevent freezing of ocular fluids, while homeoviscous adaptations allow normal cellular function, including synaptic transmission in the retina and eye muscle contraction, at subzero temperatures. Apparently light penetration in McMurdo Sound is sufficient to permit ocular function in cold adapted, but morphologically unspecialized, eyes.

Diversification of notothenioids

The Antarctic continental shelf is 400–500 m deep at the edge and also contains innershelf depressions 1000 m deep (Andriashev 1965). When compared to the depth distribution of temperate fishes, species diversity among Recent notothenioids is greatest at 300–600 m rather than 100–200 m (Andriashev 1965, 1987; DeWitt 1971). This pattern of distribution, known as glacial submergence, may be attributable to destruction of bottom habitat by continental glaciers and ice shelves (Andriashev 1987). This habitat and faunal destruction may have caused the local extinction of most of the early fish fauna, leaving an ecological void which was filled by a notothenioid fauna tolerant of deep water conditions.

Most Recent notothenioids are bottom fishes confined to waters less than 1000 m deep, although the depth range of individual species may be considerable (DeWitt 1971). There is no reason to suspect that the ancestral notothenioid stock lived any deeper than 1000 m because closely related perciform groups like blennies are coastal fishes. Notothenioids lack swim bladders, are usually denser than seawater and commonly feed and reproduce on the substrate. The midwaters of the Southern Ocean are underutilized by fishes, in an ecological sense, and could theoretically support more species. The waters south of the Antarctic Convergence are productive during the summer, but contain relatively few non-notothenioid fishes. There is evidence of a trend toward diversification, particularly of pelagic species, among notothenioids (Nybelin 1947; Andriashev 1970; DeWitt 1970; Eastman 1985a;

Hubold & Ekau 1987). For example, *Dissostichus mawsoni Pleuragramma antarcticum* and *Aethotaxis mitopteryx* are neutrally buoyant, permanent members of the midwater community (Eastman 1985a, 1988a).

Another factor that may have facilitated diversification of notothenioids is the tendency for the young of some species to pass through a distributive life history stage. Young of the bovichthyid *Bovichthys variegatus*, for example, are common near the sea surface in the outer shelf waters off New Zealand (Robertson & Mito 1979). Adults, however, are typical heavy bottom dwellers in shallow coastal and intertidal habitats. The evolution of neutrally buoyant species like *Pleuragramma antarcticum* could have been the result of neoteny.

The absence of competition and the isolation of Antarctica have provided the opportunity for speciation within this group. Thus notothenioids fill ecological roles normally occupied by taxonomically diverse fishes in temperate waters. There has obviously been evolutionary diversification in the basic notothenioid body plan (Fig. 3). Work over the last 10 years has revealed the morphological basis for the diversification in buoyancy and body types in the family Nototheniidae (Eastman & DeVries 1981, 1982, 1985, 1986a; DeVries & Eastman 1978, 1981; Clarke *et al.* 1984). This family contains 16 genera and about 50 species, including a variety of ecological types (Fig. 3).

(1) *Large, pelagic midwater predator.* Averaging 127 cm in total length and 28 kg in weight, *Dissostichus mawsoni* is six times longer and 250 times heavier than most other nototheniids. In McMurdo Sound *Dissostichus* live at 300–500 m and can be considered the ecological equivalent of a shark.

(2) *Shoaling midwater zooplanktivore.* *Pleuragramma antarcticum* has a depth range of 0–900 m and is found in both open water and beneath ice. A vital component of the food web in the Southern Ocean, it is an especially numerous and ecologically important nototheniid (Eastman 1985b). Although not a filter feeder, *Pleuragramma* may be ecologically equivalent to a herring.

(3) *Cryopelagic species. Pagothenia borchgrevinki* are specialized for swimming and feeding near the undersurface of the sea ice (Eastman & DeVries 1985).

(4) *Benthopelagic species.* Although they live close to the bottom, their streamlined appearance and lack of substrate contact adaptations indicate that they do not actually live on the substrate. In McMurdo Sound *Trematomus loennbergii* inhabits depths of at least 450 m.

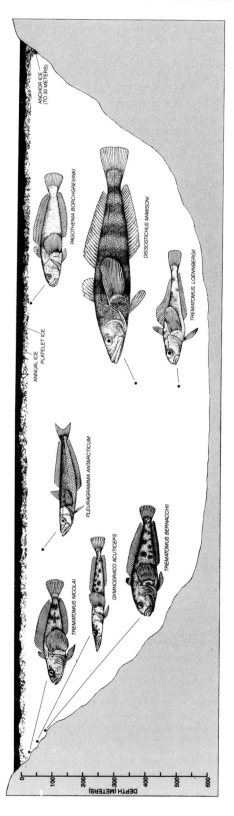

(5) *Benthic species*. Like their ancestors, most nototheniids are benthic, spending their lives on or near the substrate. There is a greater variety of food and micro-habitats available to bottom dwelling fishes than to midwater fishes, consequently bottom dwelling communities are more diverse (Roberts 1982).

(6) *Generalized species that are difficult to classify ecologically*. *Cryothenia peninsulae* exhibits morphological characteristics of both pelagic and benthic species. It may be an ecological generalist living and feeding in the mid-waters or on the bottom.

(7) *Species with an ontogenetic change in life cycle*. In *Notothenia rossii marmorata* fingerlings are pelagic, nearshore juveniles are demersal and offshore adults are both demersal and pelagic. These changes in habitat are accompanied by changes in colour, caudal fin shape and body shape (Burchett 1983).

Final Remarks

The diversification and dominance of Recent notothenioids probably indicate that the Southern Ocean is, in an ecological sense, underutilized by fishes and theoretically capable of supporting more species. Notothenioids fill roles occupied by taxonomically diverse fishes at lower latitudes. Since fossil notothenioids have not been discovered, we can say little about the prior distribution of this group. The fossil record does indicate, however, that a reasonably diverse fish fauna was present in the temperature waters near Seymour Island about the time notothenioids are suspected to have originated. We have discussed the disappearance of the Seymour Island fauna, the emergence of notothenioids and the lack of diversity among Recent non-notothenioid fishes. Further studies in Antarctic fish biology and additional palaeontological work at Seymour Island will contribute to the resolution of these intriguing issues.

Fig. 3. These seven species of notothenioids from McMurdo Sound demonstrate some of the morphological and ecological divergence mentioned in the text. Pelagic, cryopelagic, benthopelagic and benthic species are illustrated. All are members of the family Nototheniidae except *Gymnodraco acuticeps* which is a bathydraconid. Dots indicate typical habitat; however most species have considerably wider depth ranges. From Eastman & DeVries (1986*a*), copyright © 1986 by Scientific American, Inc. All rights reserved.

JTE was supported by NSF grant DPP 79–19070 and by funds from the Ohio University College of Osteopathic Medicine. Jim Eastman of Weston Printing in Minneapolis kindly converted the original Fig. 3 from colour to black and white.

References

AHLGREN, J. A. & DeVRIES, A. L. 1984. Comparison of antifreeze glycopeptides from several Antarctic fishes. *Polar Biology*, **3**, 93–97.

ANDERSEN, N. C. 1984. Genera and subfamilies of the family Nototheniidae (Pisces, Perciformes) from the Antarctic and Subantarctic. *Steenstrupia*, **10**, 1–34.

ANDRIASHEV, A. P. 1965. A general review of the Antarctic fish fauna. *In*: VAN OYE, P. & VAN MIEGHEM, J. (eds) *Biogeography and Ecology in Antarctica*, Monographiae Biologicae, **XV**, Junk, The Hague, 491–550.

— — 1970. Cryopelagic fishes of the Arctic and Antarctic and their significance in polar ecosystems. *In*: HOLDGATE, M. W. (ed.) *Antarctic Ecology*, **1**, Academic Press, London, 297–304.

— — 1987. A general review of the Antarctic bottom fish fauna. *In*: KULLANDER, S. O. & FERNHOLM, B. (eds) *Proceedings, Fifth Congress of European Ichthyologists, Stockholm 1985*, Swedish Museum of Natural History, Stockholm, 357–372.

BALUSHKIN, A. V. 1984. Morphological bases of the systematics and phylogeny of the nototheniid fishes. *USSR Academy of Sciences, Zoological Institute, Leningrad*, (In Russian.) 1–140.

BERKMAN, P. A., MARKS, D. S. & SHREVE, G. P. 1986. Winter sediment resuspension in McMurdo Sound, Antarctica, and its ecological implications. *Polar Biology*, **6**, 1–3.

BURCHETT, M. S. 1983. Morphology and morphometry of the Antarctic nototheniid *Notothenia rossii marmorata*. *British Antarctic Survey Bulletin*, **58**, 71–81.

CLARKE, A. 1983. Life in cold water: The physiological ecology of polar marine ectotherms. *Oceanography and Marine Biology: An Annual Review*, **21**, 341–453.

— — 1985. Energy flow in the Southern Ocean food web. *In*: SIEGFRIED, W. R., CONDY, P. R. & LAWS, R. M. (eds) *Antarctic Nutrient Cycles and Food Webs*, Springer-Verlag, Berlin, 573–580.

— — 1987. The adaptation of aquatic animals to low temperatures. *In*: GROUT, B. W. W. & MORRIS, G. J. (eds) *The Effects of Low Temperatures on Biological Systems*, Edward Arnold, London, 315–348.

— —, DOHERTY, N., DeVRIES, A. L. & EASTMAN, J. T. 1984. Lipid content and composition of three species of Antarctic fish in relation to buoyancy. *Polar Biology*, **3**, 77–83.

CRADDOCK, C. 1982. Antarctica and Gondwanaland. *In*: CRADDOCK, C. (ed.) *Antarctic Geoscience*, International Union of Geological Sciences, Series B, Number 4, University of Wisconsin Press, Madison, 3–13.

DAYTON, P. K. & OLIVER, J. S. 1977. Antarctic soft-bottom benthos in oligotrophic and eutrophic environments. *Science*, **197**, 55–58.

DELL, R. K. 1969. Benthic Mollusca. *In*: BUSHNELL, V. C. & HEDGPETH, J. W. (eds) *Distribution of Selected Groups of Marine Invertebrates in Waters South of 35°S Latitude, Antarctic Map Folio Series, Folio* **11**, American Geographical Society, New York, 25–26.

— — 1972. Antarctic benthos. *In*: RUSSELL, F. S. & YONGE, M. (eds) *Advances in Marine Biology*, **10**, Academic Press, London, 1–216.

DeVRIES, A. L. 1977. The physiology of cold adaptation in polar marine poikilotherms. *In*: DUNBAR, M. J. (ed.) *Polar Oceans*, Arctic Institute of North America, Calgary, 409–422.

— — 1982. Biological antifreeze agents in coldwater fishes. *Comparative Biochemistry and Physiology*, **73A**, 627–640.

— — & EASTMAN, J. T. 1978. Lipid sacs as a buoyancy adaptation in an Antarctic fish. *Nature*, **271**, 352–353.

— — & — — 1981. Physiology and ecology of notothenioid fishes of the Ross Sea. *Journal of the Royal Society of New Zealand*, **11**, 329–340.

— — & LIN, Y. 1977. The role of glycoprotein antifreezes in the survival of Antarctic fishes. *In*: LLANO, G. A. (ed.) *Adaptations Within Antarctic Ecosystems*, Smithsonian Institution, Washington, 439–458

— —, O'GRADY, S. M. & SCHRAG, J. D. 1982. Temperature and levels of glycopeptide antifreeze in Antarctic fishes. *Antarctic Journal of the United States*, **17**, 173–175.

DeWITT, H. H. 1970. The character of the midwater fish fauna of the Ross Sea, Antarctica. *In*: HOLDGATE, M. W. (ed.) *Antarctic Ecology*, **1**, Academic Press, London, 305–314.

— — 1971. Coastal and deep-water benthic fishes of the Antarctic. *Antarctic Map Folio Series*, **15**, 1–10.

DUNBAR, M. J. 1968. *Ecological Development in Polar Regions: A Study in Evolution*. Englewood Cliffs, NJ, Prentice-Hall.

EAKIN, R. R. 1981. Osteology and relationships of the fishes of the Antarctic family Harpagiferidae (Pisces, Notothenioidei). *In*: KORNICKER, L. S. (ed.) *Antarctic Research Series, Volume 31, Biology of the Antarctic Seas IX*, Washington, American Geophysical Union, 81–147.

EASTMAN, J. T. 1985a. The evolution of neutrally buoyant notothenioid fishes: Their specializations and potential interactions in the Antarctic marine food web. *In*: SIEGFRIED, W. R., CONDY, P. R. & LAWS, R. M. (eds) *Antarctic Nutrient Cycles and Food Webs*, Springer-Verlag, Berlin, 430–436.

— — 1985b. *Pleuragramma antarcticum* (Pisces,

Nototheniidae) as food for other fishes in McMurdo Sound, Antarctica. *Polar Biology*, **4**, 155–160.

—— 1988*a*. Lipid storage systems and the biology of two neutrally buoyant Antarctic notothenioid fishes. *Comparative Biochemistry and Physiology*, **90B**, 529–537.

—— 1988*b*. Ocular morphology in Antarctic notothenioid fishes. *Journal of Morphology*, **196**, 283–306.

—— & DeVries, A. L. 1981. Buoyancy adaptations in a swim-bladderless Antarctic fish. *Journal of Morphology*, **167**, 91–102.

—— & —— 1982. Buoyancy studies of notothenioid fishes in McMurdo Sound, Antarctica. *Copeia*, **2**, 385–393.

—— & —— 1985. Adaptations for cryopelagic life in the Antarctic notothenioid fish *Pagothenia borchgrevinki*. *Polar Biology*, **4**, 45–52.

—— & —— 1986*a*. Antarctic fishes. *Scientific American*, **254**, 106–114.

—— & —— 1986*b*. Renal glomerular evolution in Antarctic notothenioid fishes. *Journal of Fish Biology*, **29**, 649–662.

FORDYCE, E. 1982. The fossil vertebrate record of New Zealand. *In*: RICH, P. V. & THOMPSON, E. M. (eds) *The Fossil Vertebrate Record of Australasia*, Monash University Offset Printing Unit, Clayton, 629–698.

GOSLINE, W. A. 1968. The suborders of perciform fishes. *Proceedings of the United States National Museum*, **124**, 1–78.

GRANDE, L. & CHATTERJEE, S. 1987. New Cretaceous fish fossils from Seymour Island, Antarctic Peninsula. *Palaeontology*, **30**, 829–837.

—— & EASTMAN, J. T. 1986. A review of Antarctic ichthyofaunas in the light of new fossil discoveries. *Palaeontology*, **29**, 113–137.

HEDGPETH, J. W. 1969. Introduction to Antarctic zoogeography. *In*: BUSHNELL, V. C. & HEDGPETH, J. W. (eds) *Distribution of Selected Groups of Marine Invertebrates in Waters South of 35°S Latitude, Antarctic Map Folio Series, Folio* **11**, American Geographical Society, New York, 1–9.

HOCHACHKA, P. W. & SOMERO, G. N. 1984. *Biochemical Adaptation*. Princeton University Press, Princeton.

HUBOLD, G. & EKAU, W. 1987. Midwater fish fauna of the Weddell Sea, Antarctica. *In*: KULLANDER, S. O. & FERNHOLM, B (eds) *Proceedings, Fifth Congress of European Ichthyologists, Stockholm 1985*, Swedish Museum of Natural History, Stockholm, 391–396.

IWAMI, T. 1985. Osteology and relationships of the family Channichthyidae. *Memoirs of National Institute of Polar Research, Tokyo, Series E*, **36**, 1–69.

JOHNSON, G. L., VANNEY, J. R. & HAYES, D. 1982. The Antarctic continental shelf. *In*: CRADDOCK, C. (ed.) *Antarctic Geoscience*, International Union of Geological Sciences, Series B, Number 4. Madison, University of Wisconsin Press, 995–1002.

KENNETT, J. P. 1977. Cenozoic evolution of Antarctic glaciation, the Circum-Antarctic Ocean, and their impact on global paleoceanography. *Journal of Geophysical Research*, **82**, 3843–3860.

—— 1978. The development of planktonic biogeography in the Southern Ocean during the Cenozoic. *Marine Micropaleontology*, **3**, 301–345.

—— 1980. Paleoceanographic and biogeographic evolution of the Southern Ocean during the Cenozoic, and Cenozoic microfossil datums. *Palaeogeography, Palaeoclimatology, Palaeoecology*, **31**, 123–152.

KERR, R. A. 1984. Ice cap of 30 million years ago detected. *Science*, **224**, 141–142.

—— 1987. Ocean drilling details steps to an icy world. *Science*, **236**, 912–913.

LONG, J. 1982. The history of fishes on the Australian continent. *In*: RICH, P. V. & THOMPSON, E. M. (eds) *The Fossil Vertebrate Record of Australasia*, Monash University Offset Printing Unit, Clayton, 53–85.

MACDONALD, J. A. 1981. Temperature compensation in the peripheral nervous system: Antarctic vs temperate poikilotherms. *Journal of Comparative Physiology A*, **142**, 411–418.

MOYLE, P. B. & CECH, J. J., Jr. 1988. *Fishes: An Introduction to Ichthyology (Second Edition)*. Prentice-Hall, Englewood Cliffs, NJ.

MILLER, R. G. 1987. Origins and pathways possible for the fishes of the Antarctic Ocean. *In*: KULLANDER, S. O. & FERNHOLM, B. (eds) *Proceedings, Fifth Congress of European Ichthyologists, Stockholm 1985*, Swedish Museum of Natural History, Stockholm, 373–380.

MONTGOMERY, J. C. MACDONALD, J. A. 1984. Performance of motor systems in Antarctic fishes. *Journal of Comparative Physiology A*, **154**, 241–248.

NORMAN, J. R. 1937. Coast fishes. Part II. The Patagonian region. *Discovery Reports*, **16**, 1–150.

—— 1938. Coast fishes. Part III. The Antarctic zone. *Discovery Reports*, **18**, 1–104.

NYBELIN, O. 1947. Antarctic fishes. *Scientific Results of the Norwegian Antarctic Expeditions 1927–1928*, **26**, 1–76.

PEZZETTI, T. F. & KRISSEK, L. A. 1986. Re-evaluation of the Eocene La Meseta Formation of Seymour Island, Antarctic Peninsula. *Antarctic Journal of the United States*, **21**, 75.

PROSSER, C. L. 1986. *Adaptational Biology: Molecules to Organisms*. John Wiley, New York.

REGAN, C. T. 1914. Fishes. *British Antarctic ("Terra Nova") Expedition 1910, Natural History Report, Zoology*, **1**, 1–54.

ROBERTS, T. R. 1982. Unculi (horny projections arising from single cells), an adaptive feature of the epidermis of ostariophysan fishes. *Zoologica Scripta*, **11**, 55–76.

ROBERTSON, D. A. & MITO, S. 1979. Sea surface ichthyoplankton off southeastern New Zealand, summer 1977–78. *New Zealand Journal of Marine & Freshwater Research*, **13**, 415–424.

ROMER, A. S. 1966. *Vertebrate Paleontology* (Third Edition). University of Chicago Press, Chicago.

SCHAEFFER, B. 1972. A Jurassic fish from Antarctica. *American Museum Novitates*, **2495**, 1–17.

SCHOPF, T. J. M. 1980. *Paleoceanography*. Harvard University Press, Cambridge.

SCOTT, G. K., FLETCHER, G. L. & DAVIES, P. L. 1986. Fish antifreeze proteins: Recent gene evolution. *Canadian Journal of Fisheries and Aquatic Sciences*, **43**, 1028–1034.

STINTON, F. C. 1957. Teleostean otoliths from the Tertiary of New Zealand. *Transactions of the Royal Society of New Zealand*, **84**, 513–517.

TASCH, P. 1977. Ancient Antarctic freshwater ecosystems. *In*: LLANO, G. A. (ed.) *Adaptations Within Antarctic Ecosystems*, Smithsonian Institution, Washington, 1077–1089.

THOMSON, K. S. 1977. The pattern of diversification among fishes. *In*: HALLAM, A. (ed.) *Patterns of Evolution as Illustrated by the Fossil Record*, Elsevier, Amsterdam, 377–404.

VOGEL, S. 1981. *Life in Moving Fluids*. Willard Grant, Boston.

WOODBURNE, M. O. & ZINSMEISTER, W. J. 1984. The first land mammal from Antarctica and its bio-geographic implications. *Journal of Paleontology*, **58**, 913–948.

WOODWARD, A. S. 1908. On fossil fish-remains from Snow Hill and Seymour Islands. *Wissenschaftliche Ergebnisse der Schwedische Südpolar-Expedition 1901–1903*, **3**, 1–4.

YOUNG, G. C. 1987. Devonian vertebrates of Gondwana. *In*: MCKENZIE, G. D. (ed) *Gondwana Six: Stratigraphy, Sedimentology, and Paleontology*, Geophysical Monograph **41**, American Geophysical Union, Washington, 41–50.

—— 1989. The Aztec fish fauna (Devonian) of southern Victoria Land: evolutionary and biogeographical significance. *In*: CRAME, J. A. (ed.) *Origins and evolution of the Antarctic Biota*. Geological Society, London, Special Publication, **47**, 43–62.

ZINSMEISTER, W. J. 1987. Cretaceous paleogeography of Antarctica. *Palaeogeography, Palaeoclimatology, Palaeoecology*, **59**, 197–206.

—— & FELDMANN, R. M. 1984. Cenozoic high latitude heterochroniety of Southern Hemisphere marine faunas. *Science*, **224**, 281–283.

The origin of the Southern Ocean marine fauna

ANDREW CLARKE & J. ALISTAIR CRAME

British Antarctic Survey, NERC, High Cross, Madingley Road, Cambridge CB3 OET,
UK

Abstract: Current knowledge of the break-up of Gondwana during the Tertiary indicates that shallow water marine habitats may have been present continuously, and on occasions were considerably more extensive than at present. Although direct fossil evidence is sparse after the Eocene, geophysical evidence suggests that shallow waters have been present since the late Mesozoic, and possibly much longer. The break-up of Gondwana was accompanied by a more or less steady lowering of both surface and bottom temperatures in the Southern Ocean from about 15°C in the Late Cretaceous to the present range of roughly +2 to −1.8°C. Microfossils in deep-sea drilling cores indicate that temperature drops were particularly sharp in the early Oligocene (c. 38 Ma), mid-Miocene (10−14 Ma) and Pliocene (c. 4 Ma BP). Geological evidence suggests that the Drake Passage opened, and the present oceanographic regime established, about 25−30 Ma BP. This is now known to be about the time of full-scale development of the East Antarctic ice cap. Subsequently ice sheets extended across, and deeply eroded, the continental shelves but the effects of these glacial maxima on the marine biota are not fully understood. Late Cretaceous/early Tertiary marine fossils from the James Ross Island group indicate a diverse shallow water marine fauna, including two groups notably lacking in diversity in the living fauna: decapods and teleost fish. In several genera occurrences in this fauna predate first occurrences in lower latitudes by as much as 40 Ma, suggesting the possibility that a number of groups originated at high southern latitudes.

The living fauna exhibits a high biomass in many areas, and within-site diversity can be as high as anywhere in the world. Some individual taxonomic groups, however, (notably bivalves and gastropods) have a lower diversity than in the tropics, supporting the concept of a latitudinal cline in diversity. Studies of physiological adaptation to temperature suggest that the decline in seawater temperature during the Cenozoic has not presented a particularly severe evolutionary problem. The reasons for the absence of large decapods and the low diversity of fish in the present fauna are unclear. Most of the biological features of the modern fauna are more likely a response to the seasonality of the ecosystem rather than low temperature *per se*. Overall the evidence suggests that the present Southern Ocean shallow water marine fauna largely evolved in situ, having been present since at least the Late Cretaceous, and possibly much longer. Some groups have invaded, for example along the Scotia arc, but the isolation of the Southern Ocean by the present oceanographic regime and the limited dispersal ability of many forms means that exchange with lower latitudes is very slow.

Until relatively recently there has been a general impression amongst many palaeontologists and biologists that the Southern Ocean marine fauna is impoverished and consists largely of stress-tolerant organisms that are resistant to extinction but little prone to speciation (Vermeij 1978). Indeed the high latitudes have sometimes been regarded as evolutionary backwaters. However we now know that this is a gross oversimplification; palaeontological investigations have revealed a dynamic history, with several periods of diversification and change, and physiological studies have revealed subtle and complex adaptations to low temperature. These new data suggest that the Southern Ocean marine fauna may have evolved largely in situ over a long period of time, and that evolutionary processes at high latitudes are every bit as complex as those elsewhere (Lipps & Hickman 1982).

The origin of the Southern Ocean marine fauna has long intrigued biologists. Hypotheses have included colonization from the deep sea, colonisation via South America, and evolution in situ (see Lipps & Hickman 1982 for a recent summary). In this paper we have attempted to combine information from recent advances in marine geology, palaeontology and physiology to understand the evolution of the Antarctic marine fauna. We will deal with each area of evidence in turn, but first it is necessary to outline possible hypotheses to explain the origin of the fauna.

From Crame, J. A. (ed.), 1989, *Origins and Evolution of the Antarctic Biota*,
Geological Society Special Publication No. 47, pp. 253−268.

Possible origins of the fauna

Although much of the Southern Ocean consists of deep water, these habitats have been poorly sampled and are little known. Most discussion of the fauna has, therefore, been concerned with the shallow water organisms of the continental shelves. These shelves have been subject to a varying degree of glacial influence for approximately the past 35 million years (Ma). There are thus two major classes of hypothesis to explain the evolution of the present fauna, namely that the shelves have always been available despite glaciation, or that the shelves have been periodically cleared of life. In the first case the fauna may have evolved in situ to cope with the low temperature, or it may have become extinct and colder water forms invaded from elsewhere (or perhaps a mixture of both). In the second case the modern fauna would of necessity invaded from elsewhere. However there is unlikely to be a single origin for a fauna as diverse as that of the Southern Ocean. All faunas contain a mixture of species with different histories; some species are recent arrivals, others have existed for a long time, and most are probably somewhere in between. Nevertheless the above hypotheses provide a useful framework within which to discuss the origin of the Southern Ocean shallow water fauna. First, however, the geological and thermal history of the Southern Ocean will be examined.

Geological history

Distribution of the major landmasses

The broad features of the development of the Southern Ocean are shown in Fig. 1. Although the history of dispersal of the component Gondwana continents is not yet completely understood, we do know enough to be able to draw a broad picture (see e.g. Lawver et al. 1985 and Lawver & Scotese 1987 for recent summaries).

Many of the precise details of the breakup of Gondwana have still to be unravelled. It is not clear when the separation of western Antarctica and South America was complete, but the best current estimate for the opening of the Drake Passage is 25–30 Ma BP (Oligocene: Barker & Burrell 1977). The final deep-water separation of Australia from eastern Antarctica occurred shortly afterwards, and this enabled the establishment of the circum-Antarctic current which isolates the Southern Ocean marine fauna to this day (Kennett 1977). Particular difficulty

surrounds the establishment of the detailed history of the Scotia arc/Antarctic Peninsula region (Dalziel & Elliot 1982; Storey in press), which is unfortunate since this is an area of particular importance in the understanding of the history of the Southern Ocean marine fauna.

Distribution of shallow water

The large scale distribution of continental landmass gives only limited information to the marine biologist. In order to understand the evolution of the marine fauna we need to know the distribution of shallow water habitats through time (for it is here that most of the biomass and species are concentrated). The presence of shallow water has to be inferred from the distribution of the major landmasses and the presence of shallow water deposits in the fossil record. Present knowledge is summarized in Fig. 1.

At the present time the area of shallow water available to marine fauna is at a minimum, due mainly to the extensive ice-cap (Fig. 1c). Prior to widespread glaciation there appear to have been large areas of shallow water. In particular there were a number of major embayments situated between eastern and western Antarctica. Even further back in time there were also extensive areas of shallow water associated with the widespread encroachment of Cretaceous epicontinental seas (Webb et al. 1984).

Evidence for the former existence of extensive seaways across Antarctica comes from the presence of reworked marine microfossils in the Sirius Formation of the Transantarctic Mountains. The latter formation comprises glacial sediments that were eroded from the former Wilkes and Pensacola basins and transported across the Transantarctic Mountains in the mid-late Pliocene (Webb et al. 1984). These fossils imply that open marine conditions existed, at least intermittently, in these basins from the late Cretaceous to the early Pliocene; indeed it is even possible that there was a complete trans-Antarctic seaway for part of this time.

Even after the main Antarctic ice-cap formed in the late Oligocene (approximately 30 Ma BP: Barrett et al. 1987; Leg 113 shipboard scientific party 1987), there is evidence of substantial climatic warming and deglaciation in the early Pliocene (4–5 Ma BP). This comes both from stable isotope analysis of deep sea foraminifera (Hodell & Kennett 1986) and the presence of widespread fossiliferous deposits around the Antarctic continental margins. The Pecten

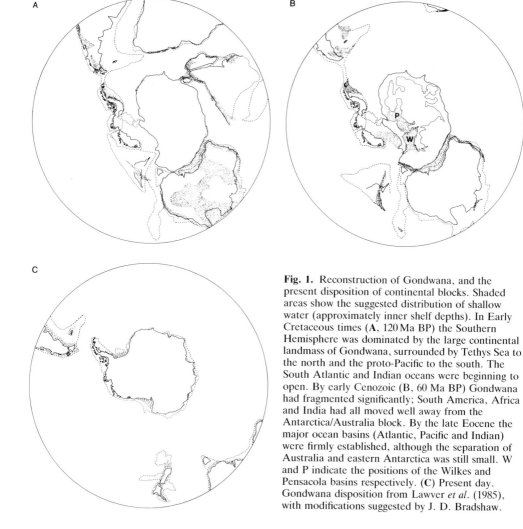

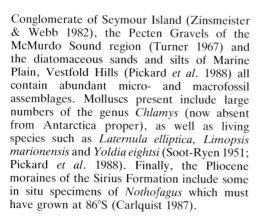

Fig. 1. Reconstruction of Gondwana, and the present disposition of continental blocks. Shaded areas show the suggested distribution of shallow water (approximately inner shelf depths). In Early Cretaceous times (**A**, 120 Ma BP) the Southern Hemisphere was dominated by the large continental landmass of Gondwana, surrounded by Tethys Sea to the north and the proto-Pacific to the south. The South Atlantic and Indian oceans were beginning to open. By early Cenozoic (**B**, 60 Ma BP) Gondwana had fragmented significantly; South America, Africa and India had all moved well away from the Antarctica/Australia block. By the late Eocene the major ocean basins (Atlantic, Pacific and Indian) were firmly established, although the separation of Australia and eastern Antarctica was still small. W and P indicate the positions of the Wilkes and Pensacola basins respectively. (**C**) Present day. Gondwana disposition from Lawver *et al.* (1985), with modifications suggested by J. D. Bradshaw.

Conglomerate of Seymour Island (Zinsmeister & Webb 1982), the Pecten Gravels of the McMurdo Sound region (Turner 1967) and the diatomaceous sands and silts of Marine Plain, Vestfold Hills (Pickard *et al.* 1988) all contain abundant micro- and macrofossil assemblages. Molluscs present include large numbers of the genus *Chlamys* (now absent from Antarctica proper), as well as living species such as *Laternula elliptica*, *Limopsis marionensis* and *Yoldia eightsi* (Soot-Ryen 1951; Pickard *et al.* 1988). Finally, the Pliocene moraines of the Sirius Formation include some in situ specimens of *Nothofagus* which must have grown at 86°S (Carlquist 1987).

It seems reasonable to conclude that there have been extensive areas of shallow water associated with Gondwana (and the subsequent isolated continental blocks) throughout the Mesozoic and up to the development of extensive glaciation in the late Oligocene. Since then the area of shallow water has fluctuated widely in response to the volume of the continental ice-cap.

Glaciological history

As emphasized earlier, in understanding the history of the Southern Ocean fauna it is clearly important to know whether the ice shelves have at any time scraped the continental shelves free of biota. Although evidence from sediment structure, ice rafted debris and continental shelf topography indicates that ice has extended over many areas of continental shelf (Anderson, in press), it is not clear whether these extensions were simultaneous or whether they would have been sufficient to eradicate the biota. A major extension of the East Antarctic ice sheet occurred in the late Miocene (Kennett 1977) but some significant extension in both western Antarctica and the central Transantarctic Mountains may have begun prior to this.

The weight of the ice-cap has depressed the continental landmass, leading to unusually deep continental shelves (as at the present day). There is also seismic evidence that previous extensions of the ice-cap may have physically eroded parts of the continental shelf (Anderson, in press). Although benthic fauna have been recovered from beneath George VI Sound and the Ross ice-shelf (Lipps et al. 1977, 1979) it is not clear whether the few species recovered are indicative of a depauperate fauna or merely difficulties of sampling. Overall, therefore, we are not yet in a position to decide conclusively whether periodic extension of the ice-shelves will have eradicated the shallow water Antarctic marine fauna, either locally or over extensive areas.

Thermal history of the Southern Ocean

Evidence for thermal history comes from essentially two sources. These are studies of the stable isotope composition of foraminifera recovered from sediment cores, and shifts in biological communities. Oxygen isotope studies provide the least equivocal data, although care must be taken in making correction for the volume of isotopically light water contained in ice caps. It is also necessary to ensure that accurate segregation into benthic and planktonic species has been accomplished, for the thermal histories of these two environments are slightly different. The thermal history of the Southern Ocean has been discussed in detail by Savin (1977), Shackleton & Kennett (1975), Frakes (1979), Shackleton (1982) and Hodell & Kennett (1986); its broad features are summarized in Fig. 2.

There has been a general cooling of both surface and bottom temperatures in the Southern Ocean from warm Cretaceous temperatures of about 15–20°C, to present temperatures of between +2°C and −1.8°C. However, this general trend has been interrupted by periods of more rapid temperature

change at the Eocene/Oligocene boundary (35–40 Ma BP) and in the late Pliocene (4–5 Ma BP). There was also a less dramatic period of cooling in the middle Miocene (15–20 Ma BP). Although these changes in temperature are rapid on a geological time scale (perhaps up to 8°C in 3 Ma), they are very slow in relation to typical oceanographic changes. Normal geological climatic change is usually much slower than this, by several orders of magnitude (Schopf 1984). Neither would seem to pose particularly severe physiological problems (Clarke, in press).

Several pieces of biological evidence have been used in support of physical data for changes in seawater temperature. These include changes in overall taxonomic diversity, switches in the balance of different taxonomic groups, changes in shell structure and ornamentation, and changes in reproductive biology and/or larval type. None of these can safely be taken in isolation as conclusive evidence for a change in seawater temperature. This is because in many (but not all) cases the interpretation depends upon a previously established correlation between temperature and the particular biological feature in question. It is now becoming clear that such correlations may be spurious, being related instead to features such as the degree of seasonality of the environment, which co-varies with temperature along a latitudinal gradient from the tropics to the poles (Clarke 1988).

Biological features which do seem to be somehow linked to temperature and which do lend support to the pattern of Southern Ocean temperature change shown in Fig. 2, are a switch from calcareous to largely siliceous microfossil taxa (Kennett 1977) and sedimentation, and a general reduction in shell size and ornamentation in calcareous macrofossils as seawater temperature decreases (Nicol 1967). The relationship between biological features and temperature is discussed further below, and in Clarke (in press).

Palaeontological evidence

The terrestrial flora and fauna

Although it may seem somewhat incongruous to begin a review of the marine fauna by examining the terrestrial record, this is nevertheless where some striking evidence regarding the origins of austral biotas can be found. Geological and palaeontological evidence has established that Southern Hemisphere high latitude terrestrial connections were possible, at least

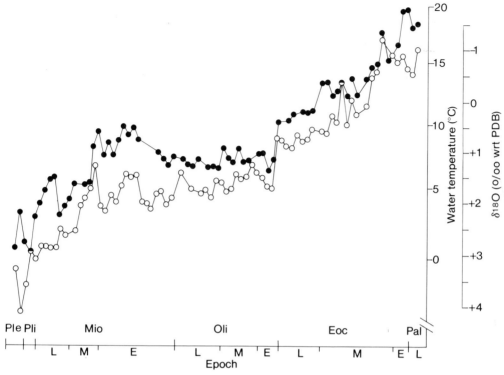

Fig. 2. Temperature of the Southern Ocean during the Tertiary as inferred from oxygen isotope composition of foraminifera recovered from sediment cores obtained by the Deep Sea Drilling Project. δ ^{18}O shows the raw data which is isotope fractionation (ppt) with respect to a standard (PDB). The water temperatures are calculated after making corrections for factors such as salinity and ice caps. Data from benthic ($\bigcirc$) and planktonic ($\bullet$) species shown separately. Pal, Paleocene; Eoc, Eocene; Oli, Oligocene; Mio, Miocene; Pli, Pliocene; Ple, Pleistocene. Redrawn from Kennett (1977).

intermittently, throughout the Cretaceous (144−65 Ma BP). Indeed they probably started as early as the Triassic and continued until at least the end of the Eocene (Raven & Axelrod 1974; Woodburne & Zinsmeister 1982; Briggs 1987).

The classic view of the southern Gondwana margins is that they formed a major migration route for taxa that originated elsewhere (Darlington 1965). However, this view may now need to be modified in the light of recent palaeontological discoveries, notably that certain key elements of the southern cool-temperate flora may have originated in situ. The earliest occurrences of *Nothofagus*, for example, are now known to be in the Upper Cretaceous of the Antarctic Peninsula region (Dettmann 1989) and this genus clearly spread outwards from this area throughout the Late Cretaceous and Cenozoic. Similar adaptive radiations from Cretaceous high latitude centres of origin have been postulated for several other

early angiosperm groups (Moore 1972; Johnson & Briggs 1975; Dettmann 1986, 1989; Askin 1989). Bryophytes are an important component of living *Nothofagus* communities and may have similarly disseminated northwards (Schuster 1969, 1979). Such radiations are inferred to have occurred over similar time periods as their associated higher plants. Finally it is also possible to view the present day distribution of terrestrial arthropods in the Antarctic and sub-Antarctic as at least a partial product of Cenozoic high latitude radiations (Wallwork 1973).

Certain terrestrial groups may have radiated from the southern high latitudes in the late Mesozoic/Cenozoic, for a vast land area was available for evolutionary innovation at this time (Fig. 1). A general cooling of the climate probably served to push new terrestrial communities northward and interestingly a number of lineages apparently diversified as they did so. Temperature fluctuations may well have been an important aid to this process of diversification (Valentine 1968, 1984a).

The marine fauna

The processes which appear to have operated in late Mesozoic/Cenozoic terrestrial communities at high southern latitudes may also have influenced marine environments. Certainly a vast area of continental shelf was available for marine organisms at this time (Fig. 1) and this would have included sites of active speciation. Examination of the living shallow water Antarctic and sub-Antarctic marine fauna reveals several groups that comprise a large number of closely related species; these are almost certainly the product of adaptive radiations in situ. Examples include pycnogonids, gastropods (especially trochids, littorinids, trichotropids and buccinids), echinoderms (ophiuroids, ctenocidarid cidaroids and schizasterid spatangoids), ascidians and notothenioid fish (Dell 1972). All of these radiations point to a long period of evolution in the Antarctic or sub-Antarctic, perhaps in some cases stretching back as far as the late Mesozoic.

For example the Antarctic and sub-Antarctic molluscan fauna comprises a comparatively small number of cosmopolitan families, and Powell (1951) reasoned that it could therefore be of no great antiquity. By far the strongest taxonomic links are with the Patagonian region and immigration into the Antarctic was apparently achieved by using the islands of the Scotia arc as a series of stepping stones (Powell 1951; Dell 1972). However this conventional view may now need to be modified in the light of important new palaeontological evidence.

As a result of recent advances in both offshore drilling and onshore mapping, a more detailed Cenozoic biostratigraphy of the Antarctic continent is beginning to emerge (Webb, in press). A series of drillholes in the Ross Sea, Weddell Sea and Prydz Bay (Amery Ice Shelf, East Antarctica) has clearly established the presence of a thick sequence of late Oligocene/early Miocene (approx. 30–21 Ma BP) glacial sediments, and these provide the earliest physical record of an extensive East Antarctic ice cap (Barrett et al. 1987; Leg 113 shipboard scientific party 1987; Leg 119 shipboard scientific party 1988). In addition early Oligocene (35 Ma BP) glacial sediments are known from both the Ross and Weddell seas, and diamictites recovered from one of the Prydz Bay boreholes may even extend back to the middle Eocene (42.5 Ma BP: Leg 119 shipboard scientific party 1988). It is possible, however, that these earlier deposits

may be merely the products of local glaciations.

Interbedded with the glacial diamictite beds in the drill cores are finer-grained marine sandstones and mudstones that have yielded both micro- and macrofossils. The former provide good age control for the latter, which include a range of molluscs, polychaetes and bryozoa. Of these, the molluscs are particularly important as they contain a number of forms that are either identical with, or very close to, living species (Dell & Fleming 1975). Bivalves such as *Nuculana* cf. *inaequisculpta*, *Yoldiella sabrina*, *Yoldia eightsi*, *Ennucula* aff. *grayi* and *Limopsis* aff. *marionensis*, together with gastropods such as *Amauropsis* sp. and *Proneptunea* aff. *duplicarinata*, indicate very close links with the American sub-Antarctic and Antarctic recent faunas. Unfortunately the palaeontological record of forms such as these is still not good enough to say whether immigration may have occurred from South America to Antarctica, or vice versa. However, their presence does indicate that elements of the Antarctic molluscan fauna have been in place for at least 25 Ma.

Extensive late Oligocene–early Miocene glaciomarine strata are also known from onshore exposures at King George Island, South Shetland Islands (Birkenmajer 1987). At the Lions Rump group of localities they are intimately associated with a suite of both intrusive and extrusive igneous rocks, and it is these which have been used to provide age control (by K-Ar whole-rock dating). Two principal glaciomarine sequences can be recognised: the Polonez Cove Formation, which is dated by overlying lavas at approximately 24 Ma (i.e. latest Oligocene), and the Cape Melville Formation, which is dated by cross-cutting dykes at approximately 20 Ma (i.e. early Miocene) (Birkenmajer et al. 1986). One tillite on King George Island may be as old as middle Eocene (50 Ma BP: Birkenmajer et al. 1986).

The Polonez Cove Formation contains worms (*Serpula* and *Spirorbis*), abundant bryozoans (49 species), occasional brachiopods, abundant bivalves (28 species), gastropods (11 species), ostracodes, crinoids and occasional ophiuroids and echinoids (Gazdzicki & Pugaczewska 1984). Although no precise matches can be made between these groups and their recent counterparts at the species level, a number of general similarities exist. These are again apparent in the molluscan fauna, where there are species of the bivalve genera *Nuculana*, *Limopsis*, *Perrierina* (cyamiid), *Eucrassatella* and *Panopea*, and the gastropods *Polinices*, *Struthiolaria*, *Trophon*, *Nassa* and *Acteon* (Gazdzicki & Pugaczewska 1984).

The gastropod assemblage is very similar to that described from the late Eocene of Seymour Island (Wilckens 1912; Zinsmeister & Camacho 1982).

The fauna of the Cape Melville Formation is particularly rich in solitary corals, polychaetes, bivalves, gastropods, scaphopods, crabs, echinoids, asteroids and ophiuroids, and it too bears a close resemblance to the living marine biota. The gastropods provide perhaps the best example here, with the thirty species so far recognised including four naticids, the struthiolariid *Perissodonta*, eight whelks (Buccinacea) two volutids, two cancellariids and two turrids (*Austrotoma* and *Aforia*) (Karczewski 1987).

These compositional traits can be recognised further back in the palaeontological record of the southern high latitudes. For example, both the late Eocene La Meseta Formation of Seymour Island and the Wangaloan Fauna of New Zealand (Danian or Paleocene) contain a strong representation of nuculid, nuculanid, malletiid and limopsid bivalves, and naticid, struthiolariid, buccinacean and turrid gastropods (Finlay & Marwick 1937; Zinsmeister & Camacho 1982; Zinsmeister 1984). The apparent radiation of the latter four gastropod groups at this time is particularly striking.

Traditionally, the struthiolariid gastropods are regarded as the only high-latitude molluscan group with an ancestry that can be traced back to the Cretaceous (Powell 1965). Forms such as the Antarctic *Perissodonta*, the New Zealand *Struthiolaria* and *Pelicaria*, and the Australian *Tylospira*, can be traced back with some certainty to the genus *Conchothyra* from the latest Cretaceous of New Zealand (Zinsmeister 1980). Nevertheless other distinctive elements of the fauna, and in particular other gastropod groups, may have a similar ancestry. Although a thorough taxonomic revision of Southern Hemisphere Late Cretaceous gastropod faunas is needed, whelks in, particular would seem to be well represented in Patagonia (Wilckens 1907), Antarctica (Wilckens 1910) and New Zealand (Wilckens 1922). There would also appear to be turrids and muricids, and perhaps even volutids and cancellariids too.

The Southern Ocean gastropod fauna would thus seem to be the product of a long period of evolution in situ, perhaps stretching back to the Late Cretaceous (that is, at least 65 Ma BP). At least in the initial stages of development of this fauna there would have been no shortage of available habitat on and around the Antarctic continent. Prior to the development of the vast continental ice sheet in the mid-Miocene it is almost certain that there would have been nu-

merous incursions of shallow seas across the continental shield (Fig. 1). Largely in response to climates that fluctuated between temperate and glacial, and the high palaeolatitude position, communities evolved that were dominated by generalist feeding types. These included suspension feeders such as the struthiolariids as well as predators able to take a very wide variety of prey types (principally whelks and naticids; e.g. Taylor 1981). Such a development is extremely interesting, for it may turn out to be a microcosm of the Antarctic marine fauna as a whole (where generalist feeding strategies appear to predominate: Arnaud 1974).

High latitude heterochroneity

One major development in Southern Hemisphere palaeontology has been the discovery in the rich and diverse early Tertiary marine invertebrate and fish fauna of the La Meseta Formation (Seymour Island) of a number of elements previously known only from the late Cenozoic or Recent at lower latitudes (Zinsmeister & Feldmann 1984; Wiedman et al. 1986). In the early Tertiary, high latitude climates were considerably milder, probably cool-temperate, and the tropics somewhat cooler. The latitudinal temperature gradient was thus far less extreme than today (Valentine 1984b) and it would seem that many groups evolved at relatively high latitudes. Only later did they extend their range into lower latitudes, possibly as a consequence of the general cooling of global seawater. Some groups also extended their range into deeper waters (Wilson & Feldmann 1986).

This demonstration that some important speciation events occurred at high latitudes should not be a cause for surprise. There is no evidence that evolutionary process are any different in high latitudes or cold temperatures from elsewhere (Clarke, in press), and several groups appear to be undergoing active speciation in the high latitudes at present. It does, however, demonstrate that the tropics have not acted as the sole source of evolutionary novelty (Crame 1986). In many ways the most interesting question posed by the concept of high latitude heterochroneity is why certain groups have subsequently died out in high Southern latitudes.

The modern fauna

Of particular relevance to the current debate on the origin of the biota is knowledge of the biomass, diversity, affinities and physiological adaptation of the fauna.

Biomass

In contrast to the earlier impression of a sparse, species-poor fauna (based in part on comparison with the more thoroughly documented and genuinely depauperate Arctic fauna), the Southern Ocean is now known to contain areas of seabed rich in biomass. For example the eastern side of McMurdo Sound has an infaunal density comparable with soft substrates anywhere in the world (Dayton & Oliver 1977).

Although single estimates like this are subject to bias by factors such as recent spatfall (which will elevate the biomass) or recovery from iceberg scour, there are sufficient data from other sites to conclude that the Southern Ocean benthic environment can support a high biomass and a large number of individuals (Andriashev 1968; Hardy 1972; White & Robins 1972; Lowry 1975; Richardson & Hedgpeth 1977; Arnaud et al. 1986; Jazdzewski et al. 1986). Since many Southern Ocean species are small, numbers of individuals will clearly be higher for a given biomass. As in other areas, biomass decreases from shallow to deeper waters (Uschakov 1963).

Diversity

The measurement of diversity indices in a living marine assemblage involves a considerable amount of time-consuming work, and in the Southern Ocean data are available only for a soft-bottom community in Arthur Harbour, Anvers Island (Richardson 1976; Richardson & Hedgpeth 1977). Here the values of the Shannon-Weaver index of diversity, H', and species richness, SR, at 14 stations were comparable with similar studies from temperate and warmer waters; indeed they were higher than in many (Fig. 3). This high within-habitat diversity is matched by the high species richness of many invertebrate groups in the Southern Ocean. These include amphipods (Barnard 1969; Watling & Thurston 1989), polychaetes (Hartmann 1964, 1966), sponges (Koltun 1970) and bryozoans (Bullivant 1969).

There are, however, several taxonomic groups with a strikingly low species richness in the Southern Ocean, such as decapods (Yaldwyn 1965) and fish (Eastman & Grande 1989). Other groups appear to have an intermediary level of species richness; not as depauperate as decapods, but well below levels found in tropical waters. Examples of these are bivalve and gastropod molluscs, two groups

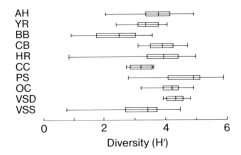

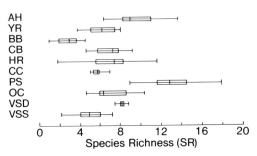

Fig. 3. Diversity and species richness in a soft-bottom marine community at Arthur Harbour, Anvers Island, in comparison with similar communities elsewhere in temperate and warmer waters. Diversity was expressed as Shannon-Weaver H'. Species richness SR, was calculated as SR = $(S - 1)/\ln N$, where S = the number of species and N is the total number of individuals (Margalef 1958). The data shown are mean (vertical line), standard error (bar) and range (horizontal line). Redrawn from Richardson (1976). (Note that Margalef's formula for SR is quoted incorrectly in this reference, although the computation is correct.) Abbreviations: AH, Arthur Harbour, Antarctica; YR, York River, Virginia; BB, Bramble Bay, Australia; CB, Chesapeake Bay; HR, Hampton Roads; CC, Cape Cide Bay; PS, Puget Sound; OC, Columbia River, Oregan Coast; VSD, VSS, Virginia Shelf Deep and Shallow respectively.

which have been instrumental in the establishment of the concept of a *latitudinal diversity gradient* in shallow marine waters.

Latitudinal diversity gradients

In several groups of marine organism (such as bivalves, gastropods, corals, decapods, foraminifera) tropical communities contain many more species than temperate or polar ones. This is a clear feature of terrestrial communities, and the concept of latitudinal gradient in diversity has been extended to the marine environment by Stehli et al. (1967, 1969) and

Stehli (1968). For example, data for bivalves at species, genus and family level exhibit a trend from peak diversity in the tropics to lower values at the poles (Fig. 4).

The possible reasons for this latitudinal gradient in diversity have attracted considerable debate (see for example reviews by Sanders 1968 and Pianka 1966), but the underlying mechanism(s) remain unclear. Certainly the cause cannot be simply temperature, for some groups are actually more diverse at the poles (see above). Although the younger age of the Arctic may explain, in part, the lower general diversity of this region compared with the Antarctic, age cannot explain global latitudinal diversity gradients. In terms of temperature the tropics are just as much a youthful ecosystem as the Southern Ocean, low latitude sea surface temperature having increased from roughly 15°C to present values (roughly 25 to 30°C) over approximately the last 30 Ma (Shackleton 1979; Valentine 1984b). The present severe latitudinal cline in mean seawater temperature is not typical of most of the history of the marine biosphere.

Whatever the underlying explanation of the latitudinal diversity gradient in shallow water taxa, it does seem to be particularly striking in some groups with calcareous skeletons. The small size, poor calcification and low incidence of ornamentation in cold water molluscs has frequently been commented upon (see, for example, Nicol 1967). Clarke (1983) suggested that this may reflect directly the increased metabolic cost of precipitating calcium carbonate from solution in sea water, but there is at present no information with which to judge whether this increase in metabolic cost is significant to the organisms concerned. Nevertheless there is a parallel decrease from tropics to poles in both the species richness and the degree of calcification in some taxa with calcareous skeletons.

If the explanation of both these trends *in some taxa with calcareous skeletons* is indeed the increased energetic cost of producing a shell in colder water then the use of diversity trends in the interpretation of palaeocommunities must be undertaken with care.

Where are the fish and decapods?

Although several taxa are species-rich in the Southern Ocean, the relatively poor fish fauna and the highly impoverished decapod fauna have long attracted attention. However these may only be extreme (albeit well documented) cases of a more general phenomenon. Gastropods and bivalve molluscs for example also

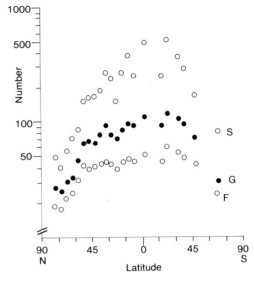

Fig. 4. The relationships between bivalve diversity and latitude. Individual points are the number of species (S, ○), genera (G, ●) or families (F, ○), pooled into blocks of 5° of latitude. Redrawn from Stehli *et al.* (1967).

show a much reduced species richness in the Southern Ocean, though not as severe as in fish or decapods.

Most attempts to explain the present low diversity of the fish fauna have involved consideration of the break-up of Gondwana (Anderson 1984; Grande & Eastman 1986; Miller 1987; Eastman & Grande 1989). However tectonic events cannot be the complete explanation, for all groups of shallow water invertebrates will presumably also have been subject to the same geographical constraints, and yet many show high species richness. Clearly it is necessary to consider the evolution of the Southern Ocean fish and decapod fauna in relation to that of the fauna as a whole.

Certainly cold water *per se* cannot be the explanation. Both teleost fish and decapods are well represented in the Arctic and in bathyal regions, where the temperatures are also very low. Teleost fish have a particular problem in being hyposmotic to seawater, and therefore liable to freeze at low seawater temperatures. However the development of antifreeze does not seem to have been a difficult evolutionary problem for there is evidence that this has occurred perhaps four or five times (Scott *et al.* 1986).

It is interesting that the Antarctic fish fauna is dominated by members of one group, the notothenioids. Although lacking a swim-bladder they have undergone a remarkable adaptive radiation, which has included the evolution of neutral buoyancy to exploit the midwater (Eastman & DeVries 1982). This bout of speciation suggests strongly that the notothenioids are expanding to fill 'vacant' ecological niches, created by the extinction of most of the previous fish fauna (which is well represented in the early Tertiary: Eastman & Grande 1989). Similarly the expansion of isopods in the Southern Ocean may represent the filling of an ecological vacuum left by the absence or extinction of the decapods.

We are not yet able to explain why some groups have all but vanished from the Southern Ocean and others not. Perhaps the explanation lies in severe habitat reduction caused by periodic extensions of the ice-shelves over the continental shelf with a purely stochastic loss of some groups. This will be difficult to demonstrate unless we can locate good fossiliferous deposits of ages later than early Tertiary. In particular it would be valuable to have a detailed sequence which would illuminate how the shallow water marine fauna of the Southern Ocean has responded to the periodic glacial maxima of the past few million years. Until then (and probably afterwards, should such deposits be found) we will need to rely on both palaeontological and biological evidence for reconstructing the history of the Southern Ocean marine fauna.

Affinities of the modern fauna

An enormous amount of descriptive and taxonomic work has been undertaken on the Southern Ocean marine fauna (see summary by Dell 1972). Although a great deal still needs to be done, especially on the smaller groups, the major taxonomic affinities of the Southern Ocean fauna are now broadly understood. Knox & Lowry (1977) have suggested that the Southern Ocean fauna as a whole is composed of:

(1) A relict autochthonous fauna
(2) A fauna derived from adjacent deep-water basins
(3) A fauna dispersing from South America along the Scotia arc
(4) A fauna which has spread in the opposite direction from Antarctica northwards along the Scotia arc.

The Southern Ocean fauna thus consists of a mixture of taxa with differing biogeographic affinities. This suggests that, as would be expected, there is unlikely to be a single 'origin'

for the overall fauna. Rather there are a variety of different origins, depending on the taxonomic group or level in question.

One striking feature of the Southern Ocean fauna is the very high level of endemism in many taxonomic groups (Table 1). This undoubtedly reflects the relative isolation of the fauna since the inception of the Polar Frontal Zone (or Antarctic Convergence) which has limited gene exchange with lower latitudes.

Physiology of cold adaptation

The way marine organisms cope with the differing seawater temperatures in the varying parts of the globe has long attracted attention. For the endothermic (warm-blooded, or homoiothermic) mammals and birds the key factor is to maintain a constant internal body temperature, which is generally between 35 and 40°C depending on the species. This is usually achieved in polar regions by increasing insulation (fur, feathers, blubber), as well as by behavioural means such as huddling.

Ectothemic (cold-blooded, or poikilothemic) organisms in general have only limited capabilities for regulating their internal temperature and marine ectotherms virtually none at all. This means, for example, that tropical species must function at about 25 to 30°C, many temperate species anywhere between 5 and 25°C depending on the location and season, and some polar species within a range of no more than 0.1°C either side of −1.8°C. These temperatures apply to fish, all marine invertebrates, marine algae, together with all prokaryotes and unicellular eukaryotes.

Living in the sea can present two distinct problems in relation to temperature. The first is evolving to operate most effectively at a more or less constant annual temperature, the second is evolving to cope with a seasonally (or in some cases daily) varying temperature. Adapting to a stable low temperature (as in the Southern Ocean) and adapting to seasonally varying temperature (such as is found in many cool temperate shallow-water marine environments) are two vary different problems physiologically. We should not expect organisms necessarily to have evolved the same means for solving these two problems.

Apart from the special case of antifreeze in fish (DeVries 1984), recent physiological studies of polar marine organisms have concentrated on molecular mechanisms of compensation for temperature in fish white muscle (Johnston 1985) and fish brain tubulin (Dietrich et al. 1987). These, and other, studies (summarized

Table 1. *Incidence of endemism in selected groups of Southern Ocean marine invertebrates (−, no data). Data extracted from Picken (1985) and White (1984)*

Taxon (Class)	Number of species recorded	Percentage of endemic	
		genera	species
PORIFERA	300	−	50
COELENTERATA			
Scleractinia	10	−	60
Actinaria	31	−	74
BRACHIOPODA	16	−	69
PYCNOGONIDA	100	14	90
ANNELIDA			
Polychaeta	650	5	57
BRYOZOA	321	−	58
CRUSTACEA			
Cirripedia	37	−	76
Isopoda + Tanaidacea	229	10	66
ECHINODERMATA		27	73
Echinoidea	44	25	77
Holothuroidea	38	5	58
CHORDATA			
Ascidia	129	−	27

recently by Clarke 1987*a*) have shown that compensation for temperature is generally good, though not always perfect, and it involves subtle adaptations at the molecular level as well as at the physiological and structural levels. These results have so far been confined to fish. However since the Southern Ocean fauna contains representatives of all taxonomic groups (at least down to the level of class and probably order as well), it is likely that broadly similar compensation has been achieved in most marine organisms.

It has long been felt that the slow rates of growth, extended development times and low metabolic rates of most cold-water ectotherms were all due to a direct rate-limiting effect of low temperature (for example, that marine invertebrates grow slowly at high latitudes simply because the water is cold). This explanation leads to something of a paradox: namely that evolution would appear to have been able to produce compensation for temperature in some processes (locomotion, for example) but not others (such as growth or 'metabolism'). A more detailed examination reveals that growth in many species is slow not because of temperature limitation but because of a seasonal shortage of food (Clarke 1988).

The low rates of basal (that is maintenance, resting or 'standard') metabolism at low temperatures are not so easy to explain. Here more than anywhere the explanation of a direct rate-limitation by temperature is intuitively appealing. In fact the explanation is complex and is likely to be related to the low requirements for protein turnover and osmotic work in ectotherms living at low temperatures (Clarke 1983, 1987*a*). Viewed in this light the low rates of basal metabolism in polar organisms are a positive energetic advantage, for they lead to an increased ecological growth efficiency because of lower maintenance costs (Clarke 1987*b*). Contrary to the traditional and anthropocentric picture of a cooling of the seas representing a climatic 'deterioration', a lowered seawater temperature confers a positive energetic advantage. The disadvantage (if indeed it is one) of living in polar seas comes from the severe seasonality of resource availability.

The relationship between physiology and the evolution of a cold-water fauna at high latitudes is examined in more detail in a companion paper elsewhere (Clarke, in press). However the diversity and biomass of the modern fauna and evidence from recent detailed studies of adaptation to cold would suggest that living at low temperatures poses no insuperable evolutionary problems for marine organisms. Indeed from an energetic viewpoint it may be more of a problem to live in tropical waters.

Discussion and Conclusions: the origin of the Southern Ocean marine fauna

A single origin for the fauna?

All present faunas consist of a mixture of species

with different histories. Some have been present a long time, others are newly arrived. Some groups may be undergoing speciation, others in the process of becoming extinct; some ranges are expanding, others contracting, and some will be more or less stable. Clearly there can be no such thing as an 'origin' for such a fauna; the origin will depend on the particular species, or group of species, in question.

However in one sense we can enquire about the origin of the Southern Ocean marine fauna, and that is in regard to its age. In essence we are asking whether the whole fauna is a new arrival, colonising from elsewhere a shallow water environment only recently made available by the retreat of the ice-cap from the previous glacial maximum. Unfortunately the evidence from different taxonomic groups is conflicting. Teleost fish suggest the expansion of a single group into an ecological vacuum, whereas the gastropods indicate a long history of evolution in situ.

The origin of the Southern Ocean marine fauna

The evidence bearing on the origin of the Southern Ocean marine fauna may be summarized as follows.

(1) Throughout the Mesozoic and Tertiary (Paleogene/Neogene) there have been extensive areas of shallow water associated with Gondwana, and subsequently with the different continental blocks produced by its break-up.

(2) The faunas inhabiting these areas would have had affinities with both the Tethys and Pacific basins (Fig. 1).

(3) Fossil remains of Late Cretaceous and early Tertiary marine faunas indicate a relatively rich and diverse fauna. These faunas included several groups which are only poorly represented in the living fauna (notably decapods and fish), and several groups which predate their appearance in low latitude fossil communities by 10–15 Ma.

(4) Both surface and bottom waters have cooled, more or less steadily, since the Cretaceous to the present. There were three

periods of more rapid cooling (roughly 35–40, 15–20 and 4–5 Ma BP).

(5) During the current glaciation there have been four (excluding the present) periods when extension of the ice-caps reduced greatly the living space available for shallow water marine organisms. Whether this living space was at any time reduced to zero throughout the Southern Ocean is not presently known.

(6) The present fauna contains representatives of virtually all marine groups, frequently at high biomass, and is generally diverse. Some groups show a low diversity, notably gastropods, bivalves, decapods and fish.

(7) Physiological research indicates that, although some of the rate-limiting effects of temperature apparently cannot be circumvented, in general adaptation to live in seawater at temperatures close to zero presents no insuperable evolutionary problem.

The generally high diversity (species-richness) of many groups in the Southern Ocean, the wide range of affinities demonstrated by biogeographic studies (Dell 1972; Knox & Lowry 1977), the subtle molecular adaptation to cold revealed by physiological studies and the likely presence of suitable shallow-water habitat since at least the early Mesozoic all point to a long evolutionary history in situ for much of the Southern Ocean fauna.

Although a number of groups may show particular biogeographical affinities (for example certain isopods have clearly colonised the shallow-waters from the deep-sea: Hessler & Thistle 1975), other groups appear to be in the process of speciating (for example: amphipods: Watling & Thurston 1989; littorinids: Arnaud & Bandel 1976; buccinids: Dell 1972), and others to have all but vanished (for example decapods and oysters). Overall the fauna clearly has had a long evolutionary history and thus has no single origin. In a sense, it has always been there.

We thank P. F. Barker, J. H. Lipps and M. R. A. Thomson for useful discussions and critically reading the final manuscript. J. D. Bradshaw (University of Canterbury) is thanked for his assistance with the Gondwana reconstruction presented in Fig. 1.

References

ANDERSON, J. B. In press. Glacial erosion of the Antarctic shelf, shelf sedimentation, and deep sea fan evolution. *In*: THOMSON, M. R. A., CRAME, J. A. & THOMSON, J. W. (eds) *Geological evolution of Antarctica*. Cambridge University press, Cambridge.

ANDERSON, N. C. 1984. Genera and subfamilies of the family Nototheniidae (Pisces, Perciformes) from the Antarctic and Subantarctic. *Steenstrupia*, **10**, 1–34.

ANDRIASHEV, A. P. 1968. The problem of the life community associated with Antarctic fast ice. *In*:

CURRIE, R. I. (ed.) *Symposium on Antarctic oceanography, Santiago, Chile, 13–16 September 1966 (SCAR/SCOR/IAPO/IUBS)*. Scott Polar Research Institute, Cambridge, 147–155.

ARNAUD, P. M. 1974. Contribution a la bionomie marine benthique des regions Antarctiques et Subantarcticques. *Tethys*, **6**, 467–653.

— & BANDEL, K. 1976. Comments on six species of marine Antarctic Littorinacea (Mollusca, Gastropoda). *Tethys*, **8**, 213–230.

—, JAZDZEWSKI, K., PRESLER, P. & SICINSKI, J. 1986. Preliminary survey of benthic invertebrates collected by Polish Antarctic expeditions in Admiralty Bay (King George Island, South Shetland Islands, Antarctica). *Polish Polar Research*, **7**, 7–24.

ASKIN, R. A. 1989. Endemism and heterochroneity in the Seymour Island. Campanian to Paleocene palynofloras: implications for origins, dispersal and palaeoclimates of southern floras. *In*: CRAME, J. A. (ed.) *Origins and Evolution of the Antarctic Biota*. Geological Society, London, Special Publication, **47**, 107–120.

BARKER, P. F. & BURRELL, J. (1977). The opening of Drake Passage. *Marine Geology*, **25**, 15–34.

BARNARD, J. L. 1969. The families and genera of marine gammaridean Amphipoda. *Bulletin of the United States National Museum*, **271**.

BARRETT, P. J., ELSTON, D. P., HARWOOD, D. M., McKELVEY, B. C. & WEBB, P. N. 1987. Mid-Cenozoic record of glaciation and sea-level change on the margin of the Victoria Land basin, Antarctica. *Geology*, **15**, 634–637.

BIRKENMAJER, K. 1987. Oligocene-Miocene glaciomarine sequences of King George Island (South Shetland Islands), Antarctica. *Palaeontologia Polonica*, **49**, 9–36.

—, DELITALIA, M. C., NAREBSKI, W., NICOLETTI, M. & PETRUCCIANI, C. 1986. Geochronology of Tertiary island-arc volcanics and glacigenic deposits, King George Islands, South Shetland islands (West Antarctica). *Bulletin of the Polish Academy of Sciences, Earth Sciences*, **34**, 257–273.

BRIGGS, J. C. 1987. *Biogeography and plate tectonics*. Elsevier, Amsterdam.

BULLIVANT, J. S. 1969. Bryozoa. *American Geographical Society Antarctic Map Folio Series*, **11**, 22–23.

CARLQUIST, S. 1987. Pliocene *Nothofagus* wood from the Transantarctic Mountains. *Aliso*, **11**, 571–583.

CLARKE, A. 1983. Life in cold water: the physiological ecology of polar marine ectotherms. *Oceanography and Marine Biology: an Annual Review*, **21**, 241–453.

— 1987a. The adaptation of aquatic animals to low temperatures. *In*: GROUT, B. W. W. & MORRIS, G. J. (eds) *The effects of low temperatures on biological systems*. Edward Arnold, London, 315–348.

— 1987b. Temperature, latitude and reproductive effort. *Marine Ecology Progress Series*, **38**, 89–99.

— 1988. Seasonality in the Antarctic marine environment. *Comparative Biochemistry and Physiology* **90B**, 461–473.

— In press. Temperature and evolution: Southern Ocean cooling and the Antarctic marine fauna. *In*: KERRY, K. R. & HEMPEL, G. (eds) *Antarctic ecosystems – change and constants*. Springer-Verlag, Berlin.

CRAME, J. A. 1986. Polar origins of marine invertebrate faunas. *Palaios*, **1**, 616–617.

DALZIEL, I. W. D. & ELLIOT, D. H. 1982. West Antarctica: problem child of Gondwanaland. *Tectonics*, **1**, 3–19.

DARLINGTON, P. J. 1965. *Biogeography of the Southern End of the World*. Harvard University Press.

DAYTON, P. K. & OLIVER, J. S. 1977. Antarctic soft-bottom benthos in oligotrophic and eutrophic environments. *Science*, **197**, 55–58.

DELL, R. K. 1972. Antarctic benthos. *In*: RUSSELL, F. S. & YONGE, M. (eds) *Advances in marine biology. Volume 10*. Academic Press, London, 1–216.

— & FLEMING, C. A. 1975. Oligocene-Miocene bivalve Mollusca and other macrofossils from sites 270 and 272 (Ross Sea), DSDP, Leg 28. *In*: HAYES, D. E., FRAKES, L. A. et al. (eds) *Initial Reports of the Deep Sea Drilling Project*. U.S. Government Printing Office, Washington, **28**, 693–703.

DETTMANN, M. E. 1986. Significance of the Cretaceous-Tertiary spore genus *Cyatheacidites* in tracing the origin and migration of *Lophosoria* (Filicopsida). *Special papers in Palaeontology* **35**, 63–94.

— 1989. Antarctica: Cretaceous cradle of austral temperate rainforests. *In*: CRAME, J. A. (ed.) *Origins and evolution of the Antarctic Biota*. Geological Societies, London. Special Publication, **47**, 89–106.

DEVRIES, A. L. 1984. Role of glycopeptides and peptides in inhibition of crystallisation of water in polar fishes. *Philosophocal Transactions of the Royal Society of London, Series B*, **304**, 575–588.

DIETRICH, H. W., PRASAD, V. & LUDUENA, R. F. 1987. Cold-stable microtubules from Antarctic fishes contain unique alpha tubulins. *Journal of Biological Chemistry*, **262**, 8360–8366.

EASTMAN, J. T. & DEVRIES, A. L. 1982. Buoyancy studies of notothenioid fishes in McMurdo Sound, Antarctica. *Copeia*, **1982**, 385–393.

— & GRANDE, L. 1989. Evolution of the Antarctic fish fauna with emphasis on the Recent notothenioids. *In*: CRAME, J. A. (ed.) *Origins and Evolution of the Antarctic Biota*. Geological Society, Special Publication, **47**, 241–252.

FINLAY, H. J. & MARVICK, J. 1937. The Wangaloan and associated molluscan faunas of Kaitangata-Green Island subdivision. *New Zealand Geological Survey Palaeontological Bulletin*, **15**.

FRAKES, L. A. 1979. *Climates through geologic time*. Elsevier, Amsterdam.

GAZDZICKI, A. & PUGACZEWSKA, H. 1984. Biota of the 'Pecten Conglomerate' (Polonez Cove formation, Pliocene) of King George Island (South Shetland Islands, Antarctica). *Studia Geologica Polonica*, **79**, 59–120.

GRANDE, L. & EASTMAN, J. T. 1986. A review of Antarctic ichthyofaunas in the light of recent fossil discoveries. *Palaeontology*, **29**, 113–137.

HARDY, P. 1972. Biomass estimates for some shallow-water infaunal communities at Signy Island, South Orkney Islands. *British Antarctic Survey Bulletin*, **31**, 93–106.

HARTMANN, O. 1964. Polychaeta Errantea of Antarctica. *American Geophysical Union Antarctic Research Series*, **3**.

—— 1966. Polychaeta Myzostomidae and Sedentaria of Antarctica. *American Geophysical Union Antarctic Research Series*, **7**.

HESSLER, R. R. & THISTLE, D. 1975. On the place of origin of deep-sea isopods. *Marine Biology* **32**, 155–165.

HODELL, D. A. & KENNETT, J. P. 1986. Late Miocene–Early Pliocene stratigraphy and paleoceanography of the south Atlantic and southwest Pacific Oceans: a synthesis. *Paleoceanography*, **1**, 285–311.

JAZDZEWSKI, K., JURASZ, W., KITTEL, W., PRESLER, E., PRESLER, P. & SICINSKI, J. 1986. Abundance and biomass estimates for benthic fauna of the Admiralty Bay, King George Island, South Shetland Islands. *Polar Biology*, **6**, 5–16.

JOHNSON, L. A. S. & BRIGGS, B. G. 1975. On the Proteaceae: the evolution and classification of a southern family. *Botanical Journal of the Linnaean Society*, **70**, 83–182.

JOHNSTON, I. A. 1985. Temperature adaptation of enzyme function in fish muscle. In: LAVERACK, M. S. (ed.) *Physiological adaptations of marine animals*. Society for Experimental Biology (The Company of Biologists), Cambridge. *Symposia of the Society for Experimental Biology*, **34**, 95–122.

KARCZEWSKI, L. 1987. Gastropods from the Cape Melville Formation (Lower Miocene) of King George Island, West Antarctica. *Palaeontologica Polonica*, **49**, 127–145.

KENNETT, J. P. 1977. Cenozoic evolution of Antarctic glaciation, the circum-Antarctic Ocean, and their impact on global paleoceanography. *Journal of Geophysical Research*, **82**, 3843–3860.

KNOX, G. A. & LOWRY, J. K. 1977. A comparison between the benthos of the Southern Ocean and the North Polar Ocean with special reference to the amphipods and the Polychaeta. *In*: DUNBAR, M. J. (ed.) *Polar oceans*. Arctic Institute of North America, Calgary, 423–462.

KOLTUN, V. M. 1970. Sponges of the Arctic and Antarctic: a faunistic review. *Symposia of the Zoological Society of London*, **25**, 285–297.

LAWVER, L. A. & SCOTESE, C. R. 1987. A revised reconstruction of Gondwanaland. *In*: MCKENZIE, G. D. (ed.) *Gondwana six: structure, tectonics and geophysics*. American Geophysical Union, Washington, D.C. *Geophysical Monograph* **40**, 17–23.

LAWVER, L. A., SCLATER, J. G. & MEINKE, L. 1985.

Mesozoic and Cenozoic reconstructions of the South Atlantic. *Tectonophysics*, **114**, 233–254.

LEG 113 SHIPBOARD SCIENTIFIC PARTY. 1987. Glacial history of Antarctica. *Nature*, **328**, 115–116.

LEG 119 SHIPBOARD SCIENTIFIC PARTY. 1988. Early glaciation of Antarctica. *Nature*, **333**, 303–304.

LIPPS, J. H. & HICKMAN, C. S. 1982. Origin, age and evolution of Antarctic and deep-sea faunas. *In*: ERNST, W. G. & MORIN, J. G. (eds) *The environment of the deep sea (Rubey Volume II)*. Prentice Hall, Englewood Cliffs, New Jersey, 325–356.

——, KREBS, W. N. & TEMNIKOW, N. K. 1977. Microbiota under Antarctic ice shelves. *Nature*, **265**, 232–233.

——, RONAN, T. E. & DELACA, T. E. 1979. Life below the Ross Ice Shelf, Antarctica. *Science*, **203**, 447–449.

LOWRY, J. K. 1975. Soft bottom macrobenthic community of Arthur Harbour, Antarctica. *Antarctic Research Series (American Geophysical Union)*, **23**, 1–19.

MARGALEF, D. R. 1958. Information theory in ecology. *General Systems*, **3**, 36–71.

MILLER, R. G. 1987. Origins and pathways possible for the fishes of the Antarctic Ocean. *Proceedings of the Vth Congress of European Ichthyologists, Stockholm 1985*, 373–380.

MOORE, D. M. 1972. Connections between cool temperate floras, with particular reference to southern South America. *In*: VALENTINE, D. H. (ed.) *Taxonomy, phytogeography and evolution*. Academic Press, London, 115–138.

NICOL, D. 1967. Some characteristics of cold-water marine pelecypods. *Journal of Paleontology*, **41**, 1330–1340.

PIANKA, E. R. 1966. Latitudinal gradients in species diversity: a review of the concepts. *American Naturalist*, **100**, 33–46.

PICKARD, J. ADAMSON, D. A., HARWOOD, D. M. MILLER, G. H., QUILTY, P. G. & DELL, R. K. 1988. Early Pliocene marine sediments, coastline and climate of East Antarctica. *Geology*, **16**, 158–161.

PICKEN, G. B. 1985. Marine habitats — benthos. *In*: BONNER, W. N. & WALTON, D. W. H. (eds) *Antarctica*, Pergammon Press, Oxford, 154–172.

POWELL, A. W. B. 1951. Antarctic and Subantarctic Mollusca: Pelecypoda and Gastropoda. *Discovery Reports*, **26**, 47–196.

—— 1965. Mollusca of Antarctic and Subantarctic seas. *In*: WEISBACH, W. W. & VAN OYE, P. (eds) *Biogeography and ecology in Antarctica*. Dr W Junk Publishers, The Hague, 333–80.

RAVEN, P. H. & AXELROD, D. I. 1974. Angiosperm biogeography and past continental movements. *Annals of the Missouri Botanic Garden* **61**, 539–673.

RICHARDSON, M. D. 1976. *The classification and structure of marine macrobenthic assemblagies at Arthur Harbour, Anvers Island, Antarctica*. PhD thesis, Oregon State University.

—— & HEDGPETH, J. W. 1977. Antarctic soft-bottom macrobenthic community adaptations to a cold, stable, highly productive, glacially affected environment. *In*: LLANO, G. A. (ed.) *Adaptations within Antarctic ecosystems*. The Smithsonian Institution, Washington D.C., 181–196.

SANDERS, H. L. 1968. Marine benthic diversity: a comparative study. *American Naturalist*, **102**, 243–282.

SAVIN, S. M. 1977. The history of the earth's surface temperature during the past 100 million years. *Annual Review of Earth & Planetary Science*, **5**, 319–355.

SCHUSTER, R. M. 1969. Problems of antipodal distribution in lower land plants. *Taxon*, **18**, 46–91.

—— 1979. On the persistence and dispersal of transantarctic Hepaticae. *Canadian Journal of Botany*, **57**, 2179–2225.

SCHOPF, T. J. M. 1984. Climate is only half the story in the evolution of organisms through time. *In*: BRENCHLEY, P. J. (ed.) *Fossils and climate*. John Wiley, Chichester, 278–289.

SCOTT, G. K., FLETCHER, G. L. & DAVIES, P. L. 1986. Fish antifreeze proteins: recent gene evolution. *Canadian Journal of Fisheries and Aquatic Science*, **43**, 1028–1034.

SHACKLETON, N. J. 1979. Evolution of the earth's climate during the Tertiary era. *In*: GANTIER, D. (ed.) *Evolution of planetary atmospheres and climatology of the earth*. Centre Nationale d'Etudes Spatiale, Toulouse, 49–58.

—— 1982. The deep-sea sediment record of climatic variability. *Progress in Oceanography*, **11**, 199–218.

—— & KENNETH, J. P. 1975. Palaeotemperature history of the Cenozoic and the initiation of Antarctic glaciation: oxygen and carbon isotope analyses in DSDP sites 277, 279 and 281. *In*: KENNETT, J. P., HOUTZ, R. E. *et al.* (eds) *Initial Reports of the Deep Sea Drilling Project*. U.S. Government Printing Office, Washington, **29**, 743–755.

SOOT-RYEN, T. 1951. Antarctic pelecypods. *Scientific results of the Norwegian Antarctic Expedition 1927–28*, **32**, 1–46.

STEHLI, F. G. 1968. Taxonomic diversity gradients in pole location: the recent model. *In*: DRAKE, E. T. (ed.) *Evolution and environment*. Yale University Press, New Haven, pp. 163–227.

——, DOUGLAS, R. G. & NEWELL, N. D. 1969. Generation and maintenance of gradients in taxonomic diversity. *Science*, **164**, 947–949.

——, MCALESTER A. L. & HELSLEY, C. E. 1967. Taxonomic diversity of recent bivalves and some implications for geology. *Geological Society of America Bulletin* **78**, 455–466.

STOREY, B. C. In press. The crustal blocks of West Antarctica within Gondwana: reconstruction and break-up model. *In*: THOMSON, M. R. A., CRAME, J. A. & THOMSON, J. W. (eds) *Geological evolution of Antarctica*. Cambridge University Press, Cambridge.

TAYLOR, J. D. 1981. The evolution of predators in the late Cretaceous and their ecological significance. *In*: FOREY, P. L. (ed.). *The evolving biosphere*. British Museum (Natural History) and Cambridge University Press, Cambridge,

229–240.

TURNER, R. D. 1967. A new species of fossil *Chlamys* from Wright Valley, McMurdo Sound, Antarctica. *New Zealand Journal of Geology and Geophysics*, **10**, 446–454.

USCHAKOV, P. V. 1963. Quelques particularites de la bionomie benthique de l'Antarctique de l'Est. *Cahiers de Biologie*, **4**, 81–89.

VALENTINE, J. W. 1968. Climatic regulation of species diversification and extinction. *Geological Society of America Bulletin*, **79**, 273–276.

—— 1984a. Neogene marine climate trends: implications for biogeography and evolution of the shallow-sea biota. *Geology*, **12**, 647–650.

—— 1984b. Climate and evolution in the shallow sea. *In*: BRENCHLEY, P. J. (ed.) *Fossils and climate*. John Wiley, Chichester, 265–277.

VERMELJ, G. J. 1978. *Biogeography and adaptation: patterns of marine life*. Harvard University Press, Cambridge, Mass.

WALLWORK, J. A. 1973. Zoogeography of some terrestrial micro-Arthropoda in Antarctica. *Biological Reviews*, **48**, 233–259.

WATLING, L. & THURSTON, M. H. Antarctica as an evolutionary incubator: evidence from the cladistic biogeography of the amphipod family Iphimediidae. *In*: CRAME, J. A. (ed.) *Origins and Evolution of the Antarctic Biota*. Geological Society, London, Special Publication, **47**, 297–313.

WEBB, P. N. In press. A review of the Cenozoic stratigraphy and palaeontology of Antarctica. *In*: THOMSON, M. R. A., CRAME, J. A. & THOMSON, J. W. (eds). *Geological evolution of Antarctica*. Cambridge University Press, Cambridge.

WEBB, P. N., HARWOOD, D. M., MCKELVEY, B. C., MERCER, J. H. & STOTT, L. D. 1984. Cenozoic marine sedimentation and ice-volume variation on the East Antarctic craton. *Geology*, **12**, 287–291.

WHITE, M. G. 1984. Marine Benthos. *In*: LAWS, R. M. (ed.) *Antarctic ecology*, Volume 2. Academic Press, London, 421–461.

WHITE, M. G. & ROBINS, M. W. 1972. Biomass estimates from Borge Bay, Signy Island, South Orkney Islands. *British Antarctic Survey Bulletin*, **31**, 45–50.

WIEDMAN, L. A., FELDMANN, R. M., ZULLO, V. A. & MCKINNEY, M. L. 1986. Antarctic Eocene marine macroinvertebrates: ecological pioneers. *Geological Society of America, Abstracts with programs*, **18**, 231.

WILCKENS, O. 1907. Die Lamellibranchiaten, Gastropoden, &c., der oberen Kreide Südpatagoniens. *Berichte der Naturforschen den Gesellschaft zu Frieberg*, **15**, 97–166.

—— 1910. Die Anneliden, Bivalven und Gastropoden der antarktischen Kreidenformation. *Wissenschaftliche Ergebnisse der Schwedischen Südpolar-Expedition 1901–1903*. **3**, No. 12, 1–132.

—— 1912. Die Mollusken der antarktischen Tertiaformation. *Wissenschaftliche Ergebnisse der Schwedischen Südpolar-Expedition 1901–1903*,

3, No. 13, 1−42.

—— 1922. The Upper Cretaceous gastropods of New Zealand. *New Zealand Geological Survey Palaeontological Bulletin* **9**, 42 pp.

WILSON, M. T. & FELDMANN, R. M. 1986. Comparative functional morphology of fossil and Recent species of the brachyuran decapod genus *Lyreidus*. *Geological Society of America, Abstracts with Programs*, **18**, 331.

WOODBURNE, M. O. & ZINSMEISTER, W. J. 1982. Fossil land mammal from Antarctica. *Science*, **218**, 284−286.

YALDWYN, J. C. 1965. Antarctic and Subantarctic decapod Crustacea. *In*: VAN MEIGHEM, J. & VAN OYE, P. (eds) *Biogeography and ecology in Antarctica. Monographiae Biologicae*, **15**, Junk, The Hague, 323−332.

ZINSMEISTER, W. J. 1980. Marine terraces of Seymour Island. *Antarctic Journal of the United States*, **15** (No 5), 25−26.

—— 1984. Late Eocene bivalves (Mollusca) from the La Meseta formation, collected during the 1974−1975 joint Argentine-American expedition to Seymour Island, Antarctic Peninsula. *Journal of Paleontology*, **58**, 1497−1527.

—— & CAMACHO, H. H. 1982. Late Eocene (to possibly earliest Oligocene) molluscan fauna of the La Mesata Formation, Antarctic Peninsula. *In*: CRADDOCK, C. (ed.) *Antarctic geoscience.* University of Wisconsin Press, Madison, Wisconsin, pp. 299−304.

—— & FELDMANN, R. M. 1984. Cenozoic high latitude heterochroneity of southern hemisphere marine faunas. *Science* **224**, 281−283.

—— & WEBB, 1982. Cretaceous-Tertiary geology and paleontology of Cockburn Island. *Antarctic Journal of the United States*, **17**(No 5), 41−42.

Origins and evolution of Antarctic marine mammals

R. EWAN FORDYCE

Department of Geology, University of Otago, P.O. Box 56, Dunedin, New Zealand

Abstract: Austral marine mammals evolved in a Southern Ocean influenced by changing post-Gondwana geography and consequent new water-masses, progressive cooling, increased thermal gradients and physical and biological seasonality. The biology of extant species and temperate- to high-latitude fossils suggests, but does not confirm, that such factors have governed evolution, and that details of marine mammal evolution therefore differed from those of the north. Cladistic analyses may eventually allow better interpretation of the record.

Cetacea were present in high southern latitudes by the late Eocene. Mysticetes appeared in the south, where they may have originated, about the early Oligocene. They were diverse by the late Oligocene. Early records of Odontoceti are more ambiguous, but they were diverse by the latest Oligocene, when some taxa had circum-Antarctic distributions. The Neogene records of Austral Odontoceti and Mysticeti broadly complement the taxonomic and ecological diversification seen in the north. Amongst largely southern seals, lobodontine Phocidae radiated in the south about the late Miocene, and perhaps ecologically excluded the later arctocephaline Otariidae from high-latitude pagophilic lifestyles. There are no reports of the important northern groups Odobenidae, Sirenia or Desmostylia. Overall, southern marine mammal faunas were ecologically and/or taxonomically similar to extant faunas by the later Miocene to Pliocene.

The extensive Southern Ocean (Figs 1 and 2) has existed since the later Paleogene when the final separation of Australia and South America from Antarctica induced gradual Antarctic cooling, stronger latitudinal thermal gradients, and the development of modern southern oceanic patterns. Some characteristics of the Southern Ocean are therefore more ancient than those of the Arctic. The climate and oceanic conditions of the south, which reflect major geographical change early in the Cenozoic, appear to govern the distribution and abundance of extant southern cetaceans (whales, dolphins), otariids (fur seals, sea lions) and phocids (true seals). These are amongst the most unusual mammals: the marine mammals, an ecologically unified group of otherwise diverse taxa (Table 1) whose distant ancestors returned to the sea earlier in the Cenozoic perhaps to exploit food resources untapped since the demise of Cretaceous large marine reptiles.

Current interpretations of the phylogeny of marine mammals reasonably suggest origins from terrestrial ancestors in temperate to warm northern latitudes (e.g. Barnes *et al.* 1985). Some interpretations have also viewed most later phases of marine mammal evolution as northern, a notion not embraced here. Rather, the Austral fossil record and the adaptations and distribution of extant Austral marine mammals indicate that the south may have been more important in evolution than acknowledged traditionally. This article reviews the evidence for the origins and evolution of marine mammals in, and in response to the development of, the Southern Ocean: the extensive cool-temperate to polar southern waters.

Evolutionary biology

Relevant aspects of the evolutionary biology of marine mammals were summarized by, e.g., Barnes *et al.* (1985), Fordyce (1980), Gaskin (1982), de Muizon (1982), and Ray (1977, and other contributors in Repenning 1977). Most mammals are well-adapted terrestrial organisms, and a secondary aquatic lifestyle is associated with profound changes in structure. Structures in marine mammals might be viewed as constrained (e.g. Lauder 1981) in the following ways.

(1) Historical factors. Despite adaptations to a marine lifestyle (2 and 3 below), marine mammals have inherited structures that indicate both distant terrestrial ancestry, and groups within groups at lower ranks. Interrelationships, reflected in such groups, are best assessed cladistically (e.g. Fig. 3).

(2) Functional factors. Marine mammals differ markedly from their terrestrial relatives in functional complexes adapted for aquatic life, e.g. feeding, locomotion (both limbs and body profiles), respiration and acoustics. When identified in fossils, these may indicate the minimum times of evolution of new lifestyles, per-

From Crame, J. A. (ed.), 1989, *Origins and Evolution of the Antarctic Biota,*
Geological Society Special Publication No. 47, pp. 269–281.

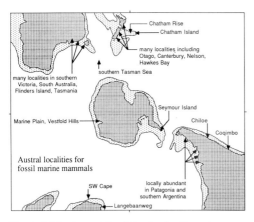

Fig. 1. Main Austral localities for fossil marine mammals, shown on modern geography. Based on references cited in text.

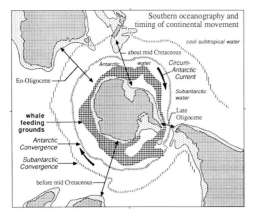

Fig. 2. Main features of the modern Southern Ocean and timing of break up of the southern continents, shown on modern geography.

haps in response to environmental influence.

(3) Structural or fabricational factors. Marine mammals are relatively large, with a lower surface area to volume ratio than usual for terrestrial mammals. Such a ratio perhaps reflects

constraints of thermoregulation and/or buoyancy. Other structural ratios warrant study in an evolutionary context, e.g. the relative surface area of limbs, used in thermoregulation and/or propulsion, and the surface area of the feeding apparatus relative to body mass.

The marine habits and large size of marine mammals lead to problems of study which in turn constrain interpretations of history. Extant and fossil specimens are difficult to handle because of large size. Studies may be limited because of legal restraints and/or public pressure. These points compound problems of rarity; many taxa are poorly known and some fossils are known from single specimens. Perhaps because good fossils are rare, many structures are, or have been, of uncertain homology and function, and many incomplete and debatably diagnostic elements have been used as type specimens. The broader evolution of marine mammals has largely been interpreted stratophenetically (e.g. the oldest fossils have often been identified as ancestors and/or as evidence for a centre of origin for a group). Such an approach may suit abundant organisms from successive close-spaced horizons, but seems dubious for fossil marine mammals. Stratophenetics has not dealt well with 'relict' taxa or the question of centres of origin, and a cladistic approach would seem to deal better with the sparse marine mammal fossils and their palaeozoogeography. All other things being equal, geographically restricted branches in a clade are assumed here to indicate a restricted area of radiation, with early branches indicating possible centres of origin. There is not the space to consider little-discussed broader issues of speciation (anagenetic or cladogenetic, gradualistic or punctuated, via dispersal or vicariance).

The distant ancestors of marine mammals returned to the sea earlier in the Cenozoic presumably to exploit food resources, rather than through ecological displacement by other terrestrial groups. Later major evolutionary and ecological changes amongst marine mam-

Table 1. *Main taxa of marine mammals*

Taxon (clade)	Global record	Southern fossil records
Cetacea (whales, dolphins)	early Eocene–Recent	late Eocene and younger
Phocidae (earless seals)	early Miocene–Recent	mid-Miocene and younger
Odobenidae (walruses)	middle Miocene–Recent	none
Otariidae (fur seals, sea lions)	latest Oligocene–Recent	early Pliocene
Lutridae (otters)	Quaternary (marine spp)	none
Ursidae (bears)	Quaternary (marine spp)	none
Sirenia (sea cows)	early Eocene–Recent	none
Desmostylia (desmostylians)	Oligocene–Miocene	none

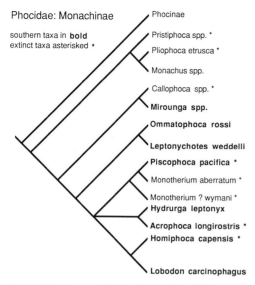

Phocidae: Monachinae

southern taxa in **bold**
extinct taxa asterisked *

Phocinae
Pristiphoca spp. *
Pliophoca etrusca *
Monachus spp.
Callophoca spp. *
Mirounga spp.
Ommatophoca rossi
Leptonychotes weddelli
Piscophoca pacifica *
Monotherium aberratum *
Monotherium ? wymani *
Hydrurga leptonyx
Acrophoca longirostris *
Homiphoca capensis *
Lobodon carcinophagus

Fig. 3. Cladogram of interrelationships of Recent and fossil Phocidae: Monachinae, based on de Muizon 1982. See Wyss (1988, fig. 9) for an alternative.

mals also presumably related to food resources. For example, present and therefore past cetacean distributions relate to sea temperature zones which govern food distribution and abundance (Brownell 1974, p. 15; Laws 1977a, b and references therein). Seal behaviour and ecology reflects an interplay between the need to exploit offshore resources and breed onshore (Bartholomew 1970, contributors in Repenning 1977), although little is known of the factors which constrain specific distributions. Major evolutionary and ecological changes should have occurred in areas of high and/or localized productivity (the much-discussed 'centres of origin' or of radiation), and the Southern Ocean was probably one such area for some groups (Fordyce 1980, and below). In practice centres of origin have been difficult to recognise (e.g. Fordyce 1985, p. 96), since they rely on sound taxonomy and a good idea of geographical and stratigraphic distribution. Is absence of specimens from one area real (taxa never there) or apparent (taxa not found)? Most described fossils are from the Northern Hemisphere, a bias which may explain why many groups are regarded as of northern origin. Current knowledge of southern fossil marine mammals reflects a shorter history of serious study, a relatively smaller published record, and relatively less land (and hence potentially fossiliferous outcrop) in cool temperate latitudes.

Geographic and taxonomic scope

This article concerns taxa and events mostly from latitudes south of 45°S (the Southern Ocean) which is dominated by both circumpolar and meridional features (e.g. Circum-Antarctic Current, Antarctic Convergence). Fossil marine mammals usually come from shallow marine sediments exposed on land; little is known of marine mammal history on continental shelves, and less about that in the broad ocean basins. The Southern Ocean between 50°S and 65–70°S is largely barren (but see Whitmore et al. 1986), although fossil material is known from mid latitude shelves (e.g. Fordyce 1984b). Two key fossil localities on land in Antarctica (Vestfold Hills and Seymour Island) are reviewed below. Fossils from modern cool to warm temperate southern regions (Figs 1 and 4) provide some idea of ancient Austral faunas, since all extant Southern Ocean Cetacea and occasionally Austral pagophilic seals also occur further north. Of the Austral fossil marine mammals covered below, the Cetacea were reviewed recently by Fordyce (1985), and comments on southern taxa were made by e.g. Barnes et al. (1985), Berta & Deméré (1986), de Muizon (1982) and Wyss (1988). Phocid and otariid seals are considered separately herein, although Wyss has suggested that they may represent a monophyletic Pinnipedia. Other articles are cited below.

Stratigraphy

The stratigraphic record of Austral marine mammals (Tables 1, 2) extends back to the Eocene. Successions of local stages are generally reliable, but it is difficult to correlate from higher southern latitudes to the European Tertiary type-sections. Heterochrony may complicate high latitude correlation (Zinsmeister & Feldmann 1984), and biostratigraphic indicators otherwise regarded as reliable may have diachronous ranges (e.g. *Globoquadrina dehiscens* in New Zealand, cf. further north). Thus, some ages cited below are vague (e.g. latest Oligocene–earliest Miocene). Such problems emphasize the possible evolutionary role of high latitudes. This aside, Table 2 shows the minimum times of occupation of Austral waters by the taxa listed.

High latitude fossils

'Subfossil' seals (within age limits of radiocarbon dating) occur in Antarctica, but only Cetacea are known before the Holocene. There are two localities (Fig. 1).

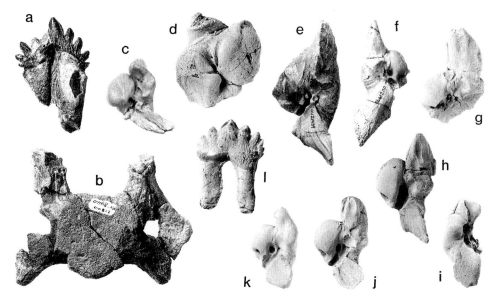

Fig. 4. Selected marine mammal fossils from mid to high southern latitudes: (**a**) middle or posterior cheek-tooth (lingual face), undescribed species of? Mysticeti, late Eocene−earliest Oligocene, Seymour Island, UCR uncatalogued (field number REF S1−67), × 0.39; (**b**) cervical vertebra of undescribed species of Cetacea (posterior face), late Eocene−earliest Oligocene, Seymour Island, Antarctica, UCR 21800 (field number RV 8440), × 0.39; (**c**) left periotic of undescribed species of Delphinidae (ventral face), early Pliocene, Marine Plain, Vestfold Hills, Antarctica, CPC 25730, × 0.78; (**d**) right temporal of undescribed species of Phocidae (ventral face), late Miocene, Beaumaris, MV P160399, × 0.51; (**e**) lef periotic of undescribed species of Mysticeti (ventral face), late early Oligocene, Hakataramea, New Zealand, OU21939, × 0.39; (**f**) right periotic of undescribed species of Mysticeti (ventral face), late Oligocene, Kokoamu, New Zealand, OU21975, × 0.39; (**g**) left periotic of undescribed species of Odontoceti (ventral face), late Oligocene, Kokoamu, New Zealand, OU11518, × 0.78; (**h**) left periotic of undescribed species of squalodont Odontoceti (ventral face), early or late Oligocene, Aorere, New Zealand, OU21803, × 0.78; (**i**) right periotic of Kogiidae species (ventral face), about late Miocene, Chatham Rise, New Zealand, OU12089, × 0.78; (**j**) left periotic of undescribed species of kentriodontid Odontoceti (ventral face), latest Oligocene or earliest Miocene, Port Waikato, New Zealand, AUGD V9, × 0.78; (**k**) left periotic of Phocoenidae species (ventral face), about late Miocene, Chatham Rise, New Zealand, OU12087, × 0.78; (**l**) middle or posterior cheek-tooth (lingual face), undetermined species of ? Mysticeti, early Oligocene, Weston, New Zealand, GS10897 CD39/2, × 0.51.

Seymour Island

Fossil Cetacea were collected from Seymour Island (64° 15′S) during the Nordenskjöld expedition of 1901−1903. Wiman (1905) described two incomplete vertebrae as those of *Zeuglodon* [=*Basilosaurus*] sp. There have been other comments on these specimens; e.g. Abel (1914) referred them unjustifiably to the monotypic *Kekenodon onamata*, a late Oligocene relict archaeocete from New Zealand for which vertebrae are unknown. Kellogg (1936) listed them as Archaeoceti *incertae sedis*. More recently Elliot *et al.* (1975) reported the discovery of a 'small whale skeleton (*Zeuglodon*) (?)', fragments of which were recovered and have since been under study by E. D. Mitchell.

A concerted effort by Fordyce and others (December 1986−January 1987) located sparse cetacean fossils, mostly from the upper part of the late Eocene to perhaps earliest Oligocene La Meseta Formation (see Woodburne & Zinsmeister 1984 for stratigraphy). One specimen, collected by Zinsmeister from the supposedly Paleocene Cross Valley Formation, is a heavy large (*c* 75 mm diameter) pachyostotic fragment of rib. A possible identification is a species of Basilosaurinae (specialized late middle and late Eocene Archaeoceti). If correctly identified as cetacean, and if in situ, it suggests a middle Eocene or later age for part of the Cross Valley Formation. Otherwise, the Cetacea are currently of no particular stratigraphic value.

Table 2. *Approximate timing of key events in the evolution of marine mammals, oceans, southern climates and southern continents. Key references for the oceanic, climatic and geographic events are: Kennett & von der Borch 1985, Murphy & Kennett 1985, and contributors in Tsuchi 1987.*

million years before present	Epoch		Approximate timing of some events in marine mammal evolution	Approximate timing of oceanic, climatic and geographic events
0	QUAT.		widespread occurrence of neospecies in south earliest cool stenothermal neospecies in north cool stenothermal phocid neospecies (NZ)	
	PLIO-CENE	l		increasing Milankovitch cycle amplitude terrestrial glaciation in S temperate regions first? major buildup of Arctic ice (~2.4 Ma) Panama seaway closed marked increase in Antarctic glaciation
		e	Vestfold Hills fauna oldest southern otariids lobodontine phocids widespread in south	Messinian crisis
	MIOCENE	late	extant genera of Cetacea present globally ancestors to some extant cool stenothermal Arctic neospecies present in north low latitudes last Squalodontidae (worldwide) early Delphinidae (NE & ?SW Pacific), Phocoenidae, Monodontidae	major southen cooling; regression; C13 shift; increased upwelling? increased gyral circln., latitudinal thermal gradients (~8.8 Ma) cooling, sea level drop (~10 Ma)
10		mid	early phocids in S (SW Atlantic) early Balaenopteridae (SW Pacific) diverse faunas in N Atlantic and NE Pacific extinction of Squalodelphidae	cooling; E Antarctic ice sheet buildup
20		early	rather poor published southern record early Balaenidae and Physeteridae (SW Atlantic) and other faunas of modern aspect	global thermal maximum; high sea level warming
			Circum-Antarctic distribution in some odontocetes Otariidae (N. Pacific) and Phocidae (N. Atlantic) radiate diverse cetacean faunas of Neogene aspect (SW Pacific; N Atlantic)	reduction in Antarctic ice? Drake Passage open; probable full Circum Antarctic Current flow
	OLIGOCENE	late	latest occurrences of Archaeoceti (?SW Pacific) diverse early Odontoceti (NE Pacific, NW Atlantic) diverse Mysticeti (toothed & baleen) (SW Pacific - NZ)	sea level glaciation in W Antarctica
30		early	some of globally earliest Odontoceti (SW Pacific - NZ) large Mysticeti with baleen (SW Pacific - NZ) marine mammal fossils rare globally, but probably time of major radiation of Mysticeti and Odontoceti toothed ?early Mysticeti (SW Pacific - NZ) Seymour Island fauna	major sea level drop major cooling and/or accumulation of Antarctic ice; increased latitudinal thermal gradients with subantarctic but not temperate cooling Tasmanian seaway deepens
40	EOCENE	late		first major flow from Indian to Pacific Ocean; cooling of surface and bottom waters; psychrosphere developed; Antarctic Convergence developed (~37 Ma)
		mid	Cetacea (Archaeoceti?) present in SW Pacific ecological diversification reflected in northern faunas of Cetacea	warm equable climates; low pole-tropics surface water temp gradient; Antarctic & Subantarctic Convergences not developed; Antarctica relatively warm.
50		early	earliest Cetacea (northern? Tethys) early Sirenia (Tethys and other northern areas) oceans unoccupied by marine mammals before about Early Eocene	NZ and Africa separated from Antarctica about mid Cretaceous

The relationships of the new fossils are uncertain, since they are still being prepared at University of Otago on loan from University of California (Riverside). If the La Meseta Formation is Eocene, only Archaeoceti would be expected, since the oldest known Mysticeti and Odontoceti are Oligocene. However, the constituent species of the paraphyletic Archaeoceti are identified by their lack of diagnostic cranial features of Mysticeti and Odontoceti. Thus, the postcranial elements (e.g. Fig. 4b), which are relatively common on Seymour Island, cannot be identified reliably to suborder.

A more informative specimen is an incomplete large skull and associated teeth, parts of mandibles, vertebrae and ribs. The specimen is still under preparation, and relationships are uncertain. Its large size (exoccipital width c. 630 mm, length from condyles to narial opening c. 1330 mm) suggests an overall skull length of 2 m (bigger than any described archaeocete) and a body length of 8–10 m. The size and apparent age originally suggested a species of Basilosaurinae, the only large Cetacea hitherto known before the later early Oligocene. This identification now seems unlikely, since the middle to posterior of the rostrum is wide and flat, the posterior cheek-teeth have wide diastemata, and the cheek-teeth are delicate with complex ornament and gracile denticles (Fig. 4a). Such dental features are seen in fragmentary Oligocene fossils from New Zealand and others from the NE Pacific which may be primitive Mysticeti. The flat rostrum on the large skull suggests a species of Mysticeti. Isolated large dense pachyostotic vertebrae from elsewhere in the La Meseta Formation may represent the same taxon. None has the markedly elongate centrum of basilosaurines, and this subfamily may well not be present. Small vertebrae from mature animals represent a small undetermined cetacean species. Odontoceti have not been recognized.

Marine Plain, Vestfold Hills

A myriad of small fragments collected by P. G. Quilty from the marine Pliocene at Marine Plain, Vestfold Hills (Fig. 1; 68.5°S) have been reconstructed to reveal the skull and mandibles of a dolphin (Odontoceti). Part of the rostrum of a second specimen is known. The skull represents a new species, with a high-domed cranium and long apparently toothless jaws. Earbones (Fig. 4c) associated with the skull show synapomorphies of Delphinidae. These fossils, to be described formally by Fordyce and Quilty, appear to be the only higher vertebrates from the Oligocene–Pleistocene interval on Antarctica. Marine Plain is a potentially important locality for other high latitude fossil marine mammals.

The broader record and its interpretation

Record of southern Phocidae: earless seals

Four extant pagophilous species, each in a monotypic genus (e.g. *Lobodon*, crabeater seal), inhabit high southern latitudes (e.g.

Bonner 1982; Laws 1977*a*, *b*). All are placed in the tribe Lobodontini, a clade within the probably paraphyletic subfamily Monachinae (*sensu* Wyss 1988). They are less diverse than northern polar phocids, and are markedly larger (Laws 1977*b*). The relatively few fossil Austral phocids are geologically young, and have been taken to support the idea of a late invasion of the south (e.g. Repenning *et al.* 1979).

Amongst the globally oldest phocids are middle Miocene taxa from around the N Atlantic (Barnes *et al.* 1985, Ray 1977). The southern fossil record includes lower latitude taxa: *Properiptychus argentinus* (middle Miocene, Argentina, about 32°S; de Muizon & Bond 1982) and two monotypic genera of monachines described by de Muizon (1981) from the early Pliocene of Peru (15.5°S), *Acrophoca longirostris* and *Piscophoca pacifica*. *Homiphoca capensis* is known from many specimens from the latest Miocene or earliest Pliocene of Langebaanweg (33°S), South Africa (de Muizon & Hendey 1980; see also Wyss 1988 for cautionary comments on relationships). At higher latitudes, latest Miocene phocid fossils (Fig. 4d; probably lobodontine) are known from Victoria, Australia (37–38°S; Fordyce & Flannery 1983). These specimens are under study by Fordyce and M. P. Beentjes. King (1973) described a fossil jaw of the pagophilous neospecies *Ommatophoca rossi* (Ross seal) from the latest Pliocene of Hawkes Bay (39.5°S). Other undescribed fragments, apparently phocid, are known from the late Pliocene or Pleistocene of New Zealand (Fleming 1968; Fordyce 1982).

Origins of southern Phocidae

De Muizon (1982) followed others (e.g. Ray 1977; Repenning *et al.* 1979) in suggesting that the monachines, otherwise viewed as largely southern, had a northern origin. The oldest and most primitive fossils indicated to de Muizon a European 'original homeland' for monachines (but cf. Wyss 1987, 1988). If this is correct, distributions of southern phocids arose presumably through successive invasions from the north. Such interpretations of palaeozoogeography rely on explicit statements of relationship, such as the cladistic interpretation of fossil and Recent southern Phocidae shown in Fig. 3 (derived from de Muizon 1982, fig. 8). As is usual for cladograms, Fig. 3 does not show ancestors (e.g. *Callophoca* as possibly ancestral to *Mirounga*, elephant seals), and other limits to the use of such published studies must also be recognized.

(1) Approaches to characters and analysis.

Some supposed apomorphies used in de Muizon's cladogram could equally be plesiomorphies or homoplasies, e.g. sexual dimorphism in *Callophoca* (used to substantiate a *Callophoca–Mirounga* link), and proportions of feeding apparatus, likely to have stronger functional and fabricational than historical constraints. Wyss (1988; see also Wyss 1987) in particular recently questioned the polarity of some characters used in both traditional and cladistic analyses of Phocidae. De Muizon's cladogram of 15 taxa is also based on only 20 characters; a larger matrix processed by computer might provide a more robust cladogram. Critical reanalysis and expansion of characters, though, is inappropriate here.

(2) Problems inherent in fossils. Species of *Callophoca* and *Monotherium* are less well known (e.g. Ray 1977) than other taxa listed in Fig. 3; they are represented by incomplete fossils that might not show many apomorphies useful at low ranks, and thus will plot as higher-level relatively plesiomorphic taxa (perhaps strictly, plesions) as far as can be told from the literature. Whether they are demonstrably ancestral to other taxa is debatable.

(3) The incomplete record. The fossil record of southern phocids outside South America and South Africa is poor, with only one Australian specimen described (Fig. 4d; Fordyce & Flannery 1983). Since there has not been a long history of southern interest in fossil marine mammals it would be imprudent to assume that the lack of fossils indicates a real absence of phocids older than latest Miocene.

The cladogram of monachines (Fig. 3) shows mostly southern taxa. The radiation of de Muizon's *Mirounga* to *Lobodon* clade of monachines (Fig. 3) might be viewed most parsimoniously as a southern radiation. Wyss (1988) also interpreted the extant Antarctic lobodontines as forming a clade, which supports the idea of a southern radiation for this group; however, he did not detail the relationships of fossil taxa to the extant lobodontines, to the other members of what he identified as the paraphyletic Monachinae or to the extant wholly northern clade Phocinae. The otherwise prominent role of incomplete northern fossils in the interpretation shown in Fig. 3 is questioned here, since the lack of southern fossils plausibly reflects the intensity of work on southern marine mammals. Indeed, cladistic analysis of extant Phocidae (Wyss 1988, fig. 9) could be taken as evidence of a Pacific–Southern Ocean initial radiation of phocids.

Record of southern Otariidae: fur seals and sea lions

Of eight species of extant fur seal, *Arctocephalus* spp. (Arctocephalinae), only one is wholly northern. Another, *A. gazella*, is circumpolar south of the Antarctic Convergence, and overlaps little in range with the six other cool temperate southern species (Bonner 1982, fig. 4.1). Three of the five extant monotypic genera of sea lions (*Otaria*, *Neophoca*, *Phocarctos*) occupy cool temperate southern waters. There seems to be no southern ecological equivalent to the large bottom-feeding northern walrus, *Odobenus*.

The globally oldest otariid appears to be the late Oligocene–early Miocene *Enaliarctos*, from the NE Pacific (Barnes *et al.* 1985). This area has yielded other important fossils from the Mio-Pliocene (Berta & Deméré 1986). The poor southern fossil record includes *Hydrarctos lomasiensis* (early Pliocene, Peru, 15.5°S; de Muizon 1978), *Neophoca palatina* (early Pleistocene, New Zealand, 38°S; King 1983), and undescribed material (early Pleistocene, New Zealand, 40°S; Fleming 1968). Southern fossil neospecies (Repenning & Tedford 1977) include the fur seal *Arctocephalus pusillus* (late Pleistocene, S Africa, about 33°S), and the sea lions *Neophoca cinerea* (late Pleistocene, Victoria, 38°S), and *Otaria byronia* (late Pleistocene, Argentina, latitude uncertain).

Origins of southern Otariidae

Otariids are thought to have arisen in the N Pacific, whence they ultimately crossed the tropical barrier to colonise the south (e.g. Repenning *et al.* 1979, fig. 1). The oldest southern fur seals are early Pliocene (at low latitudes), and sea lions early Pleistocene. Repenning *et al.* (1979) suggested that otariids invaded high southern latitudes via the Peru Current, and then the Circum-Antarctic Current.

A cladistic interpretation of the taxonomy of the Otariidae is shown in Fig. 5 (derived from Berta & Deméré 1986, fig. 7; but cf. Wyss 1987). Figure 5 does not show ancestors, and the same broad limitations apply to interpretations as for the phocids. In this case, no northern fossil has been identified clearly as the ancestor for a southern group. The simplest interpretation of the three *Arctocephalus* clades is that they represent a southern radiation. As shown, the southern *A. pusillus* group represents a polychotomy of relatively plesiomorphic and perhaps, therefore, early-branching

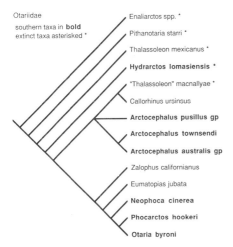

Otariidae

southern taxa in **bold**
extinct taxa asterisked *

Enaliarctos spp. *

Pithanotaria starri *

Thalassoleon mexicanus *

Hydrarctos lomasiensis *

"Thalassoleon" macnallyae *

Callorhinus ursinus

Arctocephalus pusillus gp

Arctocephalus townsendi

Arctocephalus australis gp

Zalophus californianus

Eumatopias jubata

Neophoca cinerea

Phocarctos hookeri

Otaria byroni

Fig. 5. Cladogram of interrelationships of Otariidae, based on Berta & Deméré 1986.

species. The sea lions form a clade, with two earlier-branched northern taxa perhaps suggesting a northern origin for the group, and three southern taxa marking a southern radiation. Hypotheses of cladistic relationships are volatile, and those here are likely to change. In particular, a single well-preserved southern fossil could both usefully calibrate some of the cladogenetic events and allow reinterpretation of relationships, and Wyss (1988) has re-emphasized that alternative assignments of character polarities amongst seals can fundamentally change inferred relationships.

Record of southern Cetacea

Austral Cetacea are much more diverse but perhaps known less perfectly than the seals, and their origins and evolution have not been discussed as widely. Southern biomasses are high amongst extant species, and there is sophisticated ecological partitioning between roughly sympatric species in a superficially homogeneous ocean (e.g. Laws 1977b, fig. 2), but there is apparently no high latitude breeding (Brownell 1974). Highly mobile large Mysticeti (baleen whales) occupy high southern latitudes beyond the Antarctic convergence, mostly for summer feeding. They include four species of rorqual (*Balaenoptera*), the humpback (*Megaptera*) and the right whale (*Eubalaena*), all of which have globally broad antitropical distributions which include mid-to warm-latitude waters (winter breeding) (Laws 1977a, b). The pygmy right whale (*Caperea*) appears restricted to subantarctic and southern cooler subtropical

waters (Ross *et al.* 1975). Odontocetes (toothed whales, dolphins and porpoises) include the cosmopolitan large sperm whale (*Physeter*), which concentrates around the Antarctic Convergence, and nine or more smaller odontocetes which are cosmopolitan or are Austral endemics: the killer, *Orcinus* and smaller Delphinidae, porpoises, Phocoenidae, and beaked whales, Ziphiidae. It seems that few odontocetes are common south of the Convergence, and most may be associated with the Circum-Antarctic Current (Brownell 1974; Laws 1977a, b; Nishiwaki 1977).

The Austral fossil cetacean record is more complex than for seals. Fordyce (1985) covered points not referenced below in the family-level synopsis of the southern marine record, and listed valuable earlier works not cited here. Barnes *et al.* (1985, fig. 1) provided general references to taxonomy and stratigraphy.

Of the Archaeoceti, Basilosauridae (zeuglodons) are known provisionally from the late Eocene. *Kekenodon onamata* (late Oligocene, New Zealand) may be a relict basilosaurid and thus one of the last Archaeoceti (*sensu stricto*).

The oldest Mysticeti, apart from the possible records from Seymour Island, appear to be fragmentary skulls and teeth (currently Family *incertae sedis*) from the earliest Oligocene of New Zealand. A late Oligocene toothed mysticete (currently Family *incertae sedis*, cf. the NE Pacific Family Aetiocetidae), *Mammalodon colliveri*, occurs in Victoria (Fordyce 1984a). The oldest of the Cetotheriidae appear to be specimens from the late early Oligocene of New Zealand (Fig. 4e; Fordyce 1987) which include a large edentulous skull. Other Austral cetotheres are known from the late Oligocene (Fig. 4; New Zealand) and Miocene (Australia, Argentina). Balaenopteridae (rorquals) include indeterminate middle Miocene species (Australia, New Zealand) which appear to be the earliest records (Bearlin 1988), late Miocene and Pliocene Megapterinae, humpbacks (Bearlin 1985; Dathe 1983), and a range of other balaenopterines (e.g. Bearlin 1985, 1988). Austral Balaenidae (right whales) include the oldest (early Miocene) record of *Morenocetus parvus* (Patagonia), and later Miocene and younger specimens not yet described formally. The debatably distinct Neobalaenidae (includes the extant pygmy right whale) is not known from clearly identified fossils. Austral Eschrichtiidae (grey whales) are unknown.

None of the most primitive odontocetes (Agorophiidae *sensu lato*) are known. Squalodontidae include latest Oligocene–earliest Miocene taxa, amongst the oldest in the world, with nominally but not clearly confamilial older specimens (e.g. *Austrosqualodon*, and Figs 4g, h) from the late Oligocene of New Zealand. Earliest Miocene *Prosqualodon* spp. appear to have been circumpolar. Despite reports to the contrary, there are no Austral squalodontids younger than middle Miocene; some northern late Miocene taxa are known. *Notocetus vanbenedeni* (Patagonia) is one of two species of Squalodelphidae (early Miocene; also in Italy). Eurhinodelphidae [=Rhabdosteidae] include early Miocene species (Patagonia; New Zealand, undescribed); early and/or middle Miocene taxa occurred in central Australia, the N Atlantic and N Pacific, but no late Miocene species are known. Ziphiidae, beaked whales (middle Miocene−Recent), include later Miocene and Pliocene Austral specimens (e.g. Australia). Physeteridae (sperm whales) include some of the earliest global records (early Miocene, Patagonia) and fragmentary specimens, some named, from later in the Neogene (e.g. Australia). An isolated earbone from Chatham Rise associated with a shelf assemblage of ill-constrained late Miocene age (Fig. 4i; Fordyce 1984*b*) gives a possible early record of the otherwise poorly-represented Kogiidae, pygmy sperm whales (the oldest firm record is otherwise early Pliocene, NE Pacific). Among the Delphinoidea, there is an undescribed latest Oligocene−earliest Miocene specimen (Fig. 4j; from New Zealand) of the paraphyletic Kentriodontidae, otherwise known from the later early Miocene to early Pliocene in the Northern Hemisphere and perhaps by the poorly known *Incacetus broggi* (Pliocene?, Peru). Delphinidae (dolphins) for which the oldest global records are late Miocene, include presumed early specimens of possible late Miocene age (Fordyce 1984*b*), and widespread later records around southern temperate latitudes. Austral Phocoenidae (porpoises) include an earbone from Chatham Rise associated with a shelf assemblage of ill-constrained late Miocene age (Fig. 4k; Fordyce 1984*b*) and the early Pliocene *Piscolithax longirostris* (Peru). Pontoporiidae are represented by *Pliopontos littoralis*, also from Peru (de Muizon 1984). No Austral fossil Monodontidae (beluga, narwhal) have been described and the family is not certainly extant in the south. Iniidae, Lipotidae and Platanistidae are neither extant in southern seas nor represented by Austral marine fossils.

Origins of southern Cetacea

Discussion of origins is hampered by a lack of published cladograms on the bewildering but stimulating diversity of Cetacea, and it is inappropriate to present and discuss original cladograms here. There appear to be no detailed accounts of the origins of family-level taxa relevant to the Southern Ocean, although cladograms have been presented for some more-northern taxa. In a general account, Fordyce (1980) suggested that the Southern Ocean was important in the origins of Mysticeti and Odontoceti, and reviewed some other concepts of centres of origin (Fordyce 1985).

General discussion

The following interpretive synopsis of Austral marine mammal history is necessarily narrative (see Table 2 for summary). By some time in the late Eocene, archaeocetes had dispersed from warm northern Tethyan origins to equable mid-southern latitudes. Good specimens are unknown, so there is little idea of taxa, adaptive types and likely resource partitioning in the south, but 'large' and 'small' presumed Archaeoceti occur in New Zealand. If archaeocetes were seal-like amphibious land-breeders (Fordyce 1980), they may have been mainly shelf animals which might have coped well with an archipelagic pre-glacial West Antarctica. There is no evidence that archaeocetes could echolocate or filter-feed, unlike their extant descendants. Sirenians and desmostylians, recorded in the north about now, are not known from southern fossils except for a few young low latitude sirenians.

In the latest Eocene and early Oligocene, when southern climates changed significantly, global records of Cetacea are poor. Ironically, this was probably the time when filter-feeding Mysticeti and echolocating Odontoceti evolved from archaeocetes (Fordyce 1980), since these groups are diverse by the early late Oligocene but absent in the late Eocene. These taxa, with new feeding strategies, arose as the psychrosphere, Circum-Antarctic Current, Antarctic Convergence and southern glaciation developed (Kennett & von der Borch 1985; Murphy & Kennett 1985) in response to increasing physical and thermal isolation of Antarctica induced by the final fragmentation of Gondwana.

Seymour Island whales indicate early occupation of high latitudes about the time climatic deterioration started; they may include one of the oldest (latest Eocene or earliest Oligocene) Mysticeti. Fragmentary toothed Cetacea, probably mysticetes (e.g. Fig. 4l), are known from the earlier Oligocene (early Whaingaroan Stage) of what is now E New Zealand, from localities which bounded the Southern Ocean

(reconstruction: Kamp 1986). If correctly identified, these are the oldest mysticetes, animals
which probably filter-fed with multicusped teeth
as does the extant Austral crab-eater seal
Lobodon. Such toothed mysticetes persisted
until the late Oligocene in the SW Pacific
(Victoria) and early Miocene in the NE Pacific.
The globally oldest toothless (baleen-bearing)
mysticetes are large and small specimens from
the New Zealand later early Oligocene (late
Whaingaroan Stage), and diverse late Oligocene
mysticetes also occur there (Fig. 4e, f; Fordyce
1987). Fordyce (1980) suggested that such
records support the idea of a Southern Ocean
origin for Mysticeti, an idea not yet refuted by
fossils from elsewhere and reinforced by continued expansion of the southern Oligocene
record of Mysticeti (Fordyce 1987). Features
other than the early southern records attest to
the role of the Southern Ocean in the origin and
later radiation of Mysticeti. For example, many
aspects of the biology of extant Mysticeti are
adaptations to exploit high-latitude food, the
seasonal abundance of which reflects attributes
of polar climate and oceanography that can be
traced back to the earlier Oligocene. Evolution
occurred in a Southern Ocean that was probably
more physically diverse than today, with a more
heterogeneous distribution of land to sea (e.g.
Kennett & von der Borch 1985, fig. 3), and thus
many ecological opportunities. Cladistic analysis of early Mysticeti, beyond the scope of this
article, will allow critical reanalysis of such
interpretations. Late Oligocene mysticetes also
occur in the NE Pacific and eastern Atlantic,
whence they may have dispersed from the south.

Late Oligocene Cetacea from the SW Pacific
(New Zealand) which give insight into the
Southern Ocean fauna encompass an archaeocete (thus dispelling the idea of a late Eocene
mass extinction involving Archaeoceti), diverse
Mysticeti (perhaps nine species), and rare
squalodontid-like Odontoceti (Fordyce 1987).
The late early to late Oligocene fauna occurs
above a widespread facies boundary (the
'Marshall Unconformity') which is similar in
age and perhaps causally related to major
southern cooling and a marked eustatic sea
level change. The localized high diversity in the
SW Pacific may reflect further ecological partitioning in the Southern Ocean, induced by
increased thermal gradients.

Elsewhere in the Southern Hemisphere the
published late Oligocene record is poor.
Toothed mysticetes (SW Pacific−Victoria)
had not been displaced by seals and/or baleen
whales. No cetacean genera are clearly shared
with the Northern Hemisphere late Oligocene

faunas, which include primitive odontocetes
('Agorophiidae', NE Pacific and NW Atlantic)
and more modern forms (e.g. Squalodontidae,
N Atlantic). The complex stratigraphic geographic patterns of early odontocetes, combined
with uncertain higher level relationships, mean
that areas of radiation are uncertain. Cladistic
analyses, presently unavailable for early
Odontoceti, are beyond the scope of this article;
they will probably show early branches in both
the NE Pacific and NW Atlantic. Echolocation,
inferred from skull structures to have developed
by the late Oligocene (Fordyce 1980), is not
only an adaptation to allow exploitation of food
below the photic zone but would allow high
latitude feeding during polar winter (J. D.
Bradshaw, pers. comm.).

By the earliest Miocene much of Antarctica
was probably modern in aspect, with narrow
continental shelves and surrounded by extensive Southern Ocean with a well developed
Antarctic Convergence (see Murphy & Kennett
1985), Circum-Antarctic Current, and deep-
water circulation. Circum-polar distributions
were attained in the early Miocene in some
Cetacea (*Prosqualodon*), and many modern
circum-polar distributions may have arisen then
(e.g. penguins; kelp (*Durvillea*) apparently
never cropped in the south by ecological equivalents of northern sea cows such as *Hydro-
damalis*: see also Fleming 1979, p. 102). Cetacea
of 'modern' aspect, which appeared at least as
early in mid southern latitudes as elsewhere,
included dolphins (Kentriodontidae), sperm
whales (Physeteridae) and right whales
(Balaenidae); all probably had earlier origins.
Early otariids and phocids radiated in the north,
but whether in response to circulation changes
induced by the south, to earlier Miocene warming, or to some other trigger(s) is uncertain.
Other cetacean groups, less well known,
also appeared in temperate sequences; short-
ranged Squalodelphidae (early Miocene),
Eurhinodelphidae (early−middle Miocene),
and Acrodelphidae (a dubious clade; early−
middle Miocene) disappeared during the
later Miocene, as did the longer-ranged Squalodontidae, perhaps because of climatic
deterioration.

No certain Archaeoceti are known from the
Neogene, and it is possible that they were ecologically displaced by the early seals (amphibious breeders, shelf feeders) and odontocetes
and penguins (competitors for food). Extinctions such as these are difficult to deal with,
since they assume that absence of evidence
(marine mammal fossils, rare at the best of
times) provides evidence of absence.

The rather poor published Austral middle and earlier late Miocene record (and the good northern record) gives a few tantalising insights into the anticipated influence on marine mammals of successive high latitude southern coolings and the evolution of the E Antarctic ice sheet. Many taxa of 'modern' aspect appeared now. The middle and late Miocene were times of radiation in temperate latitude N Pacific otariids, especially those that gave rise by the Quaternary to the northern pagophilous walrus, *Odobenus*. Similarly, phocids radiated in warm to mid-northern latitudes in the Atlantic but diverse northern pagophilous phocines did not appear until the Quaternary. Early balaenopterids occur in the SW Pacific. Early members of the extant cetacean dolphin (delphinoid) families Delphinidae, Phocoenidae and Monodontidae are known from the late Miocene of the temperate latitude N Pacific, and roughly contemporaneous delphinids and a phocoenid occur in the SW Pacific. No evolutionary centre has been identified clearly. Delphinids include many highly pelagic extant species which have characteristically well developed acoustic apparatuses, and their evolution may mark better exploitation of oceanic resources. The monodontids include the extant beluga (*Delphinapterus*) and narwhal (*Monodon*), both Quaternary obligate Arctic taxa descended from warmer water ancestors. The identification of the extant Indopacific *Orcaella* as a southern monodontid (e.g. Barnes *et al.* 1985) is unsubstantiated.

Lobodontine phocids occur around the edges of the Southern Ocean (Australia, South Africa, Argentina) in the latest Miocene, fortuitously at a time of a cooling, regression, and phosphatisation. It is not clear how the southern fossils, especially the earlier *Properiptychus* (middle Miocene), are related precisely to the extant endemic southern lobodontines, but the radiation of the lobodontine clade seems to be southern. The four extant monotypic genera may be ancient, perhaps at least of late Miocene

origin. Latest Miocene Southern Cetacea are largely of modern aspect, suggesting no profound changes in ecological partitioning of mid-latitude southern waters since.

Amongst otariids, *Arctocephalus* and sea lion clades, poorly represented by fossils, probably represent southern radiations no younger than Pliocene. These taxa perhaps were excluded from the far south by the presence of pagophilic phocids (and other marine homoiotherms such as smaller Cetacea and penguins), and the relative lack of extensive circum-Antarctic continental shelf and suitable breeding grounds on Antarctica. The observation that 'the sea lions... have not completed the encircling of Antarctica' (Repenning *et al.* 1979) might best be explained by competitive exclusion.

Austral Pliocene marine mammals include some extinct taxa (the Vestfold Hills dolphin, the sea lion *Neophoca palatina*) and early records of neospecies (the pagophilous *Ommatophoca rossi*). No extinct Austral taxa have been identified in the Pleistocene. By this time obligate high latitude Arctic species had arisen from warmer water ancestors presumably following the formation of sea-level ice in the near land-locked Arctic Ocean, in turn perhaps related to the closure of the Panama Seaway and northward deflection of the Gulf Stream. The Arctic marine mammal fauna, so often regarded as the standard for cold water stenothermal higher vertebrates, perhaps arose some 30 Ma after its southern high latitude equivalents.

This review is based on results of work supported by the University of Otago Research Committee and the National Geographic Society, an invitation for field work from M. Woodburne and through him a subcontract (from U.S. National Science Foundation; Seymour Island Cetacea) and the cooperation of P. Quilty and the Australian Antarctic Division (Vestfold Hills Cetacea). I thank preparators A. Grebneff and C. Jones for their valuable efforts with often-intractable specimens, and other colleagues, particularly M. Armstrong, D. Campbell, and J. Hooker, for discussion and comments on the text.

References

ABEL, O. 1914. Die Vorfahren der Bartenwale. *Denkschriften Mathematisch-naturwissenschaftlichen Klasse, Academie der Wissenschaften, Wien*, **90**, 155−244.

BARNES, L. G., DOMNING, D. P. & RAY, C. E. 1985. Status of studies on fossil marine mammals. *Marine Mammal Science*, **1**, 15−53.

BARTHOLOMEW, G. A. 1970. A model for the evolution of pinniped polygyny. *Evolution*, **24**, 546−59.

BEARLIN, R. K. 1985. The morphology and systematics of Neogene mysticetes (baleen whales)

from Australia and New Zealand — a progress report. *New Zealand Geological Survey Record*, **9**, 11−13.

−− 1988. The morphology and systematics of Neogene Mysticeti from Australia and New Zealand. [Abstract.] *New Zealand Journal of Geology and Geophysics*, **31**, 257.

BERTA, A. & DEMÉRÉ, T. 1986. *Callorhinus gilmorei* n.sp., (Carnivora: Otariidae) from the San Diego Formation (Blancan) and its implications for otariid phylogeny. *Transactions of the San Diego*

Society of Natural History, **21**, 111–126.

BONNER, W. N. 1982. *Seals and man. A study of interactions*. University of Washington, Seattle.

BROWNELL, R. 1974. Small odontocetes of the Antarctic. *In*: BUSHNELL, Y. C. (ed) *Antarctic Map Folio Series, Folio*, **18**. American Geographical Society, New York, 13–19.

DATHE, F. 1983. *Megaptera hubachi* n.sp., ein fossiler Bartenwal aus marinen Sandsteinschichten des tieferen Pliozäns Chiles. *Zeitschrift für Geologische Wissenschaften*, **11**, 813–848.

ELLIOT, D., RINALDI, C., ZINSMEISTER, W. J., TRAUTMAN, T. A., BRYANT, W. A. & DEL VALLE, R. 1975. Geological investigations on Seymour Island, Antarctic Peninsula. *Antarctic Journal of the United States*, **10**, 182–186.

FLEMING, C. A., 1968. New Zealand fossil seals. *New Zealand Journal of Geology and Geophysics*, **11**, 1184–1187.

—— 1979. *The geological history of New Zealand and its life*. Auckland University Press. Auckland.

FORDYCE, R. E. 1980. Whale evolution and Oligocene Southern Ocean environments. *Palaeogeography, Palaeoclimatology, Palaeoecology*, **31**, 319–336.

—— 1982. The fossil vertebrate record of New Zealand. *In*: RICH, P. V. & THOMPSON, E. M. (eds) *The fossil vertebrate record of Australasia*. Monash University Offset Printing Unit, Clayton, Vic., 629–698.

—— 1984a. Evolution and zoogeography of cetaceans in Australia. *In*: ARCHER, M. & CLAYTON, G. (eds) *Vertebrate zoogeography and evolution in Australasia*. Hesperion, Perth, 929–948.

—— 1984b. Preliminary report on cetacean bones from Chatham Rise (New Zealand). *Geologisches Jahrbuch*, **D 65**, 117–120.

—— 1985. The history of whales in the Southern Hemisphere. *In*: LING, J. K. & BRYDEN, M. M. (eds), *Studies of sea mammals in south latitudes*. South Australian Museum, Adelaide, 79–104.

—— 1987. New finds of whales from the Kokoamu Greensand. [Abstract.] *Geological Society of New Zealand Miscellaneous Publication*, **37A**.

——, & FLANNERY, T. F. 1983. Fossil phocid seals from the Late Tertiary of Victoria. *Proceedings of the Royal Society of Victoria*, **95**, 99–100.

GASKIN, D. E. 1982. *The ecology of whales and dolphins*. Heinemann, London.

KAMP, P. J. J. 1986. The mid-Cenozoic Challenger Rift system of western New Zealand and its implications for the age of Alpine fault inception. *Bulletin of the Geological Society of America*, **97**, 255–281.

KELLOGG, A. R. 1936. A review of the Archaeoceti. *Carnegie Institute of Washington Publication*, **482**.

KENNETT, J. P. & VON DER BORCH, C. C. 1985. Southwest Pacific Cenozoic paleoceanography. *In*: KENNETT, J. P. & VON DER BORCH, C. C. *et al.* (eds) *Initial Reports of the Deep Sea Drilling Project*, **90**, 1493–1517.

KING, J. E. 1973. Pleistocene Ross Seal (*Ommatophoca rossi*) from New Zealand. *New Zealand Journal of Marine and Freshwater Research*, **7**, 391–397.

—— 1983. The Ohope skull – a new species of Pleistocene sea lion from New Zealand. *New Zealand Journal of Marine and Freshwater Research*, **17**, 105–120.

LAWS, R. M. 1977a. The significance of vertebrates in the Antarctic marine ecosystem. *In*: LLANO, G. (ed.) *Adaptations within Antarctic ecosystems*. Smithsonian Institution, Washington D.C., 411–438.

—— 1977b. Seals and whales of the Southern Ocean. *Philosophical Transactions of the Royal Society, London*, **B279**, 81–96.

LAUDER, G. V. 1981. Form and function: structural analysis in evolutionary morphology. *Paleobiology*, **7**, 430–442.

MUIZON, C. DE 1978. *Arctocephalus* (*Hydrarctos*) *lomasiensis*, subgen. et. nov. sp., un nouvel Otariidae du Mio-Pliocène de Sacaco (Perou). *Bulletin de l'Institut Français d'Etudes Andines*, **7**. 169–188.

—— 1981. Les vertébrés fossiles de la Formation Pisco (Pérou). Première partie: Deux nouveaux Monachinae (Phocidae, Mammalia) du Pliocène de Sud-Sacaco. *Institut Français d'Études Andines, Memoire*, **6**, [=*Travaux de l'Institut Français d'Études Andines (Lima)*. **22**].

—— 1982. Phocid phylogeny and dispersal. *Annals of the South African Museum*, **89**, 175–213.

—— 1984. Les vertébrés fossiles de la Formation Pisco (Pérou). Deuxieme partie: Les odontocetes (Cetacea, Mammalia) du Pliocène inferieur de Sud-Sacaco. *Institut Français d' Études Andines, Memoire*, **50** [=*Travaux de l'Institut Français d'Études Andines (Lima)*, **27**].

—— & BOND, M. 1982. Le Phocidae (Mammalia) miocène de la formation Paraná (Entre Rios, Argentine). *Bulletin du Muséum National d'Histoire Naturelle, Paris, série 4*, **4**, **C**, 165–207.

—— & HENDEY, Q. B. 1980. Late Tertiary seals of the South Atlantic Ocean. *Annals of the South African Museum*, **82**, 91–128.

MURPHY, M. G. & KENNETT, J. P. 1985. Development of latitudinal thermal gradients during the Oligocene: oxygen-isotope evidence from the Southwest Pacific. *In*: KENNETT, J. P. & VON DER BORCH, C. C. *et al.* (eds) *Initial Reports of the Deep Sea Drilling Project*, **90**, 1347–1360.

NISHIWAKI, M. 1977. Distribution of toothed whales in the Antarctic Ocean. *In*: LLANO, G. (ed.) *Adaptations within Antarctic ecosystems*. Smithsonian Institution, Washington D.C., 783–791.

RAY, C. E. 1977. Geography of phocid evolution. *Systematic Zoology*, **25**, 391–406.

REPENNING, C. (ed.) 1977. Advances in systematics of marine mammals. *Systematic Zoology*, **25**, 301–436.

—— & TEDFORD, R. H. 1977. Otarioid seals of the Neogene. *U.S. Geological Survey Professional Paper*, **992**, 1–93.

——, RAY, C. E. & GRIGORESCU, D. 1979. Pinniped biogeography. *In*: GRAY, J. & BOUCOT, A. J. (eds), *Historical biogeography, plate tectonics, and the changing environment*. Oregon State

University Press, Corvallis, 357–369.

ROSS, G. J. B., BEST, P. B. & DONELLY, B. G. 1975. New records of the pygmy right whale (*Caperea marginata*) from South Africa, with comments on distribution, migration, appearance, and behaviour. *Journal of the Fisheries Research Board of Canada*, **32**, 1005–1017.

TSUCHI, R. (ed.) 1987. *Pacific Neogene event studies.* 1GCP-246, Shizuoka University, Shizuoka.

WHITMORE, F. C., MOREJOHN, G. Y. & MULLINS, H. T. 1986. Fossil beaked whales — *Mesoplodon longirostris* — dredged from the ocean bottom. *National Geographic Research*, **2**, 47–56.

WIMAN, C. J. E. 1905. Über die altertertiären Vertebraten der Seymourinsel. *Wissenschaftliche Ergebnisse der Schwedischen Südpolar-Expedition 1901–1903, Stockholm*, **3**(1), 1–37.

WOODBURNE, M. O. & ZINSMEISTER, W. J. 1984. The first land mammal from Antarctica and its biogeographic implications. *Journal of Paleontology*, **58**, 913–948.

WYSS, A. R. 1987. The walrus auditory region and the monophyly of pinnipeds. *American Museum Novitates*, **2871**, 1–31.

— — 1988. On "retrogression" in the evolution of the Phocinae and phylogenetic affinities of the monk seals. *American Museum Novitates*, **2924**, 1–38.

ZINSMEISTER, W. & FELDMANN, R. 1984. Cenozoic high latitude heterochroneity of Southern Hemisphere marine faunas. *Science*, **224**, 281–283.

Development of Cenozoic deep-sea benthic foraminiferal faunas in Antarctic waters

ELLEN THOMAS

Department of Earth and Environmental Sciences, Wesleyan University, Middletown, CT 06457, USA

Abstract: Upper abyssal to lower bathyal benthic foraminifera from Ocean Drilling Program Sites 689 (present water depth 2080 m) and 690 (present water depth 2914 m) on Maud Rise (Antarctica) recorded changes in deep-water characteristics at high southern latitudes during the Cenozoic. The benthic foraminiferal faunas show only minor differences as a result of the difference in water depths between the sites, and changes in faunal composition were coeval. These changes occurred at the early/late Paleocene boundary (± 61.6 Ma), in the latest Paleocene (± 57.5 Ma), in the middle early Eocene (± 55.0 Ma), in the middle middle Eocene (± 46.0 Ma), in the earliest Oligocene (± 36.5 Ma) and in the early middle Miocene (± 14.5 Ma). The faunal change at the end of the Paleocene was the most important and has been recognized world-wide. On Maud Rise, the diversity decreased by 50% and many common species became extinct over a period of less than 20 000 years. Diversity increased again during the early Eocene, and reached the same values as in the Paleocene by the middle Eocene. In the middle Eocene the diversity started to decrease, and continued to decrease until the middle Miocene. From the beginning of the middle Miocene until today biosiliceous oozes accumulated and calcareous benthic foraminifera were generally absent, with the exception of part of the late Miocene ($\pm 8.5-7.5$ Ma) and the Quaternary.

Changes in composition of the benthic foraminiferal faunas over a wide depth range (upper abyssal−lower bathyal) probably indicate periods of major changes in the formational processes of the deep waters in the oceans. The earliest Eocene faunas, living just after the major extinction at the end of the Paleocene, are characterized by low diversity and high relative abundance of small species that probably migrated downslope into the deep waters. These faunas, and to a lesser degree those in the early middle Eocene, are characterized by high relative abundance of biserial and triserial species. In contrast, older and younger faunas have high relative abundances of spiral species. This suggests that bottom waters on Maud Rise were poor in dissolved oxygen in the latest Paleocene through early middle Eocene, and that the major extinction of benthic foraminifera at the end of the Paleocene might have resulted from a decrease in availability of dissolved oxygen as a result of warming of the deep waters. Warming might have been caused by a change in sources of deep waters, possibly as a result of plate-tectonic activity. The overall decrease in diversity from middle Eocene through Miocene probably reflects continual cooling of the deep waters.

Benthic foraminiferal faunas thus indicate that Cenozoic changes in the deep oceanic waters at high latitudes did not consist of gradual progression from Cretaceous circulation to the present-day patterns of formation of deep water: the benthic faunal changes occurred in discrete steps. Benthic faunal composition indicates that deep water most probably formed at high latitudes during the Maastrichtian−early Paleocene, and from the middle Eocene to Recent, with episodes of deep water formation at low latitudes (warm, salty deep water) during the latest Paleocene and early Eocene.

Goals of the research

The Earth's climate and the temperature-structure of the oceans have changed considerably during the Cenozoic, from a warm climate with equable temperatures in the Cretaceous to the present-day situation with extensive polar ice caps and strong latitudinal and vertical temperature gradients in the oceans (e.g. Douglas & Savin 1975; Savin 1977; Barron 1985). The world may not have been completely ice-free in the Cretaceous (Frakes & Francis 1988), but ice-caps were certainly either absent or much smaller than they presently are (e.g., Frakes 1979; Schnitker 1980; Berger *et al.* 1981). The changes in climate were largest at high latitudes: cooling was of little importance in the tropics, but it was considerable in Antarctica. Beech-type (*Nothofagus*) forests were growing on the Antarctic Peninsula during the Eocene, where vegetation is much more limited today (e.g., Frakes 1979; Mohr in Barker *et al.* 1988;

From Crame, J. A. (ed.), 1989, *Origins and Evolution of the Antarctic Biota,*
Geological Society Special Publication No. 47, pp. 283−296.

283

Birkenmayer & Zastawniak 1989; Chaloner & Creber 1989). The decrease in polar temperatures during the Cenozoic probably did not occur gradually, but proceeded stepwise (e.g., Kennett 1977; Berger *et al.* 1981; Miller *et al.* 1987*a*).

Polar cooling probably caused major changes in the formation processes of deep waters in the oceans. At present the deep waters of the oceans are all formed at high latitudes, in the Norwegian–Greenland Sea (North Atlantic) and the Weddell Sea (South Atlantic) (e.g. Tolmazin 1985). In those areas the surface waters are very cold, and salinities increase as a result of extensive winter freezing (e.g. Weddell Sea; Foster & Carmack 1976) or through advection of lower latitude surface waters (e.g. Norwegian-Greenland Sea; Worthington 1972). In the absence of large ice caps deep water may still have formed as a result of cooling at high latitudes (e.g. Schnitker 1980, figs 1 & 2; Wilde & Berry, 1982; Manabe & Bryan 1985; Barrera *et al.* 1987). Alternatively, deep waters might have formed at lower latitudes because of strong evaporation and formation of dense, salty, warm waters (Brass *et al.* 1982; Prentice & Matthews 1988).

Thus during the Cenozoic the sources of deep waters in the oceans changed considerably, leading to changes in the physical and chemical properties of the deep waters because these are dependent upon the properties of the surface waters in the area of deep-water formation (e.g. Schnitker 1980; Manabe & Bryan 1985). Deep-sea benthic foraminiferal faunas are known to reflect changes in deep-water characteristics, and therefore it should be possible to trace major changes in deep water formation by studying changes in faunal composition of these organisms (e.g. Douglas & Woodruff 1981). Study of deep-sea benthic foraminiferal faunas from high latitudes is of particular interest for reconstructing the deep-water formation patterns of the past, because there the environmental changes are expected to have been greatest. Until recently the southernmost sections of Paleogene deep-sea calcareous oozes (containing a record of calcareous deep-sea benthic foraminifera) were recovered from the Falkland Plateau area (Dailey 1983). More southerly sites in the Pacific sector of the Southern Oceans and in the Ross Sea do not contain deep-water calcareous sediments older than Oligocene (e.g. Leckie & Webb 1985).

On Ocean Drilling Program Leg 113, Upper Cretaceous through upper Miocene calcareous and mixed siliceous/calcareous biogenic oozes were recovered at two sites on the Maud Rise, constituting the southernmost record of calcareous deep-sea benthic foraminiferal faunas through the Cenozoic. In this paper preliminary data are presented of the changes in benthic foraminiferal faunas through the Cenozoic, with emphasis on the Paleogene.

Material and methods

Ocean Drilling Program (ODP) Sites 689 and 690 were drilled on Maud Rise, Antarctica (**Site 689:** 64°31.01′S, 03°05.99′E, present water depth 2080 m; **Site 690:** 65°09.63′S, 01°12.30′E, present water depth 2914 m; Fig. 1; Barker *et al.* 1988). At both sites siliceous, mixed siliceous-calcareous, and calcareous biogenic oozes were recovered spanning the lowermost Maastrichtian through Pleistocene. Upper Eocene and older sediments are dominantly calcareous, upper Eocene through lower Miocene sediments mixed siliceous-calcareous with increasing amounts of siliceous oozes in younger sediments, and middle Miocene through Pleistocene sediments are dominantly siliceous.

Ages were estimated from the diatom zonation by Burckle & Gersonde and the calcareous nannofossil zonation by Pospichal & Wise in Barker *et al.* in press, in combination with data on the palaeomagnetic records at the sites (Spiess, in press), and using the time scale of Berggren *et al.* 1985. At Site 689 there are prominent unconformities in the lower Eocene and Paleocene, at Site 690 the upper Paleocene record is expanded, but there are prominent unconformities in the upper Eocene. Recovery at both sites was very good with only minor core disturbance. Palaeodepths cannot easily be estimated by backtracking (using simple thermal subsidence models) because of the location of the sites on Maud Rise. Maud Rise is an aseismic ridge with basement consisting not of typical mid-ocean ridge basalts (MORBs), but of alkali basalts of the ocean island basalt type (Barker *et al.* 1988) Depth estimates from benthic foraminiferal faunas for Site 689 suggest a depth of 1000–1500 m in the Paleocene, increasing to 1500–2000 m in the Eocene and later. For 690 estimates are 1500–2000 m in the Paleocene, increasing to more than 2500 m in the Eocene and later (Thomas in Barker *et al.* 1988).

For a preliminary overview of the material two samples per core (9.6 m) were studied, with additional samples (one per section of 1.5 m) in intervals where faunas showed changes, to a total of 100 samples. Samples of 15 cm^3 were dried at about 75°C, soaked in Calgon, washed

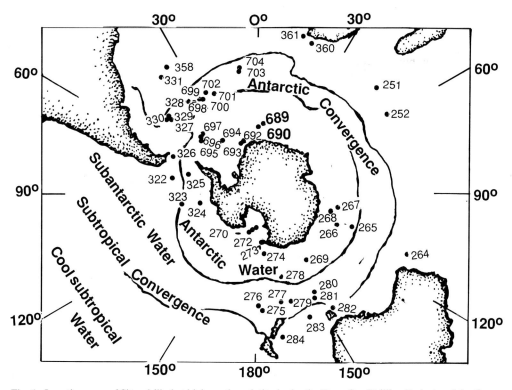

Fig. 1. Location map of Sites drilled at high southern latitudes by the Deep Sea Drilling Project and the Ocean Drilling Program. Sites 689 through 697 were drilled on Leg 113; only at Sites 689 and 690 were Upper Cretaceous through Paleogene calcareous oozes recovered. At most other sites of this leg the oldest sediment recovered was Neogene, with the exception of Sites 692 (the only sediment recovered was Lower Cretaceous), Site 693 where lower Oligocene biosiliceous hemipelagic muds were in contact with upper Albian 'black shales', and Site 696, where Oligocene (?) biosiliceous hemipelagic muds rested on barren glauconitic sands, which lay on upper–middle Eocene inner neritic sediments. Sites 608–704 were drilled on ODP Leg 114.

through sieves with openings of 63 μm, and residues dried at 75°C. Splits were made of such a size that they contained at least 300 specimens. Middle Miocene, most of the upper Miocene, and Pliocene samples consist of radiolarian-diatom oozes and are barren of benthic foraminifera. All other samples contained enough specimens for study. Specimens were picked from the greater than 63 μm size fraction to make sure that small specimens were well-represented (Thomas 1985; Schroeder *et al.* 1987). At least 300 specimens per sample were picked, because rarefaction curves (plots of number-of-species versus number-of-specimens recorded) showed that this number was required to give a good representation of the species in the most diverse samples from the sites. All specimens of benthic foraminifera in the sample splits were picked and mounted in cardboard slides; the counts and a discussion of the taxonomy will be fully presented in Thomas (in

press). The taxonomy is after Morkhoven *et al.* 1986.

Counts were completed on the 100 samples selected, and data are shown in Table 1 and Fig. 2. In addition, benthic foraminifera were picked from samples at intervals of 1.5 m over the whole section and qualitative data on these samples are included in this report; counts, however, are not yet complete.

Results

The faunas of the two sites are similar, with differences in relative abundance of species but overall the same species composition. The faunas resemble coeval lower bathyal to upper abyssal faunas from lower latitudes, as described by Vincent *et al.* (1974), Proto-Decima & Bolli (1978), Schnitker (1979), Tjalsma & Lohmann (1983), Boersma (1984), Wood *et al.* (1985), Miller & Katz (1987b), and Katz & Miller (1988).

Table 1. *Assemblages at Site 689 and Site 690.*

	Site 690	Site 689	Abundant species-morphology
Assemblage 1:	nr. of samples: 5	nr. of samples: 3	*E. exigua*-spiral
	nr. of species: 31−41	nr. of species: 21−31	*Oridorsalis* spp.-spiral
Quaternary and late Miocene (8.5−7.5 Ma)			
Assemblage 2:	nr. of samples: 9	nr. of samples: 11	*N. umbonifera*-spiral
	nr. of species: 31−48	nr. of species: 21−46	*Stilostomella* spp.-cylindrical
Oligocene-early Miocene (36.5−14.5 Ma)			*C. mundulus*-spiral
Assemblage 3:	nr. of samples: 7	nr. of samples: 10	*N. umbonifera*-spiral
	nr. of species: 31−57	nr. of species: 37−52	*Stilostomella* spp.-cylindrical
late middle Eocene-late Eocene (46.0−36.5 Ma)			*B. elongata*-buliminid
Assemblage 4:	nr. of samples: 5	nr. of samples: 5	*N. truempyi*-spiral
	nr. of species: 40−65	nr. of species: 37−46	*S. eleganta*-buliminid
4a: early middle Eocene (51.8−46.0 Ma)			*B. semicostata*-buliminid
4b: late early Eocene (55.0−51.8 Ma)			*S. brevispinosa*-buliminid
Assemblage 5:	nr. of samples: 8	nr. of samples: 5	*S. brevispinosa*-buliminid
	nr. of species: 28−49	nr. of species: 27−47	*Tappanina selmensis*-buliminid
early Eocene (57.5−55.0 Ma)			*B. simplex*-buliminid
Assemblage 6:	nr. of samples: 12	nr. of samples: 3	*B. thanetensis*-buliminid
	nr. of species: 52−67	nr. of species: 32−70	*Gavelinella* spp.-spiral
late Paleocene (61.6−57.5 Ma)			*S. brevispinosa*-buliminid
Assemblage 7:	nr. of samples: 7	nr. of samples: 4	*Gavelinella* spp.-spiral
	nr. of species: 58−79	nr. of species: 35−50	*O. navarroana*-spiral
late Maastrichtian-early Paleocene (68.9−61.6 Ma)			diverse lagenids-cylindrical
Assemblage 8:	nr. of samples: 3	nr. of samples: 5	*Gavelinella* spp.-spiral
	nr. of species: 48−62	nr. of species: 32−52	*Gyroidinoides* spp.-spiral
Maastrichtian (or older: sedimentary units rest on basement)			

Ages were derived from correlation with the diatom zonation by Gersonde & Burckle and the calcareous nannofossil zonation of Pospichal & Wise in Barker *et al.*, in press, using the time scale of Berggren *et al.* 1985. Cretaceous assemblage 8 is not discussed in detail in this paper.

There is a particularly great resemblance between Eocene−Oligocene faunas from Maud Rise and those from the Eocene of the northern Atlantic (Berggren & Aubert 1976*a*, 1976*b*; Murray 1984) because of the dominance of *Stilostomella* species.

At each site eight assemblages could be recognized, one of which was subdivided into two sub-assemblages (Table 1). The assemblages were distinguished using a combination of data on presence−absence of species with a short range, and relative abundances of the most common or dominant species. Species used to distinguish between assemblages are listed in Table 1, and most are illustrated in Fig. 3.

The transitions between the assemblages are coeval within the resolution of the diatom, calcareous nannofossil and planktonic foraminiferal zonations (Barker *et al.* 1988) and preliminary data on carbon and oxygen isotopic, composition of bulk carbonate (N. J. Shackleton, pers. comm. 1987) and benthic foraminifera (L. D. Stott, pers. comm. 1988), in combination with data on the palaeomagnetic record at the sites (Spiess, in press).

The diversity was higher at the deeper Site 690 at all times (Fig. 2), as was expected because of the commonly observed increase in diversity towards deeper water (e.g. Douglas & Woodruff 1981). There is an overall decrease in diversity during the Cenozoic, but the exception to this overall trend is the high diversity in middle Eocene assemblage 4. The diversity of lower Eocene assemblage 5 is exceptionally low as compared with that of upper Paleocene assemblage 6 and middle Eocene assemblage 4, as a result of the disappearance of many species at the end of the Paleocene. Similar disappearances occurred at the end of the Paleocene at sites in the Caribbean and Atlantic (Tjalsma & Lohmann 1983) and the Pacific (Miller *et al.* 1987*b*) and southernmost Atlantic (Katz & Miller 1988). Little detailed information was available on this major extinction event of deep-sea benthic foraminifera because few Paleocene/Eocene boundary sections were re-

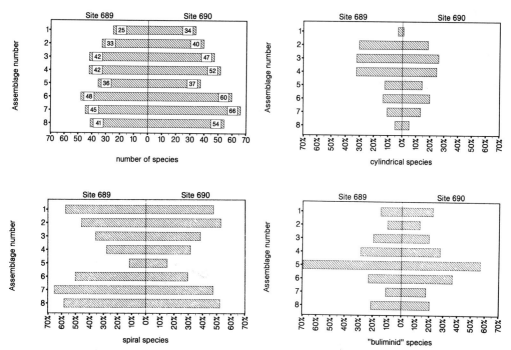

Fig. 2. Comparison of the species diversity and composition of the benthic faunal assemblages at Sites 689 and 690. Each bar indicates the average value for one assemblage (see Table 1 for number of samples in each assemblage). Assemblages are shown from 1 (youngest, see Table 1) on top to oldest (assemblage 8, see Table 1) on the bottom, but they are not spaced according to a linear time scale (see Table 1 and Fig. 4 for ages). 'Spiral species' include trochospiral and planispiral species; 'cylindrical species' include uniserial lagenids and *Stilostomella* spp., and 'buliminid' species include biserial and triserial species. The three groups do not add up to 100% because unilocular and agglutinated taxa were not included.

covered and sedimentation rates were usually low in this interval. At Site 690 the sedimentation rate for the uppermost Paleocene–lowermost Eocene palaeomagnetic Chrons 24 and 25 was about 14.4 m Ma^{-1}; at that site the major part of the faunal change occurred over less than 35 cm (equivalent to about 20 000 years). The diversity remained very low (25–30 species per 300 specimens) for about 260 000 years.

Placement of the Paleocene/Eocene boundary with regard to biostratigraphic zonations is being discussed (Berggren *et al.* 1985; M. P. Aubry *et al.* pers. comm.). Data on the calcareous nannofossil zonation at Site 690 (Pospichal & Wise in press) and the palaeomagnetic record (Spiess in press) are in better agreement with Aubry *et al.* (pers. comm.) than with Berggren *et al.* (1985). At the Maud Rise sites the extinction event of benthic foraminifera occurred in the middle of calcareous nannofossil zone CP8, after the first appearance (FA) of *Discoaster multiradiatus*, and before the FA of *Tribrachiatus bramlettei* and the LA of *Fasciculithus* spp., i.e. before the end of the Paleocene as recognized by Berggren *et al.* 1985. The last appearance (LA) of the planktonic foraminiferal species *Morozovella velascoensis*, used to recognize the boundary by Berggren *et al.* (1985), could not be determined at Site 690 because of the absence of this low-latitude species. The FA of the planktonic foraminiferal taxon *Pseudohastigerina* spp., used to mark the boundary at higher latitudes, occurs higher in the section than the boundary as defined by the nannofossils (FA of *T. bramlettei*). Tjalsma & Lohmann (1984) placed the extinction at the end of planktonic foraminiferal Zone P5 in the uppermost Paleocene, whereas Miller *et al.* (1987*b*) placed it within or at the end of planktonic foraminiferal Zone P6a (the Paleocene/Eocene boundary as defined in Berggren *et al.* 1985). At Site 690 the distinction between zones P5 and P6a is difficult to make because of the absence of marker species at high latitudes, which makes recognition of these planktonic

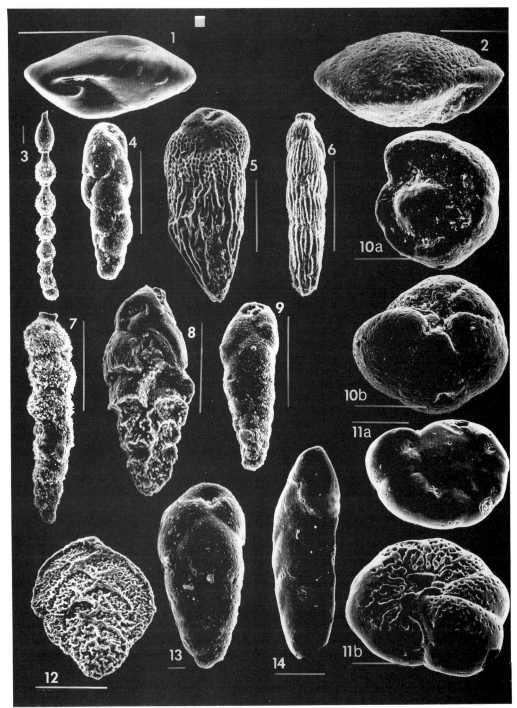

Fig. 3. The scale bar at (13) corresponds to 10 μm, all other scale bars correspond to 100 μm. (**1**) *Epistominella exigua*, side view; sample 690B-1H-1,0−2 cm. (**2**) *Nuttallides umbonifera*, side view; sample 690B-4H,CC. Note the traces of dissolution in the corrugation of the surface of the test. (**3**) *Stilostomella subspinosa*; sample 690B-12H-6,45−47 cm. (**4**) *Bulimina elongata*; sample 690B-12H-4,41−45 cm. (**5**) *Bulimina semicostata*; sample 690B-12H,CC. (**6**) *Siphogenerinoides eleganta*; sample 690B-12H,CC. (**7**) *Siphogenerinoides brevispinosa*; sample 690B-18H-4,40−44 cm. (**8**) *Tappanina selmensis*; sample 689B-22X,CC. (**9**) *Bulimina thanetensis*; sample 690C-12X-2,40−42 cm. (**10**) *Gyroidinoides quadrata*; sample 690C-15X,CC. (10a) dorsal view; (10b) ventral view. (**11**) *Stensioina beccariiformis*; sample 690C-15X,CC. (11a) dorsal view; (11b) ventral view. (**12**) *Aragonia aragonensis*; sample 689B-22X,CC. (**13**) *Bulimina simplex*; sample 690B-13H-3,40−42 cm. (**14**) *Coryphostoma incrassata*; sample 690C-15X,CC.

foraminiferal zones impossible.

The extinction of benthic taxa occurred exactly during a major decrease in the $\delta^{13}C$-values and $\delta^{18}O$ values of bulk carbonate and benthic foraminifera (Shackleton, pers. comm. 1987; Stott, pers. comm. 1988). These isotopic changes have been correlated with the Paleocene/Eocene boundary (Shackleton 1986; Miller *et al.* 1987*b*), but data from the high-sedimentation rate section at Site 690 suggest that the isotopic changes occurred during the latest Paleocene, and definitely before the Paleocene/Eocene boundary as defined by Aubry *et al.* (pers. comm.)

Concise overviews of changes in composition of benthic foraminiferal faunas are difficult to give, because faunas are very diverse and many species need to be discussed to give a complete picture of faunal events. The most distinctive species in the assemblages are listed in Table 1 and shown in Fig. 3.

In general, assemblages 8 through 6 (Maastrichtian–Paleocene) have much in common, as do assemblages 3 through 1 (Oligocene–Recent). There are large differences in faunal composition of the older assemblages 6 through 8 and the younger assemblages 1 through 3, but these two groups of assemblages have strong similarities as to the morphology of the dominant taxa (see below). Assemblages 4 (early–middle Eocene) and especially 5 (earliest Eocene), however, strongly differ from both older and younger faunas (Fig. 2).

The three oldest assemblages (covering Maastrichtian through Paleocene) are very diverse and usually dominated by trochospiral and planispiral species. They contain many large, heavily calcified species, with *Stensioina beccariiformis*, the *Nuttallides truempyi* group, *Gyroidinoides* spp., *Gavelinella* spp., *Alabamina creta*, and *Pullenia* spp. as the most common trochospiral and planispiral species, and diverse lagenids and small agglutinated forms as common constituents. There are fluctuations in faunal composition in all these assemblages, with varying amounts of biserial and triserial taxa ('buliminid' species). In assemblage 8 the biserial/triserial group consists mainly of *Coryphostoma incrassata* and *Praebulimina reussi* with *Reussella szajnochae* in some samples. In assemblage 7 the 'buliminid' group consists of *Coryphostoma* spp. with rare *P. reussi*. In assemblage 6 *Bulimina thanetensis*, *Rectobulimina carpentierae* and *Siphogenerinoides brevispinosa* are the dominant buliminids. There are minor faunal changes at the Cretaceous/Tertiary boundary within assemblage 7 (Thomas 1988), but no major faunal

turnover occurred in the benthic foraminiferal faunas, in agreement with Tjalsma & Lohmann (1983), Dailey (1983), and Widmark & Malmgren (1988).

Assemblage 5, present after the major extinction at the end of the Paleocene, differs considerably from the older assemblages in the following ways: the faunas are less diverse, dominated by 'buliminid' species, and contain predominantly small, thin-walled forms. There are no indications that the decrease in diversity resulted from dissolution. A similar decrease in size of benthic taxa at the end of the Paleocene was noted by Boersma (1984). There is little fluctuation from sample to sample in the relative abundance of spiral as compared with 'buliminid' species. Of the spiral species present, *Nuttallides truempyi* occurs in most samples but constitutes less than 10% of the fauna. Other spiral species are rare, with the exception of *Abyssamina quadrata*, *Clinapertina* spp., and small specimens of *Anomalinoides spissiformis* and *Nonion havanense*. The most common 'buliminid' forms are *Tappanina selmensis*, *Siphogenerinoides brevispinosa*, small, smooth-walled *Bulimina* species such as *B. simplex* and *B. trihedra*, small *Bolivinoides* species, and in a few samples *Aragonia aragonensis*. The fauna has a low diversity, partially caused by the absence or rare occurrence of lagenids. The big difference between assemblages 6 and 5 is the loss of the many common *spiral* species, which had existed for a long time and had given the older assemblages their overall character. Species such as *Stensioina beccariiformis*, *Alabamina creta*, *Pullenia coryelli* and *Gavelinella* spp. rapidly (within 20 000 years) disappeared from the fauna, in addition to the rare large biserial species *Bolivinoides delicatulus*.

From the base of the Eocene the diversity increases upwards (through the middle Eocene), and assemblage 4 is once again very diverse, with large, heavily calcified species, and common, diverse lagenids. In the spiral species group *N. truempyi* is common, as are *Gyroidinoides* spp., and *Oridorsalis umbonatus*; in all samples of assemblage 4a the large species *Bulimina semicostata* is common (4–35%), in addition to *Siphogenerinoides eleganta*. Assemblage 4b is transitional between 5 and 4a, with common *S. eleganta*, and *S. brevispinosa*, but rare *B. semicostata*.

Assemblage 3 (middle–late Eocene) forms a transitional assemblage between the high-diversity assemblage 4 and the lower-diversity younger assemblages. Faunal changes taking place during this transition have been described from many areas (Tjalsma & Lohmann 1983;

Corliss & Keigwin 1986). The diverse lagenid species (uniserial and lenticulinid) gradually decreased in relative abundance during the middle Eocene (assemblage 3), and disappeared at the end of the Eocene. In assemblage 3 there are once again fluctuations in relative abundance of spiral and 'buliminid' species, with *Bulimina elongata* as the dominant buliminid. Assemblages 2 and 3 contain few lagenids, and common to abundant *Nuttallides umbonifera* (with common *N. truempyi* in the lower part of assemblage 3 at Site 690), *Oridorsalis umbonatus*, *Pullenia* spp., and *Gyroidinoides* spp. as spiral species. In addition, *Stilostomella* spp. are very common to abundant (15–45%).

Assemblage 1 has a composition similar to that of assemblage 2, with the addition of *Epistominella exigua*, and the disappearance of *Stilostomella* spp. at Site 690, the very strong decrease in relative abundance of *Stilostomella* spp. at Site 689, and overall decrease in diversity.

Assemblages 1 through 3 have lower diversities than assemblages 6 through 8, and contain more small, smooth, thin-walled species. In this group of assemblages, as in the older group of assemblages 6 through 8, there are fluctuations in relative abundance, with triserial/biserial species common in some samples. In assemblage 3 the triserial/biserial group ('buliminid' species) consists mainly of the small, smooth-walled species *Bulimina elongata*, in assemblage 2 of *Bolivina* spp., *Uvigerina* spp., and *Turrilina alsatica*, and in assemblage 1 of *Bolivina decussata* and *Fursenkoina* species.

In conclusion, we see relatively long-lived assemblages of deep-sea benthic foraminifera in the Paleocene and older sediments, and in the upper middle Eocene and younger sediments at Maud Rise. These long-lived assemblages are characterized by the dominance of spiral species, with fluctuations in relative abundance of species of a triserial-biserial group. The species in the latter group tend to have shorter ranges than the spiral species. In contrast, there were short-lived assemblages in the early through early middle Eocene. These assemblages are characterized by high relative abundances of 'buliminid' species, and are very different from both younger and older faunas because of the low relative abundance of spiral species.

First and last appearances of deep-sea benthic foraminiferal species can probably not be used to correlate over large distances; the easily recognized species *Bulimina semicostata*, for instance, has a much shorter range at Sites 689 and 690 (limited to assemblage 4) than in the

rest of the Atlantic Ocean (Tjalsma & Lohmann 1983; Morkhoven *et al*. 1986). Periods of faunal change in deep-sea benthic foraminifera, however, can probably be correlated well from one basin to another: the boundaries between the faunal assemblages at Sites 689 and 690 appear to be well-correlated with the zonal boundaries of the proposed deep-sea benthic foraminiferal zones (Berggren & Miller, in press), although specific first and last appearances are not synchronous (see also Thomas 1986).

Discussion

To evaluate the environmental significance of changes in deep-sea benthic foraminiferal assemblages, the faunal records were compared with the records of carbon and oxygen isotopic values for bulk carbonate and benthic foraminifera for several sites on the Walvis Ridge–Rio Grande Rise, southern Atlantic Ocean (Shackleton 1987; Fig. 3). Data on the isotopic record for the Maud Rise sites have not yet been published, but major features of isotopic records are similar at different places in the oceans (e.g. Berger *et al*. 1981; Miller *et al*. 1987*a*), and the record at the sites on Maud Rise is in its major features similar to that collected at other sites (N. J. Shackleton, pers. comm. 1987; L. D. Stott, pers. comm. 1988; Kennett & Stott 1988).

The isotopic records from the Walvis Ridge sites show the major features of the global record: the oxygen isotopic curve shows a decrease in bottom water temperature and surface water temperature just after the Eocene/Oligocene boundary (e.g. Keigwin & Corliss 1986), highest temperatures of bottom waters during the middle Eocene with a strong decrease during the middle Eocene, and an increase in bottom water temperatures during the latest Paleocene/earliest Eocene (Oberhaensli 1986; Oberhaensli & Toumarkine 1985). The temperature difference between deep and surface waters was lowest in the early to early middle Eocene. The carbon isotope record of the surface waters shows a major positive excursion during the late Paleocene (see Miller & Fairbanks 1985, and Shackleton 1987, for a discussion of the Cenozoic carbon isotope record). Carbon isotopic values during the Late Cretaceous were about as high as values during the late Paleocene (Zachos & Arthur 1986).

Shackleton (1987, Fig. 4) suggested that the major decrease in $\delta^{13}C$ at the end of the Paleocene resulted from a drop in primary productivity. During a collapse of primary productivity, however, carbon isotopic ratios of

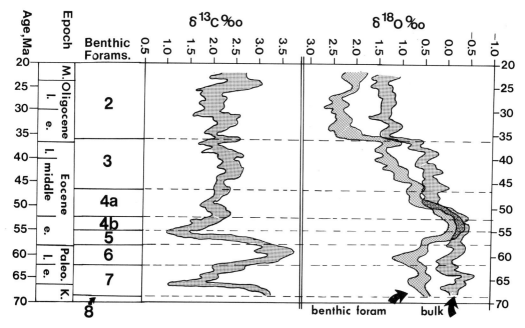

Fig. 4. Paleogene benthic foraminiferal assemblages plotted against the time scale of Berggren *et al.* 1985, and compared with the oxygen and carbon isotope records for sites on the Walvis Ridge (Shackleton 1987). The carbon isotope record and the oxygen isotope record on the right are for bulk-carbonate; the oxygen isotope record on the left is for benthic foraminifera. Data on benthic foraminifera can be used to derive estimates of bottom water temperature, but the bulk record cannot be used to estimate the temperature of the surface waters, because the 'vital effect' of the organisms forming the bulk carbonate is not known. The difference between bulk and benthic record does give an estimate of changes in the difference in temperature of deep and bottom waters.

deep-sea benthic foraminifera should become heavier while the ratios in bulk carbonate (mainly consisting of surface-dwellers) should become lighter, as was observed in many sections across the Cretaceous/Tertiary boundary (Zachos & Arthur 1986). Curves of carbon isotopic ratios of bulk carbonate and benthic foraminifera for the uppermost Paleocene, however, are parallel (Miller *et al.* 1987*b*), indicating a change in $\delta^{13}C$ of mean ocean water (probably as a result of changes in the input or output ratio of organic carbon to carbonate carbon; Miller & Fairbanks 1985).

There is agreement in timing of changes in deep-sea benthic foraminiferal assemblages and global isotopic changes, but the exact relation cannot yet be established because the isotopic record from Sites 689 and 690 is not yet available. Ocean-wide faunal events in the middle Miocene, for instance, generally predated the middle Miocene oxygen isotopic shift (Thomas & Vincent 1987; Miller & Katz 1987*b*; Berggren & Miller, in press). The boundary between assemblages 2 and 3 appears to be coeval with

this major decrease in temperature and probably an increase in ice volume (e.g. Corliss & Keigwin 1986; Miller *et al.* 1987*a*), but there is a short hiatus in the records across this boundary at the Maud Rise sites making a precise correlation impossible. The boundary between assemblages 3 and 4 and the more gradual changes within assemblage 3 occur during a long-term decrease in bottom water temperature (see Corliss & Keigwin 1986, fig. 5). The short-lived assemblages 4b and 5 occur during a period of maximum bottom water temperatures and minimum differences in temperature between deep and surface waters. The major extinction at the end of the Paleocene (assemblages 5 and 6) occurred at the same time as a major decrease in $\delta^{13}C$ and an increase in temperature. The boundary between assemblages 6 and 7 occurred during a strong increase in $\delta^{13}C$, but no large change in temperature.

Deep-sea benthic foraminifera are conservative organisms with long ranges, living in a conservative environment (e.g. Boltovskoy 1987). Many deep-sea species have cosmopolitan distributions (e.g. Douglas & Woodruff, 1981) and the Paleogene faunas at Sites 689 and 690 are similar to coeval faunas described from sites at high northern latitudes (Murray 1984; Berggren & Aubert 1976a, b) and at lower latitudes throughout the Caribbean and South Atlantic (Tjalsma & Lohmann 1983), the New Jersey continental margin (Miller & Katz 1987a), the Bay of Biscay (Schnitker 1979; Miller et al. 1984), Barbados (Wood et al. 1985), and the central Pacific Ocean (Miller et al. 1987b). The short-lived assemblages characterized by the presence of several species with ranges of a few millions of years indicate a period of unusual variability in the deep oceans (Wilde & Berry 1982). The major extinction event at the end of the Paleocene is especially remarkable: during other extinctions in deep-sea benthic foraminifera, the extinction process appears to have had a duration of millions of years (Miller & Katz 1987b; Thomas 1985, 1986; Thomas & Vincent 1987) as opposed to the estimate in this study of less than 20 000 years for the Paleocene event in this paper. The species that became extinct at this time were almost all survivors from the Cretaceous, and had survived the collapse of primary productivity at the Cretaceous/Tertiary boundary (Zachos & Arthur 1986).

During the Cretaceous/Tertiary boundary extinction planktonic taxa were very severely affected, which probably means that this was mainly a disturbance of the surface of the Earth with only secondary effects felt at greater depths in the oceans (Thomas 1988). The extinction event at the end of the Paleocene, a period characterized by few extinctions overall (Raup & Sepkoski 1986), must thus probably be seen as an event rooted in the deep waters of the oceans. Present biology of deep-sea benthic foraminifera is not well known, and thus it is difficult to evaluate the meaning of fluctuations in species composition. Corliss (1985) and Corliss & Chen (1988), however, suggest that the morphology of the tests of benthic foraminifera reflects the microhabitat in which the organisms live. Specifically, Corliss & Chen (1988) suggest that many trochospiral species have an epifaunal way of living, whereas biserial and triserial species live infaunally; these authors show that faunas dominated by infaunal species are associated with relatively high organic carbon contents of the sediments, whereas epifaunal species live in microhabitats with lower amount of organic carbon. Similarly, Lutze & Coulbourn (1984) and Bernard (1988) found that biserial/triserial species are typical of high organic carbon or low O_2 contents, and Caralp (1984) and Sen Gupta

et al. (1981) demonstrated that high percentages of triserial and biserial species in benthic foraminiferal faunas reflect high primary productivity in the surface waters.

Using the morphology of recent taxa as a guide, it appears that the group of spiral taxa is indicative of what we call 'normal bottom water conditions' in the present oceans, i.e., well-oxygenated water with fairly low levels of organic carbon. The 'buliminid' species, on the other hand, may indicate lower levels of oxygen (although by no means anoxic waters; see also Lutze & Coulbourn 1984) and possibly a higher supply of nutrients. This explanation suggests that the faunal fluctuations within assemblages 8 through 6 as well as in assemblages 3 through 1 indicate fluctuating primary planktonic productivity, resulting in fluctuating levels of nutrients and oxygen available to benthic organisms.

The species adapted to 'normal' (i.e., well-oxygenated and low in nutrients) waters have very long ranges; possibly such species could survive the collapse of the surface productivity at the end of the Cretaceous because they were well-adapted to low nutrient levels. The few species becoming extinct at Sites 689 and 690 at the end of the Cretaceous (e.g., Coryphostoma incrassata and Bolivinoides draco) belong to the 'buliminid' group thought to indicate a high level of nutrient supply, and these species probably could not survive this strong drop in their food supply.

The lack of a large extinction event of deep-sea benthic foraminifera, the extinction process Tertiary boundary at Sites 689 and 690 suggests that these organisms were well-suited to survive strong fluctuations in primary productivity. However, those species that survived the end of the Cretaceous were strongly affected by events at the end of the Paleocene, and many of them became extinct; therefore the author concludes that the extinction at the end of the Paleocene was not caused by a change in surface productivity. It seems more probable that the extinction of benthic faunas was caused by changes in the deep waters, probably in bottom water chemistry. Such a change in chemistry could well be caused by a switch of the sources of deep waters.

Miller et al. (1987b) suggested that deep waters were formed at the poles at least for some time during the late Paleocene, in agreement with Barrera et al. (1987), but that waters might have formed at lower latitudes later on. These authors place the change in circulation to deep-water sources at lower latitudes in the early Eocene, but indicate that the change might have occurred at the Paleocene/Eocene boundary (Miller *et al.* 1987*b*, fig. 6b). More recent data (Katz & Miller 1988; Miller & Katz 1988) on material recovered from ODP Sites 698–702 (Fig. 1) suggest that the Southern Oceans were filled with nutrient-depleted bottom water during the late Paleocene and early Eocene, but that there might have been an interval of reduction or shutdown of bottom water formation in the Southern Oceans at the time of the extinction of deep-sea benthic foraminifera. A change from formation of deep waters somewhere close to Sites 689 and 690 to deep-water formation at lower latitudes would explain the changes in the benthic faunas. The faunas had been able to adapt to fluctuations in productivity over a long period, but could not handle an overall change in bottom water characteristics to warmer deep waters with less oxygen. If deep waters formed in the Weddell Sea region then deep waters over Maud Rise would be 'young', cold and well-oxygenated even in the absence of large ice caps. In contrast, deep waters formed at low latitudes from warm, salty water with a high density as a result of evaporation (Brass *et al.* 1982; Wilde & Berry 1982) would be not only lower in oxygen content because of the higher temperature, but also old, and thus even lower in oxygen-content and higher in nutrients by the time that they arrived at the Maud Rise sites.

Such deep waters would be unsuitable for the faunas dominated by epifaunal species, but 'buliminid'-dominated faunas with a high abundance of small species (assemblage 5) might be expected to do well. These deep waters from another source region might bring with them juvenile mobile stages of benthic foraminifera (Berggren & Aubert 1975). The fact that the major extinction of deep-sea benthic foraminifera at the end of the Paleocene occurred at many sites in all oceans, and the short time scale (less than 20 000 years) imply a major change in circulation as the cause of the event.

The change in deep-water source might have resulted from a global increase in temperature (Owen & Rea 1985; Barron 1987), which caused stronger evaporation in low-latitude basins. It might also have been caused by changes in plate tectonic arrangements: at the end of the Paleocene large changes in plate arrangement occurred as a result of the collision of India and the Asiatic land mass (Williams 1986), as well as in the North Atlantic (Berggren & Schnitker 1985). These two possibilities might have been related: plate tectonic activity could have resulted in warmer climates because of high production of CO_2 at mid oceanic ridge systems (Owen & Rea 1985). Barron (1985), however,

denies that plate tectonic rearrangement and the resultant circulation changes could have caused the warm and very equable Eocene climate (Shackleton & Boersma 1984; Boersma *et al.* 1987).

Deep-sea benthic faunal compositions suggest that warm salty bottom waters formed during the late Paleocene through early Eocene only, in disagreement with the conclusion by Prentice & Matthews (1988) that low-latitude production of such waters dominated the thermohaline circulation during the Tertiary. More detailed research is necessary to determine whether the formation of such waters occurred continuously during the late Paleocene and early Eocene, or whether there were several short periods (less than 1 million years) of formation of warm, salty bottom water during that time interval. The deep-sea benthic foraminiferal faunas indicate that formation of warm, salty deep water probably stopped at some time in the middle Eocene, a period characterized by a very long-term decrease in temperature (Keigwin & Corliss 1986; Boersma *et al.* 1987) and global faunal changes in the deep-sea benthic assemblages. The benthic faunal changes from assemblage 4 to assemblage 3 and within assemblage 3 might reflect such a change in bottom water sources. These faunal changes were not as rapid as the latest Paleocene extinction, suggesting that these circulation changes were more gradual or possibly occurred in several steps. The overall decrease in diversity from the middle Eocene onwards probably reflects continuing cooling of the deep waters. The faunas also show the decrease in availability of $CaCO_3$ (colder waters will take up more CO_2 from the atmosphere), reflected in decrease in size and wall-thickness of the benthic faunas. The end of the Eocene is marked by the disappearance of the last species of *Bulimina* that is common at Sites 689 and 690, *B. elongata*. The fauna in the earliest Oligocene (assemblage 2) strongly resembles younger faunas (assemblage 1), which differ by lower diversities, and especially by the decrease in relative abundance of *Stilostomella* spp. at the end of the early Miocene (similar to the situation in equatorial Pacific sites; Thomas & Vincent 1987).

Conclusions

(1) Cenozoic deep-sea benthic foraminiferal faunas reflect changes in the physicochemical character of the deep waters and thus reflect processes of deep-water formation.

(2) Changes occurred at the early/late Paleocene boundary, in the latest Paleocene, in the early Eocene, in the earliest middle Eocene,

in the middle middle Eocene, in the earliest Oligocene, and in the middle Miocene.

(3) Benthic foraminiferal assemblages indicate that deep-water formation in the early and early middle Eocene was very different from the periods before and after that. Possibly deep water during this period formed by evaporation at low latitudes ('warm saline bottom water'), whereas deep water formed at high latitudes earlier and later.

(4) Many benthic foraminiferal species do not appear to have globally synchronous first and last appearances, but periods of faunal change appear to be synchronous.

The author thanks M. Katz, M. Leckie, K. Miller, and an anonymous reviewer for comments on the manuscript, and NSF-USSAC for financial support. N. Siegal and R. Garniewicz assisted in sample preparation, and N. Siegal made the SEM micrographs.

References

BARKER, P. F., KENNETT, J. P. et al. (eds), 1988, *Proceedings of the Ocean Drilling Program*, **113A**. Ocean Drilling Program, College Station (USA).

BARRERA, E., HUBER, B. T., SAVIN, S. M. & WEBB, P. N. 1987. Antarctic marine temperatures: late Campanian through early Paleocene. *Paleoceanography*, **2**, 21−48.

BARRON, E. J. 1985. Explanations of the Tertiary global cooling trend. *Palaeogeography, Palaeoclimatology, Palaeoecology*, **50**, 45−61.

−− 1987. Eocene equator-to-pole surface ocean temperatures: a significant climate problem? *Paleoceanography*, **2**, 729−740.

BERGER, W. H., VINCENT, E. & THIERSTEIN, H. R., 1981. The deep-sea record: major steps in Cenozoic ocean evolution. *Society of Economic Paleontologists and Mineralogists Special Publication*, **32**, 489−504.

BERGGREN, W. A. & AUBERT, J., 1975. Paleocene benthonic foraminiferal biostratigraphy, paleobiogeography and paleoecology of Atlantic-Tethyan regions: Midway-type fauna. *Palaeogeography, Palaeoclimatology, Palaeoecology*, **18**, 73−192.

−− & −− 1976a. Late Paleogene (late Eocene and Oligocene) benthonic foraminiferal biostratigraphy and paleobathymetry of Rockall Bank and Hatton-Rockall Basin. *Micropaleontology*, **22**, 307−326.

−−, & −−, 1976b. Eocene benthonic foraminiferal biostratigraphy and paleobathymetry of Orphan Knoll (Labrador Sea). *Micropaleontology*, **22**, 327−346.

−− & MILLER, K. G., in press. Cenozoic bathyal and abyssal calcareous benthic foraminiferal zonations. *Micropaleontology*

−− & SCHNITKER, D., 1983. Cenozoic marine environments in the North Atlantic and Norwegian-Greenland Sea. *In*: BOTT, M. H. P., SAXOV, S., TALWANI, M. & THIEDE, J. (eds), *Structure and development of the Greenland-Scotland Ridge*. NATO Conference Series, Plenum Press, New York, 495−548.

−− KENT, D. V., FLYNN, J. J. & VAN COUVERING, J. A., 1985. Cenozoic Geochronology. *Geological Society of America Bulletin*, **96**, 1407−1418.

BERNARD, J. M. 1986. Characteristic assemblages and morphologies of benthic foraminifera from

anoxic, organic-rich deposits: Jurassic through Holocene. *Journal of Foraminiferal Research*, **16**, 207−215.

BIRKENMAJER, K. & ZASTAWNIAK, E. 1989. Late Cretaceous−early Tertiary floras of King George Island, West Antarctica: their stratigraphic distribution and palaeoclimatic significance. *In*: CRAME, J. A. (ed.) *Origins and Evolution of the Antarctic Biota*. Geological Society, London, Special Publication, **47**, 227−240.

BOERSMA, A. 1984. Oligocene and other Tertiary benthic foraminifera from a depth traverse down Walvis Ridge, Deep Sea Drilling Project Leg 74, Southeast Atlantic. *In*: HAY, W. W., SIBUET, J. C., *et al.* (eds), *Initial Reports DSDP*, **74**, 1273−1300.

−−, PREMOLI-SILVA, I. & SHACKLETON, N. J. 1987. Atlantic Eocene planktonic foraminiferal paleohydrographic indicators and stable isotope paleoceanography. *Paleoceanography*, **2**, 287−331.

BOLTOVSKOY, E., 1987. Tertiary benthic foraminifera in bathyal deposits of the Quaternary world ocean. *Journal of Foraminiferal Research*, **17**, 279−285.

BRASS, G. W., SOUTHAM, J. R. & PETERSON, W. H. 1982. Warm saline bottom water in the ancient ocean. *Nature*, **296**, 620−623.

CARALP, M. H. 1984. Impact de la matiere organique dans des zones de forte productivite sur certains foraminiferes benthiques. *Oceanologica Acta*, **7**, 509−15.

CHALONER, W. G. & CREBER, G. T. 1989. The phenomenon of forest growth in Antarctica: a review. *In*: CRAME, J. A. (ed.) *Origins and evolution of the Antarctic Biota*. Geological Society, London, Special Publication, **47**, 85−88.

CORLISS, B. H. 1985. Microhabitats of benthic foraminifera within deep-sea sediments. *Nature*, **314**, 435−438.

−− & CHEN, C. 1988, Morphotype patterns of Norwegian deep-sea benthic foraminifera and ecological implications. *Geology*, **16**, 716−719.

−− & KEIGWIN, L. D. 1986. Eocene−Oligocene Paleoceanography. *In: Mesozoic and Cenozoic Oceans*, Geodynamics Series, **15**, AGU, 101−118.

DAILEY, D. H. 1983. Late Cretaceous and Paleocene benthic foraminifers from Deep Sea Drilling Project Site 516, Rio Grande Rise, western South

Atlantic. *In*: BARKER, P. F., CARLSON, R. L., JOHNSON, D. A. *et al.* (eds), *Initial Reports DSDP*, **74**, 757–82.

DOUGLAS, R. G. & SAVIN, S. M. 1975. Oxygen and carbon isotope analyses of Tertiary and Cretaceous microfossils from Shatsky Rise and other sites in the North Pacific Ocean. *In*: LARSON, R. L., MOBERLY, R. *et al.* (eds), *Initial Reports DSDP*, **32**, 509–20.

— — & WOODRUFF, F. 1981. Deep sea benthic foraminifera. *In*: EMILIANI, C. (ed.), *The Oceanic Lithosphere, The Sea*, **7**, 1233–1327.

FOSTER, T. D. & CARMACK, E. C. 1976. Temperature and salinity structure in the Weddell Sea. *Journal of Physical Oceanography*, **6**, 36–44.

FRAKES, L. A. 1979. *Climates throughout geologic time*, Elsevier Scientific Publishing Company, Amsterdam (Netherlands).

— — & FRANCIS, J. E., 1988. A guide to Phanerozoic cold polar climates from high-latitude ice-rafting in the Cretaceous. *Nature*, **333**, 547–549.

GERSONDE, R. & BURCKLE, L. in press. Neogene diatom biostratigraphy of ODP Leg 113. *In*: BARKER, P. F., KENNETT, J. P. *et al.* (eds), *Proceedings ODP*, **113B**.

KATZ, M. R. & MILLER, K. G. 1988. Paleocene to Eocene benthic foraminiferal turnover, Atlantic Sector Southern Oceans. *Geological Society of America, Programs & Abstracts*, **20**, 7, A251.

KEIGWIN, L. D. & CORLISS, B. H. 1986. Stable isotopes in late middle Eocene through Oligocene foraminifera. *Geological Society of America Bulletin*, **97**, 335–345.

KENNETT, J. P. 1977. Cenozoic evolution of Antarctic glaciation, the circum-Antarctic Ocean, and their impact on global paleoceanography. *Journal of Geophysical Research*, **82**, 3843–3860.

— — in press. Oxygen isotopes, Sites 689 and 690, Maud Rise. *In*: BARKER, P. F., KENNETT, J. P. *et al.* (eds), *Proceedings ODP*, **113B**

— — & SHACKLETON, N. J. in press. Carbon isotopes, Sites 689 and 690, Maud Rise. *In*: BARKER, P. F., KENNETT, J. P. *et al.* (eds), *Proceedings ODP*, **113B**

— — & STOTT, L. D. 1988. Antarctic Paleogene oxygen isotopic and climatic history, Maud Rise, Antarctica. *Geological Society of America, Programs & Abstracts*, **20**, 7, A251.

LECKIE, R. M. & WEBB, P.-N., 1985. Late Paleogene and early Neogene foraminifers of Deep Sea Drilling Project Site 270, Ross Sea, Antarctica. *In*: KENNETT, J. P., VON DER BORCH, C. C. *et al.* (eds), *Initial Reports DSDP*, **90**, 1093–142.

LUTZE, G. F. & COULBOURN, W. T. 1984. Recent benthic foraminifera from the continental margin of Northwest Africa: community structure and distribution. *Marine Micropaleontology*, **8**, 361–401.

MANABE, S. & BRYAN, K. 1985. CO_2-induced change in a coupled ocean–atmosphere model and its paleoclimatic implications. *Journal of Geophysical Research*, **90, C6**, 11689–11707.

MILLER, K. G. & FAIRBANKS, R. G. 1985. Oligocene–Miocene global carbon and abyssal circulation

changes. *In*: SUNDQUIST, E., & BROECKER, W. S. (eds), *The carbon cycle and atmospheric CO_2: natural variations, Archean to Present*. American Geophysical Union Monograph Series. **32**, 469–486.

— — & KATZ, M. E. 1987a. Eocene benthic foraminiferal biofacies of the New Jersey transect. *In*: POAG, C. W., WATTS, A. B. *et al.* (eds), *Initial Reports DSDP*, **95**, 267–298.

— — & — — 1987b. Oligocene to Miocene benthic foraminiferal and abyssal circulation changes in the North Atlantic. *Micropaleontology*, **33**, 97–149.

— — & — — 1988. Deep-water changes near the Paleocene-Eocene boundary: stable isotope results from the Southern Ocean. *Geological Society of America, Programs & Abstracts*, **20**, 7, A251.

— — CURRY, W. B. & OSTERMANN, D. R. 1984. Late Paleogene (Eocene to Oligocene) benthic foraminiferal oceanography of the Goban Spur Region, Deep Sea Drilling Project Leg 80. *In*: DE GRACIANSKY, P. G., POAG, C. W. *et al.* (eds), *Initial Reports DSDP*, **80**, 505–38.

— —, FAIRBANKS, R. G. & MOUNTAIN, G. S. 1987a. Tertiary isotope synthesis, sea level history, and continental margin erosion. *Paleoceanography*, **2**, 1–20.

— —, JANECEK, T. R., KATZ, M. E., & KEIL, D. J. 1987b. Abyssal circulation and benthic foraminiferal changes near the Paleocene/Eocene boundary. *Paleoceanography*, **2**, 741–61.

MORKHOVEN, F. P. C. M. van, BERGGREN, W. A. & EDWARDS, A. S. 1986. Cenozoic cosmopolitan deep-water benthic Foraminifera. *Bulletins Centres Recherches Exploration-Production Elf-Aquitaine, Memoirs*, **11**, Pau, France.

MURRAY, J. W., 1984. Paleogene and Neogene benthic foraminifers from Rockall Plateau. *In*: ROBERTS, D. G., SCHNITKER, D. *et al.* (eds), *Initial Reports DSDP*, **81**, 503–34.

OBERHAENSLI, H., 1986. Latest Cretaceous-early Neogene oxygen and carbon isotopic record at DSDP sites in the Indian Ocean. *Marine Micropaleontology*, **10**, 91–115.

— — & TOUMARKINE, M. 1985. The Paleogene oxygen and carbon isotope history of Sites 522, 523, and 524 from the central South Atlantic. *In*: HSU, K. H. & WEISSERT, H. J. (eds), *South Atlantic Paleoceanography*, Cambridge University Press (Cambridge, U.K.), 125–147.

OWEN, R. M. & REA, D. K. 1985. Sea-floor hydrothermal activity links climate to tectonics: the Eocene carbon dioxide greenhouse. *Science*, **227**, 166–169.

PRENTICE, M. L. & MATTHEWS, R. K. 1988. Cenozoic ice-volume history: development of a composite oxygen isotope record. *Geology*, **16**, 963–966.

POSPICHAL, J. & WISE, S. W. in press. Cretaceous-Eocene calcareous nannofossils, ODP Sites 689 and 690, Maud Rise, Antarctica, *In*: BARKER, P. F., KENNETT, J. P. *et al.* (eds), *Proceedings ODP*, **113B**

PROTO-DECIMA, F. & BOLLI, H. M. 1978. Southeast

Atlantic DSDP Leg 40 Paleogene benthic foraminifers. *In*: BOLLI, H. M., RYAN, W. B. F. *et al.*
(eds), *Initial Reports DSDP*, **40**, 783–810.

RAUP, D. M. & SEPKOSKI, J. J. 1986. Periodic extinction of families and genera. *Science*, **231**, 833–
836.

SAVIN, S. M. 1977. The history of the Earth's surface
temperature during the past 100 million years.
Annual Reviews of Earth and Planetary Science,
5, 319–344.

SCHNITKER, D. 1979. Cenozoic deep-water benthic
foraminifers, Bay of Biscay. *In*: MONTADERT, L.,
ROBERTS, D. G. *et al.* (eds), *Initial Reports
DSDP*, **48**, 377–414.

SCHNITKER, D. 1980. Global paleoceanography and
its deep water linkage to the Antarctic glaciation.
Earth Science Reviews, **16**, 1–20.

SCHROEDER, C. J., SCOTT, D. B., & MEDIOLI, F. S.
1987. Can smaller benthic foraminifera be ignored in paleoenvironmental analyses? *Journal
of Foraminiferal Research*, **17**, 101–105.

SEN GUPTA, B. K., LEE, R. F. & MALLORY, S. M.
1981. Upwelling and an unusual assemblage of
benthic foraminifera on the northern Florida continental slope. *Journal of Paleontology*, **55**, 853–
857.

SHACKLETON, N. J. 1986. Paleogene stable isotope
events. *Palaeogeography, Palaeoclimatology,
Palaeoecology*, **57**, 91–102.

SHACKLETON, N. J. 1987. The carbon isotope record
of the Cenozoic: history of organic carbon burial
and of oxygen in the ocean and atmosphere.
In: BROOKS, J. & FLEET, A. J. (eds), *Marine
Petroleum Source Rocks*. Geological Society,
London, Special Publication, **26**, 423–434.

SHACKLETON, N. J. & BOERSMA, A. 1981. The climate
of the Eocene ocean. *Journal of the Geological
Society, London*, **138**, 153–157.

SPIESS, V. in press. Neogene and Paleogene magnetostratigraphy of the Maud Rise and Antarctic
continental margin, Leg 113 drill sites. *In*:
BARKER, P. F., KENNETT, J. P. *et al.* (eds),
Proceedings ODP, **113B**

STOTT, L. D. & KENNETT, J. P. in press. Paleogene
planktonic foraminifer biostratigraphy: Leg 113,
Weddell Sea, Antarctica. *In*: BARKER, P. F.,
KENNETT, J. P. *et al.* (eds), *Proceedings ODP*, **113B**

THOMAS, E., 1985. Late Eocene to Recent deep-sea
benthic foraminifers from the central equatorial
Pacific Ocean. *In*: MAYER, L., THEYER, F. *et al.*
(eds), *Initial Reports DSDP*, **85**, 655–694.

— — 1986. Changes in composition of Neogene benthic foraminiferal faunas in equatorial Pacific and
North Atlantic. *Palaeogeography, Palaeoclimatology, Palaeoecology*, **53**, 47–61.

— — 1988. Mass extinctions in the deep sea. *In:
Abstracts, Conference on Global Catastrophes in
Earth History*, Snowbird, Utah, 192–3.

— —, in press. Late Cretaceous through Neogene
benthic foraminifers from Maud Rise, Antarctica. *In*: BARKER, P. F., KENNETT, J. P. *et al.*
(eds), *Proceedings ODP*, **113B**

— — & VINCENT, E. 1987. Equatorial Pacific deep-sea
benthic foraminifera: faunal changes before the
middle Miocene polar cooling. *Geology*, **15**,
1035–1039.

TJALSMA, R. C. & LOHMANN, G. P., 1983. Paleocene-
Eocene bathyal and abyssal benthic Foraminifera
from the Atlantic Ocean. *Micropaleontology
Special Publication*, **4**, Micropaleontology Press,
American Museum of Natural History, New
York.

TOLMAZIN, D. 1985. *Elements of Dynamic Oceanography*, Allen & Unwin, London.

VINCENT, E., GIBSON, J. M. & BRUN, L. 1974.
Paleogene and early Eocene microfacies, ben-
thonic foraminifera, and paleobathymetry of
Deep Sea Drilling Project Sites 236 and 237,
Western Indian Ocean. *In*: FISHER, R. L., BUNCE,
E. T. *et al.* (eds), *Initial Reports DSDP*, **17**, 607–
71.

WEI, W. & WISE, S. W. in press. Eocene through
Miocene calcareous nannofossils, ODP Leg 113.
In: BARKER, P. F., KENNETT, J. P. *et al.* (eds),
Proceedings ODP, **113B**

WIDMARK, J. G. V. & MALMGREN, B. A. 1988.
Cretaceous/Tertiary boundary — benthonic for-
aminiferal changes in the deep sea. *Geological
Society of America, Programs & Abstracts*, **20**, 7,
A223.

WILDE, P. & BERRY, W. B. N. 1982. Progressive
ventilation of the oceans — a potential for return
to anoxic conditions in the post-Paleozoic. *In*:
SCHLANGER, S. O., & CITA, M. B. (eds), *Nature
and Origin of Cretaceous Carbon-rich facies*.
Academic Press, London, 209–224.

WILLIAMS, C. A., 1986. An oceanwide view of plate
tectonic events. *Palaeogeography, Palaeoclimatology, Palaeoecology*, **57**, 3–26.

WOOD, K. C., MILLER, K. G. & LOHMANN, G. P.
1985. Middle Eocene to Oligocene benthic Foraminifera from the Oceanic Formation, Barbados.
Micropaleontology, **31**, 181–197.

WORTHINGTON, L. V. 1972. On the North Atlantic
circulation. *Johns Hopkins Oceanographic
Studies*, **6**, 110.

ZACHOS, J. C. & ARTHUR, M. A. 1986. Paleoceanography of the Cretaceous/Tertiary boundary
event: inferences from stable isotopic and other
data. *Paleoceanography*, **1**, 5–26.

Antarctica as an evolutionary incubator: evidence from the cladistic biogeography of the amphipod Family Iphimediidae

LES WATLING[1] & MICHAEL H. THURSTON[2]

[1] Oceanography Program and Zoology Department, Darling Marine Center, University of Maine, Walpole, Maine 04573, USA

[2] Institute of Oceanographic Sciences, Wormley, Godalming, Surrey GU8 5UB, UK

Abstract: The modern Antarctic marine fauna is quite distinct, with levels of endemism reaching 90% at the species level and 70% at the generic level. Since the modern Antarctic waters are very cold and isolated from all of the worlds oceans except the deep-sea, the origin of the modern Antarctic fauna has remained in doubt. The amphipod Family Iphimediidae is represented in Antarctic and Sub-Antarctic waters by 16 of its 21 genera and 49% of its species. A phylogenetic analysis of the family indicates that the most primitive genera are distributed primarily outside Antarctic waters. It is suggested that these are relicts of a former global distribution, and that once the Antarctic began to cool (38 Ma ago), a radiation of the iphimediids occurred in the Southern Ocean. This radiation was furthered by the reorientation of the mandible from one where the incisor cut in the horizontal transverse plane to cutting in the vertical frontal plane. With this advance, the family spread outward from the Antarctic, principally through the genus *Iphimedia*, which is today represented by 35 known species worldwide. The tracking of features in *Iphimedia* suggests that its initial spread beyond Antarctic waters occurred before the continent was finally thermally isolated 23 Ma ago. It is proposed that the cooling Antarctic waters acted as an incubator for this family, producing an evolutionary advance that allowed it to be a successful colonizer of the thermally changing global ocean.

Modern Antarctic biogeography and problems of its origin

The strong isolation of the modern Antarctic marine fauna has been well-documented (Knox & Lowry 1977; White 1984). Endemism levels vary, however, depending on the taxonomic level (e.g., genus, species) and group (e.g., polychaetes, amphipods, fish) under consideration. For example, fish, amphipods, and pycnogonids show greater than 90% species endemism, but at the generic level, fish are 70%, amphipods 39%, and pycnogonids 14% endemic. At the other end of the spectrum, polychaetes are 5% and 57% endemic to the Antarctic region at the generic and specific levels, respectively (all data from summary in White 1984).

Biogeographical patterns for the benthos around the Antarctic continent have been discussed by many authors (for example, Andriashev 1965; Hedgpeth 1969; DeWitt 1971; Knox & Lowry 1977; White 1984). The exact affinities of the regions south of the Sub-Antarctic Convergence vary according to the taxon under consideration, but for many groups the patterns are quite similar. Knox & Lowry (1977) noted that the biogeographical divisions established on the basis of fish and amphipod distributions were transcended by the polychaetes. The difference, they suggested, was due to the fact that polychaetes are an old homogeneous fauna with a slow evolutionary rate. White (1984) examined data on isopods and concluded that their distribution patterns were similar to most other benthic groups. In summary, the area south of the Antarctic Convergence can be divided into two sub-regions, the Western Antarctic (extending from Marguerite Bay through the Antarctic Peninsula and Scotia arc to somewhere in the Weddell Sea) and the Eastern (or Continental) Antarctic (extending the remaining distance from the Weddell Sea around the continent through the Ross Sea to the Bellingshausen Sea). South Georgia, with its warmer summer temperatures, is a transition area between the continental and Sub-Antarctic regions. The latter is often divided into Magellanic (including Tierra del Fuego and the Falklands) and Kerguelen sub-regions.

The distinctiveness of the Antarctic fauna has lead many authors to speculate as to its origins. From palaeoceanographic evidence it is clear that the Antarctic was much warmer in the past than it is today, with cooling beginning in late Eocene, perhaps on the Weddell Sea side first, and becoming complete with the opening of Drake Passage by the end of the Oligocene

From Crame, J. A. (ed.), 1989, *Origins and Evolution of the Antarctic Biota*, Geological Society Special Publication No. 47, pp. 297–313.

(Kennett 1978; Schnitker 1980). Further, as this cold Antarctic water sank and spread throughout the world ocean basins, it very likely initiated a change in the deep water fauna (Benson 1975). In the modern Antarctic fauna there are many species whose distributions range into bathyal depths. Considering both the zoogeographical and palaeoceanographic evidence, it has been suggested that the faunal components now found in Antarctic waters originated by one or more of the following mechanisms: evolution 'in situ' from relict species; migration from deep-sea basins; migration from South America via the Scotia arc; migration of northern species through the Sub-Antarctic region.

Discerning which mechanism played the major role in the present distribution of elements of the Antarctic fauna has been very difficult. For macrobenthic species there is little fossil record more recent than the Eocene. Consequently, it is possible to know only what species existed when the Antarctic waters were warm, but not their fate once the waters began to cool. In some groups, for example fish and crabs, where there is a Cretaceous and early Tertiary fossil record, extinctions of whole Families and Orders occurred (Grande & Eastman 1986; Zinsmeister & Feldmann 1984; Crame 1986). For the micro- and meiofauna, on the other hand, records obtained from the deep-sea drilling cores suggest a gradual change in the fauna, often with little or no change in genera. It is probable that this pattern of gradual change is not applicable to most components of the macrofauna, as the evolutionary rates of the taxa are most likely quite different and we know that in some major taxa whole groups went extinct (Crame 1986).

In this paper, the origin and subsequent dispersal of the amphipod Family Iphimediidae, which has quite high endemism at both generic and species levels within the Antarctic region, will be investigated. Since there is no fossil record for this group, the study will utilize the newly developed methods of cladistic biogeography (Wiley 1981; Humphries & Parenti 1986).

Biogeography of Antarctic gammaroidean Amphipoda with special reference to the Family Iphimediiae

Gammaroidean amphipods are good candidates for biogeographical studies because they brood their young and those that inhabit soft substrata or special environments such as sponge surfaces swim only for very limited periods of time. Unless a species is associated with objects that float, such as algal thalli, it is unlikely that such a species will ever be transported over long distances. Consequently, it is quite probable that reproductive isolation can develop rather quickly in some groups of Amphipoda (Kinne 1954), as shown by the large number of species flocks in diverse areas of the world (Barnard & Barnard 1983).

The distribution of gammaroidean amphipods in Antarctic waters has been summarized by Knox & Lowry (1977), with the details of each species' distribution being given by Lowry & Bullock (1976). Within the region the Scotia arc has the greatest number of species and genera, but its level of endemism for both taxonomic ranks is no higher than for any of the other areas (40–50% and 12–15%, respectively). Highest levels of affinity occur within the Antarctic Peninsula and Scotia arc areas, and along the coast from the Ross Sea to the Davis Sea. The Sub-Antarctic islands and Magellanic region show low levels of affinity to any other area, possibly because the faunas of many of the islands have been so poorly examined.

The Iphimediidae was chosen for this study because it is widespread and diverse in Southern Ocean waters (Table 1) and has no known deep-sea representatives. This family, long referred to as Acanthonotozomatidae, should revert to the earlier name of Iphimediidae (J. L. Barnard, pers. comm.). Recently the Ochlesidae have been subsumed within the Iphimediidae at the subfamily level (Barnard & Karaman 1987) but this group will not be considered further here. Currently accepted iphimediine genera and species are listed in Appendix Tables A and B. Individuals range in size from 2–30 mm in body length. Half of the currently known species and 13 of the 21 known genera in this family are endemic to the Antarctic and Sub-Antarctic region: of these, 14% of the species and only one of the genera are endemic to the Sub-Antarctic. Several of the Antarctic genera are monotypic, perhaps reflecting the low level of sampling in specialized habitats in the waters around the continent. Outside the Antarctic and Sub-Antarctic, all genera except *Iphimedia* are relatively restricted in their distributions. Only the genera *Odius*, *Iphimedia*, and *Labriphimedia* have both Southern Ocean and more northern species. Based on the distributions of species within the Southern Ocean, Knox & Lowry (1977) suggested that the centre of speciation and radiation for the family must have been the East Antarctic area. The phylogenetic analysis

Table 1. *Distribution of species of iphimediid genera in major ocean basins*

Genus	Antarctic	Sub-Antarctic	Indian O. W. Pacific	Medit.	North Atlantic	South Atlantic	Arctic	North Pacific	South Pacific	Total Species
Acanthonotozoma							9			9
Acanthonotozomella	2	1								3
Acanthonotozomoides	1	1								2
Acanthonotozomopsis	1									1
Anchiphimedia	1									1
Coboldus				1	1			1		3
Dikwa										1
Echiniphimedia	3									3
Gnathiphimedia	5	1	1							6
Iphimedia	2	5	11	7	6			2	2	35
Iphimediella	10									10
Labriphimedia	1	1							1	3
Maxilliphimedia	1									1
Nodotergum	1									1
Odius	1				1		2	1		4
Panoploeopsis	1									1
Paranchiphimedia	1									1
Parapanoploea	2									2
Pariphimedia	2	1								3
Postodius								1		1
Pseudiphimediella		2								2
Total Species	34	12	12	8	8	—	11	5	3	93
% of Total	37	13	13	9	9	—	12	4	3	
No. of Genera	15	7	2	2	3	—	2	4	2	21

to follow will test this idea and offer an hypothesis to explain the distributional pattern for the family as a whole.

Bases for the phylogenetic analysis of Iphimediidae

The techniques of cladistic analysis used are those outlined by Wiley (1981). In the Iphimediidae, most of the characters can be represented by multistate transformation series. These were coded using the nonredundant linear coding method as recommended by O'Grady & Deets (1987). Polarity of mouth field characters was based on functional analysis (following principles outlined by Fisher 1981) and outgroup comparisons, whereas the polarity of characters not associated with the mouth-field was determined solely by outgroup analysis. The Family Paramphithoidae, which occasionally has been combined in a loose association with the Iphimediidae (Karaman & Barnard 1979), was used as the out-group for the determination of character polarity.

The head appendages surrounding the mouth field of an amphipod are, from anterior to posterior, upper lip, mandible, lower lip, maxilla 1, maxilla 2, and maxilliped. The upper and lower lips and maxilla 2 will not be dealt with in this account. For each of the other appendages, along with the gnathopods and dorsal ornamentation of the pleosome, the important morphological features will be delimited, their functions outlined (where appropriate), and the resultant coded multistate transformation series given.

Mandible. Within the Family Iphimediidae, the mandible shows an extensive range in morphology. In most families of amphipods, it consists of a short, compact body which bears a molar proximally, an incisor process distally, a row of raker setae between the two, and a three-articulate palp. Normally, the mandible is oriented such that as it moves inwards and outwards on its horizontally positioned hinge, the incisor cuts in the horizontal transverse plane of the body and in a plane perpendicular to the long axis of the mandible (Fig. 1). Among the iphimediids, there is a reduction in the size of the molar, a loss of the row of raker setae, and finally, an elongation of the mandible concomitant with a change in the incisor such that the cutting plane becomes oriented in the vertical frontal plane of the body and in the medial-lateral plane through the long axis of the mandible. A transformation series detailing these gradual changes is coded as 10 character states. These are listed in Table 2 and examples of these mandible designs are given in Fig. 2.

Angles at which iphimediine mandibles are viewed are critical to the correct interpretation of structure, and this must be borne in mind when studying published illustrations. Specimens of many species have been examined in order to enhance consistency of interpretation of structure (Appendix B).

Maxilla 1. This appendage, in its basic form, consists of two endites, known as inner and outer plates, and a palp of two articles reaching at least to the distal edge of the outer plate. In the Iphimediidae, the important morphological changes are all associated with the palp: it may remain as long as the outer plate but become uniarticulate; it may remain two-articulate but be much shorter than the outer plate; or, it may be uniarticulate and much shorter than the outer plate (Fig. 3). These apomorphic variations are coded as character states 1−3 (Table 2).

Maxilliped. The only part of the maxilliped to vary from the basic form is the palp, normally comprised of four articles. The fourth article may become minute, or be lost altogether (Fig. 3), giving only 2 apomorphic states for this character (Table 2).

Gnathopods 1 and 2. It is likely that the earliest amphipod had simple thoracic appendages. However, in most families, including the sister group of the Iphimediidae, the Paramphithoidae, the first two pereopods are sub-chelate. The latter condition, then, must be taken as plesiomorphic for the Iphimediidae. Within the family two other gnathopodal morphologies can be seen, chelate and simple (Fig. 3). The simple gnathopod can be derived from the sub-chelate condition by reducing and finally eliminating the palmar surface, whereas the minutely chelate condition is produced by a reduction of the dactyl and extension of the ventro-distal corner of the propodus. It is not known which of these character states is actually plesiomorphic to the other, so the coding of these states is somewhat arbitrary (Table 2). In addition, as will be discussed later, there is in the genus *Iphimedia* the tendency for the propodus to become so enlarged as to create a condition that appears subchelate. This condition is considered to be secondarily subchelate and is coded as apomorphic to the other states.

Pleosome dorsal ornamentation. The predominant feature of both paramphithoids and iphimediids is the presence of large mid-dorsal carinae, especially on the last pereonite and the first three somites of the pleosome. Iphimediids may have dorsolateral carinal projections as well, or in a few cases, only dorsolateral carinal projections or no projections at all. These have been coded as character states 0−3 (Table 2).

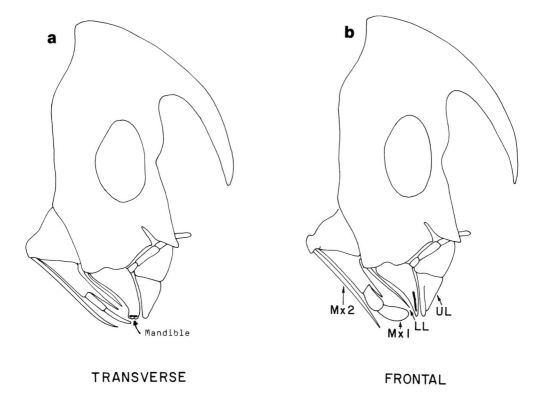

TRANSVERSE   **FRONTAL**

Fig. 1. Diagrammatic representation of the arrangement of the appendages in the mouth field of an amphipod. In (a) the mandible incisor cutting surface is oriented in the horizontal transverse plane, whereas in (b) it is oriented in the vertical frontal plane. The incisor cutting surface is indicated by the thick line on the mandible. UL, upper lip; LL, lower lip; Mxl, maxilla 1; Mx2, maxilla 2.

Cladistic analysis and biogeography

The family cladogram

The character state matrix for the genera, based on the above analysis of features, is given in Table 3. A complete list of character states as determined for all known species is included as Appendix B. The 52 of the 93 species which have been examined by us are marked by an asterisk; data for the other species were obtained from published descriptions. The cladogram summarizing the hypothesized phylogeny of this family is given in Fig. 4.

The genus retaining the greatest number of primitive features, especially in the mouth-field appendages, is *Dikwa*, an inhabitant of South African shallow subtidal waters. The other genera (*Odius* and *Postodius*) arising near the base of the cladogram are almost exclusively Northern Hemisphere in distribution. There is currently one known species of *Odius* from Antarctic waters. In *Odius* and *Postodius* the

mandible is elongate, there is a crushing molar, and at least some vestige of the raker seta row is present. There then follow three Southern Ocean genera that show only a slight modification of this mandible design, primarily in the strength of the molar. *Acanthonotozoma* is the most highly derived genus near the base of the cladogram, and is also the most diverse with nine species currently known from boreal and Arctic waters. The exact placement of this genus is troublesome; its mandible is most like that of *Iphimedia*, yet to place it later in the cladogram would require reversals of characters 3, 4, and 5. Just (1978) suggested that *Acanthonotozoma* might best be derived from an ancestor near to *Iphimedia*, most probably in the Atlantic sector of the low Arctic. It is unlikely that this issue will be resolved unless new genera exhibiting specific transitional features are found elsewhere in the Atlantic.

The next few genera in the cladogram are almost exclusively Antarctic and/or Sub-Antarctic in distribution. The mandibles of the

Table 2. *Coded values of multi-state transformation series for characters 1–6*

Character 1: Mandible

Plesiomorphic state:
    0, body compact, crushing molar, wide transverse incisor, raker setae.
Apomorphic states:
    1, elongate, crushing molar, narrow transverse incisor, raker setae;
    2, elongate, reduced molar, wide transverse incisor, raker setae;
    3, elongate, reduced molar, narrow transverse incisor, raker setae;
    4, elongate, reduced molar, narrow transverse incisor, no raker setae;
    5, elongate, molar absent, wide transverse incisor, no raker setae;
    6, elongate, reduced molar, wide oblique incisor, no raker setae;
    7, elongate, reduced molar, thick medial incisor, no raker setae;
    8, elongate, molar absent, thick medial incisor, no raker setae;
    9, elongate, reduced molar, narrow short sagittal incisor, no raker setae;
    10, elongate, reduced molar, narrow elongate sagittal incisor, no raker setae.

Character 2: Maxilla 1

Plesiomorphic state:
    0, palp of 2 articles, equal to or exceeding outer plate in length.
Apomorphic states:
    1, palp of 1 article, greater than or equal to outer plate in length;
    2, palp of 2 articles, shorter than outer plate;
    3, palp of 1 article, shorter than outer plate.

Character 3: Maxilliped

Plesiomorphic state:
    0, palp of 4 articles.
Apomorphic states:
    1, palp with minute article 4;
    2, palp of 3 articles.

Characters 4 and 5: Gnathopods 1 and 2

Plesiomorphic state:
    0, subchelate.
Apomorphic states:
    1, simple;
    2, minutely chelate;
    3, secondarily subchelate.

Character 6: Pleosome dorsal ornamentation

Plesiomorphic state:
    0, single median carina;
Apomorphic states:
    1, median carina plus dorsolateral projections;
    2, dorsolateral projections only;
    3, smooth.

first three of these genera are characterized by a transversely oriented incisor, a reduced molar, and the loss of raker setae. Furthermore, the incisor is widened in *Pseudiphimediella* and *Maxilliphimedia*. The junction in the cladogram leading to *Nodotergum* and beyond also marks a point in the evolution of the Iphimediidae where the states of characters 3 to 6 are all apomorphic.

There then occurred a significant change in the orientation of the mandible which, in our opinion, allowed the subsequent rich diversification to occur. The genus *Anchiphimedia* possesses a mandible with an obliquely sloping incisor, which is interpreted as the transition state between the plesiomorphic transversely oriented incisor and the apomorphic form, which cuts in the vertical frontal plane. There

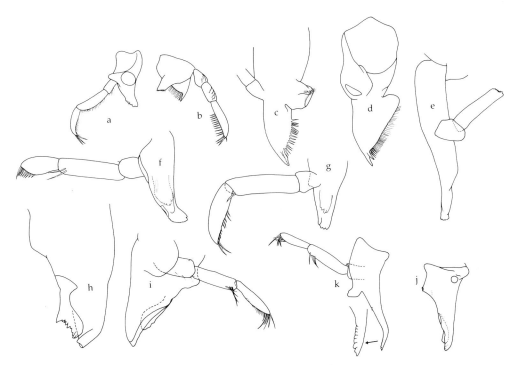

Fig. 2. The mandible in the Family Iphimediidae can take a wide variety of forms, several representative types of which are shown. For each mandible illustrated, the following information is provided: species name and citation, value of character state as coded in the transformation series for the mandible and listed in Table 2. (a) *Paramphithoe cuspidata* (Lepechin, 1780) (from Sars 1895), plesiomorphic; (b) *Dikwa acrania* Griffiths, 1974, character state 0; (c) *Odius antarcticus* Watling & Holman, 1981, character state 1; (d) *Acanthonotozomella barnardi* Watling & Holman, 1980, character state 3; (e) *Acanthonotozomoides oatesi* (K. H. Barnard, 1930) (from Bellan-Santini 1972), character state 3; (f) *Pseudiphimediella nodosa* (Dana, 1853) (from Watling & Holman 1980), character state 5; (g) *Nodotergum bicarinatum* Bellan-Santini, 1972, character state 4; (h) *Anchiphimedia dorsalis* K. H. Barnard, 1930 (from Watling & Holman 1981), character state 6; (i) *Gnathiphimedia barnardi* Thurston, 1974 (from Watling & Holman, 1981), character state 7; (j) *Iphimedia quasimodus* Ruffo & Schiecke, 1978, character state 9; (k) *Parapanoploea longirostris* Bellan-Santini, 1972, character state 10.

are two types of frontally cutting incisor, one which is thickened (possibly in response to the extreme reduction of the molar, and may therefore be a crushing surface) and one which is elongate, thin, and quite sharp along the medial edge. These two mandibles and the associated morphological features of the other mouth-field appendages characterize the last major divergence within the family. The thick-incisor group (*Gnathiphimedia*, *Labriphimedia*, and *Echiniphimedia*) is primarily Antarctic and Sub-Antarctic in distribution (with 1 species of *Labriphimedia* being found in New Zealand), whereas the narrow-incisor group, through the genus *Iphimedia*, has spread into all oceans of the world. The latter group also contains over half of the known species in the Family, both inside and outside Southern Ocean waters,

suggesting that the development of a scissors-like mandible was a very important advance. Unfortunately, nothing is known of the food of these animals so the exact significance of this mandible morphology cannot be stated.

Patterns in the distribution of the genus *Iphimedia*

Currently there are 35 known species of *Iphimedia*, of which three are still undescribed but have been examined by the authors. Of these, only seven occur in Antarctic and Sub-Antarctic waters, whereas 13 species are known from the Northeast Atlantic and Mediterranean areas (Table 4). No species have been described yet from South America (other than the Magellanic

Table 3. *Summary character matrix for genera of the Family Iphimediidae and its sister group, the Family Paramphiihoidae*

Genus	Character					
	1	2	3	4	5	6
PARAMPHITHOIDAE	0	0	0	0	0	0
Dikwa Griffiths	0	0	0	2	1	0
Odius Lilljeborg	1	1/3	0/1	2	0	0/3
Postodius Hirayama	1	3	2	2	0	0
Acanthonotozomopsis Watling & Holman	2	0	1	1	1	0
Acanthonotozomella Schellenberg	3	0	0	1	1	0
Acanthonotozomoides Schellenberg	3	0	0	1	1	0
Acanthonotozoma Boeck	9(10)	0/2	0/1	1	1	0/3
Panoploeopsis Kunkel	9	2	2	1	1	0
Nodotergum Bellan-Santini	4	0	2	1	2	1
Pseudiphimediella Schellenberg	5	0	1	2	2	1
Maxilliphimedia K. H. Barnard	5	0	2	2	2	2
Anchiphimedia K. H. Barnard	6	2	1	2	2	1
Paranchiphimedia Ruffo	?6	3	2	2	2	1
Labriphimedia K. H. Barnard	7	0	1/2	2	2	1
Gnathiphimedia K. H. Barnard	7	0	1	2	2	1/2
Echiniphimedia K. H. Barnard	8	0	1	2	2	1
Iphimediella Chevreux	9	0	1	2	2	1/2
Iphimedia Rathke	7,9(10)	0/2	2	2	2/3	(0)1/2
Coboldus Krapp-Schickel	9	3	2	2	2/3	2
Parapanoploea Nicholls	9(10)	0	1	2	2	1/2
Pariphimedia Chevreux	10	3	2	2	2	1/3

region) or the Caribbean, but it is probable that this is due to lack of collecting in the appropriate habitats.

Species in the genus *Iphimedia* are all quite similar to each other, that is, there is very little morphological diversity either in body form or in appendage shape or size. Differences between species are often delimited on the basis of setal armature, subtle changes in the shape of the mandible and other mouth-field appendages, or the shape of individual articles of the second gnathopod. While a more detailed revision of this genus will be reported elsewhere, for the present study only three appendages have been considered in any detail — maxilla 1, maxilliped, and gnathopod 2 (Table 4). On maxilla 1 the length and number of palp articles are used to define the character states. On the maxilliped it is the degree to which the second article of the palp extends along the margin of the third article. The greatest variation, however, seems to be in the shape of the propodus of gnathopod 2. In the plesiomorphic condition, the propodus is elongate with a subacute fixed finger. The propodus gradually changes in shape, becoming in some cases grossly enlarged and rounded, until the 'fixed finger' is so large that the appendage appears subchelate (Fig. 5).

As yet, a cladogram has not been generated for this genus; however, the distribution of species or species groups has been plotted on a world ocean map on the basis of the shape of gnathopod 2 article 6 (Fig. 6). All species with a 'rounded' to subchelate article 6 occur in the Indian Ocean and Western Pacific waters. Those with a linear article 6 occur from the Antarctic and New Zealand, to North America, to the Northeast Atlantic and Mediterranean. This is the only feature which shows a distinct biogeographic pattern and is taken to be of some significance.

Inferred historical biogeography of the Iphimediidae

The presence of primitive iphimediid genera in the Northern Hemisphere and southern Africa, as well as in the Antarctic, suggests that the family was widespread before the thermal isolation of the Antarctic continent began and that there are only a few remnants of that early distribution left. It is probable that the major diversification in the family occurred sometime after the Southern Ocean began to be isolated. Although it is not possible to say exactly when

Maxilla 1

Maxilliped

Gnathopods

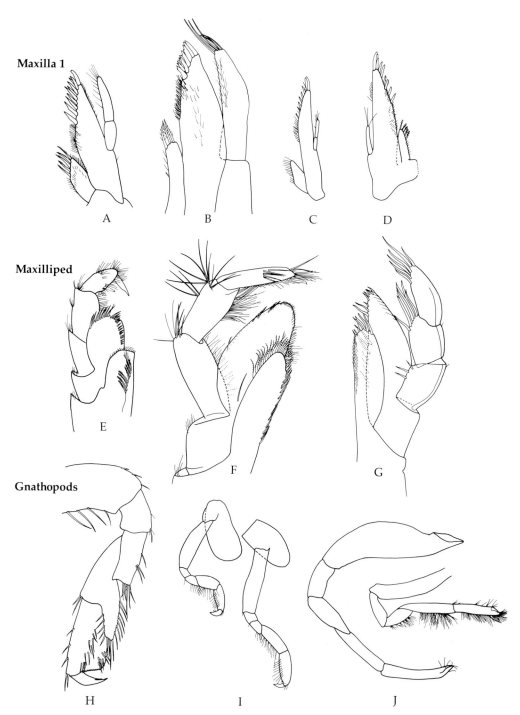

Fig. 3. Other appendages used in the phylogenetic analysis include maxilla 1, the maxilliped, and the gnathopods. The morphological variation exhibited by these appendages is shown here. Maxilla 1: (a) *Iphimedia quasimodus* Ruffo & Schiecke, 1978, plesiomorphic state; (b) *Odius antarcticus* Watling & Holman, 1981, character state 1; (c) *Iphimedia joubini* (Chevreux, 1913), character state 2; (d) *Coboldus (?) hedgpethi* (Barnard, 1969), character state 3. Maxilliped: (e) *Epimeria georgiana* Schellenberg, 1931 (from Watling & Holman 1981), plesiomorphic state; (f) *Iphimediella cyclogena* Barnard, 1930 (from Watling & Holman 1980), character state 1; (g) *Nodotergum bicarinatum* Bellan-Santini, 1972, character state 2. Gnathopods: (h) *Odius antarcticus* Watling & Holman, 1981, plesiomorphic state; (i) *Acanthonotozomopsis pushkini* (Bushueva 1978), character state 1; (j) *Iphimediella cyclogena* Barnard, 1930 (from Watling & Holman 1981), character state 2.

Table 4. *Distribution and list of character states of species of Iphimedia and Coboldus*

	Distribution	Mx 1*	Mxpd†	Gn2‡
I. joubini	Antarctic	1	0	0
I. 'walkeri' n.sp.	Antarctic	0	0	0
I. pacifica	Antarctic	0	0	0
I. imparilabia	Magellanic	1	1	0
I. magellanica	Magellanic	1	0	0
I. multidentata	Magellanic	1	0	?
I. macrocystidis	Magellanic	1	0	0
I. rickettsi	California	1	0	0
I. spinosa	New Zealand	0	0	0
I. haurakiensis	New Zealand	0	0	0
I. mala	Japan	0	0	1
I. discreta	S.E. Australia	0	0	1
I. 'Westernport' n.sp.	S.E. Australia	0	1	2
I. grossimana	Madagascar	1	1	2
I. compacta	Mauritius	1	0	1
I. orchestimana	Red Sea	0	?	2
I. gladiolus	Arabia	0	?	2
I. "Kei Islands" n.sp.	Bismark Archipelago	?	1	2
I. stegosaura	South Africa	1	1	1
I. gibba	South Africa	0	0	1
I. excisa	South Africa	1	?	3
I. capicola	South Africa	?	?	3
I. brachygnatha	Mediterranean	1	0	?
I. carinata	Mediterranean	1	?	?
I. gibbula	Mediterranean	1	?	?
I. vicina	Mediterranean	1	?	?
I. jugoslavica	Mediterranean	0	0	0
I. quasimodus	Mediterranean	0	0	?
I. serratipes	Mediterranean	0	?	?
I. minuta	W. Africa, Medit. to Norway	1	0	0
I. eblanae	Medit. to Norway	1	?	0
I. obesa	Medit. to Norway	0	0	0
I. spatula	Britain	1	0	0
I. nexa	Britain	1	0	0
I. perplexa	Britain	0	0	0
Coboldus nitior	Mediterranean	2	0	0
C. hedgpethi	California	2	0	3
C. 'laetifucatus' n.sp.	Bermuda	2	0	3

* 0 = plesiomorphic state; palp of 2 articles equal in length to outer plate
  1 = apomorphic; palp of 2 articles shorter than outer plate
  2 = apomorphic; palp of 1 article shorter than outer plate
† 0 = plesiomorphic state; palp article 2 produced only partially along inner margin of article 3
  1 = apomorphic; palp article 2 produced fully along inner margin of article 3
‡ 0 = plesiomorphic state; article 6 elongate with subacute fixed finger
  1 = apomorphic; article 6 shorter with rounded, subacute fixed finger
  2 = apomorphic; article 6 with rounded, grossly expanded fixed finger
  3 = apomorphic; article 6 slightly elongate, gnathopod sub-chelate
  Mxl = Maxilla 1   Mxpd = Maxilliped   Gn2 = Gnathopod 2

this radiation occurred, it may have coincided with the onset of cold water formation, first in the Atlantic sector of the Antarctic, and later in the Pacific sector as Australia moved northward at the end of the Eocene (38 Ma ago) (Kennett 1978; Schnitker 1980).

The evolution of the family on the Antarctic shelf is marked by a singular change in the mouth field appendages, the re-orientation of the mandible incisor from cutting in the horizontal transverse plane to cutting in the vertical frontal plane. Three of the genera with frontally cutting mandibles are found almost exclusively outside the Antarctic Convergence (Fig. 4), and another genus, *Gnathiphimedia*, has recently been found to have one species in the

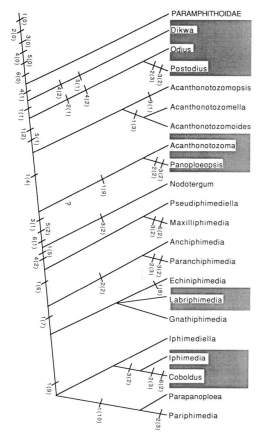

Fig. 4. Cladogram derived from the analysis of character state matrix. Cross-hatches bear character number, with coded value of transition state in parentheses. The Family Paramphithoidae is shown as the immediate sister group to the genera of Iphimediidae. Genera enclosed in dark shading are those whose distributions are entirely or nearly exclusively outside of Southern Ocean waters.

Sub-Antarctic waters of Marion Island (Bellan-Santini & Ledoyer 1986). In contrast, the species of *Labriphimedia*, the sister genus of *Gnathiphimedia*, are spread throughout the Sub-Antarctic from New Zealand to the Falklands. In the genus *Iphimedia*, only three of the 35 currently known species are found in Antarctic waters, and in the genus *Coboldus*, all known species are from waters well outside the Antarctic. [It is possible that this latter genus is not monophyletic, but that issue will have to be decided elsewhere.] There is thus a trend in this group of iphimediids toward dispersal out of the Antarctic with subsequent radiation throughout the worlds oceans.

The presence in New Zealand of the genus *Labriphimedia*, an early form with a frontally cutting mandible, and the most plesiomorphic species of the genus *Iphimedia*, suggests that one major route out of Antarctic waters was via the Macquarie Ridge as New Zealand pulled away from the developing Southern Ocean. It seems likely, then, that the dispersal of genera away from Antarctica must have occurred before the opening of the Drake Passage (23 Ma ago), the event which resulted in the complete thermal isolation of the Antarctic continent. With the consequent stabilization and homogenization of oceanographic conditions around the continent, and therefore, the loss of biogeographic provincialism, major stimuli for evolutionary innovation were removed and subsequent diversification probably occurred only at the species level.

In the approximately 25 Ma since the genus *Iphimedia* spread out of Antarctic waters, it has undergone the greatest radiation of any genus in the family. To date, 35 species are known and, since neither coast of South America has been much explored for Amphipoda, it is likely that several more will be found. The current species can be divided into two groups: those with a minutely chelate gnathopod 2 on which article 6 is linear, and those where gnathopod 2 is parachelate or subchelate and article 6 is shortened, rounded, and in many cases enlarged. The latter group is found throughout most of the Indian and western Pacific Oceans, whereas the former are known from the Southern Ocean, Magellanic region, New Zealand, California, tropical and northern east Atlantic and Mediterranean. While the northern areas are suggestive of a relict Tethyan distribution pattern, probably this is coincidental. The closing-off of Tethys at its eastern end was complete 18 Ma ago, and the Mediterranean was completely dry until about 5 Ma ago (Kennett 1982). Therefore, all the species now found in the Mediterranean must have evolved from one or more ancestors that recolonized the basin through the Straits of Gibraltar. The most likely source for these species was the North Atlantic.

Conclusions

The evidence presented in this paper, combining phylogenetic and biogeographic data, argues for considerable in situ evolution to have occurred in this family in Antarctic waters. The present day distribution of species in the genera *Dikwa*, *Odius*, and *Postodius*, suggests that the family was quite widespread and may even have

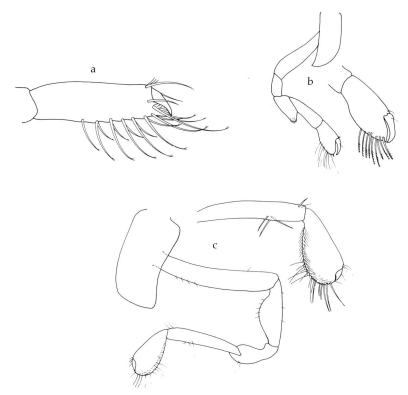

Fig. 5. Morphological variation in the shape of gnathopod 2 in the genus *Iphimedia*. (a) linear article 6 as in *Iphimedia rickettsi* (Shoemaker 1931); (b) article 6 slightly expanded as in *Iphimedia stegosaura* (Griffiths 1975); (c) article 6 greatly expanded as in *Iphimedia orchestimana* Ruffo, 1959.

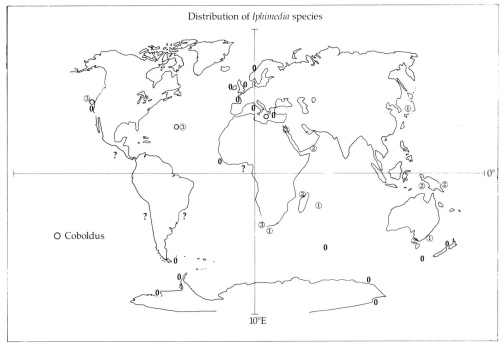

Fig. 6. Global distribution of *Iphimedia* and *Coboldus*. Numbers refer to states of gnathopod 2, article 6 as listed in Table 4.

been quite diverse. These three genera are all that remain of that earlier distribution. With the cooling of the Antarctic waters, and the concomitant increase in provinciality, there occurred an environment in which iphimediids could diversify. The radiation of iphimediids in Antarctic waters occurred as other genera in the family were most likely disappearing due to the general cooling (intermediate latitudes) or warming (low latitudes) of the world oceans (Valentine 1984). The history of the iphimediids, therefore, is not a simple one, and it is probable that similar types of patterns will be found for other taxa once their phylogenetic histories are understood. For this family, as well as for many other groups (Crame 1986) the Antarctic appears to have been both an incubator and a holding tank, supplying colonizing taxa to the world's oceans after the development of the psychrosphere. Once the cold waters of the ocean were fully developed, it is possible that there could then have been a subsequent re-colonization of Antarctic shelves by emerging deep-sea species (Hessler & Thistle 1975).

This paper is dedicated to J. Dearborn who, along with H. DeWitt, both of the University of Maine, helped to stimulate L. W.'s interest in Antarctic biology. He would also like to thank D. Schnitker for helping him understand the time sequence of changes in the Antarctic climate. Some of the taxonomic work on which this paper is based was funded by grants to L. W. from the Smithsonian Oceanographic Sorting Center. The illustrations were prepared by P. Rossi. Travel funds for L. W.'s attendance at the conference were provided by the University of Maine's Center for Marine Studies. The authors are most grateful to R. J. Lincoln for permission to study iphimediid material in the collections of the British Museum (Natural History).

Appendix A: Summary of generic additions and changes to the Family Iphimediidae since Barnard (1969a, b)

Acanthonotozomopsis Watling & Holman, 1980. Type: *Acanthonozomella pushkini* Bushueva, 1978.

Anisoiphimedia Karaman, 1980 → *Iphimedia*, herein. (The justification for this synonymy will be discussed in a forthcoming detailed examination of the genus *Iphimedia*.)

Bathypanoploea → Stilipedidae by Holman & Watling (1983).

Coboldus Krapp-Schickel, 1974. Type: *C. nitior* Krapp-Schickel, 1974.

Cypsiphimedia → *Iphimedia* by Watling & Holman (1980).

Dikwa Griffiths, 1974. Type: *D. acrania* Griffiths, 1974.

Maoriphimedia → *Labriphimedia* by Karaman & Barnard (1979).

Nodotergum Bellan-Santini, 1972. Type: *N. bicarinatum* Bellan-Santini, 1972.

Panoploea → *Iphimedia* by Karaman & Barnard (1979).

Paracanthonotozoma Bellan-Santini, 1972 → *Acanthonotozomella* by Watling & Holman (1980).

Pariphimediella → *Iphimediella* by Karaman & Barnard (1979); → (part) *Pseudiphimediella* by Watling & Holman (1980).

Postodius Hirayama, 1983. Type: *P. imperfectus* Hirayama, 1983.

Stegopanoploea Karaman, 1980 → *Iphimedia* herein. (The justification for this synonymy will be discussed in a forthcoming detailed examination of the genus *Iphimedia*.)

Appendix B: Summary table with coding of characters for all species of Iphimediidae. Species marked with an asterisk have been examined by the authors using either personal or museum material.

Genus and species	Characters					
	1	2	3	4	5	6
Acanthonotozoma Boeck, 1876	9(10)	0/2	0/1	1	1	0/3
* *cristatum* (Ross, 1835)	9	2	1	1	1	0
dunbari Just, 1978	9	0	0	1	1	3
gurjanovae Just, 1978	9	0	0	1	1	0
* *inflatum* (Kroyer, 1842)	9	0	0	1	1	0(3)
* *magnum* Just, 1978	9	0	0	1	1	0(3)
* *monodentatum* Kudrjashov, 1965	9	0	0	1	1	0(3)
* *rusanovae* Bryazgin, 1974	9(10)	2	1	1	1	0
* *serratum* (Fabricius, 1780)	9	0	0	1	1	0
* *sinuatum* Just, 1978	9(10)	2	1	1	1	0
Acanthonotozomella Schellenberg, 1926	3	0	0	1	1	0(1)
alata Schellenberg, 1926	?	0	0	1	1	0
* *barnardi* Watling & Holman, 1980	3	0	0	1	1	0
trispinosa Bellan-Santini, 1972	3	0	0	1	1	0(1)
Acanthonotozomoides Schellenberg, 1931	3	0	0	1	1	1
sublitoralis Schellenberg, 1931	3	0	0	1	1	1
* *oatesi* (Barnard, 1930)	3	0	0	1	1	1
Acanthonotozomopsis Watling & Holman, 1980	2	0	1	1	1	0
pushkini (Bushueva, 1978)	2	0	1	1	1	0
Anchiphimedia Barnard, 1930	6	2	1	2	2	1
* *dorsalis* Barnard, 1930	6	2	1	2	2	1
Coboldus Krapp-Schickel, 1974	9	3	2	2	0/2	2
nitior Krapp-Schickel, 1974	9	3	2	2	2	2
hedgpethi (Barnard, 1969)	9	3	2	2	0	2
'*laetifucatus*' n.sp. of Just	9	–	–	–	–	–
Dikwa Griffiths, 1974	0	0	0	2	1	0(1)
* *acrania* Griffiths, 1974	0	0	0	2	1	0(1)
Echiniphimedia Barnard, 1930	8	0	1	2	2	1
* *hodgsoni* (Walker, 1906)	8	0	1	2	2	1
* *echinata* (Walker, 1906)	8	0	1	2	2	1
* *scotti* Barnard, 1930	8	0	1	2	2	1
Gnathiphimedia Barnard, 1930	7	0	1	2	2	1/2
* *mandibularis* Barnard, 1930	7	0	1	2	2	2
* *barnardi* Thurston, 1974	7	0	1	2	2	2
* *fuchsi* Thurston, 1972	7	0	1	2	2	2
* *macrops* Barnard, 1932	7	0	1	2	2	2
* *sexdentata* (Schellenberg, 1926)	7	0	1	2	2	2
urodentata Bellan-Santini & Ledoyer, 1986	7	0	1	2	2	1
Iphimedia Rathke, 1843	7,9,10	0/2	2	0/2	2	0,1/2
* *obesa* Rathke, 1843	9	0	2	2	2	2
brachygnatha Ruffo & Schiecke, 1979	7	2	2	2	2	2
* *capicola* Barnard, 1932	–	0	2	2	0	2
carinata Heller, 1866	9	2	–	2	–	2
compacta Ledoyer, 1978	9	0	–	2	2	2
discreta Stebbing, 1910	9	0	2	2	2	2

Genus	Characters					
	1	2	3	4	5	6
* *eblanae* Bate, 1857	10	2	2	0	2	1
edgari Moore 1981	9	0	2	2	2	0
* *excisa* (Barnard, 1932)	–	2	–	2	0	2
gibba (Barnard, 1955)	9	0	2	2	0	2
gibbula Ruffo & Schiecke, 1979	9	2	–	–	–	1
* *gladiolus* Barnard, 1937	9	0	–	2	2	1
grossimana Ledoyer, 1972	9	0	–	2	2	2
haurakiensis Hurley, 1954	9	0	2	2	2	2(1)
* *imparilabia* Watling & Holman, 1980	9	2	2	2	2	1
* *joubini* (Chevreux, 1912)	10	2	2	2	2	1
jugoslavica Karaman, 1975	9	0	2	2	2	1
**macrocystidis* (Barnard, 1932)	9(10)	2	2	2	2	1
* *magellanica* Watling & Holman, 1980	9	2	2	2	2	1
mala (Hirayama, 1983)	9	0	2	2	2	0
* *minuta* Sars, 1882	9	2	2	2	2	2
multidentata (Schellenberg, 1931)	9	2	2	2	2	1
* *nexa* Myers & McGrath, 1987	9(10)	2	2	2	2	2
orchestimana Ruffo, 1959	9	0	2	2	2	2
* *pacifica* Stebbing, 1883	9	0	2	2	2	1
* *perplexa* Myers & Costello, 1987	9(10)	0	2	2	2	2
quasimodus Ruffo & Schiecke, 1979	9	0	2	–	–	1
rickettsi (Shoemaker, 1931)	9(10)	2	2	2	2	2
serratipes Ruffo & Schiecke, 1979	9	2	–	–	–	2
* *spatula* Myers & McGrath, 1987	10	2	2	2	2	1
spinosa (Thomson, 1880)	9	2	2	2	2	2
stegosaura (Griffiths, 1975)	9	2	2	2	2	1
vicina Ruffo & Schiecke, 1979	9	2	–	–	–	2
* 'Kei Islands' n. sp.	9	0	2	2	2	2
* 'walkeri' n. sp.	9	0	2	2	2	1
* 'Westernport' n. sp.	9	2	2	2	2	2
Iphimediella Chevreux, 1911	(7)9(10)	0	1	2	2	1/2
* *margueretei* Chevreux, 1912	(7)9	0	1	2	2	2
* *acuticoxa* Watling & Holman	9	0	1	2	2	1(2)
* *bransfieldi* Barnard, 1932	9	0	1	2	–	2
* *cyclogena* Barnard, 1930	9	0	1	2	2	2
* *georgei* Watling & Holman, 1980	9	0	1	2	2	1
imparidentata (Bellan-Santini, 1972)	9	0	1	2	2	1
* *microdentata* (Schellenberg, 1926)	9	0	1	2	2	2
octodentata (Nicholls, 1938)	9	0	–	2	2	1
* *rigida* Barnard, 1930	9(10)	0	1	2	2	1
serrata (Schellenberg, 1926)	9	0	1	2	2	2
Labriphimedia Barnard, 1932	7	0	1/2	2	2	1/2
* *vespuccii* Barnard, 1932	7	0	1	2	2	1
hinemoa (Hurley, 1954)	7	0	2	2	2	2
* *pulchridentata* (Stebbing, 1883)	7	0	2	2	2	1
Maxilliphimedia Barnard, 1930	5	0	2	2	2	2
* *longipes* (Walker, 1906)	5	0	2	2	2	2
Nodotergum Bellan-Santini, 1972	4	0	2	1	2	1
bicarinatum Bellan-Santini, 1972	4	0	2	1	2	1
Odius Lilljeborg, 1865	1	1/3	0/1	2	0	0/3
* *carinatus* (Bate, 1862)	1	3	0	2	0	0
* *antarcticus* Watling & Holman, 1981	1	1	1	2	0	3
cassigerus Gurjanova, 1972	–	3	0	2	0	0
* *kelleri* Bruggen, 1907	1	3	0	2	0	0

Genus	Characters					
	1	2	3	4	5	6
Panoploeopsis Kunkel, 1910	9	2	2	1	1	0
porta Kunkel, 1910	9	2	2	1	1	0
Paranchiphimedia Ruffo, 1949	?6	3	2	2	2	1
monodi Ruffo, 1949	?6	3	2	2	2	1
Parapanoploea Nicholls, 1938	(9)10	0	1	2	2	1/2
* *oxygnathia* Nicholls, 1938	(9)10	0	1	2	2	1
longirostris Bellan-Santini, 1972	10	0	1	2	2	2
Pariphimedia Chevreux, 1906	9/10	3	1/2	2	2	1/3
* *integricauda* Chevreux, 1906	10	3	2	2	2	1
incisa Andres, 1985	9	3	1	2	2	1
normani (Cunningham, 1871)	10	3	2(1)	2	2	3
Postodius Hirayama, 1983	1	3	2	2	0	0
imperfectus Hirayama, 1983	1	3	2	2	0	0
Pseudiphimediella Schellenberg, 1931	5	0	1	2	2	1
* *nodosa* (Dana, 1853)	5	0	1	2	2	1
glabra (Schellenberg, 1931)	5	0	1	2	2	1

Literature cited

ANDRIASHEV, A. P. 1965. A general review of the Antarctic fish fauna. In: VAN MIEGHEM, J. & VAN OYE, P. (eds) *Biogeography and Ecology in Antarctica*. Dr. W. Junk Publishers, The Hague, 491–550.

BARNARD, J. L. 1969a. Gammaridean Amphipoda of the rocky intertidal of California: Monterey Bay to La Jolla. *United States National Museum, Bulletin*, **258**, 1–230.

— — 1969b. The families and genera of marine Gammaridean Amphipoda. *United States National Museum, Bulletin*, **271**, 1–535.

— — & BARNARD, C. L. 1983. *Freshwater Amphipoda of the World. I. Evolutionary Patterns*. Hayfield Associates, Mt. Vernon, Virginia, USA.

— — & KARAMAN, G. 1987. Revisions in classification of Gammaridean Amphipoda (Crustacea), Part 3. *Proceedings of the Biological Society of Washington*, **100**, 856–875.

BELLAN-SANTINI, D. 1972. Invertébrés marins des XIIeme et XVeme expeditions Antarctiques Françaises en Terre Adélie. 10. Amphipodes Gammariens. *Tethys, Supplement*, **4**, 157–238.

— — & LEDOYER, M. 1986. Gammariens (Crustacea, Amphipoda) des Iles Marion et Prince Edward. *Bollettino del Museo civico di Storia Naturale Verona*, **13**, 349–435.

BENSON, R. H. 1975. The origin of the psychrosphere as recorded in changes of deep-sea ostracode assemblages. *Lethaia*, **8**, 69–83.

BUSHUEVA, I. V. 1978. A new amphipod species (Amphipoda, Gammaridea) from the Davis Sea (Eastern Antarctic). *Zoologicheski Zhurnal*, **57**, 450–453. (In Russian).

CHEVREUX, E. 1913. Amphipodes: Deuxieme expedition Antarctique Française (1908–1910) commandée par le Dr. Jean Charcot. *Sciences Naturelles: Documente Scientifique*, 79–186.

CRAME, J. A. 1986. Polar origins of marine invertebrate faunas. *Palaios*, **1**, 616–617.

DEWITT, H. H. 1971. Coastal and deep-water benthic fishes of the Antarctic. *Antarctic Map Folio Series, American Geographical Society*, **15**, 1–10.

FISHER, D. C. 1981. The role of functional analysis in phylogenetic inference: Examples from the history of the Xiphosura. *American Zoologist*, **21**, 47–62.

GRANDE, L. & EASTMAN, J. T. 1986. A review of Antarctic ichthyofaunas in the light of new fossil discoveries. *Palaeontology*, **29**, 113–137.

GRIFFITHS, C. L. 1974. The Amphipoda of Southern Africa. Part 4. The Gammaridea and Caprellidea of the Cape Province east of Cape Agulhas. *Annals of the South African Museum*, **65**, 251–336.

— — 1975. The Amphipoda of southern Africa. Part 5. The Gammaridea and Caprellidea of the Cape Province west of Cape Agulhas. *Annals of the South African Museum*, **67**, 91–181.

HEDGPETH, J. W. 1969. Introduction to Antarctic zoogeography. *Antarctic Map Folio Series, American Geographical Society*, **11**, 1–9.

HESSLER, R. R. & THISTLE, D. 1975. On the place of origin of deep-sea isopods. *Marine Biology*, **32**, 155–165.

HIRAYAMA, A. 1983. Taxonomic studies on the shallow water Gammaridean Amphipoda of West Kyushu, Japan. I. Acanthonotozomatidae, Ampeliscidae, Ampithoidae, Amphilochidae, Anamixidae, Argissidae, Atylidae, and Colomastigidae. *Publications of the Seto Marine Biological Laboratory*, **28**, 75–150.

HOLMAN, H. & WATLING, L. 1983. A revision of the Stilipedidae (Amphipoda). *Crustaceana*, **44**, 27–53.

HUMPHRIES, C. J. & PARENTI, L. R. 1986. *Cladistic Biogeography*. Clarendon Press, Oxford.

JUST, J. 1978. Taxonomy, biology, and evolution of the circumarctic genus *Acanthonotozoma* (Amphipoda) with notes on *Panoploeopsis*. *Acta Arctica*, **20**, 1–140.

KARAMAN, G. 1980. Revision of the genus *Iphimedia* Rathke, 1843 with description of two new genera, *Anisoiphimedia* and *Stegopanoploea*, n. gen. (fam. Acanthonotozomatidae) (Contribution to the knowledge of the Amphipoda 117). *Poljoprivreda i sumarstvo, Titograd*, **26**, 47–72.

— & BARNARD, J. L. 1979. Classificatory revisions in Gammaridean Amphipoda (Crustacea), Part 1. *Proceedings of the Biological Society of Washington*, **92**, 106–165.

KENNETT, J. P. 1978. The development of planktonic biogeography in the Southern Ocean during the Cenozoic. *Marine Micropaleontology*, **3**, 301–345.

— 1982. *Marine Geology*. Prentice-Hall, Englewood Cliffs, N.J.

KINNE, O. 1954. Die Gammarus-Arten der Kieler Bucht. *Zoologische Jarbucher, Abteilung fur Systematik Oekologie und Geographie der Tiere*, **82**, 405–424.

KNOX, G. A. & LOWRY, J. K. 1977. A comparison between the benthos of the Southern Ocean and the north polar ocean with special reference to the Amphipoda and the Polychaeta. *In*: DUNBAR, M. J. (ed.) *Polar Oceans*, Arctic Institute of North America, Calgary, 423–462.

LOWRY, J. K. & BULLOCK, S. 1976. Catalogue of the marine gammaridean Amphipoda of the Southern Ocean. *The Royal Society of New Zealand, Bulletin*, **16**, 1–187.

O'GRADY, R. T., & DEETS, G. B. 1987. Coding multistate characters, with special reference to the use of parasites as characters of their hosts. *Systematic Zoology*, **36**, 268–279.

RUFFO, S. 1959. Contributo alla conoscenza degli Anfipodi del Mar Rosso. *The Sea Fisheries Research Station Bulletin, Haifa*, **20**, 11–36.

— & SCHIECKE, U. 1978. Contributo alla conoscenza degli Acantonotozomatidi del Mediterraneo. *Bollettino del Museo Civico di Storia Naturale Verona*, **5**, 401–429.

SARS, G. O. 1895. *Amphipoda. An account of the Crustacea of Norway with short descriptions and figures of all species. Volume 1*. Alb. Cammermeyers, Christiania and Copenhagen.

SCHNITKER, D. 1980. Global paleoceanography and its deep water linkage to the Antarctic glaciation. *Earth-Science Reviews*, **16**, 1–20.

SHOEMAKER, C. R. 1931. A new species of amphipod crustacean (Acanthonotozomatidae) from California, and notes on *Eurystheus tenuicornis*. *Proceedings of the United States National Museum*, **78**(18), 1–8.

VALENTINE, J. W. 1984. Neogene marine climate trends: implications for biogeography and evolution of the shallow-sea biota. *Geology*, **12**, 647–650.

WATLING, L. & HOLMAN, H. 1980. New Amphipoda from the Southern Ocean, with partial revisions of the Acanthonotozomatidae and Paramphithoidae. *Proceedings of the Biological Society of Washington*, **93**, 609–654.

— & — 1981. Additional acanthonotozomatid, paramphithoid, and stegocephalid Amphipoda from the Southern Ocean. *Proceedings of the Biological Society of Washington*, **94**, 181–227.

WHITE, M. G. 1984. Marine benthos. *In*: LAWS, R. M. (ed.) *Antarctic Ecology*, Volume 2, Academic Press, New York, 421–461.

WILEY, E. O. 1981. *Phylogenetics, the theory and practice of phylogenetic systematics*. Wiley-Interscience, New York.

ZINSMEISTER, W. J. & FELDMANN, R. M. 1984. Cenozoic high latitude heterochroneity of southern hemisphere marine faunas. *Science*, **224**, 281–283.

Index